Sustainability in TIDES Chemistry
Green Approaches to Oligonucleotides and Oligopeptides Synthesis

Green Chemistry Series

Titles in the series:

1: The Future of Glycerol: New Uses of a Versatile Raw Material
2: Alternative Solvents for Green Chemistry
3: Eco-Friendly Synthesis of Fine Chemicals
4: Sustainable Solutions for Modern Economies
5: Chemical Reactions and Processes under Flow Conditions
6: Radical Reactions in Aqueous Media
7: Aqueous Microwave Chemistry
8: The Future of Glycerol: 2nd Edition
9: Transportation Biofuels: Novel Pathways for the Production of Ethanol, Biogas and Biodiesel
10: Alternatives to Conventional Food Processing
11: Green Trends in Insect Control
12: A Handbook of Applied Biopolymer Technology: Synthesis, Degradation and Applications
13: Challenges in Green Analytical Chemistry
14: Advanced Oil Crop Biorefineries
15: Enantioselective Homogeneous Supported Catalysis
16: Natural Polymers Volume 1: Composites
17: Natural Polymers Volume 2: Nanocomposites
18: Integrated Forest Biorefineries
19: Sustainable Preparation of Metal Nanoparticles: Methods and Applications
20: Alternative Solvents for Green Chemistry: 2nd Edition
21: Natural Product Extraction: Principles and Applications
22: Element Recovery and Sustainability
23: Green Materials for Sustainable Water Remediation and Treatment
24: The Economic Utilisation of Food Co-Products
25: Biomass for Sustainable Applications: Pollution Remediation and Energy
26: From C–H to C–C Bonds: Cross-Dehydrogenative-Coupling
27: Renewable Resources for Biorefineries
28: Transition Metal Catalysis in Aerobic Alcohol Oxidation
29: Green Materials from Plant Oils
30: Polyhydroxyalkanoates (PHAs) Based Blends, Composites and Nanocomposites

31: Ball Milling Towards Green Synthesis: Applications, Projects, Challenges
32: Porous Carbon Materials from Sustainable Precursors
33: Heterogeneous Catalysis for Today's Challenges: Synthesis, Characterization and Applications
34: Chemical Biotechnology and Bioengineering
35: Microwave-Assisted Polymerization
36: Ionic Liquids in the Biorefinery Concept: Challenges and Perspectives
37: Starch-based Blends, Composites and Nanocomposites
38: Sustainable Catalysis: With Non-endangered Metals, Part 1
39: Sustainable Catalysis: With Non-endangered Metals, Part 2
40: Sustainable Catalysis: Without Metals or Other Endangered Elements, Part 1
41: Sustainable Catalysis: Without Metals or Other Endangered Elements, Part 2
42: Green Photo-active Nanomaterials
43: Commercializing Biobased Products: Opportunities, Challenges, Benefits, and Risks
44: Biomass Sugars for Non-Fuel Applications
45: White Biotechnology for Sustainable Chemistry
46: Green and Sustainable Medicinal Chemistry: Methods, Tools and Strategies for the 21st Century Pharmaceutical Industry
47: Alternative Energy Sources for Green Chemistry
48: High Pressure Technologies in Biomass Conversion
49: Sustainable Solvents: Perspectives from Research, Business and International Policy
50: Fast Pyrolysis of Biomass: Advances in Science and Technology
51: Catalyst-free Organic Synthesis
52: Hazardous Reagent Substitution: A Pharmaceutical Perspective
53: Alternatives to Conventional Food Processing: 2nd Edition
54: Sustainable Synthesis of Pharmaceuticals: Using Transition Metal Complexes as Catalysts
55: Intensification of Biobased Processes
56: Sustainable Catalysis for Biorefineries
57: Supercritical and Other High-pressure Solvent Systems: For Extraction, Reaction and Material Processing
58: Biobased Aerogels: Polysaccharide and Protein-based Materials
59: Rubber Recycling: Challenges and Developments
60: Green Chemistry for Surface Coatings, Inks and Adhesives: Sustainable Applications
61: Green Synthetic Processes and Procedures
62: Resource Recovery from Wastes: Towards a Circular Economy
63: Flow Chemistry: Integrated Approaches for Practical Applications
64: Transition Towards a Sustainable Biobased Economy
65: Transportation Biofuels: Pathways for Production: 2nd Edition
66: Challenges in Green Analytical Chemistry: 2nd Edition
67: CO_2-switchable Materials: Solvents, Surfactants, Solutes and Solids

68: Green Toxicology: Making Chemicals Benign by Design
69: Sustainable and Functional Redox Chemistry
70: Waste Valorisation Using Ionic Liquids
71: Natural Product Extraction: Principles and Applications: 2nd Edition
72: Renewable Resources for Surface Coatings, Inks and Adhesives
73: Green Corrosion Inhibition: Fundamentals, Design, Synthesis and Applications
74: Chemical Valorisation of Carbon Dioxide
75: Green Sample Preparation Techniques: Concepts, Novel Materials and Solvents, and Applications
76: Methane Conversion Routes: Status and Prospects
77: Green Gasoline: A Green Spark Transportation Fuel
78: Agri-food Waste Valorisation
79: Bioplastics and Biocomposites: A Practical Introduction
80: Green Development of Photoluminescent Carbon Dots: For Analytical, Biomedical and Environmental Applications
81: Toward a PFAS-free Future: Safer Alternatives to Forever Chemicals
82: Green Fire Retardants for Polymeric Materials
83: Green Chemistry and Green Materials from Plant Oils and Natural Acids
84: Developments in Biodiesel: Feedstock, Production, and Properties
85: Sustainability in TIDES Chemistry

How to obtain future titles on publication:
A standing order plan is available for this series. A standing order will bring delivery of each new volume immediately on publication.

For further information please contact:
Book Sales Department, Royal Society of Chemistry, Thomas Graham House, Science Park, Milton Road, Cambridge CB4 0WF, UK
Telephone: +44 (0)1223 420066, Fax: +44 (0)1223 420247
Email: booksales@rsc.org
Visit our website at books.rsc.org

Sustainability in TIDES Chemistry

Green Approaches to Oligonucleotides and Oligopeptides Synthesis

Edited by

Alessandra Tolomelli
University of Bologna, Italy
Email: alessandra.tolomelli@unibo.it

Lucia Ferrazzano
University of Bologna, Italy
Email: lucia.ferrazzano4@unibo.it

and

Walter Cabri
University of Bologna, Italy
Email: walter.cabri@unibo.it

ROYAL SOCIETY OF CHEMISTRY

Green Chemistry Series No. 85

Print ISBN: 978-1-83767-258-5
PDF ISBN: 978-1-83767-454-1
EPUB ISBN: 978-1-83767-455-8
Print ISSN: 1757-7039
Electronic ISSN: 1757-7047

A catalogue record for this book is available from the British Library

The Royal Society of Chemistry is a charity, registered in England and Wales, Number 207890, and a company incorporated in England by Royal Charter (Registered No. RC000524), registered office: Burlington House, Piccadilly, London W1J 0BA, UK, Telephone: +44 (0)20 7437 8656.

For further information see our website at www.rsc.org

Printed in the United Kingdom by CPI Group (UK) Ltd, Croydon CR0 4YY, UK

Foreword

If Mother Nature took the form of a human being, we could all be assured that she would win all the Nobel Prizes in Chemistry for the rest of all time. And it is because of this brilliance that we are wise to look towards nature as our greatest teacher, mentor, and design partner. Throughout this book, *Sustainability in TIDES Chemistry: Green Approaches to Oligonucleotides and Oligopeptides Synthesis*, we see the wisdom of the practitioners providing us with examples of learning from and building on processes that we see in biology and applying them as best we can approximate in our laboratories. And through these examples, we see both what is possible today and also glimmers of what can be possible in the future.

The field of oligonucleotides and oligopeptides continues to grow in importance as more plentiful applications are generated for positive impact. However, the importance of this book recognizes an essential admonishment that we as chemists must always keep in mind: "it is possible to do the right things and to do them wrong". How often we have experienced throughout the history of chemistry that there have been brilliant scientists and inventors and companies that sought to advance the human condition only to find that the damage that was caused outweighed the great advance. This tragedy has been repeated in everything from transportation fuel additives, to refrigerants, to plastics, and even to the way in which we have manufactured our medicines. Many of these regrettable occurrences can be attributed to relative ignorance of the consequences at the time. Today, however, we do not have the same level of ignorance and so we must design our chemical methods with our highest level of awareness of not only trying to achieve the intended outcomes but also to avoid the unintended outcomes that can cause harm.

Green Chemistry Series No. 85
Sustainability in TIDES Chemistry: Green Approaches to Oligonucleotides and Oligopeptides Synthesis
Edited by Alessandra Tolomelli, Lucia Ferrazzano and Walter Cabri

Published by the Royal Society of Chemistry, www.rsc.org

Green chemistry is all about better chemistry. It is not about what we have to reduce, eliminate, or control, but rather what we can create, invent, and innovate. We see that spirit of creativity throughout the chapters of this book edited by Alessandra Tolomelli, Lucia Ferrazzano, and Walter Cabri. With each example, the book creates a new toolbox for us to reach into and use to make sure that we are not only doing the right things, but that we are doing the right things, right.

Paul Anastas
Yale University
New Haven, CT, USA

Preface

The increasing demand for therapeutic peptides and oligonucleotides (TIDES) has prompted a significant shift towards prioritizing their manufacture, with a focus on efficient and sustainable processes. Recognizing the expanding therapeutic application of these molecules, it was considered important to contribute to the Green Chemistry Series with an overview of *Sustainability in TIDES Chemistry*. Having devoted some years to peptide chemistry and developing green methodologies for their industrial application, we recently approached oligonucleotides with a similar strategy. As the two classes exhibit many similarities, advances in sustainable approaches can be readily adapted between them. With large-scale manufacture now imperative, sustainability and green chemistry have emerged as crucial considerations from their inception. Many papers and some interesting reviews have been published on the synthesis and purification of green oligopeptides, but a similar approach to oligonucleotides is still in its infancy. This book aims to bridge the gap between these two areas, gathering information on the green synthesis of oligopeptides – a well-established research area – and preliminary data on the sustainable synthesis of oligonucleotides.

The overview presented in this book is divided into two parts, one dedicated to oligopeptides and the other to oligonucleotides, considering each category's sustainability in productive and analytical processes. Starting from the evolution of drug design, different synthetic approaches are analysed, stressing sustainability goals and addressing critical points. Recent innovations in solvent and reagent replacement, and also technological implementations, are examined for peptides. For oligonucleotides, challenges in introducing green innovations are outlined, particularly in solid- and liquid-phase synthesis and stereocontrolled approaches. The impact of impurities on manufacturing, commercialization, and regulatory filing is

---

Green Chemistry Series No. 85
Sustainability in TIDES Chemistry: Green Approaches to Oligonucleotides and Oligopeptides Synthesis
Edited by Alessandra Tolomelli, Lucia Ferrazzano and Walter Cabri

Published by the Royal Society of Chemistry, www.rsc.org

evaluated, underscoring core aspects for synthetic chemists. The book also delves into purification processes as crucial components of TIDES green manufacturing.

The development of sustainable TIDES processes remains one of the greatest challenges in the pharmaceutical sector, requiring a robust interaction between academia and industry. Therefore, this book incorporates contributions from both academic and industrial authors, ensuring a well-rounded exploration of the topic.

Many people share our enthusiasm for the application of green chemistry principles to TIDES and have contributed to this book, and we therefore warmly thank all of them. We were fortunate to be able to collaborate with the main specialists in each category, giving us the opportunity to learn from them. We take this opportunity to stress how important it has been over the years to collaborate with colleagues from diverse backgrounds, who have enriched our understanding and fostered fundamental debates on the science involved. We also express our gratitude to all young researchers, in both academia and industry, who have contributed to sustainable TIDES chemistry research: they represent the future of green chemistry for us.

Alessandra Tolomelli
Lucia Ferrazzano
Walter Cabri

Contents

Green Chemistry Series No. 85
Sustainability in TIDES Chemistry: Green Approaches to Oligonucleotides and Oligopeptides Synthesis
Edited by Alessandra Tolomelli, Lucia Ferrazzano and Walter Cabri

Published by the Royal Society of Chemistry, www.rsc.org

CHAPTER 1

Introduction

A. TOLOMELLI, L. FERRAZZANO AND W. CABRI

P4i Laboratory, Department of Chemistry "Giacomo Ciamician", Alma Mater Studiorum – University of Bologna, via P. Gobetti 85, 40129 Bologna, Italy
E-mail: alessandra.tolomelli@unibo.it, lucia.ferrazzano4@unibo.it, walter.cabri@unibo.it

1.1 Introduction

Peptides [<40 amino acids (aa) according to the US Food and Drug Administration (FDA)][1] and oligonucleotides, including antisense oligonucleotides (ASOs), phosphorodiamidate morpholino oligomers (PMOs) and small interfering RNAs (siRNAs), are commonly referred to as TIDES.

The term "TIDES" is utilized in the pharmaceutical and research communities to categorize and describe modalities that share certain characteristics distinct from traditional small-molecule drugs and biologics. Typical positive features of these compounds are limited off-target toxicity and easy management of potential immunogenicity. In addition, TIDES have shown promise in addressing specific disease targets and potential applications in areas where other conventional drugs may not be as effective. However, historically these drugs have faced challenges related to their pharmaceutical profiles, limiting their broader application for decades. In particular, native peptides composed of coded amino acids[2] and oligonucleotides[3] are characterized by poor metabolic stability, whereby formulations are generally problematic and typically

Green Chemistry Series No. 85
Sustainability in TIDES Chemistry: Green Approaches to Oligonucleotides and Oligopeptides Synthesis
Edited by Alessandra Tolomelli, Lucia Ferrazzano and Walter Cabri

Published by the Royal Society of Chemistry, www.rsc.org

delivered through intravenous, subcutaneous, intrathecal or intravitreal routes. Advances in drug design and technology have gradually been able to overcome these limitations. Numerous strategies have been developed to improve the pharmaceutical profiles of oligonucleotides and peptides, particularly in terms of metabolic stability and drug delivery.[4,5] Whereas small molecules and biologics are highly regulated segments, specific guidance and guidelines for TIDES have only recently been introduced by different agencies.

At present, more than 80 peptides are available on the market as therapeutic drugs and >200 are progressing through clinical trials.[6] Recently approved peptides in the sectors of diabetes and obesity are the main drivers of the global market with a predicted increase from $39.3 billion in 2022 to $82.6 billion in 2032 at a compound annual growth rate (CAGR) of 7.5%.[7] Following the classification system of Lau and Dunn,[8] glucagon marketed by Novo Nordisk and approved by the FDA in 1998 is a copy of a native hormone sequence that raises blood sugar levels with a short half-life (3–6 min) (see Figure 1.1).[9] On the other hand, the blockbuster semaglutide, developed by Novo Nordisk, which was approved by the FDA in 2020, is a 31-aa homologous peptide of the GLP-1 fragment protein with a half-life of 7 days (see Figure 1.1).[10] This improvement in metabolic stability was achieved with the introduction of one non-coded amino acid and a fatty side-chain. Additionally, the drug was approved as two formulations (subcutaneous injection and oral tablet) for the treatment of type 2 diabetes and obesity. Degarelix, developed by Ferring Pharmaceuticals, a heterologous peptide characterized by the presence of seven out of ten non-coded amino acids with a median half-life of 43 days, was approved by the FDA in December 2008 (see Figure 1.1).[11] The drug is typically available as a subcutaneous formulation and displays aggregation at the site of injection.

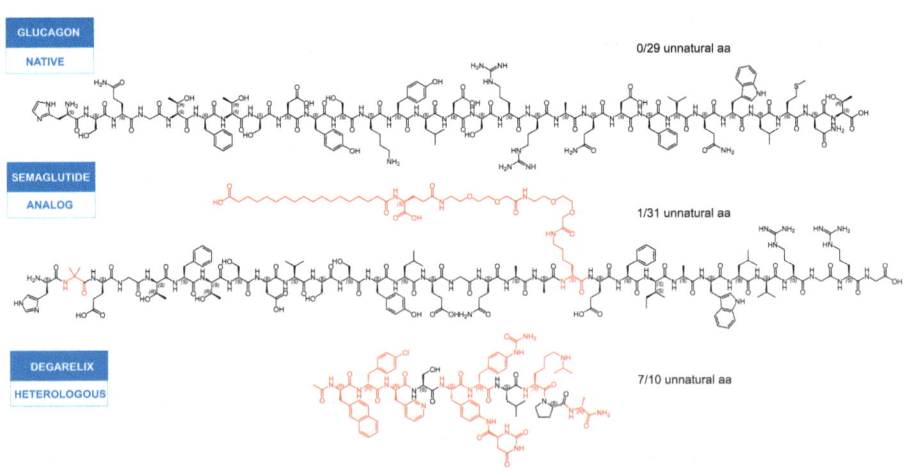

Figure 1.1 Native analogous and heterologous peptides.

Table 1.1 ASOs and siRNAs targeting lipoprotein (a).

NCE	Class	Phase	$t_{1/2}$	Delivery	Frequency of treatment	Company
Pelacarsen	ASO	III	21 days	Subcutaneous	Monthly	Novartis
Olpasiran	siRNA–GalNAc	III	6 h	Subcutaneous	3/6 monthly	Amgen
Lepodisiran	siRNA	II	1 day	Subcutaneous	Annually	Lilly

The higher success rate of peptides entering clinical phases compared with small molecules (11% *vs.* 6%) is due to their inherent high selectivity and lower toxicity and the higher success rate of native or analogous products.[8] In terms of oligonucleotides approved by the FDA, the first ASO (formivirsen) was launched in 1998, the first aptamer (pegaptanib sodium) in 2004 and the first siRNA in 2018. Eighteen oligonucleotides are currently available on the market and over the last 5 years (from 2019 to 2023), 11 have been approved as therapeutic drugs. The oligonucleotide market is expected to grow from $7.7 billion in 2022 to $16.7 billion in 2027, with a CAGR of 16.8%.[12] This estimation takes into consideration all recent approvals and also the potential impact of drugs currently in phase II/III trials on cardiovascular diseases. As shown in Table 1.1,[13] several major pharmaceutical companies are currently focused on developing molecules to lower lipoprotein(a), a risk factor for atherosclerotic cardiovascular diseases (ASCVDs).

The TIDES market is expected to increase consistently and rapidly in size and volume with higher CAGRs than all the other modalities.

The large volumes necessary to support the growing market cast a spotlight on the issues related to process sustainability. The ACS Green Chemistry Institute Pharmaceutical Roundtable identified sustainable peptide synthesis as one of the most important targets in recent years.[14,15] The application of the 12 principles of green chemistry,[16] a standard framework for small-molecule development, has been extended in recent years to TIDES.

The process design needs to guarantee not only optimal purity and yield but also sustainability of solvents, reagents, energy consumption and waste management. In addition, in TIDES chemistry, the experimental design should be tailored for each specific sequence using the most suitable technology. The purpose of this book is to serve as a valuable reference for TIDES synthesis and sustainability, with each chapter focusing on a specific topic, as outlined in the following section.

1.2 Chapter Summaries

For the rigorous assessment of process greenness, a suitable method of measurement must be selected. Green metrics are critical for guidance of process development towards sustainability. In Chapter 2, entitled *Green Metrics for a Sustainable Approach in Iterative Processes*, all the metrics are reviewed to identify the most appropriate parameters for iterative processes and TIDES in general. The first part of the book is mainly focused on sustainability challenges in peptide production.[17]

The main topic of discussion in Chapter 3 is *Recent Trends in Peptide Drug Discovery and Development*. Improvements in peptide design have promoted a steady expansion of the market from niche to widespread diseases, such as diabetes and obesity. Interestingly, the glucagon-like peptide-1 (GLP-1) receptor agonists mainly responsible for market expansion in the sectors of diabetes and obesity, including semaglutide, dulaglutide and tirzepatide, are frequently on the FDA drug shortage list.[18]

Advances in synthetic technology have been pivotal in the improvement of TIDES product designs. In view of this progress, separate chapters are dedicated to diverse chemical and biotechnological methods used within various fields. Oligopeptide synthesis has been achieved using multiple technologies, including solid-phase peptide synthesis (SPPS), liquid-phase peptide synthesis (LPPS), chemo-enzymatic peptide synthesis (CEPS) and recombinant fermentation. In Chapter 4, *Sustainable Approaches in Solid-phase Peptide Synthesis (SPPS)* are discussed. This technique involves assembling peptides on a solid support, typically a resin, allowing the stepwise addition of coded and non-coded amino acids, which results in the creation of peptide chains. On this basis, SPPS is an efficient method widely used in laboratory settings to produce both short and long peptides.

Chapter 5 provides an overview of *Sustainable Approaches in Liquid-phase Peptide Synthesis (LPPS)*. Unlike SPPS, LPPS operates in a solution phase on a growing peptide in the form of an ester or amide, which displays excellent solubility through the use of anchors (tags). This approach involves coupling of amino acids in solution and subsequent purification steps. Typically, non-toxic solvents are used and the resulting products (usually short peptides) are extremely useful for cosmetic applications.

Chapter 6 focuses on *Sustainable Approaches to Chemo-enzymatic Peptide Synthesis (CEPS)*. Using this cutting-edge technology, at least one peptide bond is created, taking advantage of the power of enzyme catalysis. The possibility of genetic modification of biocatalysts has facilitated the optimization of this technology that is effectively integrated with SPPS or LPPS for fragment synthesis.

Recombinant and Semisynthesis of Peptides is reviewed in Chapter 7. Using this method, genetically engineered organisms, such as bacteria and yeasts, can be efficiently directed to produce long peptides composed of proteinogenic amino acids.

The inclusion of multiple technologies for oligopeptide synthesis highlights the significance of combining diverse approaches to advance product design and manufacturing processes. Each method offers specific advantages and suitability for different scales of production, peptide lengths and specific requirements within the fields of biotechnology and chemical synthesis. A hybrid approach based on the integration of SPPS, LPPS, CEPS and recombinant fermentation is generally the optimal strategy for the synthesis of long peptides. Peptides such as enfuvirtide and exenatide[19] are produced using a combination of SPPS and LPPS, while semaglutide is generated *via* semisynthesis, starting with recombinant production of fragment 4–31.[20] Tirzepatide was the first double agonist of GIP and GLP-1 approved for the treatment of obesity, whereby four peptide fragments are industrially produced *via* SPPS and linked together using LPPS.[21]

As mentioned above, the recent evolution of the approach of regulatory agencies to peptides has generated critical attention to impurity control and, as a consequence, a complex qualification process. Chapter 8, *Oligopeptide Regulatory Affairs and Challenges*, discusses the impacts of impurities on the quality and safety of peptides and peptide regulatory affairs and challenges. Regulatory measures to optimize efficacy and resolve safety concerns have resulted in tightening of the impurity limits and in requiring further detailed assessment of immunogenicity and aggregation.

The second part of the book is focused on sustainability in oligonucleotide development.[22]

Chapter 9, entitled *Oligonucleotide Synthesis and State-of-the-Art Drug Discovery*, provides an overview of recent advances in oligonucleotide design and pharmaceutical profile improvements through chemical modification of ASOs and siRNAs. In contrast to peptides, oligonucleotides are only produced through chemical synthesis. A number of methodologies have been developed to date, the most notable of which are solid-phase oligonucleotide synthesis (SPOS), liquid-phase oligonucleotide synthesis (LPOS) and stereopure synthesis.

In Chapter 10, *Sustainable Approaches in Solid-phase Oligonucleotide Synthesis* are described. Similarly to SPPS, this technology involves attaching the first nucleotide to a solid support (resin or porous glass beads) and subsequently adding nucleotides step-by-step to generate the desired sequence. Protecting groups are used to ensure reaction selectivity during each coupling step. Using SPOS, moderate to large quantities of oligonucleotides are efficiently produced.

The liquid-phase oligonucleotide synthesis (LPOS) process is conducted in a solution phase using technology similar to that for LPPS. Although less commonly used than SPOS, LPOS allows the synthesis of longer oligonucleotides and involves the sequential addition of nucleotides in solution along with subsequent purification steps. Chapter 11 provides a review of *Sustainable Approaches in Liquid-phase Oligonucleotide Synthesis (LPOS)*.

Marketed oligonucleotides consist essentially of uncontrolled mixtures of diastereoisomers, since most of the phosphate backbone is generally converted to phosphorothioate or phosphorodiamidate to enhance nuclease stability and improve the pharmacokinetic profile. During SPOS of these mixtures, the P(III) → P(V) oxidation/sulfurization step is performed using a non-stereoselective reaction to generate a new phosphorus stereogenic centre. A novel approach for the stereocontrolled synthesis of the phosphorus(v) backbone has been developed to obtain this unique class of *Stereopure Oligonucleotides* that are currently in phase III trials, which is described in Chapter 12.

Similarly to that discussed for peptides, evaluation of the *Impact of Impurities on the Quality and Safety of Oligonucleotides as Drug Substances* (Chapter 13) is critical for the assessment of new technologies and process changes. Chapter 14 on *Oligonucleotide Regulatory Affairs and Challenges* highlights recent trends in this rapidly growing area. The key considerations for drug substance and drug product DMF/NDA/ANDA filing submissions are discussed in this chapter.

Finally, the importance of downstream purification[23,24] and its optimization in terms of sustainability are discussed in Chapter 15, entitled *Greening Chromatographic Purification Processes*. The technologies described in the previous TIDES chapters focus on synthesis, but the efficiency of the upstream process and the capacity to control impurity levels can simplify downstream chromatographic purification. Moreover, the purification step consistently contributes to overall process mass intensity (PMI) and sustainability. Both batch and continuous purification processes have been extensively evaluated to establish the most sustainable alternative.

1.3 Conclusion

This book comprehensively explores the design and development of TIDES, with particular focus on sustainability. For each topic, leading specialists have made significant contributions to discussions covering the design, synthesis, purification and regulatory aspects of TIDES development. This in-depth review provides a holistic perspective of the topic and represents a valuable resource for researchers interested in the systematic evolution of TIDES.

List of Abbreviations

aa	Amino acid
ACS	American Chemical Society
ASCVD	Atherosclerotic cardiovascular disease
ASO	Antisense oligonucleotide
CAGR	Compound annual growth rate
CEPS	Chemo-enzymatic peptide synthesis
DMF/NDA/ANDA	Drug master file/new drug application/abbreviated new drug application
FDA	US Food and Drug Administration
GalNAc	*N*-Acetylgalactosamine
GLP-1	Glucagon-like peptide-1
GIP	Glucose-dependent insulinotropic peptide

LPOS	Liquid-phase oligonucleotide synthesis
LPPS	Liquid-phase peptide synthesis
NCE	New chemical entity
PMI	Process mass intensity
PMO	Phosphorodiamidate morpholino oligomer
siRNA	Small interfering RNA (or short interfering RNA or silencing RNA)
SPOS	Solid-phase oligonucleotide synthesis
SPPS	Solid-phase peptide synthesis
TIDES	Oligopeptides and oligonucleotides

References

1. ANDAs for Certain Highly Purified Synthetic Peptide Drug Products That Refer to Listed Drugs of rDNA Origin – Guidance for Industry, FDA – Center for Drug Evaluation and Research, May 2021, available online at https://www.fda.gov/regulatory-information/search-fda-guidance-documents/andas-certain-highly-purified-synthetic-peptide-drug-products-refer-listed-drugs-rdna-origin.
2. C. S. B. Chia, *Int. J. Pept. Res. Ther.*, 2021, **27**, 1397.
3. H. Takakusa, N. Iwazaki, M. Nishikawa, T. Yoshida, S. Obika and T. Inoue, *Nucleic Acid Ther.*, 2023, **33**, 83.
4. A. Bak, D. Leung, S. E. Barrett, S. Forster, E. C. Minnihan, A. W. Leithead, J. Cunningham, N. Toussaint and L. S. Crocker, *AAPS J.*, 2015, **17**, 144.
5. T. C. Roberts, R. Langer, W. Langer and M. J. A. Wood, *Nat. Rev. Drug Discovery*, 2020, **19**, 673.
6. G. Rossino, E. Marchese, G. Galli, F. Verde, M. Finizio, M. Serra, P. Linciano and S. Collina, *Molecules*, 2023, **28**, 7165.
7. Global Market Insight, https://www.gminsights.com/industry-analysis/peptide-therapeutics-market.
8. J. L. Lau and M. K. Dunn, *Bioorg. Med. Chem.*, 2017, **26**, 2700.
9. A. E. Pontiroli, A. Calderara, M. G. Perfetti and S. R. Bareggi, *Eur. J. Clin. Pharmacol.*, 1993, **45**, 555.
10. S. Hall, D. Isaacs and J. N. Clements, *Clin. Pharmacokinet.*, 2018, **57**, 1529.
11. F. G. Rick, N. L. Block and A. V. Schally, *OncoTargets Ther.*, 2013, **6**, 391.
12. https://www.marketsandmarkets.com/Market-Reports/oligonucleotide-synthesis-market-200829350.html.
13. B. Sosnowska, S. Surma and M. Banach, *Pharmaceuticals*, 2022, **15**, 1573.
14. A. Isidro-Llobet, M. N. Kenworthy, S. Mukherjee, M. E. Kopach, K. Wegner, F. Gallou, A. G. Smith and F. Roschangar, *J. Org. Chem.*, 2019, **84**, 4615.
15. M. C. Bryan, P. J. Dunn, D. Entwistle, F. Gallou, S. G. Koenig, J. D. Hayler, M. R. Hickey, S. Hughes, M. E. Kopach, G. Moine, P. Richardson, F. Roschangar, A. Steven and F. J. Weiberth, *Green Chem.*, 2018, **20**, 5082.
16. P. T. Anastas and J. C. Warner, *Green Chemistry: Theory and Practice*, Oxford University Press, New York, 1998.

17. L. Ferrazzano, M. Catani, A. Cavazzini, G. Martelli, D. Corbisiero, P. Cantelmi, T. Fantoni, A. Mattellone, C. De Luca, S. Felletti, W. Cabri and A. Tolomelli, *Green Chem.*, 2022, **24**, 975.
18. https://www.accessdata.fda.gov/scripts/drugshortages/dsp_SearchResults.cfm, 17th January 2024.
19. J. R. Ferguson, M. O. Frederick, M. E. Kopach, D. Seibert and M. Strege, *Chem. Today*, 2019, **37**, 48.
20. J. Fegerman, A. S. Andersen, P. Block, P. W. Garibay, T. Kruse, I. S. N. Norby and C. U. Jessen, *US Pat.*, US9732137, 2017.
21. M. O. Frederick, R. A. Boyse, T. M. Braden, J. R. Calvin, B. M. Campbell, S. M. Changi, S. R. Coffin, C. Condon, O. Gowran, J. McClary Groh, S. R. Groskreutz, Z. D. Harms, A. A. Humenik, N. J. Kallman, N. D. Klitzing, M. E. Kopach, J. K. Kretsinger, G. R. Lambertus, J. T. Lampert, L. M. Maguire, H. A. Moynihan, N. S. Mullane, J. D. Murphy, M. E. O'Mahony, R. N. Richey, K. D. Seibert, R. D. Spencer, M. A. Strege, N. Tandogan, F. L. Torres Torres, S. V. Tsukanov and H. Xia, *Org. Process Res. Dev.*, 2021, **25**, 1628.
22. L. Ferrazzano, D. Corbisiero, A. Tolomelli and W. Cabri, *Green Chem.*, 2023, **25**, 1217.
23. C. De Luca, G. Lievore, D. Bozza, A. Buratti, A. Cavazzini, A. Ricci, M. Macis, W. Cabri, S. Felletti and M. Catani, *Molecules*, 2021, **26**, 4688.
24. M. Catani, C. De Luca, J. M. Garcia Alcântara, N. Manfredini, D. Perrone, E. Marchesi, R. Weldon, T. Müller-Späth, A. Cavazzini, M. Morbidelli and M. Sponchioni, *Biotechnol. J.*, 2020, 1900226.

CHAPTER 2

Green Metrics for a Sustainable Approach in Iterative Processes

W. CABRI* AND A. TOLOMELLI

P4i Laboratory, Department of Chemistry "Giacomo Ciamician", Alma Mater Studiorum – University of Bologna, via P. Gobetti 85, 40129 Bologna, Italy
*E-mail: walter.cabri@unibo.it

2.1 Introduction

The design and development of an industrial synthetic route for a pharmaceutical product are complex and highly regulated processes that require careful consideration of various parameters to ensure success. Planning a process that meets the basic requirements of green chemistry, while matching the needs for efficiency, is even more crucial.[1]

Usually, the introduction of environmental issues (safety, energy and waste management) during the initial stages of a project leads to a strong reduction in costs and increases the likelihood of acceptance by financial departments thanks to improved economic sustainability.

The key factors to keep in mind are as follows:

(i) Development of a patent-free process by ensuring that the chosen synthetic route does not infringe existing intellectual property (IP). In some cases, this critical factor defines limitations in the application of the basic principles of green chemistry. The presence of IP in

Green Chemistry Series No. 85
Sustainability in TIDES Chemistry: Green Approaches to Oligonucleotides and Oligopeptides Synthesis
Edited by Alessandra Tolomelli, Lucia Ferrazzano and Walter Cabri

Published by the Royal Society of Chemistry, www.rsc.org

new technologies is a limiting factor for the industrial application of sustainable processes.

(ii) High reaction yield and selectivity by limiting the formation of side products to control costs and reduce the necessity for intensive purification processes.

(iii) High productivity to increase the amount of synthesized product per unit time that can be achieved under efficient reaction conditions with the best catalysts and optimized process design. This approach also permits containing plant occupation and, therefore, the impact of fixed costs.

(iv) Limited energy requirements to reduce energy consumption by selecting reaction conditions that operate at lower temperatures and pressures.

(v) High product quality and purity throughout the synthesis. This can be achieved by implementing quality control measures to meet pharmaceutical standards.

(vi) Scalability to ensure that the synthetic route is readily transferrable from the laboratory scale to industrial-scale production. Careful monitoring of reaction kinetics, heat transfer and mixing efficiency becomes more critical if increasing the batch size.

(vii) Cost-effectiveness by taking into account the cost of raw materials, equipment, energy and labour when designing the synthetic route, without compromising quality.

(viii) Environmental, health and safety (EHS) considerations which evaluate the environmental impact of the synthetic route in terms of wastes (gas, liquid, solid) and prioritize the safety of workers and the surrounding community.

The last point requires minimizing the use of undesirable reagents and solvents, conducting thorough hazard assessments and implementing safety protocols. The generation of harmful waste is related to optimization of the reaction conditions, choice of recyclable or reusable materials and implementation of waste management practices following EHS regulations at local and international levels.

Quantitative evaluation of the environmental impact of an industrial process taking into consideration all of these issues has become a multivariable optimization process in which the definition of simple metrics represents an important challenge. For this purpose, the 12 principles of green chemistry[2] (Figure 2.1) provide a valuable framework for optimizing and guiding various aspects of the development of industrial processes for small molecules through the use of green metrics.[3–6]

The atom economy (AE)[7] and environmental impact factor (E-factor)[8] were the first green metrics, introduced by Barry Trost and Roger Sheldon, respectively, in the 1990s. Trost's AE is a theoretical measure that assesses the efficiency of a reaction by measuring atom conservation and emphasizes the importance of highly chemo-, regio-, diastereo- and enantioselective

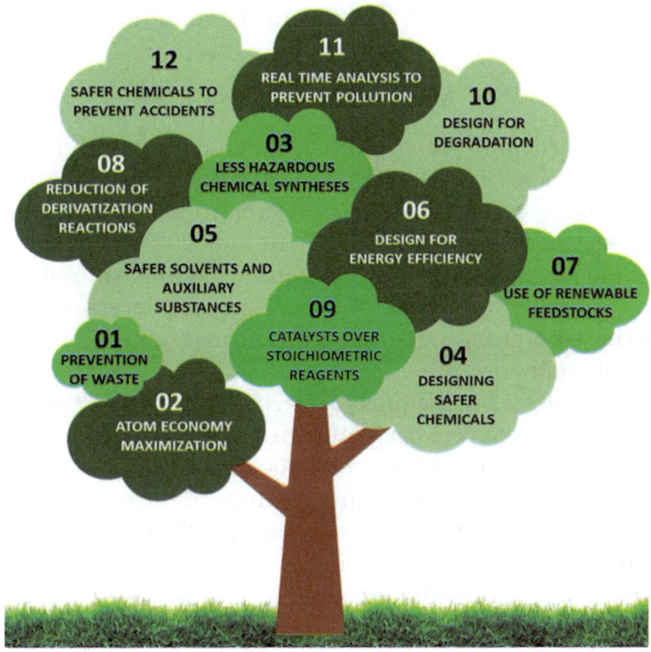

Figure 2.1 The 12 principles of green chemistry.

reactions. In the synthesis of oligopeptides and oligonucleotides (TIDES), the reactivity of functional moieties requires the unavoidable introduction of protective groups, and the AE is very low. The AE does not take into account the excess of reagents and yields. The primary role of the AE is the selection and optimization of suitable reactions for specific synthetic processes.

The E-factor promotes greener chemistry practices by measuring the ratio of the total waste mass to the mass of product. A low E-factor indicates a more environmentally friendly and efficient process because less waste is generated relative to the product. The original definition by Sheldon did not take process water into account, even though water can be a significant component of the waste in some chemical processes. Its omission from the E-factor can result in an incomplete assessment of the environmental impact. Climate change is intensifying water scarcity in various regions, thereby jeopardizing the sustainability of the use of industrial water.

To address this issue, industries are investing in water-saving technologies, recycling and long-term planning. Contaminant removal and wastewater treatment in industrial processes are critical tasks for environmental sustainability and regulatory compliance. Sheldon and co-workers recognized this issue in 2015 by introducing the "complete" environmental impact factor (cEF), which accounts also for process water.[9] However, by that time, the industry had shifted its focus to the metric of process mass intensity (PMI), which measures the number of kilograms of materials used to produce 1 kg of a product (including water).[10,11] The conversion of the PMI into the cEF is straightforward:

$$cEF = PMI - 1 \tag{2.1}$$

The PMI is favoured more readily within a company because it correlates directly with the cost of materials, and non-technical functions such as finance and sourcing can understand and support such sustainable changes. Nowadays, new processes to produce small molecules typically incorporate green metrics to assess and improve environmental sustainability. Some of them are easy to use and are part of the normal process of evaluation by chemists in the laboratory. Others require more sophisticated calculations and the work of engineers and specialists.

Introduction of the 12 principles of green chemistry by Anastas and Warner in 1998[2] marked the most significant turning point in the discussion of metrics in the pharmaceutical arena. As a result, industry and academia began to pay attention to metrics that measure and improve the environmental impact and sustainability of pharmaceutical production.

TIDES chemistry is based on the iteration of basic reaction sequences necessary for the assembly of individual amino acids into a peptide (see Scheme 2.1) or a nucleotide for antisense oligonucleotide (ASO) drugs (see Scheme 2.2). For the synthesis of a peptide of 39 amino acids, 39 coupling and 39 deprotection reactions followed by cleavage from the resin, ester or anchor are necessary. The use of a hybrid process does not change this scenario, and ~79 primary reactions are necessary. In the active pharmaceutical ingredient (API) market, the small molecule eribulin (Figure 2.2)[12] represents "Mount Everest" for a pharmaceutical synthetic chemist. The commercial synthesis, applied in the best process by Eisai, involves 62 chemical steps, fewer than the 79 needed for a 39-aa peptide.[13] Considering a standard ASO with 25 nucleotides, the number of reactions required is 99. For each nucleotide except the first one, removal of the primary alcohol protective group, followed by coupling with an additional nucleotide, capping and oxidation of the P(III) intermediate to the corresponding P(V) and the final deprotection, must be performed.[14,15]

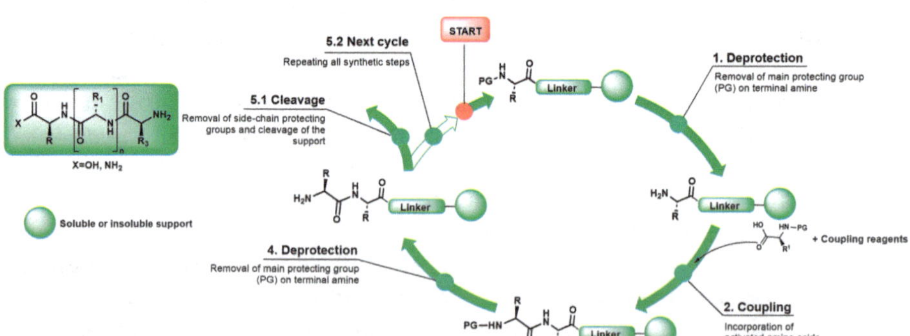

Scheme 2.1 Iterative process for oligopeptides.

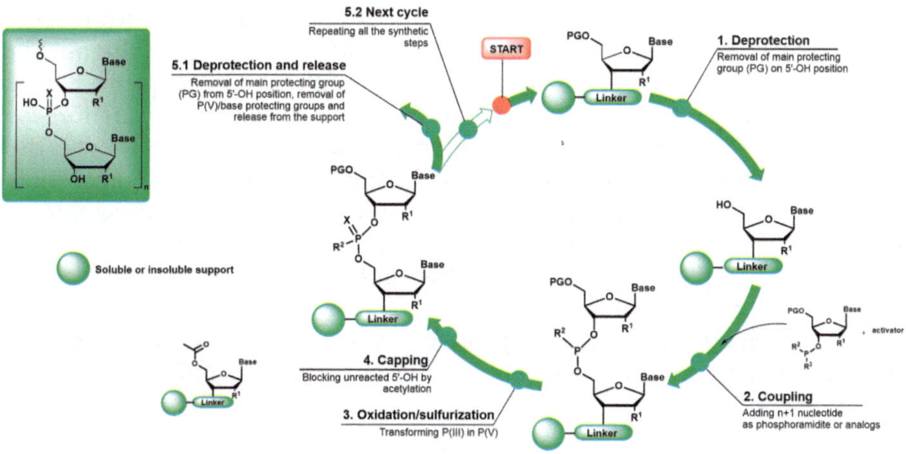

Scheme 2.2 Iterative process for oligonucleotides.

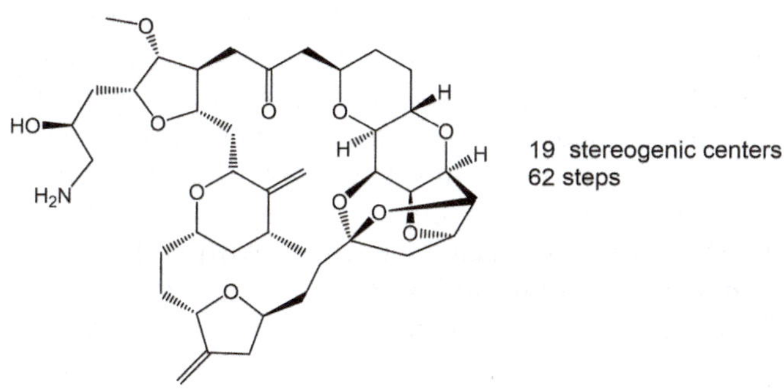

Figure 2.2 Eribulin.

2.2 Carbon Footprint

Assessment of the "carbon footprint" of chemical processes in the healthcare area is a challenge because most attention has been devoted to other economy segments (*e.g.* mining, energy, automotive industry).[16] Compared with other metrics, the carbon footprint also includes energy, which is critical in certain countries for the determination of the overall cost of a process. However, in the production of an API, this aspect is (in general) non-critical for the comparison of processes.[17] In TIDES chemistry, considering the overall carbon footprint of the entire supply chain, the main difference between peptides and ASOs is that the synthesis of proteinogenic amino acids is carried out *via* fermentation,[18] and only protective groups are added chemically.[19] Non-coded monomers, with a few exceptions,[20] are mainly produced *via* chemical syntheses.[21] For ASOs, all nucleotide building blocks are produced by more complex total syntheses or a chemo-enzymatic approach, which increases the carbon footprint of the overall production with respect to peptides.[22,23] The solid support for SPXS (SPPS or SPOS) or the anchor for LPXS (LPPS or LPOS) contributes to the overall carbon footprint, but its impact decreases with the length of TIDES. The evolution of the anchors for LPXS will not affect the carbon footprint much, whereas for SPXS there is the space to develop a new generation of solid supports based on biomaterials that do not require swelling and could allow the choice of the process solvent to be simplified. An additional issue arises from the source of the raw material and its transportation. However, this appears to be less important with respect to the quantity of waste.

This chapter critically evaluates green metrics in the chemical development of TIDES, focusing on their impact on process assessment and sustainability.

2.3 Green Metrics and Molecule Classes

Numerous metrics have been generated to guide the process development of small molecules towards sustainability. For this purpose, green metrics are employed to identify the most sustainable reaction methodologies, most efficient reaction conditions, ideal reaction sequence, least energy-demanding process and the process that produces the least waste. Matching this approach with good control of raw material costs permits the identification of the most competitive process. In this context, the collaboration between scientists from pharmaceutical companies (members) and contract manufacturing organizations (associate members) within the Green Chemical Institute Pharmaceutical Roundtable (GCI-PR) of the American Chemical Society (ACS) had a significant impact in promoting the adoption of green metrics in drug development. In addition, to facilitate the critical definition of sustainable practices, the GCI-PR sponsored opinion papers and meetings and supported research projects and educational initiatives with grants.

Several metrics are focused on specific aspects of chemical processes and others are a combination of the main ones, but not all of them suit all possible processes. For instance, Baran's ideality factor[24] is a powerful tool for small molecules but, if applied to TIDES, affords almost a fixed percentage. The value is obtained using the following calculation:

$$\text{Ideality factor} / \% = \frac{\text{number of construction reactions} + \text{number of strategic redox reactions}}{\text{number of reactions}} \tag{2.2}$$

For complex structures, the ideality factor can be used to guide strategic choices in the synthetic plan because backbone building can be approached through different routes. However, for oligomeric structures such as TIDES, the number of possible alternative reaction sequences is limited, and the protocol is necessarily iterative.

Considering only the upstream process (assembly), the ideality factor for peptides is 50% (deprotection/coupling), and also for oligonucleotides (deprotection/coupling/oxidation/capping). This is determined by the occurrence in both sequences of a protective group removal, which cannot be accounted for as a constructive step. For oligonucleotides, the capping step is required to avoid the formation of contracted sequences and to simplify the purification step.

Considering the AE in TIDES chemistry, after the first monomer (amino acid or nucleoside) has been attached to a chemical anchor or polymeric material, the process becomes iterative through transformations centred on a deprotection–coupling sequence for each added residue, which is characterized by a poor AE. In peptide synthesis, the Fmoc or Boc protective groups of the amine moiety in the added building blocks must be cleaved to start the next coupling. If amino acids bearing sensitive groups in the side-chain are involved, such as arginine, lysine, histidine, cysteine, asparagine, glutamic acid, aspartic acid, glutamine, tryptophan, serine, threonine and tyrosine, the need for additional orthogonal protective groups negatively affects green parameters. For some of these amino acids, such as serine, threonine, tyrosine and arginine, the additional protective groups can be limited by adopting special coupling protocols.[25] With regard to nucleotide chemistry, the primary alcohol is (in general) protected with a dimethoxytrityl moiety, but nucleobases also require additional orthogonal protective groups. All of these factors contribute to bringing the iterative synthesis of TIDES to the worst possible level of AE also because the size of the protective groups is not negligible with respect to the molecular weight of the single building blocks. For example, the AE for the addition of a simple glycine unit using Fmoc technology is only 19.5%. The issue is very similar in the growing oligonucleotide sequence. Using N^6-acetyl-2′-deoxy-5′-*O*-DMT-adenosine 3′-CE phosphoramidite (DMT = dimethoxytrityl; CE = cyanoethyl) as an example, the AE due to incorporation of the structure in the sequence is 40%. Even if AE is not a suitable metric for TIDES, it gives an idea of the negative impact of protecting groups on the process greenness. In summary, there are no alternatives to these iterative sequences, and the poor AE is endemic. However, it is a theoretical value and cannot be used to discriminate significantly among the different technologies in TIDES. Instead, other metrics more suited to the iterative nature of TIDES chemistry are listed in Table 2.1.

Table 2.1 Green metrics for TIDES chemistry.

Metric	Acronym	Unit	Formula
Reaction mass efficiency	RME	%	$\frac{\text{mass of product}}{\sum(\text{mass of reactants})}$
Process mass intensity	PMI	%	$\frac{\sum m(\text{input materials})\text{in kg}}{m(\text{product})\text{in kg}} \times 100$
Process mass intensity with recovery	PMIr	%	$\frac{\sum m(\text{input materials}) - \text{recovered in kg}}{m(\text{product})\text{in kg}} \times 100$

2.4 Reaction Mass Efficiency (RME)

RME is more useful than AE because it considers the reaction stoichiometry and yield.[26] In fact, in iterative processes, the excess of reagents and optimization of reaction conditions are necessary to guarantee high yields for each step. For peptides, it is necessary to consider two steps (coupling and deprotection) to incorporate each amino acid. For oligonucleotides, three steps must be taken into account: deprotection, coupling and oxidation. The capping step, which is mandatory for oligonucleotides and used in some steps for oligopeptides, must facilitate the final chromatographic purification by avoiding sequences with misincorporations and the formation of shorter truncated sequences, which would be difficult to separate from the product owing to their similar structures and behaviours.

When evaluating the efficiency of a metric and comparing it with other approaches, one must ascertain if the calculated values refer only to the upstream process or if they also consider downstream purification. The academic literature commonly reports upstream-related values. In communications from industry, the downstream stage is of fundamental importance and often guides strategic choices in the synthetic plan. In TIDES chemistry, achieving complete conversion in each step of an iterative chemical process is fundamental, as highlighted from the data reported in Table 2.2. By optimizing the conversion/yield of each step from 98 to 99.8%, the overall yield/purity of the oligomer showed a substantial improvement from 29.7 to 88.7% (see Table 2.2, entries 1–4). This value represents the limit for long sequences. An overall purity of >90% can be achieved only with shorter sequences (see Table 2.2, entry 5). Hence, in oligonucleotide synthesis where the number of units is commonly quite high, the hybrid synthetic approach is preferred.

The RME is a suitable metric to evaluate the impact of different technologies, such as solid-phase peptide synthesis (SPPS), liquid-phase peptide synthesis (LPPS) and microwave–solid-phase peptide synthesis (MW-SPPS), considering for each case the excess of amino acid required to achieve 100% conversion. The RME ranges related to the general protocol for solid and liquid phases in TIDES synthesis are reported in Table 2.3. In general, the

Table 2.2 Relationship between overall purity/yield for synthetic 30mer oligopeptides and 20mer ASO.[a]

Entry	No. of amino acids	No. of nucleotides	Single-step yield/%	Overall purity/%
1	30	20	98	29.7
2	30	20	99	54.7
3	30	20	99.5	74.0
4	30	20	99.8	88.7
5	15	10	99.8	94.2

[a]Values calculated considering two synthetic steps for the addition of a single amino acid in a peptide and cleavage, and considering three chemical steps for the addition of a nucleotide in an ASO and cleavage. The capping for ASO synthesis was not considered.

Table 2.3 Reagent excess in the upstream process.[a]

Entry	Technology	Amino acid equivalents	RME range/%	PMIaa[b]	PMIsn[b]
1	SPPS[b]	1.5–4	95–28	86	
2	LPPS[b]	1.1–1.5	88–95	200	
3	MW-SPPS[c]	3–5	37–23	180	
4	SPOS[d]	1.5	68		70–100
5	LPOS	1.05–1.4	89–72		100

[a]The reaction evaluated for the RME was Fmoc–Gly + Gly for peptides and *N*[4]-Bz-5′-*O*-DMTr-2′-*O*-(2-methoxyethyl)-5-methylcytidine-3′-CED-phosphoramidite + (2-methoxyethyl)cytidine. The solid support and the tag were not considered. Gly: Glycin.
[b]For the definition of PMIaa and PMIsn see the discussion in the text.
[c]See ref. 12.
[d]See ref. 27.

use of two equivalents is a standard procedure in SPPS technology. However, for certain difficult couplings, the reaction sometimes must be replicated (see Table 2.3, entry 1).[13] LPPS commonly requires a smaller excess of amino acid.[27] Moving to MW, the excess of amino acid is generally very high, which adversely affects the RME value.[28] For oligonucleotides, solid-phase oligonucleotide synthesis (SPOS) requires a larger amount of phosphoramidite[15,29] than liquid-phase oligonucleotide synthesis (LPOS)[30] (see Table 2.3, entries 4 and 5). As discussed above, elimination of additional protective groups for certain amino acids bearing a reactive moiety such as serine, threonine, tyrosine or arginine has a positive impact on the RME.

2.5 Process Mass Intensity (PMI)

As discussed above, the reaction stoichiometry is determined mainly by the necessity to have almost complete conversions and yields. However, considering the contributions to PMI calculations, reagents and auxiliaries are not the main components. The solvent is a critical factor, as in many other syntheses of APIs. Upstream and downstream in TIDES are closely related to each other, and both contribute consistently to the generation of waste in terms of solvent consumption. A decrease in or direct recycling of solvents is a key action to contain the PMI. In recent years, the PMI has been accepted as the metric of choice in the TIDES modality. Several calculations for oligopeptides and oligonucleotides of variable length have been reported.[14,15] However, to allow a direct comparison between different technologies, the evaluation can be done only after standardization of the PMI for incorporation in the sequence of a single amino acid (PMIaa) or a single nucleotide (PMIsn). These normalized data allow the effect due to the length of the TIDES and the size/loading of the resin/anchor to be eliminated. For the "normalization", the amount of the amino acid or the nucleoside is considered equal to 1. With respect to peptides, some data from the literature are reported in Table 2.3, but the list does not include improvements arising from industrial researchers. For SPPS, the main issue is the need to wash the resin to avoid the formation of side products coming from the residues of reagents and auxiliaries, in particular from the deprotection step. In this context, important efforts made by several groups were guided by the PMI evaluation[31] and several process analytical technologies (PATs) have been introduced.[32,33] In LPPS, the use of tags requires several water extractions and, in some cases, studies have not reported the PMI as a parameter to discuss the sustainability of the technology. For instance, the LPPS technology performances of Ajiphase were discussed in the original publication using the solvent intensity as a green metric. However, in this calculation, water was missed even if the washings necessary to remove the side products were the ones that increased the PMI consistently with respect to SPPS.[13] LPPS, anyway, can be carried out using several other coupling and purification technologies that do not require extensive washing or solvent consumption. Livingston and co-workers introduced new tags and used a membrane for diafiltration to eliminate side products and auxiliaries. However, in both versions of this approach, membrane-enhanced peptide synthesis (MEPS) and peptide synthesis *via* one-pot nanostar sieving (PEPSTAR),[34] the PMI was consistently higher than that calculated for SPPS.

For oligonucleotides, the nature of the solid phase for SPOS is crucial. Silica-based controlled-pore glass is mechanically and chemically stable towards the solvents used during oligonucleotide synthesis but is characterized by a low loading. However, the use of a polymeric support for SPOS and equipment more similar to a standard reactor for oligopeptide synthesis permits increased loading. Therefore, for large-scale production, the technology is similar to that for peptides with regard to loading, reaction time and reagent excess.

2.6 Process Mass Intensity with Recovery (PMIr)

PMIr is obtained by considering in the calculation of the PMI only the amount of new raw material and not the amount recovered. This information is useful because there is direct information on the product raw material consumption, or on the amount of waste. In TIDES, recycling of solvents by distillation is one of the most important approaches to achieving a low PMIr. For example, recycling contaminated dimethylformamide is difficult because of the side products and the tendency to decompose and be a source of nitrosamines.

2.7 From Kilograms to Large Production Volumes

The need to implement sustainable processes in TIDES production guided by green metrics is mandatory due to the shift from niche diseases to larger markets to meet an increased demand for drugs.

The gamechanger for oligopeptides was the success of this modality in the treatment of diabetes and obesity[35] and the introduction of oral formulations.[36]

Production that was focused on a few kilograms now needs to be produced at the level of tons. Interestingly, Novo Nordisk, the originator of liraglutide and semaglutide, developed a semisynthetic approach (recombinant fermentation + chemical modification).[37] The PMI remains the main green metric to manage the production of recombinant peptides according to the GCI-PR of the ACS.[38] Most companies producing generic compounds have filed processes based on SPPS, hybrid or chemo-enzymatic peptide synthesis technologies. The other blockbuster, tirzepatide from Eli Lilly, due to the presence of two non-proteinogenic amino acids in the 39-aa oligopeptide sequence, required the use of a hybrid chemical process (SPPS + LPPS).[39]

Oligonucleotides have been utilized primarily in the treatment of rare diseases owing to their ability to target the specific genetic sequences involved in these conditions.[40] However, their potential for treating more common conditions such as cardiovascular diseases, including the reduction of levels of low-density lipoprotein cholesterol, has attracted significant interest and research.[41] The launch of ASOs or small interfering RNAs (siRNAs) in the cardiovascular segment will change the production scale and primary technologies.

2.8 Conclusions

The use of green metrics is necessary to guide the process development of APIs in general. According to the Peptides Focus Team of the ACS GCI-PR,[42] and to our evaluation, the PMI is useful for the assessment of any technology. For the chemical synthesis of TIDES, the iterative nature of the sequence reduces the number of metrics that can be useful in guiding development of the process. Rapid assessment can be undertaken using the overall PMI. However, more specific parameters allow a head-to-head comparison of the different chemical technologies: the PMIaa/PMIsn allows the evaluation of the impact of solvents and chemicals normalizing for the introduction of a single amino acid or nucleotide, whereas the PMIr takes into account the recycling of some solvents or reagents, and the RME focuses on reagent excess and chemical yields.

List of Abbreviations

aa	Amino acid
ACS	American Chemical Society
AE	Atom economy
API	Active pharmaceutical ingredient
ASO	Antisense oligonucleotide
Boc	*tert*-Butoxycarbonyl
cEF	Complete E-factor
CEPS	Chemo-enzymatic peptide synthesis
E-factor	Environmental impact factor

EHS	Environmental, health and safety
Fmoc	Fluorenylmethoxycarbonyl
GCI-PR	Green Chemical Institute Pharmaceutical Roundtable
IP	Intellectual property
LPOS	Liquid-phase oligonucleotide synthesis
LPPS	Liquid-phase peptide synthesis
LPXS	LPOS and/or LPPS
MEPS	Membrane-enhanced peptide synthesis
MW	Microwave
PAT	Process analytical technology
PEPSTAR	Peptide synthesis *via* one-pot nanostar sieving
PMI	Process mass intensity
PMIaa	Process mass intensity normalized for a single amino acid addition
PMIsn	Process mass intensity normalized for a single nucleotide addition
PMIr	Process mass intensity with recovery
RME	Reaction mass efficiency
siRNA	Small interfering RNA (or short interfering RNA or silencing RNA)
SPOS	Solid-phase oligonucleotide synthesis
SPPS	Solid-phase peptide synthesis
SPXS	SPOS and/or SPPS
TIDES	Oligopeptides and oligonucleotides

References

1. M. Butters, D. Catterick, A. Craig, A. Curzons, D. Dale, A. Gillmore, S. P. Green, I. Marziano, J.-P. Sherlock and W. White, *Chem. Rev.*, 2006, **106**, 3002.
2. P. T. Anastas and J. C. Warner, *Green Chemistry: Theory and Practice*, Oxford University Press, New York, 1998.
3. F. Roschangar, Y. Zhou, D. J. C. Constable, J. Colberg, D. P. Dickson, P. J. Dunn, D. Eastgate, F. Gallou, D. Hayler, S. G. Koenig, M. E. Kopach, D. K. Leahy, I. Mergelsberg, U. Scholz, A. G. Smith, M. Henry, J. Mulder, J. Brandenburg, J. R. Dehli, D. R. Fandrick, K. R. Fandrick, F. Gnad-Badouin, G. Zerban, K. Groll, P. T. Anastas, R. A. Sheldon and C. H. Senanayake, *Green Chem.*, 2018, **20**, 2206.
4. F. Roschangar and J. Colberg, Green Chemistry Metrics, in *Green Techniques for Organic Synthesis and Medicinal Chemistry*, ed. W. Zhang, and B. W. Cue, John Wiley & Sons, Chichester, UK, 2018.
5. F. Roschangar, J. Li, Y. Zhou, W. Aelterman, A. Borovika, J. Colberg, D. P. Dickson, F. Gallou, J. D. Hayler, S. G. Koenig, M. E. Kopach, B. Kosjek, D. K. Leahy, E. O'Brien, A. G. Smith, M. Henry, J. Cook and R. A. Sheldon, *ACS Sustainable Chem. Eng.*, 2022, **10**, 5148.

6. A. P. C. Ribeiro, L. M. D. R. S. Martins, D. E. N. Bastos, A. F. Cristino and R. Galhano dos Santos, The importance of green chemistry metrics, in *Handbook of Greener Synthesis of Nanomaterials and Compounds*, ed. B. Kharisov, and O. Kharissova, Elsevier, 2021, ch. 2.
7. B. M. Trost, *Science*, 1991, **254**, 1471.
8. R. A. Sheldon, *Chem. Ind.*, 1992, 903.
9. F. Roschangar, R. A. Sheldon and C. H. Senanayake, *Green Chem.*, 2015, **17**, 752.
10. D. Hughes, *Pharma and Suppliers: Collaborating on Green Chemistry*, Launch of the PMI tool, February, 2011, http://www.acs.org/content/dam/acsorg/greenchemistry/industriainnovation/gcipr-informex-2011-pmi-tool.pdf
11. A. Borovika, J. Albrecht, J. Li, A. S. Wells, C. Briddell, B. R. Dillon, L. J. Diorazio, J. R. Gage, F. Gallou, S. G. Koenig, M. E. Kopach, D. K. Leahy, I. Martinez, M. Olbrich, J. L. Piper, F. Roschangar, E. C. Sherer and M. D. Eastgate, *Nat. Sustainability*, 2019, **2**, 1034.
12. B. C. Austad, T. L. Calkins, C. E. Chase, F. G. Fang, T. E. Horstmann, Y. Hu, B. M. Lewis, X. Niu, T. A. Noland, J. D. Orr, M. J. Schnaderbeck, H. Zhang, N. Asakawa, N. Asai, H. Chiba, T. Hasebe, Y. Hoshino, H. Ishizuka, T. Kajima, A. Kayano, Y. Komatsu, M. Kubota, H. Kuroda, M. Miyazawa, K. Tagami and T. Watanabe, *Synlett*, 2013, **24**, 333.
13. L. Ferrazzano, M. Catani, A. Cavazzini, G. Martelli, D. Corbisiero, P. Cantelmi, T. Fantoni, A. Mattellone, C. De Luca, S. Felletti, W. Cabri and A. Tolomelli, *Green Chem.*, 2022, **24**, 975.
14. L. Ferrazzano, D. Corbisiero, A. Tolomelli and W. Cabri, *Green Chem.*, 2023, **25**, 1217.
15. B. I. Andrews, F. D. Antia, S. B. Brueggemeier, L. J. Diorazio, S. G. Koenig, M. E. Kopach, H. Lee, M. Olbrich and A. L. Watson, *J. Org. Chem.*, 2021, **86**, 49.
16. L. Belkhir and A. Elmeligi, *J. Cleaner Prod.*, 2019, **215**, 185.
17. U. Onken, A. Koettgen, H. Scheidat, P. Schueepp and F. Gallou, *Chimia*, 2019, **73**, 730.
18. M. D'Este, M. Alvarado-Morales and I. Angelidaki, *Biotechnol. Adv.*, 2018, **36**, 14.
19. A. Isidro-Llobet, M. Alvarez and F. Albericio, *Chem. Rev.*, 2009, **109**, 2455.
20. P. J. Almhjell, C. E. Boville and F. H. Arnold, *Chem. Soc. Rev.*, 2018, **47**, 8980.
21. M. A. T. Blaskovich, *J. Med. Chem.*, 2016, **59**, 10807.
22. *Current Protocols in Nucleic Acid Chemistry*, ed. S. L. Beaucage, D. E. Bergstrom, G. D. Glick, and R. A. Jones, John Wiley & Sons, Inc., New York, 2000.
23. B. Roy, A. Depaix, C. Perigaud and S. Peyrottes, *Chem. Rev.*, 2016, **116**, 7854.
24. T. Gaich and P. S. Baran, *J. Org. Chem.*, 2010, **75**, 4657.
25. Y. Yang, L. Hansen and P. Ryberg, *Org. Process Res. Dev.*, 2022, **26**, 1520.
26. A. D. Curzons, D. J. C. Constable, D. N. Mortimer and V. L. Cunningham, *Green Chem.*, 2001, **3**, 1.

27. A. Sharma, A. Kumar, B. G. de la Torre and F. Albericio, *Chem. Rev.*, 2022, **122**, 13516.
28. S. L. Pedersen, A. P. Tofteng, L. Malik and K. J. Jensen, *Chem. Soc. Rev.*, 2012, **41**, 1826.
29. M. E. Kopach and B. I. Andrews, *Curr. Opin. Green Sustainable Chem.*, 2022, **37**, 100643.
30. X. Zhou, W. F. Kiesman, W. Yan, H. Jiang, F. D. Antia, J. Yang, Y. A. Fillon, L. Xiao and X. Shi, *J. Org. Chem.*, 2022, **87**, 2087.
31. R. Ravetti Duran and O. Ludemann-Hombourger, *Chem. Today*, 2021, **39**, 42.
32. B. G. de la Torre, S. Ramkisson, F. Albericio and J. Lopez, *Org. Process Res. Dev.*, 2021, **25**, 1047.
33. S. L. Broman, M. Rosenberg, F. Wojcik, A. H. Hansen, P. H. G. Egelund, J. Malmstrøm and D. S. Pedersen, *Org. Process Res. Dev.*, 2024, **28**, 666.
34. J. Yeo, L. Peeva, S. Chung, P. Gaffney, D. Kim, C. Luciani, S. Tsukanov, K. Seibert, M. Kopach, F. Albericio and A. Livingston, *Angew. Chem., Int. Ed.*, 2021, **60**, 7786.
35. L. B. Knudsen and J. Lau, *Front. Endocrinol.*, 2019, **10**, 155.
36. S. Berg, H. Edlund, W. R. F. Goundry, C. A. S. Bergstrom and N. M. Davies, *J. Pharm.*, 2022, **628**, 122238.
37. A. Staby, D. B. Steensgaard, K. F. Haselmann, J. Søndergaard Marino, C. Bartholdy, N. Videbæk, O. Schelde, H. Bosch-Traberg, L. Touborg Spang and D. J. Asgreen, *Pharm. Res.*, 2020, **37**, 120.
38. K. Budzinskia, M. Blewisb, P. Dahlinc, D. D'Aquilad, J. Esparzae, J. Gavinf, S. V. Hog, C. Hutchensg, D. Kahnh, S. G. Koeniga, R. Kottmeierg, J. Millarde, M. Snyderi, B. Stanarde and L. Sun, *New Biotechnol.*, 2019, **49**, 37.
39. M. O. Frederick, R. A. Boyse, T. M. Braden, J. R. Calvin, B. M. Campbell, S. M. Changi, S. R. Coffin, C. Condon, O. Gowran, J. McClary Groh, S. R. Groskreutz, Z. D. Harms, A. A. Humenik, N. J. Kallman, N. D. Klitzing, M. E. Kopach, J. K. Kretsinger, G. R. Lambertus, J. T. Lampert, L. M. Maguire, H. A. Moynihan, N. S. Mullane, J. D. Murphy, M. E. O'Mahony, R. N. Richey, K. D. Seibert, R. D. Spencer, M. A. Strege, N. Tandogan, F. L. Torres Torres, S. V. Tsukanov and H. Xia, *Org. Process Res. Dev.*, 2021, **25**, 1628.
40. S. J. Jo, S. U. Chae, C. B. Lee and S. K. Bae, *Int. J. Mol. Sci.*, 2023, **24**, 746.
41. K. Kyung An and P. Hun-Jun, *J. Lipid Atheroscler.*, 2023, **12**, 37.
42. I. Kekessie, K. Wegner, I. Martinez, M. E. Kopach, T. D. White, J. K. Tom, M. N. Kenworthy, F. Gallou, J. Lopez, S. G. Koenig, P. R. Payne, S. Eissler, B. Arumugam, C. Li, S. Mukherjee, A. Isidro-Llobet, O. Ludemann-Hombourger, P. Richardson, J. Kittelmann, D. S. Pedersen and L. J. van den Bos, *J. Org. Chem.*, 2024, **89**, 4261.

CHAPTER 3

Recent Trends in Peptide Drug Discovery and Development

H. Y. R. MADGE[a] AND M. MUTTENTHALER[a,b*]

[a]Institute for Molecular Bioscience, The University of Queensland, St Lucia, Brisbane, Queensland 4072, Australia; [b]Faculty of Chemistry, Institute of Biological Chemistry, University of Vienna, 1090 Vienna, Austria
*E-mail: m.muttenthaler@uq.edu.au, markus.muttenthaler@univie.ac.at

3.1 Introduction

Peptides are an important molecular drug class that have improved the lives of millions of people by offering effective treatments for metabolic disorders, cancers, and cardiovascular conditions. More than 90 peptide drugs have been approved since the first medical use of insulin in 1922,[1] accounting for ~6% (US$ 90 billion) of the total global drug market (2023; see Figure 3.1A).[2–5] Peptides are typically considered to be 4–100 amino acids in length, occupying a unique chemical and pharmacological space bridging small molecules and biologics while remaining in the realm of chemical synthesis, facilitating structure–activity relationship (SAR) studies and medicinal chemistry strategies to enhance their drug-like properties, pharmacokinetics (PK), and pharmacodynamics (PD).[1] Their chiral structure, size, and large surface area contribute to their remarkable potency and selectivity, and peptides typically display low toxicity owing to their high

Green Chemistry Series No. 85
Sustainability in TIDES Chemistry: Green Approaches to Oligonucleotides and Oligopeptides Synthesis
Edited by Alessandra Tolomelli, Lucia Ferrazzano and Walter Cabri

Published by the Royal Society of Chemistry, www.rsc.org

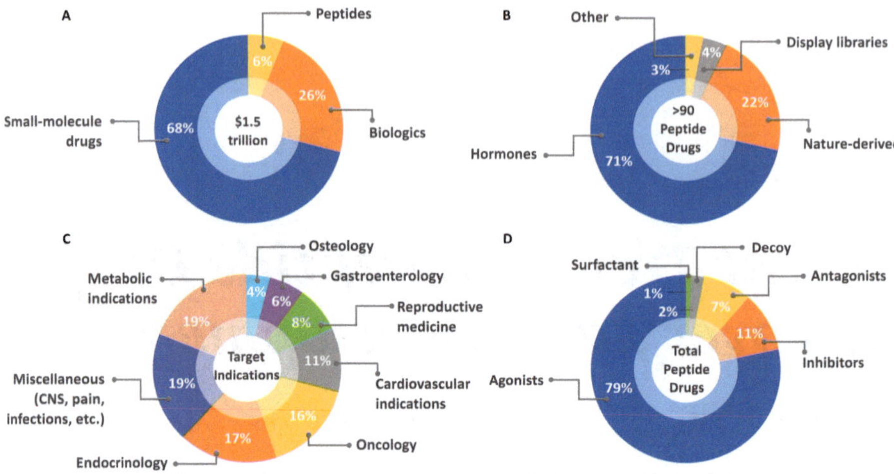

Figure 3.1 Overview of approved peptide drugs. (A) Source overview of all currently approved peptide drugs. (B) 2023 global pharmaceutical market and its distribution of small-molecule drugs, biologics, and peptide drugs. (C) Overview of distribution of function of approved peptide drugs. (D) Overview of target indications of approved peptide drugs.

target specificity.[6] Despite these advantages, some challenges and limitations remain, including difficulties crossing biological membranes, poor metabolic stability and oral bioavailability, rapid renal clearance, and the risk of immunogenicity.[7]

Peptide drugs have been discovered and developed from various sources (see Figure 3.1B), including endogenous peptide hormones and ligands [*e.g.* oxytocin, vasopressin, insulin, glucagon-like peptide 1 (GLP-1), and somatostatin analogues], natural products (*e.g.* ziconotide, exenatide, eptifibatide, and lepirudin), and, more recently, display technologies such as phage and mRNA display (pegcetacoplan and zilucoplan).[8–11]

The majority of approved peptide drugs are copies of endogenous peptide hormones or optimised mimetics; drug development efforts remain high in this area, driven by clinical success, well-understood signalling mechanisms, and broad involvement in numerous physiological and pathological processes.[1,12] Some of these have become blockbusters, with, for example, 10 insulin products in the top 200 selling pharmaceuticals, generating over $46 billion in annual revenue (2023).[13–15]

Driven by a paradigm shift in diabetes treatment, the second most successful peptide drug class is GLP-1 analogues, with several companies trying to capture a share of the ~$24 billion GLP-1 market.[5] With the recent approvals for glucagon-like peptide 1 receptor (GLP-1R) agonists also treating obesity, the total market for GLP-1-related drugs is predicted to rise to over $100 billion by 2030.[16] The range of therapeutic indications is highly diverse, with over 60% of peptide drugs equally distributed between cardiovascular indications, endocrinology, metabolic indications, and oncology (see Figure 3.1C).[1] Several somatostatin analogues (*e.g.* lanreotide and octreotide) are among the top 200 selling drugs, grossing over $1.3 billion and $1.2 billion annually, respectively.[17] Within the oncology class, gonadotropin-releasing hormone agonists goserelin and leuprolide each gross ~$1 billion annually, supporting the role of peptides in oncology.[17] Peptides are also entering the market for gastrointestinal disorders, with linaclotide leading the field as an orally administered and gut-restricted blockbuster drug for treating irritable bowel syndrome with constipation.[18]

The involvement of peptides in many disorders and the commercial successes of several first-in-class peptide drugs keep the momentum going and investors interested, with over 200 peptide drug candidates now in various stages of clinical development.[8,19]

Indications linked to a reduced physiological function of peptide hormones were the entry point for peptide drugs. Pursuing well-understood biology and potent agonists to offset the then very high manufacturing costs resulted in the first approvals and commercial successes (see Figure 3.1D).[20,21] Peptide drug manufacturing became more cost-effective with enfuvirtide, a 36-mer HIV fusion inhibitor that required a ton production scale, revolutionising the chemical production of amino acid building blocks and large-scale peptide assembly,[22] leading to a new era of peptide drug discovery and development, in both industry and academia.[21] Today, many manufacturing techniques are used to produce peptide drugs on a multikilogram scale, with linear side-chain functionalised peptides such as tirzepatide [dual GLP-1R/glucose-dependent insulinotropic polypeptide receptor (GIPR) agonist approved for type 2 diabetes and obesity] manufactured through hybrid solid-phase peptide synthesis (SPPS) and liquid-phase peptide synthesis (LPPS) in flow-based reactors.[23]

Endogenous ligands such as peptide hormones will continue to be a primary source for drug development; however, confirmational constrained and stable macrocyclic peptides discovered either from nature or through peptide display technologies (phage and mRNA) are expected to increase the number of peptide drug modalities in development as they expand the drug target space ranging from protein–protein interaction inhibitors to ion channel blockers and immune system modulators, shaping the future landscape of peptide therapeutics.

Since our latest review that covered peptide drug discovery and development up to 2019,[1] 10 new peptide drugs have been approved (see Table 3.1), with record numbers in both preclinical and clinical testing.[24–27] The following sections cover these approvals, their underlying discovery and development strategies, and the related individual fields and disease indications.

Table 3.1 FDA-approved peptide drugs from 2019 to 2023.[a]

Drug (Trade name)	Company	Mode of Action	Indication	Source	Structure	Year	Ref
Afamelanotide (Scenesse)	Clinuvel	MC1R Agonist	Erythropoietic protoporphyria	H (α-MSH)	*N*-acetylated linear 13 AA, Nle^4, D-Phe^7	2019	28-30
Bremelanotide (Vyleesi)	Palatin Technologies	MC1-4R Agonist	Women hypoactive sexual desire	H (α-MSH)	*N*-acetylated cyclic (sidechain lactam) 13 AA, Nle^4, D-Phe^7	2019	31, 32
Setmelanotide (Imcivree)	Rhythm Pharmaceuticals	MC4R Agonist	Single gene mutation obesity	H (α-MSH)	*N*-acetylated cyclic (Cys^2- Cys^8 disulfide) 9 AA, D-Ala^3, D-His^4	2019	33, 34
Bulevirtide (Hepcludex)	Gilead Sciences	LBAT Inhibitor	Antiviral hepatitis delta virus infection	N (HBV)	*N*-C_{14} acylated linear 47 AA	2020	35-37
Vosoritide (Voxzogo)	BioMarin Pharmaceutical	NPR-B Agonist	Achondroplastic dwarfism	H (CNP)	Cyclic (Cys^{23}-Cys^{29} disulfide) 37 AA	2020	38-40
Pegcetacoplan (Empaveli)	Apellis Pharmaceuticals	C3 Inhibitor	Paroxysmal nocturnal haemoglobinuria	D (Phage)	*N*-acetylated cyclic (Cys^2-Cys^{12} disulfide) 13 AA PEG linked dimer	2021	41, 42
Dasiglucagon (Zegalogue)	Zealand Pharma / Novo Nordisk	GCGR Agonist	Severe hypoglycaemia	H (GCG)	Linear 29 AA, Aib^{16}	2021	43, 44
Tirzepatide (Mounjaro)	Eli Lilly	GLP-1R/GIPR Agonist	Type 2 diabetes	H (GLP-1/GIP)	Linear 39 AA, $Aib^{2,13}$, Lys^{20}-OEG-Glu-C_{20}	2022	45-47
Motixafortide (Aphexda)	BioLineRx	CXCR4 Antagonist	Multiple myeloma	N (Horseshoe crab)	4-flurobenzyoyl *N*-acylated cyclic (Cys^4-Cys^{14} disulfide), 14 AA, Nal^3, Cit^6, D-Lys^8, Cit^{12}	2023	48, 49
Zilucoplan (Zilbrysq)	Union Chimique Belge	C5 Inhibitor	Myasthenia gravis	D (mRNA)	N-acetylated cyclic (sidechain lactam) 15 AA, cyclohexyl Gly^{14}, 7-aza-Trp^{10}, butyl-Gly^8, *N*-methyl-Asp^7	2023	50, 51

[a]AA (amino acid), H (hormone), N (nature derived), D (display technology), α-MSH (α-melanocyte-stimulating hormone), HBV (hepatitis B virus), CNP (C-type natriuretic peptide), phage (2×10^8 27-mer phage display peptide library), mRNA (co-translational enzymatic charging mRNA display), MCR (melanocortin receptor), LBAT (liver bile acid transporter), NPRB (natriuretic peptide receptor B), C3 (complement component 3), GCGR (glucagon receptor), CXCR4 (chemokine receptor 4), C5 (complement component 5), erythropoietic protoporphyria (accumulation of protoporphyrins in red blood cells causing acute, painful photosensitivity), paroxysmal nocturnal haemoglobinuria (destruction of red blood cells by the immune complement system), GLP-1 (glucagon-like peptide 1), GIP (glucose-dependent insulinotropic polypeptide), Nle (norleucine), Phe (phenylalanine), Ala (alanine), His (histidine), Cys (cysteine), Aib (aminoisobutyric acid), Lys (lysine), OEG (8-amino-3,6-dioxaoctanoic acid), Glu (glutamic acid), Gly (glycine), 7-aza-Trp (7-azatryptophan), Asp (aspartic acid), Nal (naphthylalanine), Cit (citrulline).

3.2 Peptide Drug Development Based on Peptide Hormones

3.2.1 Insulin Drug Development – Setting the Scene for Peptide Drug Development and Company Growth

Diabetes has been a major drug development target for over 100 years, with extracted animal insulin forming an effective treatment as early as 1922. The advent of recombinant peptide/protein expression technologies in the early 1980s allowed the production of human insulin (and the foundation of Genentech), followed by a suite of insulin mimetics with different PK profiles available today (see Figure 3.2A).[52,53] These rapid, short- and long-acting

A

Type	Name	Trade name	Company	Modifications	Year
Ultra-rapid-acting (onset 5-10 min, duration 3.5-4 h)	Faster insulin aspart	FiAsp	Novo Nordisk	B: P28D	2017
Rapid-acting (onset 15-20 min, duration 3.5-4.5 h)	Insulin lispro	Humalog	Eli Lilly	B: P28K, K29P	1996
	Insulin aspart	NovoRapid	Novo Nordisk	B: P28D	2000
	Insulin glulisine	Apidra	Sanofi	B: N3K, K29E	2004
Short-acting (onset 1 h, duration 6-8 h)	Neutral insulin	Actrapid	Novo Nordisk	–	2002
		Humulin R	Eli Lilly	–	1982
Intermediate-acting (onset 1-1.5 h, duration 16-24 h)	Insulin isophane	Humulin NPH	Eli Lilly	Protamine sulphate, zinc oxide crystalline suspension	1982
		Protaphane	Novo Nordisk	Protamine sulphate, zinc oxide crystalline suspension	2002
Long-acting (onset 3-4 h, duration 24-36 h)	Insulin detemir	Levemir	Novo Nordisk	B: K29K-C_{14}	2005
	Insulin glargine (U100)	Optisulin	Sanofi	A: N21G, B:+31RR	2001
		Semglee	Alphapharm	A: N21G, B:+31RR	2020
	Insulin glargine (U300)	Toujeo	Sanofi	A: N21G, B:+31RR	2000
	Insulin degludec	Tresiba	Novo Nordisk	B: ΔT30, K29E-C_{16}-COOH	2015

C_{14} (tetradecanoic acid), U100/300 (100/300 insulin units/mL), Δ (amino acid deletion), C_{16}-COOH (hexadecandioyl acid).

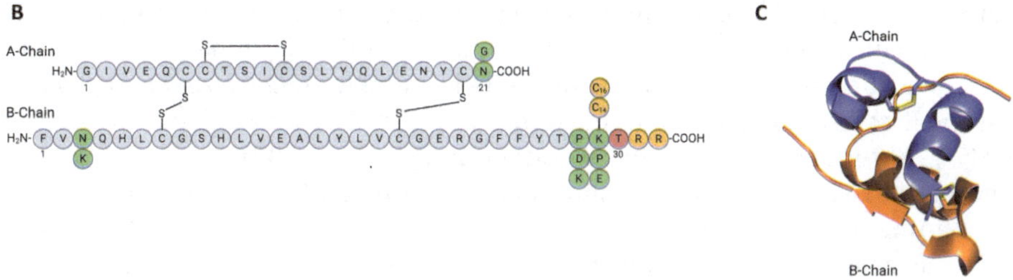

Figure 3.2 Overview of insulin peptide drugs and their structures, modifications, and applications. (A) Insulin drug analogues, their pharmacokinetics, developers, modifications and approval year. (B) Chemical structure of human insulin with residues highlighted that are modified in insulin analogues (green, substitutions; yellow, additions; red, deletions). (C) X-ray crystal structure of human insulin (PDB: 3E7Y).

insulin analogues allow patients to manage glucose levels more easily over extended periods.[54,55] Many of these modified products have been developed by simple single or double amino acid substitutions (rapid-acting) or lipidation through side-chain modifications to enhance albumin binding (long-acting) (see Figure 3.2B).[56,57] Insulin (see Figure 3.2C) drugs have seen great clinical and commercial success, generating revenue of $46 billion in 2023.[15,58]

Improving peptide drug PK and PD through small modifications to achieve prolonged circulation times, fewer injections, and enhanced PK/PD profiles for specific disease applications remains a key strategy for peptide drug development and company growth. This strategy, exemplified with insulin and later with GLP-1 analogues, represents a robust business model that results in product portfolio diversification based on leveraging existing expertise and infrastructure and regular patent extensions underpinning the market lead.

3.2.2 GLP-1 Drug Development – A New Era for Treating Diabetes, Obesity, and Cardiovascular Disorders

GLP-1 drug development goes back to the late 1980s when various companies pursued human GLP-1 as an insulin alternative for treating type 2 diabetes. Initial investigations of the therapeutic use of full-length GLP-1 (see Figure 3.3A) were unsuccessful; it was then discovered that GLP-1 is proteolytically cleaved into two active circulating forms, GLP-1(7–37) and GLP-1(7–36)-amide (see Figure 3.3A).[59] Further developments were based on discoveries that biologically active GLP-1 [GLP-1(7–37) and GLP-1(7–36)-amide] stimulated insulin secretion while also inhibiting the secretion of glucagon, lowering hepatic glucose production.[60] Intravenous infusion of GLP-1(7–37) furthermore normalises fasting hyperglycaemia in patients with type 2 diabetes, decreasing the risk of complications during hyperglycaemia.[59] However, the first attempts to develop GLP-1(7–37) into a drug failed owing to its rapid metabolism ($t_{½}$ = 1–2 min) by dipeptidylpeptidase-4 (DPP-4) and renal clearance.[61,62] Major efforts were undertaken to improve the PK of GLP-1(7–37), but it took the discovery of exenatide (Eli Lilly, 2005), a 39-residue long peptide found in the saliva of the Gila monster (*Heloderma suspectum*), for the first GLP-1R agonist to reach the clinic.[63,64] Even though exenatide has only 53% sequence homology to endogenous GLP-1(7–37), it is a potent full agonist at the human GLP-1R and stable against DPP-4 degradation due to an alanine to glycine modification at position 2 of the N-terminus. Exenatide also has a significantly slower rate of renal clearance (5–7 h) compared with GLP-1(7–37) (~5 min).[65,66] This enhanced PK profile allowed a twice-daily injection regime for treating type 2 diabetes.[67]

Following the approval of exenatide, several GLP-1R agonists based on GLP-1(7–37) and exenatide were developed and approved (see Figure 3.3B).

A

Name	Peptide Sequence	Modifications
GLP-1	HDEFERHAEGTFTSDVSSYLEGQAAKEFIAWLVKGRG	–
GLP-1(7-37)	HAEGTFTSDVSSYLEGQAAKEFIAWLVKGRG	–
GLP-1(7-36) amide	HAEGTFTSDVSSYLEGQAAKEFIAWLVKGR*	–
GIP	YAEGTFISDYSIAMDKIHQQDFVNWLLAQKGKKNDWKHNITQ	–
Exenatide	HGEGTFTSDLSKQMEEEAVRLFIEWLKNGGPSSGAPPPS*	–
Liraglutide	HAEGTFTSDVSSYLEGQAAKEFIAWLVRGRG	Lys^{20}-Glu-C_{16}
Albiglutide#	HGEGTFTSDVSSYLEGQAAKEFIAWLVKGR	C-terminal albumin
Dulaglutide#	HGEGTFTSDVSSYLEEQAAKEFIAWLVKGG	C-terminal IgG4-Fc
Lixisenatide	HGEGTFTSDLSKQMEEEAVRLFIEWLKNGGPSSGAPPSKKKKKK*	Poly-Lys C-terminus
Semaglutide	HXEGTFTSDVSSYLEGQAAKEFIAWLVRGRG	Aib^2, Lys^{20}-$(OEG)_2$-Glu-C_{18}-COOH
Tirzepatide	YXEGTFTSDYSIXLDKIAQKAFVQWLAIGGPSSGAPPPS*	Aib^2, Aib^{13}, Lys^{20}-$(OEG)_2$-Glu-C_{20}-COOH

# indicates peptide-protein conjugate

B

Name	Brand name	Company	Half-life$	Administration	Year	Ref
GLP-1(7-37)	–	–	2 min	–	–	60, 61
Exenatide	Byetta (Type 2 diabetes)	AstraZeneca, Amylin	2.4 h	2x daily (s.c.)	2005	63, 71
	Bydureon (Type 2 diabetes)	AstraZeneca, Amylin	2.4 h	1x weekly (s.c.)	2012	72, 73
Liraglutide	Victoza (Type 2 diabetes) Saxenda (Obesity)	Novo Nordisk	12 h	1x daily (s.c.)	2010 2014	69, 70, 74
Albiglutide	Tanzeum (Type 2 diabetes)	GSK	120 h	1x weekly (s.c.)	2014	75, 76
Dulaglutide	Trulicity (Type 2 diabetes)	Eli Lilly	90 h	1x weekly (s.c.)	2014	77, 78
Lixisenatide	Lixumia (Type 2 diabetes)	Sanofi	4 h	1x daily (s.c.)	2016	68, 79
Semaglutide	Ozempic (Type 2 diabetes) Wegovey (Obesity)	Novo Nordisk	160 h	1x weekly (s.c.)	2017 2021	80-82
Semaglutide	Rybelsus (Type 2 diabetes)	Novo Nordisk	5 days	1x daily (oral)	2019	83
Tirzepatide	Mounjaro (Type 2 diabetes)	Eli Lilly	5 days	1x weekly (s.c.)	2022	84, 85

$ circulating half-life

C

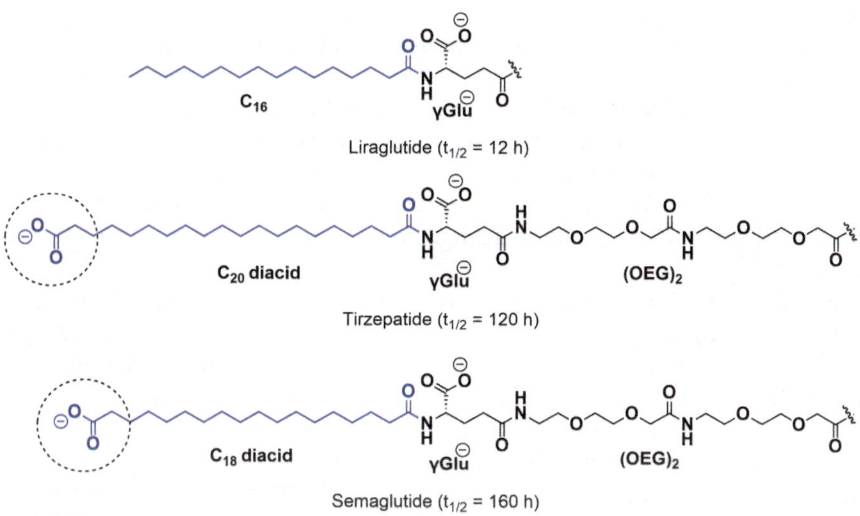

Figure 3.3 GLP-1 peptide drug development. (A) Amino acid sequence and modifications of approved GLP-1R and GLP-1R/GIPR agonists. (B) Overview of approved GLP-1 peptide drug analogues with their sequence and modification information. (C) Albumin binding fatty acids (blue) and linkers (black) in approved GLP-1R agonists. Approved GLP-1 peptide drug structures, with albumin binding fatty acids (blue) and linkers (black).

Similar approaches to those used in developing long-acting insulin analogues were used, such as lipidation and strategic amino acid substitutions, in addition to novel strategies to extend the *in vivo* half-life, for example, through covalent conjugation to IgG Fc fragments or serum proteins.[68–70]

Liraglutide (Victoza, Novo Nordisk, 2010, once-daily s.c.) was the first modified GLP-1(7–37) analogue to reach the market, representing over a decade of work by Novo Nordisk's diabetes team.[70] Simply improving the stability of GLP-1(7–37) to DPP-4 was not sufficient to extend its circulating half-life owing to rapid renal clearance; however, lipidation of the peptide for non-covalent serum albumin binding resulted in improved circulation PK.[86] Multiple acylation sites along GLP-1(7–37) were explored during development and also a range of fatty acids and hydrophilic linkers.[70] Acylation was avoided at the N-terminal part owing to participation in GLP-1R binding; positions at the C-terminal part were well tolerated, with Lys^{26} retaining the highest potency. Numerous fatty acids (monoacids and diacids) of varying lengths and chiral and achiral linkers were examined, and a combination of a negatively charged glutamic acid spacer with a C_{16} fatty monoacid provided the best outcomes regarding potency and *in vivo* half-life (see Figure 3.3C).[87,88] This resulted in a once-daily dosing schedule and improvements in glycaemic control, and liraglutide replaced exenatide as the preferred type 2 diabetes medication with annual sales of over $262 million in its first year (2010), growing to over $1 billion in 2013.[69,89] Exenatide was discontinued in 2023 owing to low prescriptions and sales.

Several companies pursued *covalent* conjugation to proteins to enhance the circulation time. Albiglutide (Tanzeum, GSK, 2014, once-weekly s.c.), for example, is a DPP-4-stable GLP-1(7–36) analogue covalently conjugated to human albumin with a half-life of 4–7 days. Dulaglutide (Trulicity, Eli Lilly, 2014, once-weekly s.c.), also a DPP-4-stable GLP-1(7–36) analogue, is C-terminally conjugated to the Fc regions of recombinant human IgG4 (one antibody with two peptide ligands) with a circulating half-life of 5 days.[77,90,91]

Semaglutide (Ozempic, Novo Nordisk, 2017, once-weekly s.c.), the follow-up drug to liraglutide, has an Aib^2 (aminoisobutyric acid) modification to prevent DPP-4 cleavage and a C_{18} fatty diacid attached to a Glu–$(OEG)_2$ linker to extend its circulation half-life *via* serum albumin binding. Several trends were observed during structure–activity relationship studies carried out on fatty acid length and monoacid *versus* diacid towards serum albumin affinity and GLP-1R potency: increasing the fatty acid length improved albumin binding and circulating half-life but reduced GLP-1R potency.[81] Albumin binding was improved by replacing the C_{16} fatty monoacid of liraglutide with a C_{18} fatty diacid. This was explained by the presence of basic amino acids (positively charged at physiological pH) at the edge of the albumin lipid binding pockets, resulting in ionic interactions with the negatively charged carboxylic head groups of the diacid linker.[92] The presence of the extended lipid moiety attenuated the potency of the compounds at GLP-1R, but incorporating a negatively charged Glu–$(OEG)_2$ linker improved potency (from 270 to 6.2 pM, even better than liraglutide).[91] Semaglutide was the first non-protein covalently conjugated GLP-1(7–37) analogue for once-weekly dosing with a half-life of 7 days.[93] Semaglutide eventually overtook dulaglutide in sales ($8.7 billion *vs.* $7.4 billion) in 2022 owing to superior performance in both glycaemic control and bodyweight reduction in a head-to-head clinical trial, even though dulaglutide entered the market three years earlier with the same dosing schedule.[17,94]

These GLP-1R agonists revolutionised type 2 diabetes treatment, becoming one of the preferred treatment options, allowing patients to achieve non-glycaemic targets that could not be obtained with regular insulin therapy, and also additional glycaemic benefits of GLP-1R agonists such as improvements in β-cell function, blood pressure, and body weight reduction. GLP-1R agonists also have a lower risk of inducing hypoglycaemia.[95,96] The declaration of obesity as a disease catalysed a new wave of research and development for GLP-1R agonists as weight-loss drugs that ultimately led to a surge in demand with global shortages.[97] GLP-1R agonists had already demonstrated weight loss in preclinical studies and clinical trials for type 2 diabetes, supporting their scope for treating obesity.[98,99]

Liraglutide (Saxenda, Novo Nordisk, 2014, once-daily s.c.) was the first GLP-1R agonist to be approved for treating obesity in combination with diet and exercise for people with a body mass index (BMI) of >30.[100] Liraglutide decreased participants' weight by 8.4 ± 7.3 kg on average, with some losing over 10% of their body weight, which was far above the gold standard of other anti-obesity drugs (typically ~3–5% of body weight).[101–103]

Semaglutide (Wegovey, Novo Nordisk, reformulated for higher dose, 2021, once-weekly s.c.) followed Saxenda based on 15% body weight reduction data among clinical trial participants, a feat that had been achieved previously only through invasive surgical action such as gastric banding.[80]

All GLP-1R agonists approved for treating obesity are formulated for subcutaneous injection; being highly potent agonists with EC_{50} below 10 pM, the doses remain incredibly small, particularly when considering once-weekly dosing schedules. Studies concluded that GLP-1R agonists are more cost-effective than their insulin counterparts in treating type 2 diabetes.[104,105]

Semaglutide (Rybelsus, Novo Nordisk, 2019, once-daily oral) is so far the first and only GLP-1R agonist developed for oral administration,[106] representing a significant landmark in peptide drug development, even though the oral bioavailability is very low (~1%) and feasible only because of the outstanding potency of semaglutide.[107] Achieving efficacious systemic levels of orally administered peptides is difficult owing to substantial obstacles such as rapid gastrointestinal degradation and restricted transepithelial uptake.[83] Nevertheless, it is still the commercial belief that oral administration is the preferred route among patients. The oral activity of semaglutide is achieved by delivering a high dose (3–14 mg once-daily oral *vs.* 0.25–1.0

mg once-weekly s.c., representing a ~100-fold higher dose) formulated with the intestinal permeation enhancer salcaprozate.[83,108–110] This can be considered a brute-force approach to oral peptide delivery, yet still breaks new ground in developing oral peptide drugs.[83] Rybelsus was benchmarked in over 10 phase III clinical trials, demonstrating equivalence and, in some cases, increased effectiveness over competitors in direct head-to-head clinical studies.[111]

Despite the extraordinary investment in clinical trials, the market uptake of Rybelsus has been lower than that of injectable semaglutide, with many patients opting for a once-weekly subcutaneous injection over a daily oral pill required to be taken on an empty stomach at least 30 min before food intake.[112] Rybelsus still brought in over $1.6 billion in 2022, becoming Novo Nordisk's second largest driver of annual growth after Ozempic (semaglutide for type 2 diabetes) in only its third year on the market.[113]

GLP-1R agonists now represent some of the most commercially successful drugs on the overall pharmaceutical market, having a market size of over $24 billion in 2023 (combined type 2 diabetes and obesity). As of 2023, there are no new GLP-1R agonist monotherapies (in contrast to dual agonists) in clinical trials; however, several approved GLP-1R agonists are under investigation for secondary indications such as heart failure, Alzheimer's disease, and Parkinson's disease.[114–116]

3.2.3 Dual GLP-1R/GIPR/GCGR Agonists – The Next Generation of Type 2 Diabetes and Obesity Treatments

Despite the success of GLP-1R agonists, many patients are still unable to meet ideal glycaemic targets, and in some the weight loss achieved is well below what can be achieved through surgical intervention.[47,117] With increasing numbers of patients suffering from obesity and/or type 2 diabetes, the need for metabolic pharmacological therapies remains high. Having made significant advances in peptide–drug modifications to improve PK properties, as evidenced by the half-life extension of the GLP-1R agonists (from 2.5 h to 7 days), the focus shifted to a novel strategy of attempting to engage additional pathways involved in nutrient and energy metabolism.[118] The glucose-dependent insulinotropic polypeptide receptor (GIPR) and glucagon receptor (GCGR) have been identified as highly promising secondary drug targets.[117,118]

Tirzepatide (Eli Lilly) was the first of these co-agonist chimera to receive regulatory approval for treating type 2 diabetes (Mounjaro, 2022, once-weekly s.c.) and for obesity (Zepbound, 2023, once-weekly s.c.).[46,84] Tirzepatide, a 39-residue long unimolecular co-agonist of GLP-1R and GIPR with a sequence comprising common motifs present from GLP-1(7–37), GIP, and exenatide (see Figure 3.3A), is one of the most advanced peptide drugs developed to date.[119,120] Unlike a dual epitope conjugate or fusion with heterobivalent pharmacophores, tirzepatide boasts a sophisticated chimeric-like design with selected residues from each of the parent peptides throughout the backbone along with the incorporation of two non-natural Aib residues (positions 2 and 13), and also a half-life-extending C_{20} fatty diacid side-chain modification that has become standard for drugs of this class (see Figure 3.3C).[47,121]

Several unimolecular co-agonists targeting GLP-1R and GIPR (VK2735 from Viking Therapeutics) or GLP-1R and GCGR (survodutide from Boehringer Ingelheim, mazdutide from Eli Lilly) have entered clinical trials; however, they are yet to receive regulatory approval.[122–125]

Retatrutide (Eli Lilly, phase III, 2023, once-weekly s.c.) is an extension of this chimera design, a unimolecular triagonist targeting GLP-1R, GIPR, and GCGR.[126] In a phase II clinical trial, over 60% of the participants lost at least 20% of their body weight, with some losing over 30%.[126] Retatrutide has the potential to induce weight loss more than any drug currently on the market.

Development of dual peptide hormone therapies for type 2 diabetes and obesity is not without challenges, and with the highly competitive nature of the field it is not surprising that a number of candidates in this class have been abandoned owing to a lack of efficacy, side effects, or lack of a suitable PK profile, limiting the commercial competitiveness of the products (*e.g.* JNJ-64565111 from Johnson & Johnson, SAR425899 from Sanofi, cotadutide from AstraZeneca). Several small-molecule GLP-1R agonists are also under clinical development for type 2 diabetes and obesity, such as danuglipron and lotiglipron from Pfizer and orforglipron from Eli Lilly.[127]

3.2.4 Melanocyte-stimulating Hormone Analogues – Treatments for an Array of Metabolic and Endocrine Disorders

The melanocortin system comprises five melanocortin receptor subtypes (MCR1–5) involved in diverse physiological functions, including hair colour and skin pigmentation (MC1R), adrenocortical steroidogenesis (MC2R), energy homeostasis (MC3R), and sexual function and energy homeostasis (MC4R).[128] The exact role of MC5R remains not fully understood, with recent evidence suggesting that it may be involved in immune and inflammatory responses.[129] The MCR system has also been identified as a valuable drug target for several indications, with ongoing clinical trials for indications such as rheumatoid arthritis (MC2R, MC3R), scleroderma (MC2R), and idiopathic membranous nephropathy (MC1R).[130] Four endogenous MCR ligands (see Table 3.2), namely adrenocorticotropic hormone (ACTH) and α-, β-, and γ-melanocyte-stimulating hormones (MSHs), activate these receptors.[42] These endogenous peptide ligands elicit activity across all MCRs with varying subtype specificity, ACTH (MC2R), α-MSH

Table 3.2 Melanocortin system ligands.

Name	Peptide Sequence	Modifications
ACTH	SYSME**HFRW**GKPVGKKRRPVKVYPNGAEDESAEAFPLEF	–
α-MSH	SYSME**HFRW**GKPV*	–
β-MSH	AEKKDEGPYRME**HFRW**GSPPKD	–
γ-MSH	YVMG**HFRW**DRF	–
Afamelanotide	Ac-SYSNleE**HfRW**GKPV*	Nle[4], D-Phe[7]
Bremelanotide	Ac-Nle-cyclo[D**HfRW**K]	Cyclic (sidechain lactam D[2]-K[7]), Nle[4], D-Phe[7]
Setmelanotide	Ac-R-cyclo[Ca**hFRW**C]*	Cyclic (Cys[2]-Cys[8] disulfide), D-Ala[3], D-His[4]

* C-terminal amide

(MC1R, MC5R), β-MSH (MC4R), and γ-MSH (MC3R).[131] These pituitary peptide hormones are produced endogenously through cleavage of the pro-opiomelanocortin hormone (POMC). Synthetic analogues of C-terminally truncated analogues of ACTH were approved for multiple sclerosis and infantile spasms as early as 1952, but most of them have seen more utility as diagnostics for pituitary and adrenal gland disorders.[132] The α-, β-, and γ-MSHs are shorter peptides, with half-lives of less than 15 min.[133] α-MSH is the best characterised and has been the focus of most clinical melanocortin drug development programmes. α-MSH has potent agonist activity (<1 nM) at MCR1, -3, -4, and -5; however, owing to its short half-life, it is not a suitable drug candidate.[42]

Afamelanotide (Scenesse, Clinuvel Pharmaceuticals, 2019, bimonthly s.c. slow-release implant) was the first α-MSH-derived drug approved for treating erythropoietic protoporphyria, an inherited disease in which the accumulation of protoporphyrins in red blood cells causes painful photosensitivity.[134] Afamelanotide is administered every two months as a subcutaneous slow-release implant during periods of high sunlight exposure. Afamelanotide activates MC1R with 4-fold greater potency than endogenous α-MSH and, upon activation, stimulates melanocytes to synthesise eumelanin, which acts as a photoprotectant by absorbing, scattering, and quenching ultraviolet light.[28] Two amino acid substitutions on the native α-MSH sequence (Met[4] to Nle and Phe[7] to D-Phe) and N-terminal capping by acetylation increased the stability of the endogenous peptide, affording a compound with increased proteolytic stability (30 min) and high potency at MC1R (0.46 nM) capable of providing photoprotection to individuals suffering from erythropoietic protoporphyria.[28,131]

Bremelanotide (Vyleesi, Palatin Technologies, 2019, 45 min s.c. prior to sexual activity) was the second α-MSH-derived drug to gain approval for treating generalised hypoactive sexual desire disorder (HSDD) in premenopausal women. Bremelanotide is a non-selective MC1R, MC3R, MC4R, and MC5R agonist; however, its effects on sexual function are driven primarily through MC3R and MC4R. Following the identification of a common melanocortin tetrapeptide His–Phe–Arg–Trp (HFRW) pharmacophore, homology modelling and docking studies indicated that this tetrapeptide motif may adopt a β-turn conformation.[135] Bremelanotide contains the same non-natural amino acid substitutions as afamelanotide, but it also has a side-chain to side-chain lactam bridge, which stabilises the conformation of the pharmacophore, leading to higher potency (0.25 nM, MC4R).[131] This cyclisation also increased the circulating half-life to 2–4 h compared with 30 min for linear afamelanotide.[136]

Owing to the role of the central melanocortin system (MC4R, CNS) in appetite control and energy homeostasis, several melanocortin peptide analogues and MCR small-molecule agonists have entered clinical development for treating obesity: LY2112688 (peptide, Eli Lilly), MC4-NN-0453 (peptide, Novo Nordisk), AZD2820 (small molecule, Astra Zeneca), and MK-0493 (small molecule, Merck). However, owing to the non-selective nature of this drug class, side effects (cardiovascular) and lack of efficacy led to all of them being abandoned.[137]

Setmelanotide (Imcivree, Rhythm Pharmaceuticals, 2020, once-daily s.c.) was the first selective MC4R (10-fold over MC3R) agonist to be approved for the treatment of severe genetic obesity caused by POMC deficiency, proprotein convertase subtilisin/kexin type 1 (PCSK1) deficiency, or leptin receptor (LEPR) dysfunction.[138,139] Structural modifications of α-MSH led to setmelenotide exhibiting 20-fold higher potency at MC4R.[140] Setmelenotide contains the same HFRW pharmacophore as bremelanotide and afamelanotide; however, compared with bremelanotide it lacks N-terminal norleucine and C-terminal lysine and instead contains D-alanine,[3] an N-terminal L-arginine,[1] and a disulfide bridge (C^2–C^8) as the conformational locking moiety.[141] It is expected that the recently resolved high-resolution structures of MC4R will accelerate the development of melanocortin agonists with improved selectivity profiles.[142]

3.2.5 Natriuretic Peptides Analogues – From Cardiovascular System to Rare Disease Treatment

Vosoritide (Voxzogo, BioMarin, 2021, once-daily s.c.) was approved for treating achondroplasia, a genetic condition causing overactive fibroblast growth factor, leading to downregulation of bone growth, which presents clinically as disproportionate short stature or dwarfism.[143] Although natriuretic peptides are traditionally known for their role in the regulation of the cardiovascular system, C-type natriuretic peptide (CNP) plays an important role in endochondral ossification.[144] In a multinational clinical trial, vosoritide, a long-acting synthetic analogue of CNP, significantly increased annual growth velocity in children following a once-daily injection.[38] Vosoritide consists of the 37 amino acid endogenous CNP with an additional Pro–Gly at the N-terminus; this minimal modification is sufficient to provide resistance to neutral endopeptidase metabolism, increasing the half-life of endogenous CNP from 2.6 to 30 min, and is suitable for once-daily subcutaneous administration.[145]

The endogenous human NP family consists of five peptides with varying activity across two major particulate guanylyl cyclase receptors (pGCR), pGC-A and pGC-B.[146] Activation of pGC-A by atrial natriuretic peptide (ANP), B-type natriuretic peptide, and urodilatin leads to natriuresis, vasodilation, renin and aldosterone inhibition, suppression of apoptosis and hypertrophy, and lipolysis, and activation of pGC-B by CNP leads to inhibition of fibrosis, anti-inflammatory action and endothelial regenerating capacity.[146] A peptide drug analogue of endogenous B-type natriuretic peptide (nesiritide from Janssen Pharmaceuticals) was approved for acute decompensated heart failure in 2001; however, owing to the increased risk of worsening renal function, it was discontinued in 2018.[147,148] Vosoritide remains the only FDA-approved natriuretic peptide drug.

3.3 Nature-derived Peptide Drugs – A Reliable and Vast Source of Therapeutic Leads

Since the first commercial use of a purified natural product as a therapeutic (morphine, Merck, 1826), small-molecule compounds derived from natural products have provided valuable drugs with sources ranging from fungi and bacteria to plants and animals.[149] Nature-derived peptides have also provided valuable and life-saving drugs. Peptide drugs derived from natural sources (excluding endogenous peptide hormones) make up the second largest proportion of approved peptide drugs (see Figure 3.1B). Snake venom peptide (eptifibatide, 1998) and blood-sucking leech salivary peptides (lepirudin, 1998) were two of the earliest approved peptide drugs derived from natural products as potent anticoagulants.[150] Other nature-derived peptide drugs include exenatide (2005) from the Gila monster (*Heloderma suspectum*) discussed earlier and ziconotide (Prialt, Elan Corporation, 2004, daily intrathecal infusion), a 25-residue long and three disulfide bond-containing peptide isolated from the venom of a marine cone snail (*Conus magus*) approved as a non-opioid analgesic for treating severe chronic pain.[151] Ziconotide is a potent antagonist of the $Ca_v2.2$, inhibiting nerve impulses and leading to its antinociceptive effects; it is ~1000-fold more potent than morphine while not developing dependence or tolerance.[152]

Despite these successes, only two nature-derived peptides have received approval in the past five years: bulevirtide (Hepcludex, Gilead Sciences, 2020), an antiviral agent for the treatment of chronic hepatitis D virus (HDV), and motixafortide (Aphexda, BioLineRx, 2023), a haematopoietic stem cell mobiliser and chemokine receptor 4 (CXCR4) antagonist for treating multiple myeloma.

Bulevirtide (Hepcludex, Gilead Sciences, 2020, once-daily s.c.) is a 47-residue long N-terminally lipidated synthetic mimetic of the pre-S1 hepatitis B virus (HBV) surface antigen (HBsAg) (C_{13}-GTNLSVPNPLGFFPDHQLDPAFGANSNNPDWDFNPNKDHWPEANKVG*), which competitively inhibits the virus's entry receptor on the hepatocyte surface.[153] The peptide inhibits both HBV and HDV entry to hepatocytes; however, it is currently approved only for HDV treatment.[154]

Motixafortide (Aphexda, BioLineRx, 2023, single-dose s.c. 10–14 h before apheresis), a 14-residue long cyclic (single disulfide) CXCR4 antagonist, was originally identified from host-defence peptides of the horseshoe crab during anti-HIV therapy development.[155] High-dose chemotherapy followed by an autologous haematopoietic stem cell transplant is a common treatment for multiple myeloma; however, in ~30% of patients the number of stem cells collected is suboptimal for transplant. Motixafortide can rapidly mobilise high numbers of haematopoietic stem cells into the bloodstream before apheresis, allowing for improved success rates in autologous stem cell transplant therapy.[156] Motixafortide (FBn-RR[Nal]CY[Cit]KKPYR[Cit]CR*) contains a number of non-natural amino acids, including naphthylalanine[3] (Nal), two citrulline[6,12] (Cit) residues, and a 4-fluorobenzoic acid (FBn)-capped N-terminus.[157] Despite the initial hit discovery from a natural product, many of these peptides required significant modifications for clinical success.

3.4 Display Library-derived Peptide Drugs – A New Era of Directed Evolution-driven Discovery

So far, only four peptide drugs have been derived from display libraries, but the number of preclinical candidates arising from these strategies is expanding rapidly.[158] Peptide display technologies allow the generation of highly diverse and complex libraries of 10^{10}–10^{15} peptides, panned against a selected drug target using affinity-based selection.[159–161] High-affinity hits are enriched and, owing to a genetic link to either encoding gene (phage) or mRNA, undergo further mutations and are screened again, allowing for multiple generations of evolutionary directed hit selection.[162]

Phage display was the first of these techniques capable of generating peptide libraries with up to 10^9 members.[163] Peptides are encoded into the phage coat protein gene, leading to the "display" of the encoded peptides on the surface of the phage.[162] The genetic link to the peptide binder allows not only the amplification of affinity-selected binders and further mutation, but also sequencing, which identifies the exact sequence of a given hit.[164] Although phage display is still commonly employed today, there are many limitations associated with this technique.

mRNA display, the most popular *in vitro* display technology, can generate libraries of 10^{13}–10^{15} peptides of increasing diversity.[165] Advances in mRNA display have led to the development of the random non-standard peptides integrated discovery (RaPID) system.[166] The flexizyme *in vitro* translation technology employed with this technique allows the incorporation of diverse non-proteinogenic amino acids.[166] Owing to the increasing interest in cyclic peptides in drug development, techniques such as the split-intein circular ligation of peptides and proteins (SICLOPPS) method have also been developed, capable of intracellular production of backbone cyclised peptide libraries with up to 10^9 unique members.[167]

3.4.1 Phage Display

Three peptides discovered by phage display have so far received regulatory approval: ecallantide (Kalbitor, Dyax Corporation, 2009) followed by peginesatide (Omontys, Affymax, 2012, discontinued) and pegcetacoplan (Empaveli, Apellis Pharmaceuticals, 2021).

Pegcetacoplan (Empaveli, Apellis Pharmaceuticals, 2021, twice-weekly s.c.) is a first-in-class drug for treating paroxysmal nocturnal haemoglobinuria, and in 2023 it was also approved under the trade name Syfovre (Apellis Pharmaceuticals, 2023, approximately once-monthly intravitreal injection) for treating geographic atrophy age-related macular degeneration.[41] Pegcetacoplan is a poly(ethylene glycol) (PEG)-linked dimer of two identical copies of a 15-residue long cyclic peptide based on a 13-residue long compstatin family peptide, originally identified during phage display screening against complement component C3b.[42] From the initial phage display hit, it took over a decade of medicinal chemistry efforts to yield the final clinical candidate.[168] Resistance to aminopeptidase activity through N-terminal acetylation, targeted *N*-methylation, and D-amino acid systemic screening were employed to provide a compound with not only improved stability towards proteolysis but also increased affinity to C3 and C3b.[169] The final modified cyclic compstatin peptide displayed high stability in human plasma but was still susceptible to renal clearance. A number of strategies were employed to improve stability and PK, including covalent conjugation to albumin and non-covalent albumin binders. Conjugation of two copies of the optimised peptide to a 40 kDa PEG linker proved the most successful PK enhancement while maintaining high affinity, resulting in the final drug candidate.[168] Although phage display is a powerful screening methodology paving the way to many approved monoclonal antibodies and this first-in-class peptide drug, it still suffers from many disadvantages that affect peptide discovery capabilities, mainly in the hit-to-lead phase exemplified by the extensive and time-consuming optimisation from first phage display hit to pegcetacoplan.

3.4.2 mRNA Display

The ability to incorporate unnatural non-proteogenic amino acids in the screening phase offers a significant advantage in the lead discovery phase over fully proteinogenic peptide hits from phage display.

Zilucoplan (Zilbrysq, Union Chimique Belge, 2023, once-daily s.c.), represents the first approved peptide drug derived from an mRNA display library.[170] Zilucoplan was developed as a complement component C5 inhibitor for treating autoimmune disease generalised myasthenia gravis.[50] Inhibition of C5 cleavage to C5a and C5b by zilucoplan ameliorates complement-mediated damage to the neuromuscular junction, improving patients' myasthenia gravis activities of living (MG-ADL) clinical scores.[50] Only limited published reports exist on the development process or chemical synthesis of zilucoplan. Analysis of the drug structure (see Figure 3.4) reveals a 15-residue long and highly modified side-chain to side-chain cyclised peptide comprised of four non-natural amino acids (*N*-methylaspartic acid,[7] L-*tert*-butylglycine,[8] L-7-azatryptophan,[10] and cyclohexylglycine[14]) conjugated to a serum albumin-binding PEG_{24}–Glu–C_{16} moiety.

MK-0616 (Merck, phase III, 2023, once-daily oral) is the second compound derived from mRNA display, currently in phase III clinical trials for treating hypercholesterolaemia.[171] An mRNA display campaign was initiated, with linear peptides of varying lengths containing two active thiols, which were converted to macrocyclic peptides with the cross-linking agent 1,3-di(bromomethyl)benzene (DBX).[11] Several hit series were initially identified and cyclised *via* non-terminal amino acids and, as such, the leads were first truncated at the N- and C-termini to afford fully cyclic peptides.[172] Owing to the goal of developing an oral drug, the compounds were assessed for their stability against gastrointestinal proteases. Both compounds were stable towards pepsin but, like many peptide-based compounds, had poor stability against intestinal proteases.[173] Identification of the major metabolites revealed five major cleavage sites

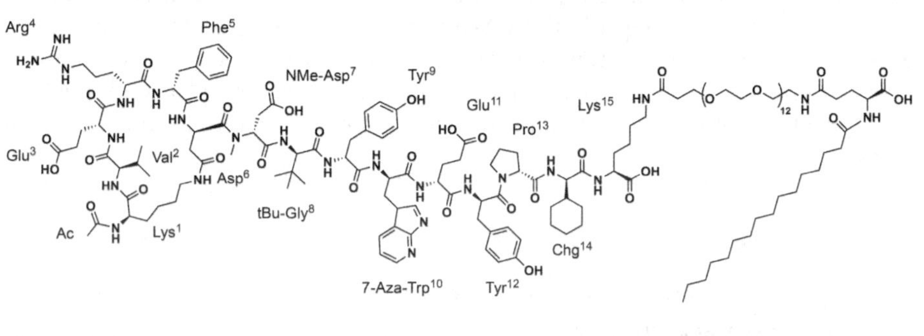

Figure 3.4 Chemical structure of zilucoplan. Ac, acetyl; NMe-Asp, *N*-methylaspartic acid; *t*Bu-Gly, L-*tert*-butylglycine; 7-Aza-Trp, L-7-azatryptophan; Chg, cyclohexylglycine.

and allowed directed modification of the peptide to improve its gastrointestinal stability.[172] D-Amino acid and *N*-methylated amino acid scanning was performed, in addition to crystal structure-guided amino acid replacement and further novel bicyclisation.[172] MK-0616 is formulated with sodium caprate permeation enhancer, achieving an oral bioavailability of 2%, and is compatible with once-daily dosing.[174] MK-0616 reinforces the potential ability of peptide-based drugs to target large, shallow binding surfaces to inhibit protein–protein interactions. This is a good example of mRNA display technology providing the initial high-affinity hits containing non-natural amino acids, followed by extensive medicinal chemistry efforts to develop the final drug candidate, a trend we expect to continue to see.

3.5 Conclusions

Peptide hormones remain one of the most successful avenues for peptide drug development and the recent successes of GLP-1R agonists have established peptide drugs as valuable treatments for patients and commercial assets for those companies leading their development (*e.g.* Novo Nordisk, Eli Lilly). Medicinal chemistry strategies played a critical part in the success of these drugs, particularly the non-covalent albumin binding half-life extension and unnatural amino acid substitutions and macrocyclisation for enhancing proteolytic stability and bioavailability. Nature-derived peptides continue to provide valuable leads for development, but at a lower rate compared with hormone mimetics and peptides identified from display libraries. Peptide display technology, in particular mRNA display, is expected to provide more and better leads for peptide drug development, with the first successful approvals of two complement binding peptides and a pipeline of clinical candidates under evaluation. The ability to incorporate non-proteogenic amino acids (which now appear common in many recently approved peptide drugs) within the screening phase has the possibility to reduce hit-to-lead development times and is predicted to continue to contribute to the future growth of the peptide field. We are in an exciting era of accelerating peptide drug development, with many pharmaceutical companies that once discounted peptide therapeutics now increasing their efforts in peptide drug development in attempts to capture a share of a rapidly growing global market.

References

1. M. Muttenthaler, G. F. King, D. J. Adams and P. F. Alewood, *Nat. Rev. Drug Discovery*, 2021, **20**, 309.
2. Statista, *Pharmaceutical Products & Market Report*, 2023.
3. Presedence-Research, *Peptide Therapeutics Market*, 2023.

4. Coherent-Market-Insights, *Biologics Market Analysis*, 2023.
5. Global Market Insights, *GLP-1 Receptor Agonist Market*, 2023.
6. M. Blanco, K. M. Gardinier and M. N. NamChuk, *ACS Med. Chem. Lett.*, 2022, **13**, 1691.
7. M. Kuna, F. Mahdi, A. R. Chade and G. L. Bidwell 3rd, *Sci. Rep.*, 2018, **8**, 7923.
8. L. Wang, N. Wang, W. Zhang, X. Cheng, Z. Yan, G. Shao, X. Wang, R. Wang and C. Fu, *Signal Transduction Targeted Ther.*, 2022, **7**, 48.
9. N. M. Tripathi and A. Bandyopadhyay, *Eur. J. Med. Chem.*, 2022, **243**, 114766.
10. Y. Zhang, J. Guo, J. Cheng, Z. Zhang, F. Kang, X. Wu and Q. Chu, *J. Med. Chem.*, 2023, **66**, 95.
11. S. E. Iskandar and A. A. Bowers, *ACS Med. Chem. Lett.*, 2022, **13**, 1379.
12. C. Lamers, *Future Drug Discovery*, 2022, **4**, 2631.
13. D. Beran, E. A. M. Gale and J. S. Yudkin, *Diabetologia*, 2022, **65**, 931.
14. I. B. Hirsch, R. Juneja, J. M. Beals, C. J. Antalis and E. E. Wright, *Endocr. Rev.*, 2020, **41**, 733.
15. Global-Information, *Insulin Market by Drug*, 2023.
16. C. Schott and J. P. Morgan, *Assest Managment Life Sciences*, 2023.
17. N. A. McGrath, M. Brichacek and J. T. Njardarson, *J. Chem. Educ.*, 2022, **87**, 1348.
18. N. Braga Emidio, H. N. T. Tran, A. Andersson, P. E. Dawson, F. Albericio, I. Vetter and M. Muttenthaler, *J. Med. Chem.*, 2021, **64**, 8384.
19. G. Rossino, E. Marchese, M. Finizio, M. Serra, P. Linciano and S. Collina, *Molecules*, 2023, **28**, 7165.
20. J. L. Lau and M. K. Dunn, *Bioorg. Med. Chem.*, 2018, **26**, 2700.
21. B. L. Bray, *Nat. Rev. Drug Discovery*, 2003, **2**, 587.
22. T. Matthews, M. Salgo, M. Greenberg, J. Chung, R. DeMasi and D. Bolognesi, *Nat. Rev. Drug Discovery*, 2004, **3**, 215.
23. M. O. Frederick, R. A. Boyse, T. M. Braden, J. R. Calvin, B. M. Campbell, S. M. Changi, S. R. Coffin, C. Condon, O. Gowran, J. McClary Groh, S. R. Groskreutz, Z. D. Harms, A. A. Humenik, N. J. Kallman, N. D. Klitzing, M. E. Kopach, J. K. Kretsinger, G. R. Lambertus, J. T. Lampert, L. M. Maguire, H. A. Moynihan, N. S. Mullane, J. D. Murphy, M. E. O'Mahony, R. N. Richey, K. D. Seibert, R. D. Spencer, M. A. Strege, N. Tandogan, F. L. Torres Torres, S. V. Tsukanov and H. Xia, *Org. Process Res. Dev.*, 2021, **25**, 1628.
24. D. Al Shaer, O. Al Musaimi, F. Albericio and B. G. de la Torre, *Pharmaceuticals*, 2019, **12**, 52.
25. D. Al Shaer, O. Al Musaimi, F. Albericio and B. G. de la Torre, *Pharmaceuticals*, 2020, **13**, 40.
26. D. Al Shaer, O. Al Musaimi, F. Albericio and B. G. de la Torre, *Pharmaceuticals*, 2021, **14**, 145.
27. D. Al Shaer, O. Al Musaimi, F. Albericio and B. G. de la Torre, *Pharmaceuticals*, 2022, **15**, 222.

28. J. G. Langendonk, M. Balwani, K. E. Anderson, H. L. Bonkovsky, A. V. Anstey, D. M. Bissell, J. Bloomer, C. Edwards, N. J. Neumann, C. Parker, J. D. Phillips, H. W. Lim, I. Hamzavi, J. C. Deybach, R. Kauppinen, L. E. Rhodes, J. Frank, G. M. Murphy, F. P. J. Karstens, E. J. G. Sijbrands, F. W. M. de Rooij, M. Lebwohl, H. Naik, C. R. Goding, J. H. P. Wilson and R. J. Desnick, *N. Engl. J. Med.*, 2015, **373**, 48.
29. D. Wensink, M. A. E. M. Wagenmakers and J. G. Langendonk, *Expert Rev. Clin. Pharmacol.*, 2021, **14**, 151.
30. T. K. Sawyer, P. J. Sanfilippo, V. J. Hruby, M. H. Engel, C. B. Heward, J. B. Burnett and M. E. Hadley, *Proc. Natl. Acad. Sci. U. S. A.*, 1980, **77**, 5754.
31. V. J. Hruby, D. Lu, S. D. Sharma, A. L. Castrucci, R. A. Kesterson, F. A. Al-Obeidi, M. E. Hadley and R. D. Cone, *J. Med. Chem.*, 1995, **38**, 3454.
32. S. Dhillon and S. J. Keam, *Drugs*, 2019, **79**, 1599.
33. A. Markham, *Drugs*, 2021, **81**, 397.
34. K. Clément, E. van den Akker, J. Argente, A. Bahm, W. K. Chung, H. Connors, K. De Waele, I. S. Farooqi, J. Gonneau-Lejeune and G. Gordon, *Lancet Diabetes Endocrinol.*, 2020, **8**, 960.
35. C. Kang and Y. Y. Syed, *Drugs*, 2020, **80**, 1601.
36. V. Soriano, V. Moreno-Torres, A. Treviño, O. Corral and C. de Mendoza, *Drug Des., Dev. Ther.*, 2023, **17**, 155.
37. A. Loglio, P. Ferenci, S. C. U. Renteria, C. Y. L. Tham, C. Scholtes, H. Holzmann, F. van Bömmel, M. Borghi, R. Perbellini and A. Rimondi, *J. Hepatol.*, 2022, **76**, 464.
38. R. Savarirayan, M. Irving, C. A. Bacino, B. Bostwick, J. Charrow, V. Cormier-Daire, K.-H. Le Quan Sang, P. Dickson, P. Harmatz, J. Phillips, N. Owen, A. Cherukuri, K. Jayaram, G. S. Jeha, K. Larimore, M.-L. Chan, A. Huntsman Labed, J. Day and J. Hoover-Fong, *N. Engl. J. Med.*, 2019, **381**, 25.
39. S. Duggan, *Drugs*, 2021, **81**, 2057.
40. E. Dardenne, N. Ishiyama, T.-A. Lin and M. C. Lucas, *Bioorg. Med. Chem.*, 2023, **87**, 117275.
41. R. P. de Latour, J. Szer, I. C. Weitz, A. Röth, B. Höchsmann, J. Panse, K. Usuki, M. Griffin, J.-J. Kiladjian and C. M. de Castro, *Lancet Haematol.*, 2022, **9**, 648.
42. D. C. Mastellos, D. Ricklin, G. Sfyroera and A. Sahu, *Clin. Immunol.*, 2022, **235**, 108785.
43. U. Hövelmann, B. V. Bysted, U. Mouritzen, F. Macchi, D. Lamers, B. Kronshage, D. V. Møller and T. Heise, *Diabetes Care*, 2017, **41**, 531.
44. T. R. Pieber, R. Aronson, U. Hövelmann, J. Willard, L. Plum-Mörschel, K. M. Knudsen, B. Bandak and R. Tehranchi, *Diabetes Care*, 2021, **44**, 1361.
45. J. P. Frías, M. J. Davies, J. Rosenstock, F. C. Pérez Manghi, L. Fernández Landó, B. K. Bergman, B. Liu, X. Cui and K. Brown, *N. Engl. J. Med.*, 2021, **385**, 503.

46. A. M. Jastreboff, L. J. Aronne, N. N. Ahmad, S. Wharton, L. Connery, B. Alves, A. Kiyosue, S. Zhang, B. Liu, M. C. Bunck and A. Stefanski, *N. Engl. J. Med.*, 2022, **387**, 205.
47. T. Coskun, K. W. Sloop, C. Loghin, J. Alsina-Fernandez, S. Urva, K. B. Bokvist, X. Cui, D. A. Briere, O. Cabrera and W. C. Roell, *Mol. Metab.*, 2018, **18**, 3.
48. Z. D. Crees, M. P. Rettig and J. F. DiPersio, *Ther. Adv. Hematol.*, 2023, **14**, 1–13.
49. Z. D. Crees, M. P. Rettig, R. G. Jayasinghe, K. Stockerl-Goldstein, S. M. Larson, I. Arpad, G. A. Milone, M. Martino, P. Stiff and D. Sborov, *Nat. Med.*, 2023, **29**, 1.
50. J. F. Howard, S. Bresch, A. Genge, C. Hewamadduma, J. Hinton, Y. Hussain, R. Juntas-Morales, H. J. Kaminski, A. Maniaol and R. Mantegazza, *Lancet Neurol.*, 2023, **22**, 395.
51. J. F. Howard Jr, J. Vissing, N. E. Gilhus, M. I. Leite, K. Utsugisawa, P. W. Duda, R. Farzaneh-Far, H. Murai and H. Wiendl, *Expert Opin. Invest. Drugs*, 2021, **30**, 483.
52. I. Vecchio, C. Tornali, N. L. Bragazzi and M. Martini, *Front. Endocrinol.*, 2018, **9**, 613.
53. E. K. Sims, A. L. J. Carr, R. A. Oram, L. A. DiMeglio and C. Evans-Molina, *Nat. Med.*, 2021, **27**, 1154.
54. M. A. Jarosinski, B. Dhayalan, Y.-S. Chen, D. Chatterjee, N. Varas and M. A. Weiss, *Mol. Metab.*, 2021, **52**, 101325.
55. T. B. Kjeldsen, F. Hubálek, C. U. Hjørringgaard, T. M. Tagmose, E. Nishimura, C. E. Stidsen, T. Porsgaard, C. Fledelius, H. H. F. Refsgaard, S. Gram-Nielsen, H. Naver, L. Pridal, T. Hoeg-Jensen, C. B. Jeppesen, V. Manfè, S. Ludvigsen, I. Lautrup-Larsen and P. Madsen, *J. Med. Chem.*, 2021, **64**, 8942.
56. H. Haahr and T. Heise, *Clin. Pharmacokinet.*, 2020, **59**, 155.
57. V. Sugumar, K. P. Ang, A. F. Alshanon, G. Sethi, P. V. C. Yong, C. Y. Looi and W. F. Wong, *Pharmaceutics*, 2022, **14**, 1406.
58. J. B. Buse, M. J. Davies, B. M. Frier and A. Philis-Tsimikas, *BMJ Open Diabetes Res. Care*, 2021, **9**, e002373.
59. M. A. Nauck, I. Weber, I. Bach, S. Richter, C. Ørskov, J. J. Holst and W. Schmiegel, *Diabetic Med.*, 1998, **15**, 937.
60. W. O. C. Creutzfeldt, N. Kleine, B. Willms, C. Ørskov, J. J. Holst and M. A. Nauck, *Diabetes Care*, 1996, **19**, 580.
61. R. Ritzel, C. Ørskov, J. J. Holst and M. A. Nauck, *Diabetologia*, 1995, **38**, 720.
62. C. F. Deacon, A. H. Johnsen and J. J. Holst, *J. Clin. Endocrinol. Metab.*, 1995, **80**, 952.
63. D. G. Parkes, K. F. Mace and M. E. Trautmann, *Expert Opin. Drug Discovery*, 2013, **8**, 219.
64. R. Göke, H. C. Fehmann, T. Linn, H. Schmidt, M. Krause, J. Eng and B. Göke, *J. Biol. Chem.*, 1993, **268**, 19650.

65. M. Yu, M. M. Benjamin, S. Srinivasan, E. E. Morin, E. I. Shishatskaya, S. P. Schwendeman and A. Schwendeman, *Adv. Drug Delivery Rev.*, 2018, **130**, 113.
66. L. L. Baggio and D. J. Drucker, *Gastroenterology*, 2007, **132**, 2131.
67. A. Bond, *Bayl. Univ. Med. Cent. Proc.*, 2006, **19**, 281.
68. M. Christensen, P. Miossec, B. D. Larsen, U. Werner and F. K. Knop, *Expert Opin. Drug Discovery*, 2014, **9**, 1223.
69. D. J. Drucker, A. Dritselis and P. Kirkpatrick, *Nat. Rev. Drug Discovery*, 2010, **9**, 267.
70. L. B. Knudsen, *ACS Pharmacol. Transl. Sci.*, 2019, **2**, 468.
71. D. J. Drucker, J. B. Buse, K. Taylor, D. M. Kendall, M. Trautmann, D. Zhuang and L. Porter, *Lancet*, 2008, **372**, 1240.
72. Y. Y. Syed and P. L. McCormack, *Drugs*, 2015, **75**, 1141.
73. E. D. Deeks, *Drugs Ther. Perspect.*, 2019, **35**, 51.
74. A. Astrup, S. Rössner, L. Van Gaal, A. Rissanen, L. Niskanen, M. Al Hakim, J. Madsen, M. F. Rasmussen and M. E. J. Lean, *Lancet*, 2009, **374**, 1606.
75. R. E. Pratley, M. A. Nauck, A. H. Barnett, M. N. Feinglos, F. Ovalle, I. Harman-Boehm, J. Ye, R. Scott, S. Johnson and M. Stewart, *Lancet Diabetes Endocrinol.*, 2014, **2**, 289.
76. E. L. St Onge and S. A. Miller, *Expert Opin. Biol. Ther.*, 2010, **10**, 801.
77. A. J. Scheen, *Expert Rev. Clin. Pharmacol.*, 2016, **9**, 385.
78. K. M. Dungan, S. T. Povedano, T. Forst, J. G. G. González, C. Atisso, W. Sealls and J. L. Fahrbach, *Lancet*, 2014, **384**, 1349.
79. J. Rosenstock, D. Raccah, L. Koranyi, L. Maffei, G. Boka, P. Miossec and J. E. Gerich, *Diabetes Care*, 2013, **36**, 2945.
80. J. P. H. Wilding, R. L. Batterham, S. Calanna, M. Davies, L. F. Van Gaal, I. Lingvay, B. M. McGowan, J. Rosenstock, M. T. D. Tran, T. A. Wadden, S. Wharton, K. Yokote, N. Zeuthen and R. F. Kushner, *N. Engl. J. Med.*, 2021, **384**, 989.
81. J. Lau, P. Bloch, L. Schäffer, I. Pettersson, J. Spetzler, J. Kofoed, K. Madsen, L. B. Knudsen, J. McGuire and D. B. Steensgaard, *J. Med. Chem.*, 2015, **58**, 7370.
82. A. Phillips and J. N. Clements, *J. Clin. Pharm. Ther.*, 2022, **47**, 184.
83. A. L. Lewis, N. McEntee, J. Holland and A. Patel, *Drug Delivery Transl. Res.*, 2022, **12**, 1.
84. J. Rosenstock, C. Wysham, J. P. Frías, S. Kaneko, C. J. Lee, L. F. Landó, H. Mao, X. Cui, C. A. Karanikas and V. T. Thieu, *Lancet*, 2021, **398**, 143.
85. J. P. Frías, *Expert Rev. Endocrinol. Metab.*, 2020, **15**, 379.
86. S. B. van Witteloostuijn, S. L. Pedersen and K. J. Jensen, *ChemMedChem*, 2016, **11**, 2474.
87. L. B. Knudsen, P. F. Nielsen, P. O. Huusfeldt, N. L. Johansen, K. Madsen, F. Z. Pedersen, H. Thøgersen, M. Wilken and H. Agersø, *J. Med. Chem.*, 2000, **43**, 1664.
88. K. Madsen, L. B. Knudsen, H. Agersoe, P. F. Nielsen, H. Thøgersen, M. Wilken and N. L. Johansen, *J. Med. Chem.*, 2007, **50**, 6126.

89. J. B. Buse, J. Rosenstock, G. Sesti, W. E. Schmidt, E. Montanya, J. H. Brett, M. Zychma and L. Blonde, *Lancet*, 2009, **374**, 39.
90. W. Glaesner, A. Mark Vick, R. Millican, B. Ellis, S. Tschang, Y. Tian, K. Bokvist, M. Brenner, A. Koester, N. Porksen, G. Etgen and T. Bumol, *Diabetes/Metab. Res. Rev.*, 2010, **26**, 287.
91. A. J. Scheen, *Expert Rev. Clin. Pharmacol*, 2016, **9**, 385.
92. A. A. Bhattacharya, T. Grüne and S. Curry, *J. Mol. Biol.*, 2000, **303**, 721.
93. C. Kapitza, L. Nosek, L. Jensen, H. Hartvig, C. B. Jensen and A. Flint, *J. Clin. Pharmacol.*, 2015, **55**, 497.
94. R. E. Pratley, V. R. Aroda, I. Lingvay, J. Lüdemann, C. Andreassen, A. Navarria and A. Viljoen, *Lancet Diabetes Endocrinol.*, 2018, **6**, 275.
95. T. Vilsbøll and A. J. Garber, *Diabetes, Obes. Metab.*, 2012, **14**, 41.
96. M. Singhal, S. Unni, M. Schauerhamer, H. Nguyen, J. Hurd and C. McAdam-Marx, *J. Manag. Care Spec. Pharm.*, 2017, **23**, 267.
97. G. Iacobucci, *Br. Med. J.*, 2023, **382**, 2019.
98. K. A. Scott and T. H. Moran, *Am. J. Physiol.: Regul., Integr. Comp. Physiol.*, 2007, **293**, 983.
99. P. K. Chelikani, A. C. Haver and R. D. Reidelberger, *Am. J. Physiol.: Regul., Integr. Comp. Physiol.*, 2005, **288**, 1695.
100. J. Crane and B. McGowan, *Ther. Adv. Chronic Dis.*, 2016, **7**, 92–107.
101. X. Pi-Sunyer, A. Astrup, K. Fujioka, F. Greenway, A. Halpern, M. Krempf, D. C. W. Lau, C. W. le Roux, R. Violante Ortiz, C. B. Jensen and J. P. H. Wilding, *N. Engl. J. Med.*, 2015, **373**, 11.
102. M. Márquez-Cruz, A. Kammar-García, J. C. Huerta-Cruz, M. D. C. Carrasco-Portugal, L. M. Barranco-Garduño, J. Rodríguez-Silverio, H. I.-R. González and J. G. Reyes-García, *Int. J. Clin. Pharmacol. Ther.*, 2021, **59**, 539.
103. E. Colman, *Circulation*, 2012, **125**, 2156.
104. C. Y. Yang, Y. R. Chen, H. T. Ou and S. Kuo, *Cardiovasc. Diabetol.*, 2021, **20**, 21.
105. C. T. Yang, W. Y. Yao, H. T. Ou and S. Kuo, *Diabetes Res. Clin. Pract.*, 2023, **198**, 110625.
106. R. V. Overgaard, A. Navarria, S. H. Ingwersen, T. A. Bækdal and R. J. Kildemoes, *Clin. Pharmacokinet.*, 2021, **60**, 1335.
107. L. Urquhart, *Nat. Rev. Drug Discovery*, 2019, **18**, 816.
108. S. L. Anderson, T. R. Beutel and J. M. Trujillo, *J. Diabetes Its Complications*, 2020, **34**, 107520.
109. R. M. Goldenberg and O. Steen, *Can. J. Diabetes*, 2019, **43**, 136.
110. V. R. Aroda, L. Blonde and R. E. Pratley, *Rev. Endocr. Metab. Disord.*, 2022, **23**, 979.
111. Y. Alhindi and A. Avery, *Contemp. Clin. Trials Commun.*, 2022, **28**, 100944.
112. B. Gallwitz and F. Giorgino, *Front. Endocrinol.*, 2021, **12**, 645507.
113. Novo Nordisk, *Annual Report*, 2022.
114. M. N. Kosiborod, S. Z. Abildstrøm, B. A. Borlaug, J. Butler, S. Rasmussen, M. Davies, G. K. Hovingh, D. W. Kitzman, M. L. Lindegaard,

D. V. Møller, S. J. Shah, M. B. Treppendahl, S. Verma, W. Abhayaratna, F. Z. Ahmed, V. Chopra, J. Ezekowitz, M. Fu, H. Ito, M. Lelonek, V. Melenovsky, B. Merkely, J. Núñez, E. Perna, M. Schou, M. Senni, K. Sharma, P. Van der Meer, D. von Lewinski, D. Wolf and M. C. Petrie, *N. Engl. J. Med.*, 2023, **389**, 1069.
115. J. Nowell, E. Blunt, D. Gupta and P. Edison, *Ageing Res. Rev.*, 2023, **89**, 101979.
116. J. R. Ussher and D. J. Drucker, *Nat. Rev. Cardiol.*, 2023, **20**, 463.
117. S. A. Sadry and D. J. Drucker, *Nat. Rev. Endocrinol.*, 2013, **9**, 425.
118. L. L. Baggio and D. J. Drucker, *Mol. Metab.*, 2021, **46**, 101090.
119. B. Manandhar and J.-M. Ahn, *J. Med. Chem.*, 2015, **58**, 1020.
120. A. Hareter, E. Hoffmann, H.-P. Bode, B. Göke and R. Göke, *Endocr. J.*, 1997, **44**, 701.
121. M. Bastin and F. Andreelli, *Diabetes, Metab. Syndr. Obes.: Targets Ther.*, 2019, **12**, 1973.
122. B. Finan, T. Ma, N. Ottaway, T. D. Müller, K. M. Habegger, K. M. Heppner, H. Kirchner, J. Holland, J. Hembree, C. Raver, S. H. Lockie, D. L. Smiley, V. Gelfanov, B. Yang, S. Hofmann, D. Bruemmer, D. J. Drucker, P. T. Pfluger, D. Perez-Tilve, J. Gidda, L. Vignati, L. Zhang, J. B. Hauptman, M. Lau, M. Brecheisen, S. Uhles, W. Riboulet, E. Hainaut, E. Sebokova, K. Conde-Knape, A. Konkar, R. D. DiMarchi and M. H. Tschöp, *Sci. Transl. Med.*, 2013, **5**, 209ra151.
123. M. M. Ali, A. Hafez, M. S. Abdelgalil, M. T. Hasan, M. M. El-Ghannam, O. M. Ghogar, A. A. Elrashedy and M. Abd-ElGawad, *BMC Endocr. Disord.*, 2022, **22**, 113.
124. J. Tillner, M. G. Posch, F. Wagner, L. Teichert, Y. Hijazi, C. Einig, S. Keil, T. Haack, M. Wagner, M. Bossart and P. J. Larsen, *Diabetes, Obes. Metab.*, 2019, **21**, 120.
125. J. W. Day, N. Ottaway, J. T. Patterson, V. Gelfanov, D. Smiley, J. Gidda, H. Findeisen, D. Bruemmer, D. J. Drucker, N. Chaudhary, J. Holland, J. Hembree, W. Abplanalp, E. Grant, J. Ruehl, H. Wilson, H. Kirchner, S. H. Lockie, S. Hofmann, S. C. Woods, R. Nogueiras, P. T. Pfluger, D. Perez-Tilve, R. DiMarchi and M. H. Tschöp, *Nat. Chem. Biol.*, 2009, **5**, 749.
126. A. M. Jastreboff, L. M. Kaplan, J. P. Frías, Q. Wu, Y. Du, S. Gurbuz, T. Coskun, A. Haupt, Z. Milicevic and M. L. Hartman, *N. Engl. J. Med.*, 2023, **389**, 514.
127. H. J. Choe and Y. M. Cho, *Endocrinol. Metab.*, 2021, **36**, 22.
128. M. D. Ericson, C. J. Lensing, K. A. Fleming, K. N. Schlasner, S. R. Doering and C. Haskell-Luevano, *Biochim. Biophys. Acta, Mol. Basis Dis.*, 2017, **1863**, 2414.
129. Y. Xu, X. Guan, R. Zhou and R. Gong, *Cell. Mol. Life Sci.*, 2020, **77**, 3831.
130. T. Montero-Melendez, T. Boesen and T. E. N. Jonassen, *Semin. Immunol.*, 2022, **59**, 101603.

131. X. C. Yuan and Y. X. Tao, *Biomolecules*, 2022, **12**, 1407.
132. G. J. Wan, J. Niewoehner and K. Hayes, *Clin. Outcomes Res.*, 2023, **15**, 499.
133. H. Ashton, J. E. Millman, R. Telford, J. W. Thompson, T. F. Davies, R. Hall, S. Shuster, A. J. Thody, D. H. Coy and A. J. Kastin, *Psychopharmacology*, 1977, **55**, 165.
134. A. M. Lane, J. T. McKay and H. L. Bonkovsky, *Appl. Clin. Genet.*, 2016, **9**, 179.
135. K. L. Chapman, G. K. Kinsella, A. Cox, D. Donnelly and J. B. C. Findlay, *J. Mol. Biol.*, 2010, **401**, 433.
136. H. M. Reddy, J. S. Poole, G. A. Maguire and S. M. Stahl, *Psychiatr. Clin. N. Am.*, 2020, **43**, 399.
137. P. Sweeney, L. E. Gimenez, C. C. Hernandez and R. D. Cone, *Nat. Rev. Endocrinol.*, 2023, **19**, 507.
138. K. Clément, E. van den Akker, J. Argente, A. Bahm, W. K. Chung, H. Connors, K. De Waele, I. S. Farooqi, J. Gonneau-Lejeune and G. Gordon, *Lancet Diabetes Endocrinol.*, 2020, **8**, 960.
139. S. H. Ju, G. B. Cho and J. W. Sohn, *Pharmacol. Res.*, 2018, **129**, 10.
140. B. A. Falls and Y. Zhang, *ACS Chem. Neurosci.*, 2019, **10**, 1055.
141. T. H. Collet, B. Dubern, J. Mokrosinski, H. Connors, J. M. Keogh, E. Mendes de Oliveira, E. Henning, C. Poitou-Bernert, J. M. Oppert, P. Tounian, F. Marchelli, R. Alili, J. Le Beyec, D. Pépin, J. M. Lacorte, A. Gottesdiener, R. Bounds, S. Sharma, C. Folster, B. Henderson, S. O'Rahilly, E. Stoner, K. Gottesdiener, B. L. Panaro, R. D. Cone, K. Clément, I. S. Farooqi and L. H. T. Van der Ploeg, *Mol. Metab.*, 2017, **6**, 1321.
142. J. Yu, L. E. Gimenez, C. C. Hernandez, Y. Wu, A. H. Wein, G. W. Han, K. McClary, S. R. Mittal, K. Burdsall, B. Stauch, L. Wu, S. N. Stevens, A. Peisley, S. Y. Williams, V. Chen, G. L. Millhauser, S. Zhao, R. Cone and R. Stevens, *Science*, 2020, **368**, 428.
143. W. A. Horton, J. G. Hall and J. T. Hecht, *Lancet*, 2007, **370**, 162.
144. Y. Komatsu, H. Chusho, N. Tamura, A. Yasoda, T. Miyazawa, M. Suda, M. Miura, Y. Ogawa and K. Nakao, *J. Bone Miner. Metab.*, 2002, **20**, 331.
145. M. L. Chan, Y. Qi, K. Larimore, A. Cherukuri, L. Seid, K. Jayaram, G. Jeha, E. Fisheleva, J. Day and A. Huntsman-Labed, *Clin. Pharmacokinet.*, 2022, **61**, 1.
146. T. Ichiki, N. Dzhoyashvili and J. C. Burnett Jr., *Int. J. Cardiol.*, 2019, **281**, 166.
147. M. M. Kittleson, *Circ.: Heart Failure*, 2018, **11**, e005440.
148. A. G. Mathur, J. K. Kairi and B. B. Nayak, *Med. J. Armed Forces India*, 2005, **61**, 375.
149. E. Muratspahić, M. Freissmuth and C. W. Gruber, *Trends Pharmacol. Sci.*, 2019, **40**, 309.

150. E. A. Nutescu, N. L. Shapiro and A. Chevalier, *Cardiol. Clin.*, 2008, **26**, 169.
151. J. G. McGivern, *Neuropsychiatr. Dis. Treat.*, 2007, **3**, 69.
152. J. A. Williams, M. Day and J. E. Heavner, *Expert Opin. Pharmacother.*, 2008, **9**, 1575.
153. V. Soriano, V. Moreno-Torres, A. Treviño, O. Corral and C. de Mendoza, *Drug Des., Dev. Ther.*, 2023, **17**, 155.
154. P. Khalfi, P. T. Kennedy, K. Majzoub and T. Asselah, *Antiviral Res.*, 2023, **209**, 105461.
155. H. Tamamura, Y. Xu, T. Hattori, X. Zhang, R. Arakaki, K. Kanbara, A. Omagari, A. Otaka, T. Ibuka, N. Yamamoto, H. Nakashima and N. Fujii, *Biochem. Biophys. Res. Commun.*, 1998, **253**, 877.
156. Z. Crees, M. P. Rettig and J. DiPersio, *Ther. Adv. Hematol.*, 2023, **14**, 1.
157. M. Rebolledo-Bustillo, D. Garcia-Gomez, E. M. Dávila, M. E. Castro, N. A. Caballero, F. J. Melendez, V. M. Baizabal-Aguirre, B. L. Sanchez-Gaytan and J. M. Perez-Aguilar, *Int. J. Mol. Sci.*, 2023, **24**, 4393.
158. X. Li, T. W. Craven and P. M. Levine, *J. Med. Chem.*, 2022, **65**, 11913.
159. W. Jaroszewicz, J. Morcinek-Orłowska, K. Pierzynowska, L. Gaffke and G. Węgrzyn, *FEMS Microbiol. Rev.*, 2022, **46**, fuab052.
160. C. Sohrabi, A. Foster and A. Tavassoli, *Nat. Rev. Chem.*, 2020, **4**, 90.
161. M. S. Newton, Y. Cabezas-Perusse, C. L. Tong and B. Seelig, *ACS Synth. Biol.*, 2020, **9**, 181.
162. J. K. Scott and G. P. Smith, *Science*, 1990, **249**, 386.
163. R. Derda, S. K. Y. Tang, S. C. Li, S. Ng, W. Matochko and M. R. Jafari, *Molecules*, 2011, **16**, 1776.
164. C.-H. Wu, I. J. Liu, R.-M. Lu and H.-C. Wu, *J. Biomed. Sci.*, 2016, **23**, 8.
165. C. Helmling and C. N. Cunningham, *Approaching the Next Inflection in Peptide Therapeutics: Attaining Cell Permeability and Oral Bioavailability*, American Chemical Society, Washington, 2022.
166. Y. Goto and H. Suga, *Acc. Chem. Res.*, 2021, **54**, 3604.
167. A. Tavassoli, *Curr. Opin. Chem. Biol.*, 2017, **38**, 30.
168. C. Lamers, D. C. Mastellos, D. Ricklin and J. D. Lambris, *Trends Pharmacol. Sci.*, 2022, **43**, 629.
169. S. M. Hoy, *Drugs*, 2021, **81**, 1423.
170. G. Q. Tang, Y. Tang, K. Dhamnaskar, M. D. Hoarty, R. Vyasamneni, D. D. Vadysirisack, Z. Ma, N. Zhu, J. G. Wang, C. Bu, B. Cong, E. Palmer, P. W. Duda, C. Sayegh and A. Ricardo, *Front. Immunol.*, 2023, **14**, 1213920.
171. C. M. Ballantyne, P. Banka, G. Mendez, R. Garcia, J. Rosenstock, A. Rodgers, G. Mendizabal, Y. Mitchel and A. L. Catapano, *J. Am. Coll. Cardiol.*, 2023, **81**, 1553.
172. C. Alleyne, R. P. Amin, B. Bhatt, E. Bianchi, J. C. Blain, N. Boyer, D. Branca, M. W. Embrey, S. N. Ha, K. Jette, D. G. Johns, A. D. Kerekes, K. A. Koeplinger, D. LaPlaca, N. Li, B. Murphy, P. Orth, A. Ricardo,

S. Salowe, K. Seyb, A. Shahripour, J. R. Stringer, Y. Sun, R. Tracy, C. Wu, Y. Xiong, H. Youm, H. J. Zokian and T. J. Tucker, *J. Med. Chem.*, 2020, **63**, 13796.

173. T. Kremsmayr, A. Aljnabi, J. B. Blanco-Canosa, H. N. T. Tran, N. B. Emidio and M. Muttenthaler, *J. Med. Chem.*, 2022, **65**, 6191.

174. D. G. Johns, L. C. Campeau, P. Banka, A. Bautmans, T. Bueters, E. Bianchi, D. Branca, P. G. Bulger, I. Crevecoeur and F. X. Ding, *Circulation*, 2023, **148**, 144.

CHAPTER 4

Sustainable Approaches in Solid-phase Peptide Synthesis (SPPS)

J. PAWLAS* AND J. H. RASMUSSEN

PolyPeptide, Limhamnsvägen 108, P.O. Box 30089, 20061 Limhamn, Sweden
*E-mail: jan.pawlas@polypeptide.com

4.1 Introduction

Since the Curtius dipeptide synthesis in 1882,[1] the advances achieved in peptide synthesis over the past 14 decades now allow peptides of practically any length and complexity imaginable to be prepared.[2] Having said that, the single discovery that altered the trajectory of peptide chemistry in the most profound way was Merrifield's solid-phase peptide synthesis (SPPS),[3] which in 1963 departed from the laborious liquid-phase peptide synthesis (LPPS) methods available at that time. Peptides could now be synthesized by attaching the first amino acid (AA) to an insoluble polymer support, carrying out all steps of peptide assembly by cycles consisting of α-AA protecting group (PG) removals, solvent washes and peptide couplings while removing excesses of all materials throughout by filtration and cleaving the peptide off the resin at the end (see Scheme 4.1).

Merrifield's original SPPS relied on carboxybenzyl (Cbz)-protected AAs, but in the following years the use of *tert*-butyloxycarbonyl (Boc)-protected AAs proved superior, with Boc SPPS emerging as an efficient method for the

Green Chemistry Series No. 85
Sustainability in TIDES Chemistry: Green Approaches to Oligonucleotides and Oligopeptides Synthesis
Edited by Alessandra Tolomelli, Lucia Ferrazzano and Walter Cabri

Published by the Royal Society of Chemistry, www.rsc.org

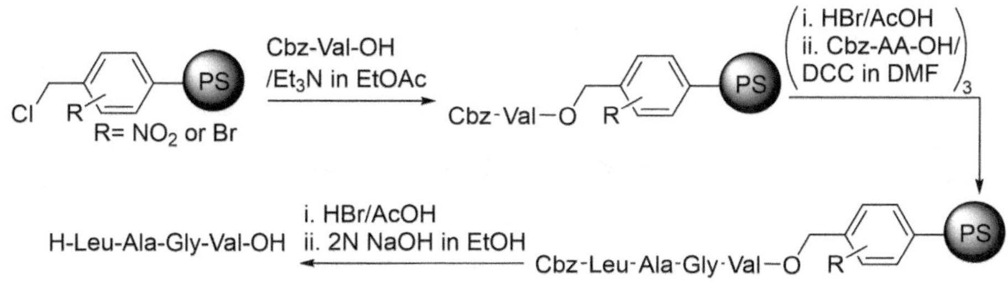

Scheme 4.1

synthesis of a wide range of peptides.[4] Nevertheless, owing to the reliance of Boc SPPS on the extremely toxic HF for resin cleavages[5] coupled with the advent of the base-labile fluorenylmethoxycarbonyl (Fmoc) amino PG by Carpino and Han,[6] the peptide field gradually shifted to Fmoc/*t*-Bu SPPS,[7,8] which is the prevalent peptide synthesis platform today. Fmoc/*t*-Bu SPPS commences by attaching an Fmoc-AA-OH moiety to a polymer support followed by alternating Fmoc removals, washes and peptide couplings until the target peptide resin is obtained, after which the peptide is cleaved off the resin using trifluoroacetic acid (TFA) in the presence of various scavengers (see Figure 4.1).

Fmoc/*t*-Bu SPPS has been used in scenarios from automated laboratory-scale synthesis[9] to industrial kilogram-scale manufacturing,[10] giving access to substrates from short[11] to highly challenging peptides[12] and even to proteins.[13] Notably, Fmoc/*t*-Bu SPPS can be exploited beyond merely assembling linear peptides. For example, protected peptide fragments **1** and **2** released from solid supports can be coupled using different coupling agents (CAs) followed by the deprotection of side-chain PGs using TFA,[14] and such hybrid SPPS approaches have been used in large-scale therapeutic peptide manufacturing.[15] Another approach exploiting SPPS is ligation, in which peptides **4** and **5** containing motifs such as a C-terminal thioester and an N-terminal Cys moiety form the peptide **6** by a native chemical ligation (NCL),[16] an approach that has also been used for other ligations.[17] Alternatively, **4** and **5** can be coupled using suitable ligases in a chemoenzymatic peptide synthesis (CEPS).[18,19] Moreover, resin **7** can be used in on-resin cyclizations, forming for example amide- or disulfide-bridged resin **8**, furnishing cyclic peptide **9** upon TFA cleavage.[20] Furthermore, peptide resin **10** can be used for PG deprotection/functionalization, for example in acylation,[21] decarboxylative arylation[22] and halogenation–functionalization.[23] Finally, peptides from SPPS can be used, for example, in conjugations at amino[24] or thiol[25] groups or in metal-aided transformations.[26]

The aforementioned SPPS and related methods provide access to peptides of almost unlimited diversity, and the most prominent area in which these molecules are being used is therapeutics.[27] Regarding numbers and quantities of therapeutic peptides synthesized from drug discovery to commercialization, owing to the wealth of drug targets[27] and the ease of SPPS

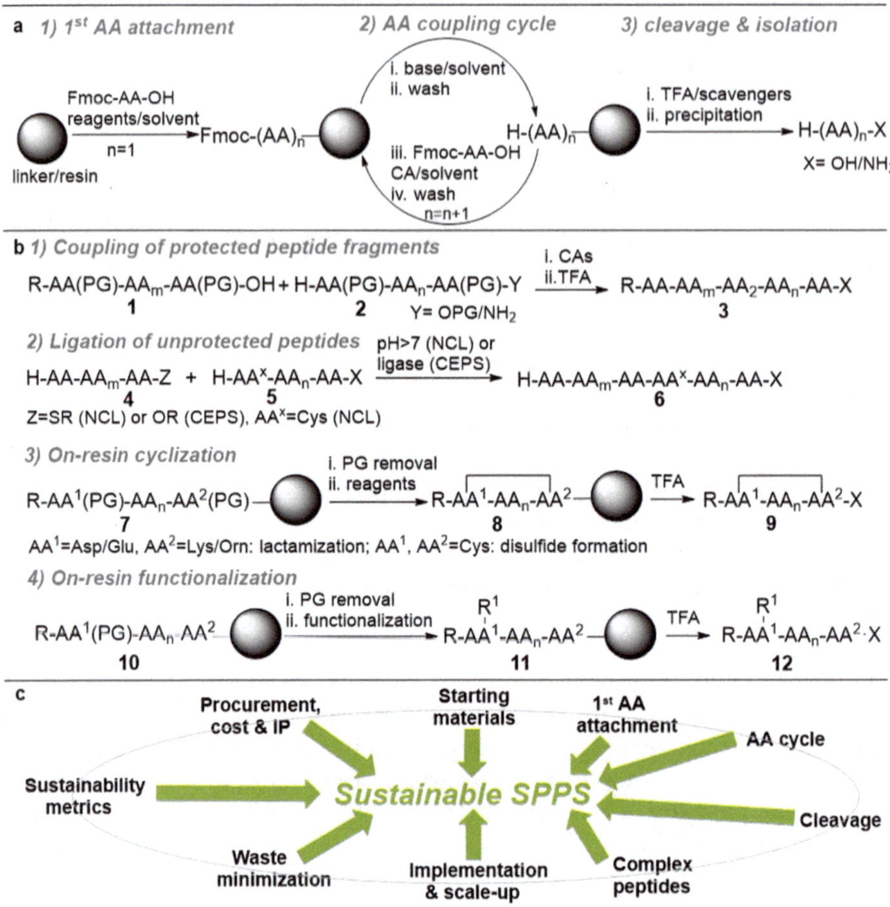

Figure 4.1 (a) Fmoc/*t*-Bu SPPS; (b) applications of Fmoc/*t*-Bu SPPS; (c) an integrated approach to sustainable SPPS.

automation,[89] thousands of peptides are synthesized in small amounts in the drug discovery phase each year. On the other hand, relatively few peptides go all the way through clinical development and approval by authorities, to be produced on a large scale, with 119 pharmaceutical peptides currently on the market.[28] From the standpoint of quality, whereas R&D peptides are produced in *ca.* 85–95% HPLC purity, commercial peptide drugs are typically required to be >98% pure, with the former using non-specific analytical methods and the latter being analyzed by the most stringent methods available.[29] Consequently, owing to the high quality demands on therapeutic peptides, highly optimized processes for the commercial phase are typically employed, *i.e.* the use of high-quality starting materials and minimized formation of impurities during synthesis ensure that the requisite quality of the peptide is met during downstream processing (DSP).[30] Producing peptides as therapeutics thus entails making small amounts of numerous peptides of modest quality in parallel with manufacturing fewer, highly pure peptides in large amounts for use in the clinic and on the market. Notably, SPPS has excelled for decades in delivering the requisite peptides at all scales with high success and in a productive manner.

There are, however, other aspects in need of consideration. Notably, producing peptides by SPPS comes at the cost of a very large footprint on the environment while potentially, as with all chemistry, exposing workers to various hazardous materials.[31] In this regard, with the 12 principles of green chemistry by Anastas and Eghbali as the guideline for greening of chemical processes,[32] the awareness of the need for more sustainable peptide synthesis and purification methods has recently increased considerably.[33,34] For example, platforms capable of manufacturing peptides with significantly lower environmental footprints than SPPS, such as tag-assisted LPPS[35] and recombinant synthesis,[36] have lately seen increased activity, although there are currently no viable alternatives to SPPS in terms of scope (molecular diversity), efficiency and scalability. In fact, with the expected growth in the therapeutic peptides sector in the coming years,[37] the need to improve sustainability of SPPS methods is particularly urgent. Consequently, various approaches towards more sustainable SPPS have been investigated by many groups.[31,33,34,38] In this chapter, advances related to sustainable SPPS are reviewed in an integrated manner, with the viewpoint that to attain a truly sustainable, industrially applicable SPPS, all relevant aspects have to be taken into consideration from the outset (see Figure 4.1c). The approaches for improving SPPS sustainability are reviewed from the starting materials to first AA attachment, AA cycles and cleavage. The SPPS of complex peptides, implementation, scale-up and waste minimization are discussed together with consideration of sustainability metrics and the impact of procurement, costs and intellectual property (IP). Finally, a perspective on future directions in sustainable SPPS is also provided.

4.2 Starting Materials

The sustainability of SPPS is affected by the starting materials, hence the routes of syntheses (RoS) for these chemicals have to be carefully evaluated. In particular, as the solvent accounts for the majority of the process mass used in SPPS,[33,34] the RoS for any SPPS solvent has to be assessed to determine its impact on sustainability.[39] Moreover, the impact of the solvent quality and stability must also be considered, as the quality of the solvent may have an impact on SPPS.[13,40] The (solid support) resin also has an impact on the SPPS sustainability and, although a variety of SPPS resins have been reported,[41] in practice mainly polystyrene (PS) resins[42] and PS resins grafted with poly(ethylene glycol) (PEG)[43] are used in manufacturing, whereas in R&D PEG resins such as ChemMatrix (CM)[44] are commonly applied. Nevertheless, the impact of resin RoS or quality[45] on SPPS sustainability still has to be evaluated.

Furthermore, the fact that large amounts of Fmoc-AA-OH derivatives are used in SPPS[46] means that significant amounts of various materials are also used during Fmoc-AA-OH manufacture. In fact, although ton quantities of highly pure Fmoc-AA-OHs are synthesized annually, there have been no overall studies evaluating the sustainability of Fmoc-AA-OH RoS, although studies on greening of some steps of Fmoc-AA-OH manufacture have been reported.[47] As impurities in Fmoc-AA-OHs can have an impact on peptide quality, these AAs must conform to high-purity requirements, in particular with regard to good manufacturing practice (GMP).[48] Nevertheless, the use of Fmoc-OSu (Fmoc-succinimide), which leads to the formation of Fmoc-β-Ala-OH/Fmoc-β-Ala-AA-OH impurities,[49] is still common, and although Fmoc reagents free of β-Ala impurities are available,[50] they have not yet been implemented in Fmoc-AA-OH manufacture. Reagents for SPPS such as bases for Fmoc deprotection and CAs for peptide couplings also require significant processing, although RoS of these reagents have not been studied in the context of SPPS sustainability. Finally, the quality of reagents for SPPS can also have an impact on sustainability; for example, impurities in hydroxybenzotriazole (HOBt)[51] and diisopropylcarbodiimide (DIC)[52] have been shown to impact peptide synthesis.

4.3 First Amino Acid Attachment

SPPS begins with the attachment of the first AA to the resin, for which various linkers are available.[53] The two most common types of C-terminus motifs are peptide acids and peptide amides, and for the latter the first AA can be coupled using linkers such as Rink amide linker (RAM)[54] or Ramage linker (RMG),[55] employing standard CAs. On the other hand, the attachment of the first AA for peptide acids requires that an ester bond between the solid support, typically Wang resin[56] or 2-chlorotrityl chloride (CTC) resin,[57] and the AA is formed.

As standard protocols for the coupling of the first AA to Wang and CTC resins proceed in hazardous solvents such as dimethylformamide (DMF) and dichloromethane (DCM), several studies aimed at making the first AA attachment more sustainable have been reported. Specifically, DIC–dimethylaminopyridine (DMAP)-mediated attachment of Fmoc-AA-OHs to CTC and Wang resins in 2-methyltetrahydrofuran (2-MeTHF) with yields comparable to those attained in DCM were reported,[58] and for racemization-prone AAs lower racemization in 2-MeTHF than in DCM was observed. A protocol for the coupling of Fmoc-AA-OHs to Wang resin in γ-valerolactone (GVL) using DIC with *N*-methylimidazole (NMI) as a catalyst has also been reported,[59] and more recently the attachment of Fmoc-AA-OHs to 4-methylbenzhydryl bromide (MBH-Br) and CTC resins in MeCN–EtOAc (1 : 1) in the presence of diisopropylethylamine (DIEA)[60] and the coupling of Fmoc-AA-OHs to CTC resin at 50 °C in *N*-butylpyrrolidinone (NBP) with DIEA were carried out.[61] It is worth noting that halogenated resins such as MBH-Br and CTC may decompose upon standing, and in fact replenishing a partially decomposed CTC resin with thionyl chloride prior to the first AA attachment increased the loading yield.[62] Finally, the preparation of halogenated resins requires harsh reagents in hazardous solvents such as DCM,[63] hence greening of the resin halogenation processes would be valuable.

4.4 Amino Acid Coupling Cycle

The "AA coupling cycle" (see Figure 4.1a) is the key part of SPPS, and in the following subsections various aspects pertaining to greening of the AA coupling cycle are discussed.

4.4.1 Solvents (see Figure 4.2a)

The most evaluated aspect of SPPS greening is replacing the standard solvent DMF with greener alternatives.[33,34,38] In fact, with DMF deemed unsustainable by regulatory agencies,[64,65] work on replacing hazardous dipolar aprotic solvents with more benign alternatives[66] has led to a variety of greener solvents for amide-forming reactions.[67,68] In SPPS, replacing DMF requires

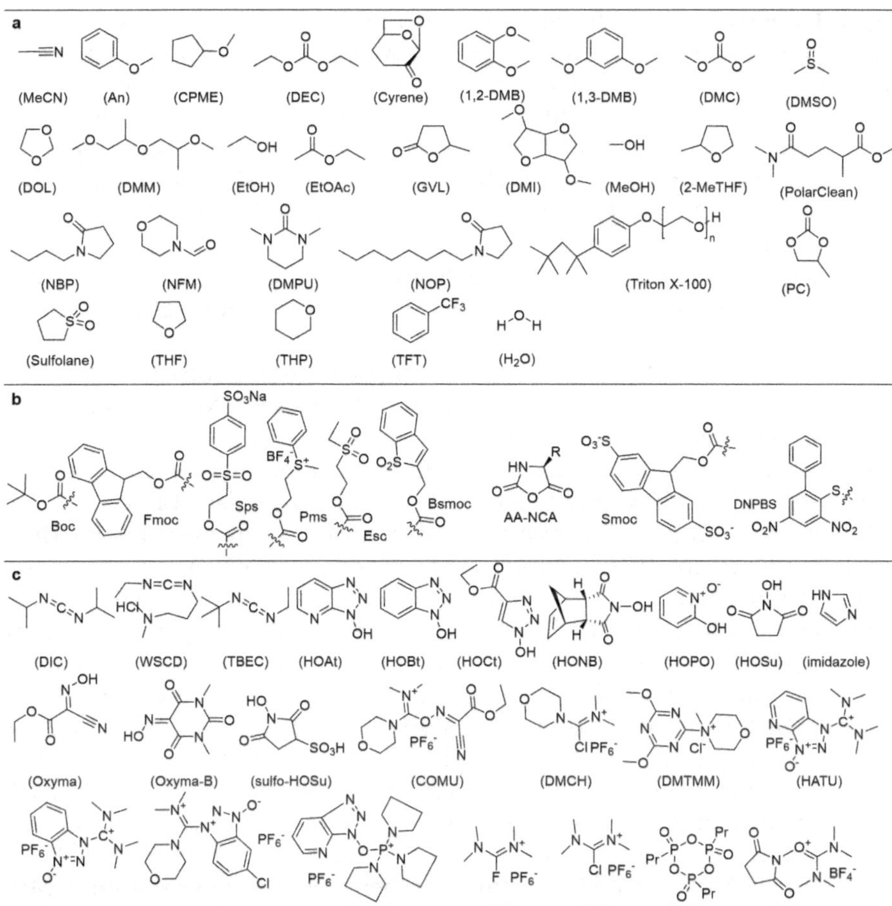

Figure 4.2 (a) Solvents; (b) α-PGs; (c) CAs reported for sustainable SPPS methods.

a holistic approach and the greenness of a solvent has to be considered[69] together with physicochemical attributes[70] such as polarity, boiling point and viscosity to ensure that adequate solubility, resin swelling and reaction kinetics are attained. Finally, the cost, life cycle[71] and security of supply of the solvent also have to be considered. Two solvent themes in sustainable SPPS have emerged: (i) Fmoc/*t*-Bu SPPS in greener organic solvents (GSPPS)[72] (see Table 4.1) and (ii) SPPS in aqueous media (ASPPS)[73] (see Table 4.2).

In early GSPPS studies, Albericio and de la Torre's group investigated acetonitrile (MeCN),[74,75] tetrahydrofuran (THF),[75] 2-MeTHF,[76,77] cyclopentyl methyl ether (CPME)[76] and ethyl acetate (EtOAc)[77] (see Table 4.1, entries 1–4), while North's group reported an SPPS in propylene carbonate (PC) on CM resin (see Table 4.1, entry 5),[78] and the first GSPPS solely on a PS resin was reported using *N*-formylmorpholine (NFM) and GVL as solvents (see Table 4.1, entry 6).[79] Microwave-assisted (MW) GSPPS in GVL was developed (see Table 4.1, entry 8),[80] followed by the report of a ring-opening side reaction involving GVL and means of alleviating it (see Table 4.1, entry 11).[81] Lopez *et al.* reported *N*-butylpyrrolidinone (NBP) as a well-suited alternative to DMF in SPPS on PS resin (see Table 4.1, entry 7)[82] and with Albericio and de la Torre's group showed that NBP is suitable for suppressing SPPS side reactions (see Table 4.1, entries 16 and 18),[83,84] while Wegner *et al.* employed NBP as solvent for coupling/deprotection and DMF as solvent for washing (see Table 4.1, entry 19).[85] MeCN and GVL were used with propylphosphonic anhydride (T_3P) as a CA (see Table 4.1, entry 20),[86] and SPPS in methyl 5-(dimethylamino)-2-methyl-5-oxopentanoate (PolarClean) (see Table 4.1, entry 24)[87] and dipropylene glycol dimethyl ether (DMM) (see Table 4.1, entry 29)[88] were also reported.

From 2019, several groups have investigated solvent mixtures in SPPS. Specifically, Schütznerová's group introduced 2-MeTHF–MeOH as a blend for Fmoc deprotections using NaOH as the base (see Table 4.1, entries 9 and 17),[89,90] also introducing anisole (An)–EtOH as an Fmoc deprotection solvent and An–dimethyl sulfoxide (DMSO) as a coupling solvent (see Table 4.1, entry 26).[91] North's group evaluated a range of binary solvent mixtures for their swelling, identifying PC–2,2,5,5-tetramethyloxolane (TMO) as a suitable solvent (see Table 4.1, entry 10),[92] while NBP–EtOAc was used in cyclic peptide SPPS (see Table 4.1, entry 12)[93] and organocatalytic SPPS (see Table 4.1, entry 28).[94] In an assessment of a range of solvent mixtures, Ferrazzano and Cabri's group identified An–dimethyl carbonate (DMC) and sulfolane–diethyl carbonate (DEC) as useful solvents for SPPS on both PS and CM resins (see Table 4.1, entry 13),[95] subsequently reporting *N*-octylpyrrolidinone (NOP)–DMC as a recyclable solvent using piperidine (pip) (see Table 4.1, entry 23)[96] and alternative Fmoc deprotection bases (see Table 4.1, entry 25).[97] DMSO–EtOAc was reported as a recyclable solvent (see Table 4.1, entry 14),[98] and NFM, NBP with anisole and 1,3-dimethoxybenzene (DMB) were used in MW SPPS (see Table 4.1, entry 15),[99] whereas Schönleber and Pedersen's group surveyed a range of binary solvents, identifying polarity and viscosity as key attributes for mimicking the characteristics of DMF in SPPS (see Table 4.1,

Table 4.1 Overview of GSPPS methods.

Entry	Solvent	Resin	CAs	AA equiv.	Fmoc removal	No. of peptides	Length (no. of AAs)	Temperature/°C	Published	References
1	MeCN	CM	DIC/Oxyma	3	20% pip	2	5–10	rt	2009	74
2	MeCN, THF	PEG	DIC/Oxyma, Oxyma-B, HOBt, HOAt	3	20% pip	4	3–10	rt	2014	75
3	2-MeTHF, CPME	CM, PS	DIC/Oxyma, Oxyma-K Oxyma-B, HOBt, HOAt, HBTU, HATU, COMU	3	20% pip	1	5	rt	2015	76
4	EtOAc, 2-MeTHF	CM, PS	DIC/Oxyma	3	20% pip	2	5–10	rt, 40	2016	77
5	PC	CM	HBTU/HOBt	3	20% pip	2	3–9	rt	2017	78
6	GVL, NFM	PS	DIC/Oxyma	3	20% pip	2	5–10	rt	2017	79
7	NBP	PS	DIC/HOBt	1.5	20% 4-mp	1	5–10	rt	2018	82
8	GVL	CM, PS	DIC/Oxyma	5	20% pip	4	10–28	90 (MW)	2018	80
9	2-MeTHF–MeOH	PS	DIC/HOBt	3.5	0.1 M NaOH	2	5–6	rt	2019	89
10	PC–TMO	PS	HBTU/HOBt	3	20% pip	1	3	rt	2019	92
11	GVL	PS	DIC/Oxyma	4–5	20% pip	1	16	90 (MW)	2019	81
12	NBP–EtOAc	PS	DIC/Oxyma	2	5% 4-mp	1	5–10	45	2019	93
13	Anisole–DMC, sulfolane–DEC	CM, PS	DIC/Oxyma	3	20% pip	3	5–10	rt, 40	2019	95
14	DMSO–EtOAc	PS	DIC/Oxyma	1.3	1–5% 4-mp	2	6–10	45–50	2019	98
15	NFM, NBP–anisole, 1,2-, 1,3-DMB	PS, PS–PEG	DIC/Oxyma	5	20% pip, 25–75% pyr	2	10–20	90–110 (MW)	2019	99
16	NBP	PS	DIC/Oxyma	1.7	20% 4-mp	1	4	45	2020	83
17	2-MeTHF–MeOH	PS	DIC/HOBt	3	0.2 M NaOH	3	5–10	rt	2020	90
18	NBP	PS	DIC/Oxyma	1.5	20% pip or 4-mp	4	3–6	rt, 45	2020	84

(*continued*)

Table 4.1 *(continued)*

Entry	Solvent	Resin	CAs	AA equiv.	Fmoc removal	No. of peptides	Length (no. of AAs)	Temperature/°C	Published	References
19	NBP, DMF	PS	COMU/Oxyma	n.r.[a]	n.r.	3	8	rt	2021	85
20	MeCN, GVL	CM	T_3P/Oxyma	5	20% pip or 4-mp	5	3–10	60	2021	86
21	DMSO–2-MeTHF, DMSO–EtOAc, DMSO–DOL	PS	DIC/Oxyma	3	20% pip	5	5–29	rt	2021	100
22	DMSO–DOL, DMSO–2-MeTHF, DMSO–EtOAc, NFM–DOL, NBP–DOL	PS	DIC/Oxyma	3	20% pip	4	3–20	rt	2021	101
23	NOP/DMC	PS	DIC/Oxyma	3	20% pip	2	5–8	rt	2021	96
24	PolarClean	PS	DIC/Oxyma	3	20% pip	3	5–6	45	2021	87
25	NOP–DMC	PS	DIC/Oxyma	1.5–3	10% DEAPA, 20% TBA or 5% TMG	3	3–8	rt, 40	2021	97
26	An, An–EtOH, An–DMSO	PS	DIC/HOBt, DIC/Oxyma	3.3	50% morph, 5% DBU, 0.2 M NaOH	3	3–6	rt, 85 (MW)	2021	91
27	DMSO–EtOAc, NBP–DOL	PS	DIC/Oxyma	3.0	20% pyr	4	5–29	rt	2021	102
28	NBP–EtOAc	PS	DIC/HOCt	1.1	15 mol% DBU	1	5	50	2022	94
29	DMM	PS, PEG	DIC/HOBt, DIC/Oxyma	2 × 3	20% pip	2	5–10	rt, 75 (MW)	2023	88

[a]n.r., not reported.

Table 4.2 Overview of ASPPS methods.

Entry	Solvent	α-PG	Resin	CAs	AA equiv.	α-PG removal	No. of peptides	Length (no. of AAs)	Temperature/°C	Published	References
1	H_2O–0.2% Triton X	Pms	PEG	WSCD/HONB	4.0	5% $NaHCO_3$	1	5	rt	2001	104
2	H_2O–0.2% Triton X	Pms	CLEAR	WSCD/HONB	4.0	0.01 N NaOH	1	5	rt	2003	105
3	H_2O–0.2% Triton X	Pms	TentaGel, CLEAR	WSCD/HONB	4.0	5% $NaHCO_3$ or 0.01 N NaOH	2	5	rt	2004	106
4	H_2O–0.2% Triton X	Esc	PEG	WSCD/HONB + DIEA	4.0	0.025 M NaOH	1	5	rt	2004	107
5	H_2O–0.2% Triton X, EtOH	Sps	TentaGel	WSCD/HONB + DIEA	4.0	0.025 M NaOH	1	5	rt	2004	108
6	H_2O–2% Triton X, EtOH	Sps	TentaGel	DMTMM	4.0	0.025 M NaOH	1	5	rt	2006	109
7	H_2O–ZrO–PEG, EtOH	Fmoc	PEG	WSCD/HONB	3.0	0.1 M NaOH	1	5	rt	2007	115
8	H_2O–0.5% Triton X-100, MeOH	Boc	CM	WSCD/HONB	5.0	1 N HCl	1	5	70, MW	2009	122
9	H_2O–0.2% Triton X-100–ZrO, EtOH	Fmoc	TentaGel	DMTMM	3.0	0.1 M NaOH	3	5–10	rt, 70	2012	116
10	H_2O–0.2% Triton X-100–ZrO, DMF	Fmoc	PEG	DMTMM	5.0	20% pip	2	3–9	rt, 70	2013	117

(continued)

Table 4.2 *(continued)*

Entry	Solvent	α-PG	Resin	CAs	AA equiv.	α-PG removal	No. of peptides	Length (no. of AAs)	Temperature/°C	Published	References
11	H_2O, DMF	Bsmoc	CM	DIC/HOBt	5.0	20% pip, 5% piz	1	10	75, MW	2013	110
12	H_2O	NCA	CM	Buffer, pH 10.4	3.0	Buffer, pH 3	1	4	rt	2013	111
13	H_2O–0.2% Triton X-100–ZrO, EtOH	Boc	PEG	DMTMM	5.0	TFA	2	5	rt	2013	120
14	H_2O–0.2% Triton X-100–ZrO, DMF	Fmoc	PEG	DMTMM	3.0	20% pip	2	3–6	70	2014	118
15	H_2O–MeCN, EtOH	Smoc	CM	WSCD/HOPO	3.0	1 M NaOH, 25% ethanolamine, 5–10% piz	22	2–10	rt	2016, 2020	112 and 113
16	H_2O–PolarClean, MeCN, NBP, DMSO	Fmoc	TentaGel	TCFH	1.3	5% 4-mp	1	5	rt, 40	2021, 2023	123 and 124
17	H_2O, DCM	Boc	PEG	DMTMM	5.0	50% TFA	1	5	rt, 75 (MW)	2022	121
18	H_2O–0.2% Triton X-100–ZrO, EtOAc	Fmoc	PEG, TentaGel	DMTMM	4.0	20% pip	2	5–10	rt, 75 (MW)	2022	119
19	H_2O, MeCN	Smoc	Amino Li	WSCD/Oxyma	4.0	1 M NaOH	4	3–5	rt	2023	114

entry 21),[100] and this approach was employed for minimizing SPPS side reactions (see Table 4.1, entry 22),[101] and also identifying solvent mixtures for use with pyrrolidine (pyr) as the base (see Table 4.1, entry 27).[102]

Whereas a key aim of GSPPS is to replace DMF adequately with greener solvents, the aim of ASPPS is to adopt water as the greenest solvent[103] in Merrifield's peptide synthesis. Nevertheless, as standard SPPS starting materials are not readily amenable for use in water, ASPPS is inherently more challenging than GSPPS. ASPPS is therefore based on ensuring compatibility with an aqueous environment while attaining suitable reaction kinetics and minimizing side reactions. The main reported ASPPS concepts are based on (i) hydrophilic α-PGs (see Table 4.2, entries 1–6, 11, 12, 15 and 19),[104–114] (ii) hydrophobic α-PG-AA-OHs formulated as nanoparticles and nanomicelles (see Table 4.2, entries 7–10, 13, 14, 17 and 18)[115–121] and (iii) enabling the use of hydrophobic α-PG-AA-OHs by reaction engineering (see Table 4.2, entries 8 and 16).[122–124] It is worth noting that only the *N*-carboxyanhydride (AA-NCA) method used pure water (see Table 4.2, entry 12), whereas in all other ASPPS methods varying amounts of organic solvents and/or Triton-X surfactants were employed.

4.4.2 α-Amino Acid Protecting Groups (see Figure 4.2b)

Despite the numerous PGs reported for α-AA protection,[125] only Boc and Fmoc have found broad use in peptide synthesis.[126] In fact, the dominance of Fmoc in GSPPS was not challenged until thiol-labile DNPBS was used as an α-AA PG for SPPS in 2-MeTHF, GVL and NFM (see Scheme 4.2),[127] and most recently by employing unprotected AAs and using DEC–sulfolane for activation and 2-MeTHF for aminolysis in ynamide-mediated inverse SPPS (ISPPS) of a 10-mer peptide.[128]

On the other hand, several α-PGs have been investigated in ASPPS, with Hojo and Kawasaki's group employing 2-[phenyl(methyl)sulfonio]ethoxycarbonyl tetrafluoroborate (Pms)-, *N*-ethanesulfonylethoxycarbonyl (Esc)- and 2-(4-sulfophenylsulfonyl)ethoxycarbonyl (Sps)-protected AAs (see Table 4.2, entries 1–6), Collins using (1,1-dioxobenzo[*b*]thiophene-2-ylmethyloxycarbonyl) (Bsmoc) (see Table 4.2, entry 11) and Gentilucci and co-workers using AA-NCAs (see Table 4.2, entry 12). In the broadest ASPPS study yet, Knauer and Kolmar's

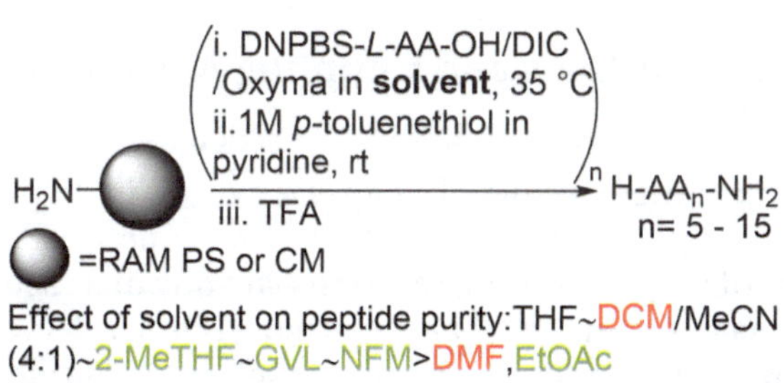

Scheme 4.2

group reported 2,7-disulfo-9-fluorenylmethoxycarbonyl (Smoc) as an α-PG for ASPPS of a wide range of peptides (see Table 4.2, entries 15 and 19). Finally, hydrophobic Fmoc and Boc were applied in ASPPS either by formulating them with zirconium oxide and a Triton X surfactant (see Table 4.2, entries 7–10, 13, 14 and 17), by MW SPPS (see Table 4.2, entry 8) or by using a water-compatible resin and CA, respectively, and leveraging increased resin swelling and AA solubility owing to the use of suitable cosolvents (see Table 4.2, entry 16). Overall, unless significant investments into some of the alternative α-PGs are made, it can be expected that GSPPS using Fmoc as the α-PG will continue to evolve at a faster pace than any other α-PG-based SPPS platform.

4.4.3 Resin Swelling and Solubility of Starting Materials and By-products

Understanding the characteristics of resins is critical[129] in SPPS, with solvation and swelling of peptides attached to solid-phase resins being particularly important,[130–132] and several studies in the area of SPPS greening reported swelling of various functionalized polymers in different solvents. In the first such swelling study, PS and CM resins in nine different solvents were evaluated,[133] followed by a comprehensive assessment of the swelling of nine resins in 25 solvents[134] and a study on the swelling of a PS resin in 36 solvents.[82] Other groups reported on the swelling of different resins in neat solvents and binary solvent mixtures,[92,93,95,98,135,136] including swelling of peptides of varying length bound to the polymer supports[100,135,136] and swelling of polymer supports in aqueous mixtures.[123,124] Most recently, swelling of amino-Li-resin in various solvents[137] and swelling of different resins in DMM[88] were reported. Taken together, these swelling studies allow the identification of resin–solvent combinations exhibiting adequate swelling in SPPS. Suitable solubility of the starting materials and by-products formed during SPPS is also important, as insolubles may impede the reaction kinetics and/or clog the reactor filters or tubing and piping. Several studies on the solubility of Fmoc-AA-OHs and common CAs and their by-products in greener solvents and mixtures have been reported,[82,87,88,95–98,100] and these studies provide a basis for selecting solvents facilitating suitable solubility of AAs and other materials used in SPPS.

4.4.4 Peptide Couplings and Coupling Agents (see Figure 4.2c)

To attain suitable kinetics and chemoselectivities in peptide couplings, choosing the right CA is important.[104,138] At the same time, with the focus on sustainability of amide-forming reactions steadily increasing,[31,33,34] CA-related risks such as allergenicity[139] and thermal stability[140] must also be considered, aided by assessments of kinetics for selected peptide couplings (see Table 4.3).

Table 4.3 Studies on peptide couplings in greener/aqueous solvents.

Entry	Solvent(s)	Coupling reagents	Coupling partners	Temperature/°C	Published	References
1	H_2O	WSCD/HOBt, HOAt, HOSu, HONB, imidazole (DIEA, pyridine)	Esc-Phe-OH + H-Leu-RAM PEG resin	rt	2004	107
2	Aq. 2% Triton X-100, aq. EtOH (5–50%)	WSCD, TNTU/HONB (DIEA), DMT-MM (NMM)	Sps-Phe-OH + H-Leu-TentaGel resin	rt	2006	109
3	Aq. 0.2% Triton X-100	WSCD, DMT-MM/HONB, HOSu, sulfo-HOSu	Fmoc-Phe-OH (ZrO) + H-Gly-RAM-TentaGel resin	70 (MW)	2012	116
4	Aq. 0.2% Triton X-100	WSCD, DMT-MM/HONB, HOSu, sulfo-HOSu	Boc-Phe-OH (ZrO) + H-Gly-RAM-PEG resin	70 (MW)	2013	120
5	NBP, NBP–EtOAc (1:1, 1:4)	DIC/Oxyma	Fmoc-Ser(*t*-Bu)-OH + H-RMG PS resin	rt	2020	142
6	H_2O, H_2O–MeCN (7:3)	WSCD/Oxyma, HOPO	Smoc-AA-OH + H-AA_n-CM H-RAM resin	rt	2020	113
7	GVL, 2-MeTHF, NBP, EtOAc, MeCN, NBP–EtOAc (3:1)	T3P/Oxyma (DIEA)	Fmoc-Tyr(*t*-Bu)-OH + H-Gly-Gly-Phe-Leu-NH-RAM resin	60	2021	86
8	NBP, DMSO, DOL, 2-MeTHF, THP and mixtures of NBP, NFM and DMSO with EtOAc, 2-MeTHF and DOL	DIC/Oxyma	Fmoc-Gly-OH + H-Pro-Phe-Ala-2CT PS resin	rt	2021	100
9	An and An–DMSO (4:1)	DIC/HOBt, Oxyma	Fmoc-Phe-OH + H-Ala-H-RAM PS resin	rt	2021	91
10	H_2O, PolarClean and mixtures thereof	COMU, HDMC, DMCH, TFFH, TCFH (collidine, DIEA, NMM, NMI)	Fmoc-Gly-OH + TentaGel S NH_2 resin	rt	2021	123
11	NBP, DMSO, PolarClean and their mixtures with EtOAc	DIC/HOCt	Fmoc-Val-OH + H-Tyr(*t*-Bu) RMG PS resin	50	2022	94
12	Aq. 0.2% Triton X-100	DMTMM, TSTU (NMM)	Fmoc-Ser/Thr/Tyr-OH (ZrO) + H-Trp(Boc)-RAM-TentaGel resin	55, 75 (MW)	2022	119
13	NBP–EtOAc (1:4)	DIC, TBEC/Oxyma	Fmoc-His(Trt)-OH + H-Phe RMG PS resin	rt	2023	147

Formulations of DIC in the presence of additives (see Table 4.1, entries 1–4, 6–9, 11–18 and 21–29) and onium salts (see Table 4.1, entries 5, 10 and 19) were the most widely used CAs in GSPPS. While DIC/ethyl (hydroxy-imino)cyanoacetate (Oxyma) has emerged as the preferred CA in GSPPS, in 2019 Kolis's group reported that DIC/Oxyma forms toxic prussic acid (HCN) during AA activation.[141] Several studies aimed at minimizing HCN formation during DIC/Oxyma couplings followed,[142–144] although ultimately, by assessing different carbodiimides with Oxyma for coupling efficiency and HCN formation, Albericio and de la Torre's group developed 1-*tert*-butyl-3-ethyl-carbodiimide (TBEC)/Oxyma as a suitable alternative, offering minimized HCN formation and lower racemization than DIC/Oxyma.[145,146] Notably, 1-*tert*-butyl-3-ethylurea (TBEU) formed during TBEC couplings was found to be markedly more soluble across a wide range of greener solvents and mixtures than diisopropylurea (DIU) formed from DIC.[147] As precipitation of DIU during SPPS in greener reaction media leads to clogging of equipment,[82,100] TBEC/Oxyma is not only a less hazardous[145,146] but also a more practical[147] choice for GSPPS than DIC/Oxyma. In fact, in a DIC-mediated solid-phase peptoid synthesis carried out in EtOAc, the post-coupling washes had to be carried out in DMSO–EtOAc owing to the poor solubility of DIU in EtOAc[148] and, owing to the higher solubility of TBEU, if TBEC was used then EtOAc could conceivably have been used throughout.

With regard to couplings in ASPPS, considering that a wide range of CAs worked well in solution-phase aqueous peptide couplings,[149] based on evaluation in selected couplings (see Table 4.3) CAs useful for ASPPS were readily identified, most notably water-soluble carbodiimide (WSCD) with various coupling additives (see Table 4.2, entries 1–5, 7, 8, 15 and 19), 4-(4,6-dimethoxy-1,3,5-triazin-2-yl)-4-methylmorpholinium chloride (DMTMM) (see Table 4.2, entries 6, 9, 10, 13, 14, 17 and 18) and chloro-*N*,*N*,*N*′,*N*′-tetramethylformamidinium hexafluorophosphate (TCFH) (see Table 4.2, entry 16). It is of note that enzymatic aqueous solid-phase couplings using Fmoc-AA-OHs have been reported,[150] although a practical ASPPS based on these biocatalytic amide formations remains to be developed.

4.4.5 Deprotection of α-Protecting Groups

To obtain peptides with satisfactory purities by SPPS, efficient removal of α-PG is essential. The standard protocol for Fmoc removal is with 20% pip in DMF,[151] with other bases such as piperazine (piz),[152] 4-methylpiperidine (4-mp),[153] 1,8-diazabicyclo[5.4.0]undec-7-ene (DBU)[154] and dipropylamine (DPA)[155] also proven to be valuable. In GSPPS, 20% pip was initially the only protocol used (see Table 4.1, entries 1–8). Nevertheless, with pip deemed a hazardous base,[34,97] alternative bases have been investigated, including NaOH,[89–91] 4-mp,[82–84,86,93,98] pyr,[102] 3-(diethylamino)propylamine (DEAPA)[97] and morpholine (morph).[91] The base-labile α-PGs in ASPPS were removed with different bases suitable for aqueous conditions such as NaOH and $NaHCO_3$[104–109,112–119] or piz,[110,112–114] while the Boc group was removed using either HCl[122] or TFA.[120,121]

4.4.6 Temperature Control and Agitation

The use of elevated temperature (ET) is an established approach for accelerating reaction rates in SPPS,[156,157] especially in the context of MW-enhanced peptide synthesis.[158,159] Although in reports on GSPPS and ASPPS ET protocols were used in some cases, most studies relied on the use of room temperature. Nevertheless, as even mild heating can accelerate the rates of reactions in greener solvents quite profoundly (see Table 4.4),[98] the effect of temperature as a means of improving rates in sustainable SPPS should always be considered.

Regarding the impact of ET on sustainability, a room temperature SPPS which requires less energy will in principle always be greener. Nevertheless, in addition to accelerating rates, an ET SPPS also allows the use of less AAs and coupling/deprotection reagents, which in turn allows for less washing.[94] Therefore, from an overall perspective, the more efficient, less wasteful ET SPPS can, in fact, be more sustainable. As mass transfer (MT) plays an appreciable role in rates of reactions on solid supports,[160] different means of agitation, such as ultrasonication,[161,162] shear mixing[163] and flow methods,[164,165] have been used for efficient MT and should be considered in the context of sustainable SPPS.

4.4.7 Chemoselectivity

Minimizing side reactions during SPPS is crucial for the efficient purification of crude peptides during DSP. As racemization during AA activation is the most common side reaction in SPPS couplings, many of the studies carried out on sustainable SPPS evaluated AA racemization. In particular, with histidine (His) being the most racemization-prone proteinogenic AA,[125] minimizing racemization during His couplings is highly important.

Several protocols for coupling of His in greener solvents have been reported (see Table 4.5), and although most of these couplings were carried out in organic solvents and mixtures, coupling of Fmoc-His(Trt)-OH ground with zirconia beads in aqueous Triton X-100 has also been reported (see Table 4.5, entry 2). Fmoc-His(Trt)-OH and Fmoc-His(Boc)-OH were evaluated as His derivatives, and in keeping with prior art,[12,166] His(Boc) afforded appreciably lower D-His product than His(Trt). Nevertheless, owing to the instability of

Table 4.4 Effects of temperature and solvent on Fmoc deprotection of Fmoc-RAM resin using 5% 4-mp.

		Conversion/%	
Entry	Solvent	25 °C	50 °C
1	DMF	**90.5**	**99.3**
2	EtOAc	5.9	36.7
3	DMSO–EtOAc (1 : 49)	13.7	62.0
4	DMSO–EtOAc (1 : 9)	28.9	**91.1**
5	DMSO–EtOAc (1 : 1)	**91.9**	**99.1**

Table 4.5 Protocols for Fmoc-His(X)-OH couplings in greener solvents.

Entry	Solvent	Coupling agents	Temperature/°C	X in His(X)	AA equiv.	D-His/%	Published	References
1	MeCN	PyAOP/DIEA	rt	Trt	5.0	28.5	2009	74
2	H_2O–0.2% Triton X-100–ZrO	DMTMM/NMM	70 (MW)	Trt	5.0	3.3	2014	118
3	NBP–EtOAc (1 : 1)	DIC/HONB/HOBt	45	Trt	2 × 2.0	0.3	2019	93
4	NFM–anisole (1 : 3)	DIC/Oxyma	90 (MW)	Boc	5.0	0.6	2019	99
5	GVL	DIC/Oxyma	rt	Trt	5.0	n.r.[a]	2020	205
6	2-MeTHF	DIC/HOBt	rt	Boc	3.3	n.d.[a]	2020	90
7	NBP	DIC/Oxyma	rt	Trt	1.5	2.5	2020	84
8	DMSO–2-MeTHF (3 : 7)	DIC/Oxyma	rt	Trt	3.0	2.0	2021	101
9	DMSO–EtOAc (1 : 9), NBP–DOL (1 : 4 and 4 : 6), NFM–DOL (1 : 4)	DIC/Oxyma	rt	Boc	3.0	<0.01	2021	101
10	NBP–EtOAc (1 : 4)	TBEC/Oxyma	rt	Trt	3.0	0.4	2023	147

[a]n.r., not reported; n.d., not determined.

His(Boc) towards prolonged pip treatments,[125] the use of His(Boc) in GSPPS should be carried out with appropriate care. For His(Trt), the lowest racemization was achieved using DIC/HONB/HOBt and TBEC/Oxyma in NBP–EtOAc (see Table 4.5, entries 3 and 10). Furthermore, minimized lactamization during Fmoc-Arg(Pbf)-OH couplings was achieved by tuning the protocol[83] and the solvent,[101] and replacing DMF with DMSO–EtOAc suppressed radical-induced oxidation during Cys couplings.[98]

Regarding side reactions that occur during α-PG removals, the formation of aspartimide (Asi)-related impurities is a daunting challenge in SPPS.[167] Several groups have demonstrated minimization of Asi formation in GSPPS, namely using pip in NBP[84] and in DMSO–2-MeTHF, DOL and EtOAc,[101] and also using NaOH,[90,91] DEAPA,[97] morph,[91] DBU[91] and pyr[102] as bases. In addition, Asi formation in Smoc ASPPS has been investigated, and it was determined that piz as the base at temperatures below room temperature provides optimal Asi minimization.[113] Diketopiperazine (DKP) formation is another side reaction that hampers basic α-PG removals[168] and, in studies aimed at minimizing pip-induced DKP formation, no correlation between the attributes of the greener solvent mixtures and the rate of DKP formation was observed.[101] In that regard, it is of note that although hampered by the use of hazardous solvents, piz–DBU,[169] DPA[155] and morph[170] were recently reported to minimize DKP efficiently during SPPS Fmoc removals.

4.4.8 Reaction Monitoring

To ensure the completion of couplings/deprotections and also removal of residual materials during washes, the progress of SPPS is customarily followed by the methods available for the monitoring of processes on solid supports.[171] Nevertheless, accurate, non-destructive analytical methods for SPPS monitoring are still desirable as a means of (i) avoiding inaccurate endpoint determinations and (ii) optimizing solvent consumption. To this end, several methods for SPPS monitoring have been reported recently, including a quantitative colorimetric assay that afforded peptides with higher purities than in syntheses monitored by the Kaiser test.[172] On the other hand, refractive index (RI) was reported as a process analytical tool (PAT) for couplings, deprotections and washes during SPPS.[173] Further, a camera-enabled, visual method was developed for monitoring coupling reactions in solution and in SPPS,[174] and a near-infrared (NIR) spectroscopic method to assess couplings in SPPS,[175] a Gaussian shape of the deprotection peaks to monitor Fmoc removals in flow SPPS[176] and the use of Oxyma anion for spectrophotometric monitoring of Fmoc deprotections in green SPPS[177] have also been reported.

4.5 Cleavage

Upon completion of SPPS, the peptide is cleaved off the resin and isolated. Although resins prepared by Fmoc/*t*-Bu SPPS can be cleaved by different means, including the use of nucleophiles[178] or photochemically,[179] the standard approach is TFA/scavengers followed by precipitation with ethereal solvents.[180,181] Nevertheless, from a sustainability standpoint TFA cleavage is hampered by the formation of various impurities,[180,181] which can be difficult to remove during the DSP. Moreover, TFA is highly corrosive and from the green chemistry perspective it should be replaced with less hazardous reagents. Furthermore, TFA-cleaved peptides are typically isolated by precipitation with flammable, potentially peroxide-containing ethers such as Et_2O and methyl *tert*-butyl ether (MTBE), the latter of which can cause the formation of *t*-Bu adducts,[182] while TFA[183] and AcOH–trifluoroethanol (TFE)[184] in the suspected carcinogen DCM are used for the cleavage of protected peptide fragments. Considering the above peptide resin cleavage-related sustainability challenges, various approaches for greening of the cleavage–peptide isolation process have been investigated.

4.5.1 Greener TFA Cleavages

To minimize the occurrence of side reactions in TFA peptide resin cleavages, the process is typically thoroughly optimized,[185] and as methods for suppressing impurities improve the sustainability of the cleavage, approaches for improving chemoselectivity during cleavages have been studied. For example, 1,4-benzenedimethanethiol (BDMT) was reported to give higher peptide purities than the classical aliphatic thiol scavengers,[186] and cleavage cocktails containing trimethylsilyl chloride and triphenylphosphine were proposed as a means of avoiding side reactions that occur on methionine.[187] In addition, replacing Boc with Trt as the Lys side-chain PG has been leveraged to suppress intramolecularly formed *t*-Bu adducts.[188]

4.5.2 TFA-free Cleavage

Based on the use of TFA for cleavages in all GSPPS approaches (see Table 4.1), avoiding TFA as the cleavage reagent is not easy, even in studies where increasing SPPS sustainability is the stated aim. Nevertheless, replacing TFA is urgent not only owing to its high corrosiveness, but also because it is classified as a per- and polyfluorinated substance (PFAS), which is of great concern.[189] Consequently, several TFA-free resin cleavages have been reported. For example, protected peptides have been cleaved off CTC resin using 1 N HCl in DMF[190] and 0.1 N HCl in TFE or hexafluoroisopropanol (HFIP) was used to cleave all common side-chain PGs and linkers,[191–193] and occasionally this protocol has actually been adopted.[113,194] Furthermore, low-concentration HCl in subcritical water has been used to cleave short peptides off Wang and RAM resins,[195] and $NiCl_2$ has been used to cleave Gly–Ser bonds in GSHHW amino acid sequence tag-containing peptides.[196] In addition, enzymatic[197] and Lewis acid[198]-mediated resin cleavages have been reported and enzymes[199] and Lewis acids[200,201] have also been used in AA side-chain PG removals. To summarize, despite the ample alternative peptide cleavage protocols that have been investigated, none has yet outcompeted TFA, and adoption in practice has therefore been elusive.

4.5.3 Precipitation in Greener Solvents

Peptide precipitations in greener ethers have been investigated, and CPME and 2-MeTHF have been used in the precipitation of ≥10-AA peptides.[202,203] These solvents could not be used for shorter peptides owing to the high peptide solubility, and in a green precipitation of a 7-AA peptide MTHP–*n*-heptane was used instead.[93] Alternatively, precipitation can be avoided by neutralizing the peptide-containing TFA solution, for example using aqueous ammonium acetate containing a suitable amount of EtOH or MeCN.[204]

4.5.4 Low-concentration TFA Cleavages in Greener Solvents

The first instance of a DCM-free, low TFA concentration peptide resin cleavage was accomplished using 2% TFA in MeCN–EtOAc in the removal of Fmoc-GG-OH from CTC resin.[93] Subsequently, 2% TFA in anisole or 1,3-DMB was used in chemoselective cleavages of a variety of peptide acids from CTC resin.[205] Furthermore, protected peptide amides were cleaved off Sieber resin using 2%TFA in toluene.[206] Finally, aiming at minimizing TFA consumption by omitting side-chain PGs for Tyr, His and Arg, Yang *et al.* used TFA–trifluorotoluene (TFT) (1 : 9) in the cleavage of an 8-mer peptide from CTC resin.[207]

4.6 Complex Peptides

Although SPPS and related methods are valuable for constructing a variety of complex peptide architectures (see Figure 4.1b), the synthesis of these molecules often requires large amounts of hazardous materials in the same way that SPPS of linear peptides does. Consequently, sustainable methods for SPPS of complex peptides are worth pursuing and, as cyclic peptides constitute 46% of the approved peptide therapeutics,[208] developing greener SPPS methods for cyclic peptides is of particular interest.

In this regard, while the method for (i) on-resin removal of side-chain PGs in MeCN–EtOAc and (ii) on-resin lactamization in NBP–EtOAc (see Scheme 4.3a) is significant in that it represents the first green SPPS of a cyclic peptide,[93] expanding further the toolbox for sustainable SPPS of complex peptides will certainly be of great value.

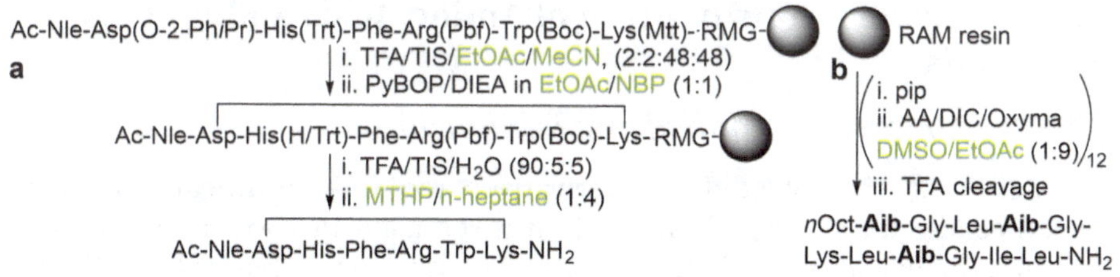

Scheme 4.3

4.7 Implementation and Scale-up

Judging by the number of reports on SPPS greening in the literature, tremendous progress towards making SPPS sustainable has been made. Nevertheless, unless these advances are broadly adopted in practice, they will not make a difference for the better, and ultimately the success of the current SPPS greening movement should be measured by the extent of implementation. In particular, as enormous amounts of waste are generated in the large-scale manufacture of peptide therapeutics, adopting sustainable tactics in industrial peptide production is of great importance. In this regard, while instances of the implementation of large-scale green SPPS have been presented at various conferences, reports in the scientific literature have been lacking, and the SPPS of anticancer peptaibols in DMSO–EtOAc by De Zotti's group thus warrants a mention as an example of using green SPPS in practice (see Scheme 4.3b).[209] Concerning scale-up, the 10 mmol assembly of Aib–ACP 10-mer (Aib = 2-aminoisobutyric acid; ACP = acyl carrier protein) in DMSO–EtOAc is the largest-scale green SPPS reported to date.[98]

4.8 Minimization of Waste

Producing synthetic peptides generates large amounts of waste, and waste trimming approaches in peptide synthesis have therefore attracted significant attention.[210] SPPS, in which substantial waste generation occurs already during the production of starting materials (see above), is regarded as particularly wasteful, as large amounts of amino acids and process aids are used to drive reactions to completion, with enormous volumes of solvents being used for washing. In fact, SPPS has been reported to be significantly more wasteful than alternative peptide synthesis platforms based on *e.g.* tag LPPS,[211] mechanochemistry[212] and recombinant production.[213] Nevertheless, with the significant potential of SPPS for waste minimization in mind (see below), if the waste reduction tactics here are deployed efficiently, SPPS can conceivably be competitive from a sustainability standpoint with the platforms currently perceived as less wasteful. Specifically, SPPS waste-reducing techniques based on (i) minimizing the consumption of starting materials, (ii) minimizing washing and (iii) recycling have been reported and are discussed in the following subsections.

4.8.1 Reducing Consumption of Amino Acids and Process Aids

4.8.1.1 Amino Acid Equivalents Minimization

The synthesis of protected AA derivatives requires significant processing, especially for the highly pure derivatives used in GMP manufacture. Although the use of these materials will be kept as low as possible, as Fmoc-AA-OHs are relatively inexpensive the use of large AA equivalents in SPPS is quite common. In fact, of all GSPPS and ASPPS approaches reviewed here, only a few used ≤1.5 equiv. of AAs in couplings (see Table 4.1, entries 7, 14, 18, 25 and 28, and Table 4.2, entry 16) and further studies aimed at reducing AA equivalents would therefore be valuable.

4.8.1.2 *Reagent Minimization*

The use of large amounts of process aids in SPPS is standard practice, exemplified by 20% pip being the norm for Fmoc deprotections. In fact, in GSPPS, only a few methods have utilized smaller amounts of a base (see Table 4.1, entries 9, 12, 14, 17, 25, 26 and 28) and 20% of a base has been used in most cases. Nevertheless, as the use of less Fmoc removal base means less washing, the base-minimizing approach[94] is beneficial from the overall waste reduction standpoint and should always be considered.

4.8.1.3 *Amino Acids with Unprotected Side-chains*

Unprotected AA derivatives require less processing than their protected counterparts, hence the use of unprotected AAs is an attractive waste-minimizing technique. In the context of ASPPS, this concept has been demonstrated by the use of Smoc-Arg-OH in the synthesis of H-Arg_7-OH[113] and the use of Fmoc-Ser/Thr/Tyr-OH ground with ZrO/aqueous Triton X-100 nanoparticles.[119] The use of side-chain unprotected Fmoc-Arg(HCl)/His/Tyr-OH has also been reported, albeit in this case the SPPS was carried out in DMF.[207]

4.8.1.4 *Catalysis*

Replacing stoichiometric reagents with catalysts is a key tenet of GC (ninth principle of green chemistry), and in the area of amide-forming reactions numerous catalytic reagents have been evaluated.[214] In SPPS, Arora's group reported diselenide catalysts for peptide couplings, the first generation relying on a stoichiometric amount of a phosphine[215] whereas in the second-generation protocol a catalytic amount of a phosphine oxide precatalyst was used.[216] Moreover, a recent study used catalysts for both Fmoc removals and couplings, revealing the possibility of a fully catalytic GSPPS.[94]

4.8.2 Reducing Washing

Solvent use during the washing stage contributes significantly to the waste generated during SPPS[33,34,217] and carrying out washes in a flow is a useful means of significantly reducing the washing.[218] On the other hand, omitting washing between SPPS steps also reduces the washing, as demonstrated by eliminating washes between couplings and Fmoc deprotections in SPPS in DMF,[159] an approach later adopted in GSPPS.[93] More recently, minimization of washing in SPPS was taken a step further by Singh and Collins, who carried out MW SPPS in DMF using one-pot couplings/Fmoc removals,[219] and this concept was refined and even adopted in GSPPS by Albericio and de la Torre's group.[220] Most recently, Collins *et al.* reported SPPS completely free of solvent washing by utilizing MW-aided, one-pot couplings/Fmoc deprotections followed by evaporation of the volatile Fmoc removal base (pyrrolidine), in which condensation of the base inside the reactor was prevented by a directed nitrogen flushing.[221] A solvent saving of up to 95% was demonstrated in the SPPS of a range of challenging peptides and even proteins (up to 89 AAs) on the 25 mmol scale, and although most of the study was carried out using DMF as the solvent, washing-free SPPS with a greener solvent (NBP) was also exemplified.

4.8.3 Recycling

As chemistry for the next decade has been defined as the science and industry of reuse,[222] adhering to principles of a circular economy has also been a point of emphasis in peptide chemistry.[223] While reports on recyclable resins[224,225] and CAs[226–228] have paved the way for more sustainable SPPS, methods for recycling of solvents, CAs and Fmoc deprotection bases in GSPPS have also been reported. In the first account of adopting a circular economy in GSPPS, EtOAc and DMSO were recycled by distillation and Oxyma was recycled by extraction/crystallization.[98] In subsequent studies by Cabri and Ferrazzano's group, recycling of NOP and DMC as solvents and pip[96] and DEAPA[97] as bases was reported, and recycling of aqueous SPPS waste by passing it through a mixed ion-exchange (IEX) stationary phase has also been reported.[124] In fact, as disposing of aqueous waste is challenging,[229] in ASPPS processes means of tackling the waste should always be considered. Most recently, recycling of DMM from SPPS waste was reported as a means of reducing the waste and compensating for the higher cost of DMM compared with DMF.[88] Taken together, the waste recycling approaches summarized here hold promise for improving the sustainability of future SPPS processes. In fact, the design of recycling strategies should be an integral part of the process design, as generating large amounts of difficult-to-recycle waste may render waste-minimizing endeavors challenging. Importantly, as recycling requires energy and processing, the sustainability and cost of recycling should be always be determined.[230] Finally, evidenced by recycled DMF causing the presence of nitrosamines in vasartan API (active pharmaceutical ingredient),[231] the use of recycled materials in GMP peptide manufacture should be approached with caution. Therefore, collaborating with recycling specialists and valorizing the waste rather than reusing it may be a sensible option. Consideration of which of the above waste-reducing tactics should be adopted has to be treated on a case-by-case basis, taking both cost and sustainability into account. Above all, as in the manufacture of peptides the quality of the product is of the utmost importance, the impact of the different waste-reducing tactics on the content of impurities should be used as the guiding principle as to which waste-reducing tactic to select.

4.9 Sustainability Metrics

To assess the sustainability of an SPPS, an analysis using suitable metrics has to be carried out. In several reports on GSPPS, complete E-factor (CE-factor)[98] and process mass intensity (PMI)[88,96,97] as sustainability metrics were calculated, proving that waste recycling leads to a significant reduction of the generated waste. Nevertheless, the impact of the RoS of the starting materials (see above) on sustainability was not considered, and future studies aiming at more adequate sustainability metrics for SPPS should take the RoS of the starting materials into account. Moreover, as the CE-factor and PMI do not determine the amount of energy used throughout the process, the use of energy consumption-quantifying metrics such as carbon intensity (CI)[232] should be considered.

4.10 Procurement, Cost and Intellectual Property (IP)

When considering the use of a new SPPS material, the procurement landscape must be thoroughly evaluated. This is particularly true for some of the green solvents that are produced by a single manufacturer using a proprietary process, which may render the devising of a sensible supply-chain strategy challenging. On the other hand, as in peptide manufacture economic and environmental sustainability considerations have to go hand in hand,[233] the impact of a new SPPS starting material on the economy of the overall process has to be determined. In fact, as some workers may be deterred by the higher cost of greener materials and protocols, to attain a truly broad adoption of sustainable SPPS cost efficiency has to be a focal point from the outset. An example of this school of thought can be found in SPPS in DMSO–EtOAc, which upon recycling of the waste resulted in both minimized waste and improved process economy *versus* a DMF-based process.[98] Finally, as some inventions in the sustainable SPPS area are covered by patents or patent applications, when considering implementation any relevant IP prior art must be taken into account. While filing a patent application is the absolute right of any applicant/inventor, it is nonetheless worth remembering that advances restricted by IP generally become less broadly adopted in the peptide synthesis field than free-to-use technologies.

4.11 Conclusion

Over the past six decades, the concept of Merrifield's SPPS has become the central platform for the assembly of a wide range of structurally diverse peptides at any scale, making it an irreplaceable tool for all peptide chemists. Although the wastefulness and reliance on hazardous materials have long hampered SPPS from the sustainability standpoint, with the advances reviewed here in mind the perception of SPPS as being an unsustainable technology is about to change. Granted, further studies will be needed to expand the still quite limited scope of sustainable SPPS, and for truly broad adoption there is still ample room for future work. Nevertheless, with the prospect of the demand for making large quantities of therapeutic peptides in a sustainable manner steadily increasing,[37] it can be expected that future advances by the peptide synthesis community will render SPPS not only the most versatile and practical approach, but also a highly sustainable peptide synthesis platform.

In fact, although literature examples of sustainable SPPS adopted in practice are currently limited, as peptide greening initiatives often take place in a confidential setting during the development of therapeutic and cosmetic peptides in the manufacturing sector, the extent of implementation is probably appreciably higher than the literature reports would suggest. Notably, as using DMF as a solvent has dominated SPPS for decades, identifying and developing less hazardous replacements for DMF are currently the greatest challenge for implementing more sustainable SPPS protocols in practice.[234] To this end, despite the numerous greener solvents and solvent mixtures that have been reported (see Tables 4.1 and 4.2), no universal replacement for DMF has emerged thus far. Although for application in peptide synthesizers in the laboratory and for deployment in large-scale reactors in manufacturing the utilization of many different solvents and mixtures would be impractical, it is nonetheless likely that several preferred greener solvents and mixtures will be necessary to fully replace DMF with efficient, safe and circular[235] reaction media in all SPPS applications. Finally, the use of DMF in the synthesis of already approved pharmaceutical products is expected to continue in the foreseeable future. Replacing DMF with a greener reaction medium constitutes a process change that will require regulatory actions, so economically viable approaches for the replacement of DMF in the manufacture of peptide drugs that have been on the market for years will still have to be established.

References

1. T. Curtius, *J. Prakt. Chem.*, 1882, **26**, 145.
2. D. M. M. Jaradat, *Amino Acids*, 2018, **50**, 39.
3. R. B. Merrifield, *J. Am. Chem. Soc.*, 1963, **85**, 2149.
4. M. Schnölzer, P. Alewood, A. Jones, D. Alewood and S. B. Kent, *Int. J. Pept. Protein Res.*, 1992, **40**, 180.
5. A. R. Mitchell, *Biopolymers*, 2008, **90**, 175.
6. L. A. Carpino and G. Y. Han, *J. Am. Chem. Soc.*, 1970, **92**, 5748.
7. G. B. Fields and R. L. Noble, *Int. J. Pept. Protein Res.*, 1990, **35**, 161.
8. R. Behrendt, P. White and J. Offer, *J. Pept. Sci.*, 2016, **22**, 4.
9. V. Mäde, S. Els-Heindl and A. G. Beck-Sickinger, *Beilstein J. Org. Chem.*, 2014, **10**, 1197.
10. M. Verlander, *Int. J. Pept. Res. Ther.*, 2007, **13**, 75.
11. H. Yan and F.-E. Chen, *Adv. Synth. Catal.*, 2022, **364**, 1934.

12. M. Paradís-Bas, J. Tulla-Puche and F. Albericio, *Chem. Soc. Rev.*, 2016, **45**, 631.
13. N. Hartrampf, A. Saebi, M. Poskus, Z. P. Gates, A. J. Callahan, A. E. Cowfer, S. Hanna, S. Antilla, C. K. Schissel, A. J. Quartararo, X. Ye, A. J. Mijalis, M. D. Simon, A. Loas, S. Liu, C. Jessen, T. E. Nielsen and B. L. Pentelute, *Science*, 2020, **368**, 980.
14. A. Okawara, K. Hada, R. Kurotani, K. Teruya and H. Konno, *Tetrahedron Lett.*, 2023, **127**, 154665.
15. H. Zhang, S. E. Schneider, B. L. Bray, P. E. Friedrich, N. A. Tvermoes, C. J. Mader, S. R. Whight, T. E. Niemi, P. Silinski, T. Picking, M. Warren and S. A. Wring, *Org. Process Res. Dev.*, 2008, **12**, 101.
16. P. Dawson, T. Muir, I. Clark-Lewis and S. Kent, *Science*, 1994, **266**, 776.
17. V. Agouridas, O. El Mahdi, V. Diemer, M. Cargoët, J.-C. M. Monbaliu and O. Melnyk, *Chem. Rev.*, 2019, **119**, 7328.
18. T. Nuijens, A. Toplak, M. Schmidt, A. Ricci and W. Cabri, *Front. Chem.*, 2019, **7**, 829.
19. T. Nuijens, A. Toplak, M. B. A. C. Van de Meulenreek, M. Schmidt, M. Goldbach, D. B. Janssen and P. J. L. M. Quaedflieg, *Chem. Today*, 2016, **34**, 16.
20. C. Bechtler and C. Lamers, *RSC Med. Chem.*, 2021, **12**, 1325.
21. D. Carbajo, A. El-Faham, M. Royo and F. Albericio, *ACS Omega*, 2019, **4**, 8674.
22. S. Pal, J. Openy, A. Krzyzanowski, A. Noisier and P. Hart, *Org. Lett.*, 2024, **26**, 2795.
23. M. Werner, J. Pampel, T. L. Pham and F. Thomas, *Chem. – Eur. J.*, 2022, **28**, e202201339.
24. A. Tantipanjaporn and M.-K. Wong, *Molecules*, 2023, **28**, 1083.
25. W. Wang, M. M. Lorion, J. Shah, A. R. Kapdi and L. Ackermann, *Angew. Chem., Int. Ed.*, 2018, **57**, 14700.
26. S. Laps, G. Satish and A. Brik, *Chem. Soc. Rev.*, 2021, **50**, 2367.
27. L. Wang, N. Wang, W. Zhang, X. Cheng, Z. Yan, G. Shao, X. Wang, R. Wang and C. Fu, *Signal Transduction Targeted Ther.*, 2022, **7**, 48.
28. http://peptherdia.herokuapp.com/, accessed on December 7, 2023.
29. S. D. Jois, *Pept. Ther.*, 2022, **47**, 287.
30. J. H. Rasmussen, *Bioorg. Med. Chem.*, 2018, **26**, 2914.
31. A. Isidro-Llobet, M. N. Kenworthy, S. Mukherjee, M. E. Kopach, K. Wegner, F. Gallou, A. G. Smith and F. Roschangar, *J. Org. Chem.*, 2019, **84**, 4615.
32. P. Anastas and N. Eghbali, *Chem. Soc. Rev.*, 2010, **39**, 301.
33. V. Martin, P. H. G. Egelund, H. Johansson, S. Thordal Le Quement, F. Wojcik and D. S. Pedersen, *RSC Adv.*, 2020, **10**, 42457.
34. L. Ferrazzano, M. Catani, A. Cavazzini, G. Martelli, D. Corbisiero, P. Cantelmi, T. Fantoni, A. Mattellone, C. De Luca, S. Felletti, W. Cabri and A. Tolomelli, *Green Chem.*, 2022, **24**, 975.
35. A. Sharma, A. Kumar, B. G. de la Torre and F. Albericio, *Chem. Rev.*, 2022, **122**, 13516.

36. B.-N. Nguyen, F. Tieves, T. Rohr, H. Wobst, F. Schöpf, J. D. M. Solano, J. Schneider, J. Stock, A. Uhde, T. Kalthof, K. E. Jaeger, L. Schmitt and C. Schwarz, *AMB Express*, 2021, **11**, 48.
37. G. Rossino, E. Marchese, G. Galli, F. Verde, M. Finizio, M. Serra, P. Linciano and S. Collina, *Molecules*, 2023, **28**, 7165.
38. Y. E. Jad, A. Kumar, A. El-Faham, B. G. de la Torre and F. Albericio, *ACS Sustainable Chem. Eng.*, 2019, **7**, 3671.
39. M. J. Raymond, C. S. Slater and M. J. Savelski, *Green Chem.*, 2010, **12**, 1826.
40. J. K. Magtaan, M. Devocelle and F. Kelleher, *J. Pept. Sci.*, 2019, **25**, e3139.
41. F. G. Martin and F. Albericio, *Chem. Today*, 2008, **26**, 29.
42. S. M. Alesso, Z. Yu, D. Pears, P. A. Worthington, R. W. A. Luke and M. Bradley, *J. Comb. Chem.*, 2001, **3**, 631.
43. W. Rapp, in *Combinatorial Peptide and Nonpeptide Libraries: A Handbook*, ed. G. Jung, VCH, Weinheim, Germany, 1996, p. 425.
44. F. Garcia-Martin, M. Quintanar-Audelo, Y. Garcia-Ramos, L. J. Cruz, C. Gravel, R. Furic, S. Cote, J. Tulla-Puche and F. Albericio, *J. Comb. Chem.*, 2006, **8**, 213.
45. I. Bouillon, M. Soural, M. J. Miller and V. Krchňák, *J. Comb. Chem.*, 2009, **11**, 213.
46. E. R. Lax, G. Stærkær, P. Möller, A. Swietlow, P. Larsson and B. Zell, *Chem. Today*, 2016, **34**, 27.
47. V. Singh, R. K. Sharma, N. Sinha and A. K. Thakur, *ChemistrySelect*, 2017, **2**, 11826.
48. I. Eggen, B. Gregg, H. Rode, A. Swietlow, M. Verlander and A. Szajek, *BioPharm Int.*, 2014, **27**, 40.
49. M. Obkircher, C. Stähelin and F. Dick, *J. Pept. Sci.*, 2008, **14**, 763.
50. A. Kuma, A. Sharma, E. Haimov, A. El-Faham, B. G. de la Torre and F. Albericio, *Org. Process Res. Dev.*, 2017, **21**, 1533.
51. D. S. Treitler, A. S. Marriott, J. Chadwick and E. Quirk, *Org. Process Res. Dev.*, 2019, **23**, 2562.
52. J. Pawlas and J. H. Rasmussen, *Org. Process Res. Dev.*, 2023, **27**, 982.
53. F. Guillier, D. Orain and M. Bradley, *Chem. Rev.*, 2000, **100**, 2091.
54. M. S. Bernatowicz, S. B. Daniels and H. Köster, *Tetrahedron Lett.*, 1989, **30**, 4645.
55. R. Ramage, S. L. Irving and C. McInnes, *Tetrahedron Lett.*, 1993, **34**, 6599.
56. S.-S. Wang, *J. Am. Chem. Soc.*, 1973, **95**, 1328.
57. P. Athanassopoulos, K. Barlos, D. Gatos, O. Hatzi and C. Tzavara, *Tetrahedron Lett.*, 1995, **36**, 5645.
58. O. Al Musaimi, Y. E. Jad, A. Kumar, A. El-Faham, J. M. Collins, A. Basso, B. G. de la Torre and F. Albericio, *Org. Process Res. Dev.*, 2018, **22**, 1809.
59. O. Al Musaimi, A. El-Faham, A. Basso, B. G. de la Torre and F. Albericio, *Tetrahedron Lett.*, 2019, **60**, 151058.
60. M. Leko, P. Filippova, M. Dorosh, K. Rustler, T. Bruckdorfer and S. Burov, *Org. Process Res. Dev.*, 2022, **26**, 144.

61. J. Lopez, J. Beck, C. Bucher, A. Berthelmann, S. Markos and S. Eissler, *Org. Process Res. Dev.*, 2022, **26**, 1450.
62. T. Román, G. Acosta, B. G. de la Torre, C. Cárdenas, F. Guzmán and F. Albericio, *Methods Protoc.*, 2023, **6**, 82.
63. G. Orosz and L. P. Kiss, *Tetrahedron Lett.*, 1998, **39**, 3241.
64. https://eur-lex.europa.eu/eli/reg/2021/2030/oj.
65. https://www.epa.gov/sites/default/files/2016-09/documents/n-n-dimethylformamide.pdf.
66. F. Gao, R. Bai, F. Ferlin, L. Vaccaro, M. Li and Y. Gu, *Green Chem.*, 2020, **22**, 6240.
67. D. Procopio, C. Siciliano, S. Trombino, D. E. Dumitrescu, F. Suciu and M. L. Di Gioia, *Org. Biomol. Chem.*, 2022, **20**, 1137.
68. S. Hazra, F. Gallou and S. Handa, *ACS Sustainable Chem. Eng.*, 2022, **10**, 5299.
69. F. P. Byrne, S. Jin, G. Paggiola, T. H. M. Petchey, J. H. Clark, T. J. Farmer, A. J. Hunt, C. Robert McElroy and J. Sherwood, *Sustainable Chem. Processes*, 2016, **4**, 7.
70. C. Reichardt, *Org. Process Res. Dev.*, 2007, **11**, 105.
71. C. Jimenez-Gonzales, *Curr. Opin. Green Sustainable Chem.*, 2019, **18**, 66.
72. F. Albericio, O. Al Musaimi and B. G. de la Torre, *Green Chem.*, 2020, **22**, 996.
73. D. M. M. Jaradat, O. Al Musaimi and F. Albericio, *Green Chem.*, 2022, **24**, 6360.
74. G. A. Acosta, M. del Fresno, M. Paradis-Bas, M. Rigau-DeLlobet, S. Côté, M. Royo and F. Albericio, *J. Pept. Sci.*, 2009, **15**, 629.
75. E. Jad, G. A. Acosta, S. N. Khattab, B. G. de la Torre, T. Govender, H. G. Kruger, A. El-Faham and F. Albericio, *Org. Biomol. Chem.*, 2015, **13**, 2393.
76. Y. E. Jad, G. A. Acosta, S. N. Khattab, B. G. de la Torre, T. Govender, H. G. Kruger, A. El-Faham and F. Albericio, *Amino Acids*, 2016, **48**, 419.
77. Y. E. Jad, G. A. Acosta, T. Govender, H. G. Kruger, A. El-Faham, B. G. de la Torre and F. Albericio, *ACS Sustainable Chem. Eng.*, 2016, **4**, 6809.
78. S. B. Lawrenson, R. Arav and M. North, *Green Chem.*, 2017, **19**, 1685.
79. A. Kumar, Y. E. Jad, A. El-Faham, B. G. de la Torre and F. Albericio, *Tetrahedron Lett.*, 2017, **58**, 2986.
80. A. Kumar, Y. E. Jad, J. M. Collins, F. Albericio and B. G. de la Torre, *ACS Sustainable Chem. Eng.*, 2018, **6**, 8034.
81. A. Kumar, A. Thompson-Adewumi, K. P. Nandhini, J. M. Collins, F. Albericio and B. G. de la Torre, *Org. Process Res. Dev.*, 2019, **23**, 1096.
82. J. Lopez, S. Pletscher, A. Aemissegger, C. Bucher and F. Gallou, *Org. Process Res. Dev.*, 2018, **22**, 494.
83. B. G. de la Torre, A. Kumar, M. Alhassan, C. Bucher, F. Albericio and J. Lopez, *Green Chem.*, 2020, **22**, 3162.
84. A. Kumar, M. Alhassan, J. Lopez, F. Albericio and B. G. de la Torre, *ChemSusChem*, 2020, **13**, 5288.

85. K. Wegner, D. Barnes, K. Manzor, A. Jardine and D. Moran, *Green Chem. Lett. Rev.*, 2021, **14**, 153.
86. O. Al Musaimi, R. Wisdom, P. Talbiersky, B. G. de La Torre and F. Albericio, *ChemistrySelect*, 2021, **6**, 2649.
87. A. Kumar, A. Sharma, B. G. de La Torre and F. Albericio, *Green Chem. Lett. Rev.*, 2021, **14**, 545.
88. G. Vivenzio, M. C. Scala, P. Marino, M. Manfra, P. Campiglia and M. Sala, *Pharmaceutics*, 2023, **15**, 1773.
89. A. Přibylka, V. Krchňák and E. Schütznerová, *Green Chem.*, 2019, **21**, 775.
90. A. Přibylka, V. Krchňák and E. Schütznerová, *J. Org. Chem.*, 2020, **85**, 8798.
91. A. Přibylka, M. Pastorek, M. Grepl and E. Přibylka Schütznerová, *Tetrahedron*, 2021, **97**, 132452.
92. Y. Ran, F. Byrne, I. D. V. Ingram and M. North, *Chem. – Eur. J.*, 2019, **25**, 4951.
93. J. Pawlas, B. Antonic, M. Lundqvist, T. Svensson, J. Finnman and J. H. Rasmussen, *Green Chem.*, 2019, **21**, 2594.
94. J. Pawlas and J. H. Rasmussen, *Org. Lett.*, 2022, **24**, 1827.
95. L. Ferrazzano, D. Corbisiero, G. Martelli, A. Tolomelli, A. Viola, A. Ricci and W. Cabri, *ACS Sustainable Chem. Eng.*, 2019, **7**, 12867.
96. G. Martelli, P. Cantelmi, A. Tolomelli, D. Corbisiero, A. Mattellone, A. Ricci, T. Fantoni, W. Cabri, F. Vacondio, F. Ferlenghi, M. Mor and L. Ferrazzano, *Green Chem.*, 2021, **23**, 4095.
97. G. Martelli, P. Cantelmi, C. Palladino, A. Mattellone, D. Corbisiero, T. Fantoni, A. Tolomelli, M. Macis, A. Ricci and W. Cabri, *Green Chem.*, 2021, **23**, 8096.
98. J. Pawlas and J. H. Rasmussen, *Green Chem.*, 2019, **21**, 5990.
99. J. M. Collins and S. K. Singh, *WO pat.*, WO2019/241586, 2019.
100. V. Martin, S. Jadhav, P. H. G. Egelund, R. Liffert, H. J. Castro, T. Krüger, K. F. Haselmann, S. T. Le Quement, F. Albericio, F. Dettner, C. Lechner, R. Schönleber and D. S. Pedersen, *Green Chem.*, 2021, **23**, 3295.
101. S. Jadhav, V. Martin, P. H. G. Egelund, H. J. Castro, T. Krüger, F. Richner, S. T. Le Quement, F. Albericio, F. Dettner, C. Lechner, R. Schönleber and D. S. Pedersen, *Green Chem.*, 2021, **23**, 3312.
102. P. H. G. Egelund, S. Jadhav, V. Martin, H. J. Castro, F. Richner, S. T. Le Quement, F. Dettner, C. Lechner, R. Schoenleber and D. S. Pedersen, *ACS Sustainable Chem. Eng.*, 2021, **9**, 14202.
103. F. Zhou, Z. Hearne and C.-J. Li, *Curr. Opin. Green Sustainable Chem.*, 2019, **18**, 118.
104. K. Hojo, M. Maeda and K. Kawasaki, *J. Pept. Sci.*, 2001, **7**, 615.
105. K. Hojo, M. Maeda, Y. Takahara, S. Yamamoto and K. Kawasaki, *Tetrahedron Lett.*, 2003, **44**, 2849.
106. K. Hojo, M. Maeda and K. Kawasaki, *Tetrahedron*, 2004, **60**, 1875.
107. K. Hojo, M. Maeda, T. J. Smith, E. Kita, F. Yamaguchi, S. Yamamoto and K. Kawasaki, *Chem. Pharm. Bull.*, 2004, **52**, 422.

108. K. Hojo, M. Maeda and K. Kawasaki, *Tetrahedron Lett.*, 2004, **45**, 9293.
109. K. Hojo, M. Maeda, N. Tanakamaru, K. Mochida and K. Kawasaki, *Protein Pept. Lett.*, 2006, **13**, 189.
110. J. M. Collins, *WO pat.*, WO20137115813, 2013.
111. R. De Marco, A. Tolomelli, A. Greco and L. Gentilucci, *ACS Sustainable Chem. Eng.*, 2013, **1**, 566.
112. S. Knauer, T. M. L. Roese, O. Avrutina, H. Kolmar and C. Uth, *WO pat.*, WO20167050764A1, 2016.
113. S. Knauer, N. Koch, C. Uth, R. Meusinger, O. Avrutina and H. Kolmar, *Angew. Chem., Int. Ed.*, 2020, **59**, 12984.
114. C. Uth, S. Englert, O. Avrutina, H. Kolmar and S. Knauer, *J. Pept. Sci.*, 2023, e3527.
115. K. Hojo, H. Ichikawa, M. Maeda, S. Kida, Y. Fukumori and K. Kawasaki, *J. Pept. Sci.*, 2007, **13**, 493.
116. K. Hojo, H. Ichikawa, A. Hara, M. Onishi, K. Kawasaki and Y. Fukumori, *Protein Pept. Lett.*, 2012, **19**, 1231.
117. K. Hojo, N. Shinozaki, A. Hara, M. Onishi, Y. Fukumori and H. Ichikawa, *Protein Pept. Lett.*, 2013, **20**, 1122.
118. K. Hojo, N. Shinozaki, K. Hidaka, Y. Tsuda, Y. Fukumori, H. Ichikawa and J. D. Wade, *Amino Acids*, 2014, **46**, 2347.
119. K. Hojo, Y. Manabe, T. Uda and Y. Tsuda, *J. Org. Chem.*, 2022, **87**, 11362.
120. K. Hojo, N. Shinozaki, Y. Nozawa, Y. Fukumori and H. Ichikawa, *Appl. Sci.*, 2013, **3**, 614.
121. K. Hojo, S. Fujiwara, H. Inai, Y. Manabe and Y. Tsuda, *Int. J. Pept. Res. Ther.*, 2022, **28**, 51.
122. A. S. Galanis, F. Albericio and M. Grøtli, *Org. Lett.*, 2009, **11**, 4488.
123. J. Pawlas and J. H. Rasmussen, *ChemSusChem*, 2021, **14**, 3231.
124. J. Pawlas and O. Ludemann-Hombourger, *WO pat.*, WO2023/287345, 2023.
125. A. Isidro-Llobet, M. Álvarez and F. Albericio, *Chem. Rev.*, 2009, **109**, 2455.
126. A. El-Faham and F. Albericio, *Pept. Sci.*, 2020, **112**, e24164.
127. Y. Zhou, H. Li, Y. Huang, J. Li, G. Deng, G. Chen, Z. Xi and C. Zhou, *Nat. Commun.*, 2023, **14**, 5324.
128. T. Liu, Z. Peng, M. Lai, L. Hu and J. Zhao, *J. Am. Chem. Soc.*, 2024, **146**, 4270.
129. A. R. Vaino and K. D. Janda, *J. Comb. Chem.*, 2000, **2**, 579.
130. V. K. Sarin, S. B. H. Kent and R. B. Merrifield, *J. Am. Chem. Soc.*, 1980, **102**, 5463.
131. G. B. Fields and C. G. Fields, *J. Am. Chem. Soc.*, 1991, **113**, 4202.
132. R. Santini, M. C. Griffith and M. Qi, *Tetrahedron Lett.*, 1998, **39**, 8951.
133. Y. E. Jad, T. Govender, H. G. Kruger, A. El-Faham, B. G. De la Torre and F. Albericio, *Org. Process Res. Dev.*, 2017, **21**, 365.
134. S. Lawrenson, M. North, F. Peigneguy and A. Routledge, *Green Chem.*, 2017, **19**, 952.

135. C. Amadi-Kamalu, H. Clarke, M. McRobie, J. Mortimer, M. North, Y. Ran, A. Routledge, D. Sibbald, M. Tickias, K. Tse and H. Willway, *ChemistryOpen*, 2020, **9**, 431.
136. J. K. Magtaan, M. Devocelle and F. Kelleher, *J. Pept. Sci.*, 2020, **26**, e3250.
137. D. C. Akintayo, B. G. de la Torre, Y. Li and F. Albericio, *Polymers*, 2022, **14**, 928.
138. A. El-Faham and F. Albericio, *Chem. Rev.*, 2011, **111**, 6557.
139. J. C. Graham, A. Trejo-Martin, M. L. Chilton, J. Kostal, J. Bercu, G. L. Beutner, U. S. Bruen, D. G. Dolan, S. Gomez, J. Hillegass, J. Nicolette and M. Schmitz, *Chem. Res. Toxicol.*, 2022, **35**, 1011.
140. J. B. Sperry, C. J. Minteer, J. Tao, R. Johnson, R. Duzguner, M. Hawksworth, S. Oke, P. F. Richardson, R. Barnhart, D. R. Bill, R. A. Giusto and J. D. Weaver III, *Org. Process Res. Dev.*, 2018, **22**, 1262.
141. A. D. McFarland, J. Y. Buser, M. C. Embry, C. B. Held and S. P. Kolis, *Org. Process Res. Dev.*, 2019, **23**, 2099.
142. M. Erny, M. Lundqvist, J. H. Rasmussen, O. Ludemann-Hombourger, F. Bihel and J. Pawlas, *Org. Process Res. Dev.*, 2020, **24**, 1341.
143. S. R. Manne, A. El-Faham, B. G. de la Torre and F. Albericio, *Tetrahedron Lett.*, 2021, **85**, 153462.
144. J. Wang, P. Agrawal, J. Y. Buser, M. C. Embry, A. D. McFarland, M. R. Berglund, J. McClary Groh and S. K. Viswanath, *Ind. Eng. Chem. Res.*, 2023, **62**, 14986.
145. S. R. Manne, O. Luna, G. A. Acosta, M. Royo, A. El-Faham, G. Orosz, B. G. de la Torre and F. Albericio, *Org. Lett.*, 2021, **23**, 6900.
146. S. R. Manne, D. C. Akintayo, O. Luna, A. El-Faham, B. G. de la Torre and F. Albericio, *Org. Process Res. Dev.*, 2022, **26**, 2894.
147. J. Pawlas, J.-H. Choi, C. von Bargen, S. Maibom-Thomsen, J. H. Rasmussen and O. Ludemann-Hombourger, *Org. Process Res. Dev.*, 2023, **27**, 1348.
148. A. M. Clapperton, K. N. M. Vera, J. Babi and H. Tran, *ACS Sustainable Chem. Eng.*, 2023, **11**, 13024.
149. M. Badland, R. Crook, B. Delayre, S. J. Fussell, I. Gladwell, M. Hawksworth, R. M. Howard, R. Walton and G. A. Weisenburger, *Tetrahedron Lett.*, 2017, **58**, 4391.
150. R. V. Ulijn, B. Baragaña, P. J. Halling and S. L. Flitsch, *J. Am. Chem. Soc.*, 2002, **124**, 10988.
151. G. B. Fields, *Methods Mol. Biol.*, 1994, **35**, 17.
152. J. D. Wade, M. N. Mathieu, M. Macris and G. W. Tregear, *Pept. Sci.*, 2000, **7**, 107.
153. J. Hachmann and M. Lebl, *J. Comb. Chem.*, 2006, **8**, 149.
154. J. D. Wade, J. Bedford, R. C. Sheppard and G. W. Tregear, *Pept. Res.*, 1991, **4**, 194.
155. H. Personne, T. N. Siriwardena, S. Javor and J.-L. Reymond, *ACS Omega*, 2023, **8**, 5050.
156. M. P. Souza, M. F. M. Tavares and M. T. M. Miranda, *Tetrahedron*, 2004, **60**, 4671.

157. M. A. Fara, J. J. Díaz-Mochón and M. Bradley, *Tetrahedron Lett.*, 2006, **47**, 1011.
158. S. L. Pedersen, A. P. Tofteng, L. Malik and K. J. Jensen, *Chem. Soc. Rev.*, 2012, **41**, 1826.
159. J. M. Collins, K. A. Porter, S. K. Singh and G. S. Vanier, *Org. Lett.*, 2014, **16**, 940.
160. S. Pinzón-López, M. Kraume, J. Danglad-Flores and P. H. Seeberger, *React. Chem. Eng.*, 2023, **8**, 2951.
161. G. Wołczański, H. Płóciennik, M. Lisowski and P. Stefanowicz, *Tetrahedron Lett.*, 2019, **60**, 1814.
162. F. Merlino, S. Tomassi, A. M. Yousif, A. Messere, L. Marinelli, P. Grieco, E. Novellino, S. Cosconati and S. Di Maro, *Org. Lett.*, 2019, **21**, 6378.
163. J. N. Naoum, I. Alshanski, G. Mayer, P. Strauss and M. Hurevich, *Org. Process Res. Dev.*, 2022, **26**, 129.
164. N. Hartrampf and Z. P. Gates, *Pept. Sci.*, 2020, **112**, e24198.
165. K. Ferentzi, D. Nagy-Fazekas, V. Farkas and A. Perczel, *React. Chem. Eng.*, 2023, **9**, 58.
166. J. Pawlas, in *Proceedings of the 34th European Peptide Symposium and the 8th International Peptide Symposium*, ed. A. Beck-Sickinger, K. Mörl, K. Bellmann-Sickert, and S. Els-Heindl, European Peptide Society, Leipzig, Germany, 2016, p. 60.
167. R. Subiros-Funosas, A. El-Faham and F. Albericio, *Tetrahedron*, 2011, **67**, 8595.
168. E. Pedroso, A. Grandas, X. de las Heras, R. Eritja and E. Giralt, *Tetrahedron Lett.*, 1986, **27**, 743.
169. Y. Yang and L. Hansen, *ACS Omega*, 2022, **7**, 12015.
170. S. N. Mthembu, A. Chakraborty, R. Schönleber, F. Albericio and B. G. de la Torre, *J. Pept. Sci.*, 2023, **28**, e3538.
171. G. Sabatino, M. Chelli, A. Brandi and A. M. Papini, *Curr. Org. Chem.*, 2004, **8**, 291.
172. T. Umeno, M. Fujihara, S. Matsumoto, N. Iizuka, K. Usui and S. Karasawa, *Anal. Chem.*, 2023, **95**, 15803.
173. B. G. de la Torre, S. Ramkisson, F. Albericio and J. Lopez, *Org. Process Res. Dev.*, 2021, **25**, 1047.
174. C. Yan, C. Fyfe, L. Minty, H. Barrington, C. Jamieson and M. Reid, *Chem. Sci.*, 2023, **14**, 11872.
175. K. Wang, X. Wang, W. Zhang, W. Zhang, Y. Dua, M. Xu, L. Wang, L. Song and H. Xu, *Infrared Phys. Technol.*, 2023, **136**, 105026.
176. N. Hartrampf, B. Tamás, P. L. Willi and H. Bürgisser, *React. Chem. Eng.*, 2024, **9**, 825.
177. S. L. Broman, M. Rosenberg, F. Wojcik, A. H. Hansen, P. H. G. Egelund, J. Malmstrøm and D. S. Pedersen, *Org. Process Res. Dev.*, 2024, **28**, 666.
178. J. Hansen, F. Diness and M. Meldal, *Org. Biomol. Chem.*, 2016, **14**, 3238.
179. Y. Bakhatan, I. Alshanski, D. Grunhaus and M. Hurevich, *Org. Biomol. Chem.*, 2020, **18**, 4183.

180. C. A. Guy and G. B. Fields, *Methods Enzymol.*, 1997, **289**, 67.
181. F. Dick, *Methods Mol. Biol.*, 1994, **35**, 63.
182. B. G. de la Torre and D. Andreu, *J. Pept. Sci.*, 2008, **14**, 360.
183. K. Barlos, D. Gatos and W. Schäfer, *Angew. Chem., Int. Ed. Engl.*, 1991, **30**, 590.
184. R. Bollhagen, M. Schmiedberger, K. Barlos and E. J. Grell, *J. Chem. Soc., Chem. Commun.*, 1994, 2559.
185. J. Wang, A. Niemoeller and S. Viswanath, *Chem. Eng. J.*, 2023, **476**, 146928.
186. J. Pawlas, T. Svensson and J. H. Rasmussen, *RSC Adv.*, 2019, **9**, 38928.
187. K. P. Nandhini, M. Alhassan, C. G. L. Veale, F. Albericio and B. G. de la Torre, *ACS Omega*, 2023, **8**, 15631.
188. C. Melander, L. Malik, J. Pawlas and S. Hansen, *WO pat.*, WO2023/118416A1, 2023.
189. N. D. Tyrrell, *Org. Process Res. Dev.*, 2023, **27**, 1422.
190. K. S. Srivastava, *WO pat.*, WO2007/130275A2, 2007.
191. P. Palladino and D. A. Stetsenko, *Org. Lett.*, 2012, **14**, 6346.
192. D. Stetsenko and P. Palladino, *WO pat.*, WO2014/033466A1, 2014.
193. D. A. Stetsenko, V. S. Apukhtina, B. P. Chelobanov and P. Palladino, *Russ. J. Bioorg. Chem.*, 2016, **42**, 143.
194. Y. Feng, S. Liang, J. Langille, B. S. Pierce, S. Chung, J. Szeliga, G. Wilcox, P. Simonds, K. A. Farley, B. Li, C. Garcia-Irizarry, P. Jones and R. Lira, *Org. Process Res. Dev.*, 2023, **27**, 2010.
195. A. Rees, *WO pat.*, WO2015/028599, 2015.
196. D. Cardella, Y.-H. Tsai and L. Y. P. Luk, *Org. Biomol. Chem.*, 2023, **21**, 966.
197. X. Cheng, T. Zhu, H. Hong, Z. Zhou and Z. Wu, *Org. Chem. Front.*, 2017, **4**, 2058.
198. A. Sewing and D. Hilvert, *Angew. Chem., Int. Ed.*, 2001, **40**, 3395.
199. M. Schmidt, E. Barbayianni, I. Fotakopoulou, M. Höhne, V. Constantinou-Kokotou, U. T. Bornscheuer and G. Kokotos, *J. Org. Chem.*, 2005, **70**, 3737.
200. R. S. Giri, S. R. Manne, G. Dolai, A. Paul, T. Kalita and B. Mandal, *ACS Omega*, 2017, **2**, 6586.
201. R. S. Giri, S. Roy, G. Dolai, S. R. Manne and B. Mandal, *ChemistrySelect*, 2020, **5**, 2050.
202. O. Al Musaimi, Y. E. Jad, A. Kumar, J. M. Collins, A. Basso, B. G. de la Torre and F. Albericio, *Curr. Opin. Green Sustainable Chem.*, 2018, **11**, 99.
203. O. Al Musaimi, Y. E. Jad, A. Kumar, A. El-Faham, J. M. Collins, A. Basso, B. G. de la Torre and F. Albericio, *Org. Process Res. Dev.*, 2018, **22**, 1809.
204. H. Zhang, J. Fomsgaard and G. Stærkær, *WO pat.*, WO2010/121835, 2010.
205. O. Al Musaimi, J. M. Collins, F. Albericio and B. G. de la Torre, *Green Chem.*, 2020, **22**, 2840.
206. O. Al Musaimi, V. Gavva and D. R. Williams, *ChemistryOpen*, 2022, **11**, e202200236.
207. Y. Yang, L. Hansen and P. Ryberg, *Org. Process Res. Dev.*, 2022, **26**, 1520.

208. L. Costa, E. Sousa and C. Fernandes, *Pharmaceuticals*, 2023, **16**, 996.
209. N. Casagrande, C. Borghese, L. Gabbatore, L. Morbiato, M. De Zotti and D. Aldinucci, *Int. J. Mol. Sci.*, 2021, **22**, 8362.
210. M. E. Kopach and B. I. Andrews, *Curr. Opin. Green Sustainable Chem.*, 2022, **37**, 100643.
211. D. Takahashi, T. Inomata and T. Fukui, *Angew. Chem., Int. Ed.*, 2017, **56**, 7803.
212. O. Maurin, P. Verdié, G. Subra, F. Lamaty, J. Martinez and T.-X. Métro, *Beilstein J. Org. Chem.*, 2017, **13**, 2087.
213. K. Yap, J. Du, F. Y. Looi, S. R. Tang, S. J. de Veer, A. R. Bony, F. B. H. Rehm, J. Xie, L. Y. Chan, C. K. Wang, D. J. Adams, L. H. L. Lua, T. Durek and D. J. Craik, *Green Chem.*, 2020, **22**, 5002.
214. M. Todorovic and D. M. Perrin, *Pept. Sci.*, 2020, **112**, e24210.
215. S. Satishkumar, N. R. Panigrahi and P. S. Arora, *J. Am. Chem. Soc.*, 2019, **141**, 15977.
216. Handoko, N. R. Panigrahi and P. S. Arora, *J. Am. Chem. Soc.*, 2022, **144**, 3637.
217. D. J. C. Constable, C. Jimenez-Gonzalez and R. K. Henderson, *Org. Process Res. Dev.*, 2007, **11**, 133.
218. O. Ludemann-Hombourger and R. R. Duran, *WO pat.*, WO2023068987A1, 2023.
219. S. K. Singh and J. M. Collins, *Methods Mol. Biol.*, 2020, **2103**, 95.
220. F. Albericio, A. Kumar, A. Sharma and B. G. de la Torre, *Green Chem.*, 2022, **24**, 4887.
221. J. M. Collins, S. K. Singh, T. A. White, D. J. Cesta, C. L. Simpson, L. J. Tubb and C. L. Houser, *Nat. Commun.*, 2023, **14**, 8168.
222. J. Garcia-Martinez, *Angew. Chem., Int. Ed.*, 2021, **60**, 4956.
223. O. Al Musaimi, D. Al Shaer, B. G. De de la Torre and F. Albericio, *Chem. Today*, 2021, **39**, 18.
224. J. K. Chang, M. Shimizu and S.-S. Wang, *J. Org. Chem.*, 1976, **41**, 3255.
225. K. Barlos and B. Knipp, *WO pat.*, WO2004/056883A2, 2004.
226. H. Bukya, K. Nayani, P. Gangireddy and P. S. Mainkar, *Eur. J. Org Chem.*, 2020, **2020**, 5358.
227. Y. Li, J. Li, G. Bao, C. Yu, Y. Liu, Z. He, P. Wang, W. Ma, J. Xie, W. Sun and R. Wang, *Org. Lett.*, 2022, **24**, 1169.
228. D. Uehara, S. Adachi, A. Tsubouchi, Y. Okada, V. V. Zhdankin, A. Yoshimura and A. Saito, *Chem. Commun.*, 2024, **60**, 956.
229. C. Krell, R. Schreiber, L. Hueber, L. Sciascera, X. Zheng, A. Clarke, R. Haenggi, M. Parmentier, H. Baguia, S. Rodde and F. Gallou, *Org. Process Res. Dev.*, 2021, **25**, 900.
230. P. T. Anastas, *ACS Sustainable Chem. Eng.*, 2020, **8**, 13111.
231. D. P. Elder, G. E. Johnson and D. J. Snodin, *J. Pharm. Sci.*, 2021, **110**, 2311.
232. J. Pawlas, T. Nuijens, J. Persson, T. Svensson, M. Schmidt, A. Toplak, M. Nilsson and J. H. Rasmussen, *Green Chem.*, 2019, **21**, 6451.

233. E. R. Lax and T. Shah, in *Peptide Therapeutics: Strategy and Tactics for Chemistry, Manufacturing and Controls*, ed. V. Srivastava, Royal Society of Chemistry, London, UK, 2019, p. 151.
234. J. Sherwood, F. Albericio and B. G. de la Torre, *ChemSusChem*, 2024, **17**, e202301639.
235. H. Flerlage and J. C. Slootweg, *Nat. Rev. Chem*, 2023, **7**, 593.

CHAPTER 5

Third Wave for Peptide Synthesis: Liquid-phase Peptide Synthesis (LPPS)

A. SHARMA[a,b], A. KUMAR[a,c,d], B. G. DE LA TORRE[c*] AND F. ALBERICIO[a,e*]

[a]Peptide Science Laboratory, School of Chemistry and Physics, University of KwaZulu-Natal, Westville, Durban 4000, South Africa; [b]Department of Natural Products and Medicinal Chemistry, CSIR–Indian Institute of Chemical Education (IICT), Tarnaka, Hyderabad, Telangana 500007, India; [c]KwaZulu-Natal Research Innovation and Sequencing Platform (KRISP), School of Laboratory Medicine and Medical Sciences, College of Health Sciences, University of KwaZulu-Natal, Durban 4041, South Africa; [d]Anthem Biosciences Pvt. Ltd., 49 Canara Bank Road, Bommasandra Industrial Area, Phase I, Bommasandra, Bangalore, Karnataka 560099, India; [e]CIBER-BBN, Networking Centre on Bioengineering, Biomaterials and Nanomedicine, and Department of Organic Chemistry, University of Barcelona, Martí i Franqués 1–11, 08028 Barcelona, Spain

*E-mail: garciadelatorreb@ukzn.ac.za, albericio@ukzn.ac.za

5.1 Introduction

Peptides have witnessed impressive progress as therapeutics in recent years,[1–5] hence there has been a strong surge in their demand.[6–9] Their recent growing interest is evidenced by the fact that around 100 peptide

Green Chemistry Series No. 85
Sustainability in TIDES Chemistry: Green Approaches to Oligonucleotides and Oligopeptides Synthesis
Edited by Alessandra Tolomelli, Lucia Ferrazzano and Walter Cabri

Published by the Royal Society of Chemistry, www.rsc.org

drugs have been approved by the US Food and Drug Administration (FDA) and are currently on the market. In addition, around 170 peptides are in clinical trials and >500 are in the preclinical stage.[10,11] Several approaches are currently used to prepare peptides, the initial one being the classical solution peptide synthesis (CSPS).[12,13] Although desirable from green and financial perspectives, CSPS is a tedious and time-consuming process as all of the intermediates require isolation. The pioneering work of Nobel Laureate Bruce Merrifield simplified peptide synthesis by introducing heterogeneous synthesis, commonly known as solid-phase peptide synthesis (SPPS). This methodology greatly simplified the work-up by removing the need to isolate the intermediates,[14] and allows efficient and rapid multi-kilogram production. However, SPPS involves the use of large amounts of reagents and solvents, which from a green perspective is not acceptable.[15]

The process mass intensity (PMI) and complete E-factor (cEF) for pharmaceuticals are calculated by measuring the amount of waste generated in the synthetic process.[16,17] The PMI in the case of SPPS is 100 times higher than that in the synthesis of small molecules.[16] Around 80–90% of this higher PMI is due to the high solvent consumption, and the rest (around 10–20%) is attributed to excess reagents and resin waste.

Owing to the limitations of the peptide synthesis protocols mentioned above, recent years have witnessed increasing interest in the development of a third approach, referred to as liquid-phase peptide synthesis (LPPS), that combines the best attributes of CSPS and SPPS. LPPS involves growing a peptide chain supported on a soluble tag, wherein the synthesis is carried out similarly to CSPS, and the by-products and excess reagents are removed by precipitation, extraction, or nanofiltration. This approach significantly reduces the large volumes of solvents and reagents used in SPPS, thus facilitating the synthesis of peptide material on the kilogram scale. The issue of excess reagent/solvent usage can be significantly reduced in LPPS protocols.[18,19]

The idea behind the LPPS strategy evolved from the pioneering work of Bayer and Mutter, who replaced solid polystyrene, used in the SPPS approach of Merrifield, with soluble poly(ethylene glycol) (PEG), so that the synthesis can be carried out in the liquid phase instead of the solid phase.[20–24] The concept of tags was introduced in the early 1970s from polydisperse PEG to monodisperse PEG. Since then, enormous growth has been witnessed in the past few decades from PEG-based tags to perfluoroalkyl, ionic liquid, PolyCarbon, silicon-based, hydrophobic polymer, and phosphorus-containing tags.[25] This chapter aims to cover these recent approaches in LPPS strategies, which have found applications in large-scale peptide production in industry. The chapter is broadly divided into the following sections: (i) membrane-enhanced peptide synthesis (MEPS), (ii) PolyCarbon, (iii) silicon-based tags, (iv) phosphorus-containing tags, and (v) miscellaneous. All other tags such as PEGs, ILs, and perfluoroalkyl tags are mentioned briefly in the miscellaneous section.

5.2 Membrane-enhanced Peptide Synthesis (MEPS)

Livingston and co-workers developed MEPS, which is a PEG tag-based technology involving organic solvent nanofiltration (OSN) as a diafiltration technique.[26–30] This filtration method separates molecules (<2000 kDa) in solution under pressure conditions.[26] The MEPS strategy implements PEG (molecular weight 2000–8000) as a tag and uses a single reactor during synthesis, which is a positive feature in greening the synthetic process.[31–33] Two types of tags, *i.e.* linear and globular, are used based on the OSN membrane. Figure 5.1 illustrates the tags developed based on the MEPS strategy for peptide synthesis. However, linear tags (M1 and M2) result in a loss of polymer and therefore are not considered further here, and only the globular type (M3–M5) is discussed.[33,34]

In general, the MEPS strategy involves the following four steps: (a) amide coupling, (b) washing using constant-volume diafiltration (CVD), (c) deprotection, and (d) washing again *via* diafiltration as shown in Figure 5.2. The cycle is repeated until the required peptide has been assembled on the tag. The peptide is recovered after cleavage and global deprotection. Figure 5.2 shows two syntheses performed with the MEPS strategy. In the first case, depicted in Figure 5.2(I), a standard four-step process, as described above,

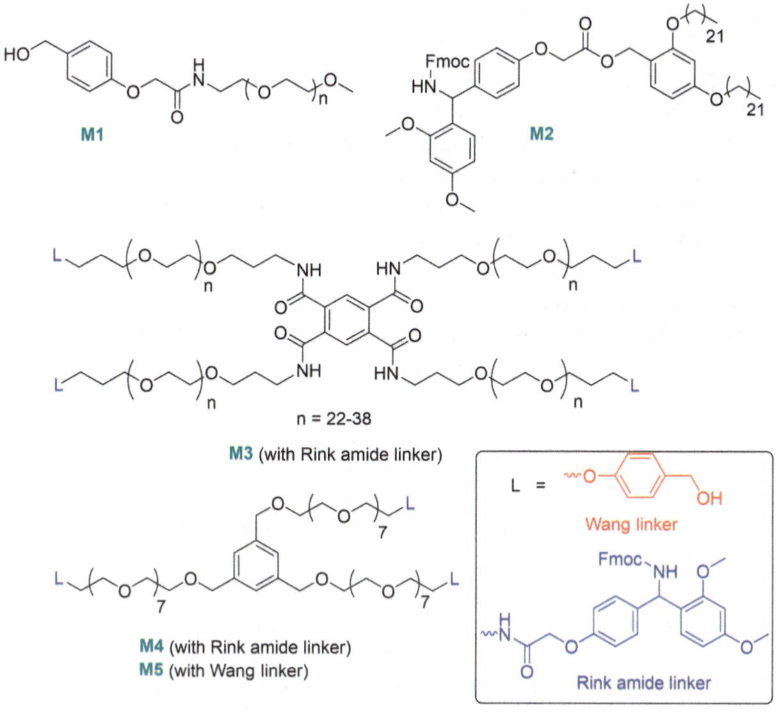

Figure 5.1 Representative structures of MEPS tags. Adapted from ref. 25 with permission from American Chemical Society, Copyright 2022.

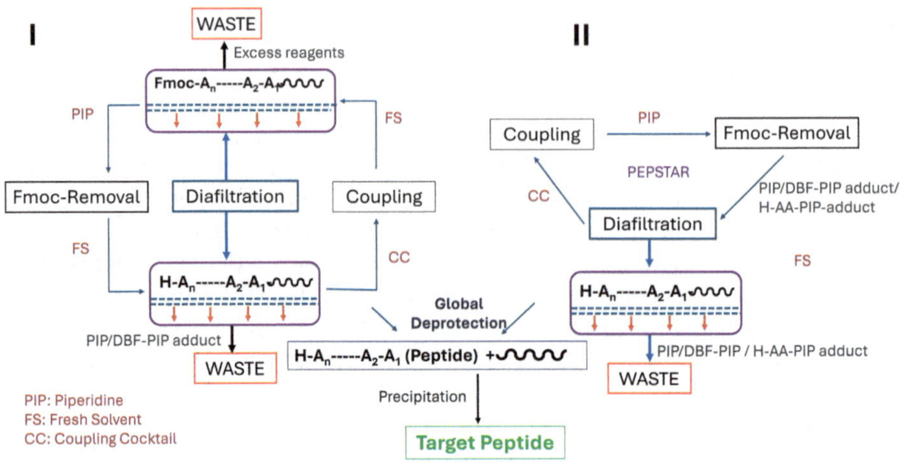

Figure 5.2 Representation of the MEPS strategy. (I) Two-times diafiltration and (II) one-time diafiltration (PEPSTAR strategy). Adapted from ref. 25 with permission from American Chemical Society, Copyright 2022.

was performed wherein defiltration was performed twice (after coupling and after deprotection). In the second case (PEPSTAR strategy), shown in Figure 5.2(II), diafiltration was performed only after deprotection.

M3, a four-armed tag, was used to synthesize a tetrapeptide (Fmoc–Arg–Ala–Asp–Ala–NH_2) at the 0.64 mmol scale.[33] Removal of the Fmoc (fluorenylmethoxycarbonyl) protecting group was performed through a standard protocol, using 20% piperidine (PIP) in dimethylformamide (DMF). Washing was then performed using 40 CVDs. This was followed by coupling for 2 h at room temperature using HBTU–DIEA (*O*-benzotriazol-1-yltetramethyluronium hexafluorophosphate–*N*,*N*-diisopropylethylamine) in dichloromethane (DCM)–DMF as solvent. Peptide M3 was obtained after performing 10 CVDs. The crude peptide was then recovered by global deprotection using a standard protocol, followed by precipitation using diethyl ether (DEE) and purification by preparative high-performance liquid chromatography (HPLC).

A compact monodisperse PEG nanostar (called PEPSTAR) was also developed.[35] The use of monodisperse PEG (PEPSTAR) offers the advantage of ease of monitoring the reaction progress by HPLC. PEPSTAR can be used in fully automated peptide manufacture. To demonstrate its applicability in one-pot peptide synthesis, two nanostars [one containing Fmoc–Rink amide linker (M4) and the other containing Wang linker (M5)] were used to synthesize a pentapeptide (Ser–enkephalin; H–Tyr–Ser–Ser–Phe–Leu–NH_2), a decapeptide (H–Tyr–Ser–Ser–Phe–Leu–Tyr–Ser–Ser–Phe–Leu–NH_2), and an octreotate peptide [H–D-Phe–Cys(PG)–D-Phe–Trp–Lys–Thr–Cys(PG)–Thr–XH, where X = NH or O and protecting group PG = acetamidomethyl (Acm)]. Fmoc removal from M4 (1.46 mmol) and esterification of M5 (1.39 mmol) were performed before using them for further coupling in a synthesizer *via* a feed tank. In this context, Fmoc was removed from M4 using 1,8-diazabicyclo[5.4.0]undec-7-ene (DBU)–thiomalic acid (26 : 10) in $CHCl_3$ at room temperature for 1 h. By-products were eliminated by subsequently washing the reaction mixture with CH_3COOK and K_2CO_3 solution. The combined $CHCl_3$ solution was collected, dried over Na_2SO_4, and concentrated to afford free amine M4. In the case of M5, esterification was performed overnight at room temperature by adding excess Fmoc–Thr(*t*Bu)–OH–2,6-dichlorobenzoyl chloride (Dcb-Cl)–1-hydroxybenzotriazole (HOBt) (six fold) in DCM. After completion, DCM was removed and the residue was fractionated through a silanized silica column using acetonitrile (ACN)–H_2O as eluting agent. ACN was then removed and the aqueous layer was extracted using $CHCl_3$. The combined $CHCl_3$ solution was dried over Na_2SO_4 and concentrated to afford Fmoc–Tyr(*t*Bu)–O–M5. Fmoc was then removed as explained earlier. Both nanostars [free amine M4 and H–Tyr(*t*Bu)–O–M5] were subjected to peptide synthesis in the feed tank. A coupling cocktail containing Fmoc–AA–OH where AA = amino acid, *N*,*N*′-diisopropylcarbodiimide (DIC), and HOBt was added to the tank. Coupling was confirmed by UHPLC-MS and then Fmoc was removed using 10% v/v PIP. Diafiltration was performed to remove excess reagents and by-products. This cycle of coupling → deprotection → diafiltration was repeated until complete peptide synthesis. For the Cys residue, the degree of epimerization was found to be insignificant (<0.1%). The peptides were cleaved using trifluoroacetic acid (TFA)–triisopropylsilane (TIS)–H_2O (95 : 2.5 : 2.5) and TFA–H_2O (95 : 5) for peptide M4 and peptide M5, respectively. Treatment with cold DEE provided decapeptide (75% using M4) and octreotate peptide (80% using M4 and 71% using M5) as precipitates.

Elimination of washing after coupling reduced the PMI by around threefold in comparison with previous LPPS methods, which is an added advantage of PEPSTAR. According to reports, the PMI of PEPSTAR is similar to that of SPPS. With further slight fine tuning, a significant reduction in PMI can be expected in the case of PEPSTAR.

Recently, Eli Lilly filed a patent for the synthesis of the blockbuster tirzepatide through the coupling in solution of protected peptide fragments (convergent approach), which in turn were prepared by LPPS using a poly-AEEA {2-[2-(2-aminoethoxy)ethoxy]acetyl} as monodisperse PEG.[36]

5.3 Group-assisted Purification (GAP) for LPPS

A group-assisted purification (GAP) strategy was proposed by Li and co-workers to simplify the purification of small molecules and, later, of peptides.[37–39] The GAP strategy is defined as "a purification technique that avoids traditional methods such as chromatography, recrystallization, *etc.*, by covalently introducing a functionalized group in either the starting material or the product".[40,41]

An *N*-phosphoryl group is used as a tag in the GAP strategy to increase the hydrophobicity, thereby offering distinct solubility. In addition to improving the solubility, it also makes monitoring of the reaction feasible by ^{31}P NMR spectroscopy in each step.[39,40,42,43] Overall, GAP tags have the following characteristics:[37,40,41,44–49] (a) soluble in solvents such as tetrahydrofuran (THF), DCM, *etc.*, and precipitating in ethers, hexanes, *etc.*, thus making purification easy; (b) easily cleavable; (c) recyclable, hence in alignment with green principles;[15] and (d) they conserve stereoselectivity. Purification with the GAP strategy is performed by precipitation [also known as support-aided purification (SAP)] wherein the tagged product is dissolved in the minimum amount of ethyl acetate and the product is precipitated by adding petroleum ether (PE), as shown in Figure 5.3. The solution contains excess reagents and by-products, which can be easily separated by filtration. This approach has been successfully demonstrated in peptide synthesis using Fmoc/Boc (*tert*-butoxycarbonyl) strategies.

GAP tags can be used with a single loading (G1) or with multiple loadings (G2–G4). Figure 5.4 illustrates reagents used to incorporate GAP tags. Owing to the ease of isolation of the tags, they can be reused further after activation, hence they are also known as greener peptide supports (GPSs).[50–54]

G1A was used at the 5 mmol scale for the synthesis of plecanatide. Coupling was achieved using ethylene dichloride (EDC)–HOBt–DIEA, and Fmoc

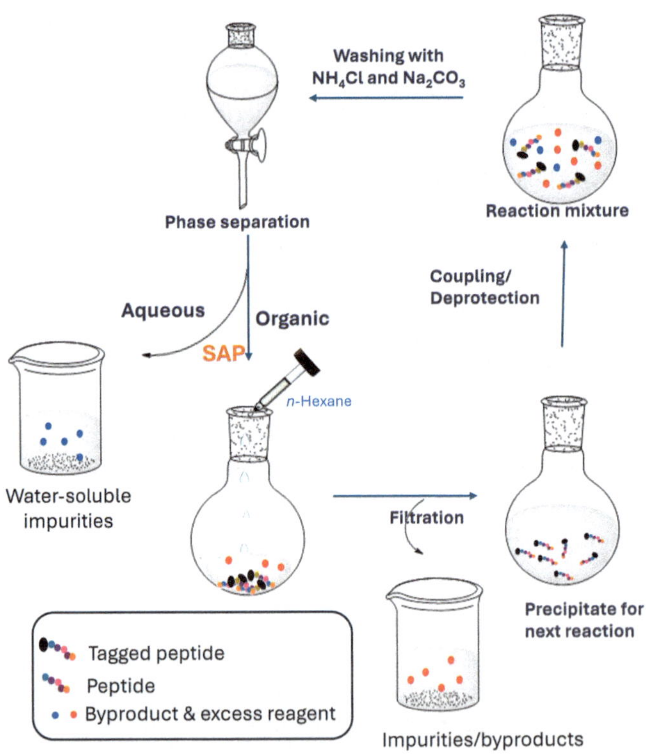

Figure 5.3 Schematic representation of the GAP strategy. Adapted from ref. 25 with permission from American Chemical Society, Copyright 2022.

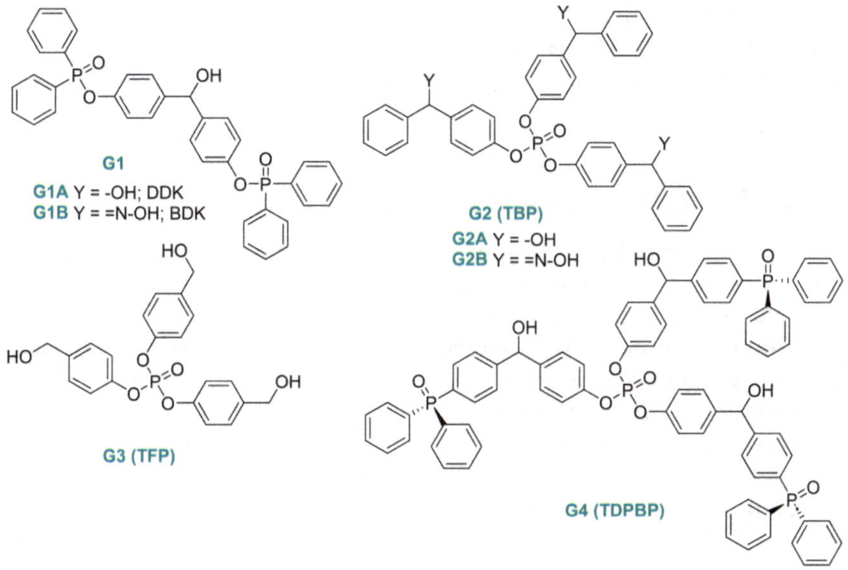

Figure 5.4 Representative structures of GAP tags. Adapted from ref. 25 with permission from American Chemical Society, Copyright 2022.

was removed using 25% diethanolamine (DEA) in ACN in accordance with green principles. After the complete synthesis, the overall yield obtained was 53%.[52] Similarly, G1B (HO-BDK) was used for the synthesis of cyclic mortiamide analogues. The peptide synthesis was carried out using Boc chemistry, while at the N-terminus Fmoc–D-Val–OH was used. The obtained linear peptide Fmoc–D-Val–L-X–D-Val–D-Val–D-Val–L-Phe–D-Phe–O–G1B (X = Leu, Phe, Ile, Val, and Pro to obtain different mortiamide analogues) was treated with 10% DEA in ACN for 2–3 h. The other cyclic mortiamide analogues were obtained in 85–92% yield.[53] The GAP tag G1B was recovered for reuse in peptide synthesis.

G1 is limited to only one loading site. Therefore, to improve the efficiency further, G2–G4 containing three anchoring sites were prepared.[50,51,54] The increased loading further aids in better and more efficient precipitation, thus making these tags suitable for peptides of various lengths. The tags were removed using standard global deprotection in the case of the Fmoc strategy whereas excess LiOH in THF–H_2O was used for Boc chemistry. In this context, Fmoc–AA–O–G2A was prepared using standard coupling conditions and Fmoc was removed by using 25% DEA in ACN (in the case of the Boc strategy using G2B, 25% TFA in DCM was used).[41] This cycle of Fmoc elimination and coupling was continued until the final peptide was assembled on the tag. In the case of G3, dicyclohexylcarbodiimide–*N*-hydroxysuccinimide (DCC–HOSu) (1 : 1) was used for coupling. All the peptides were obtained in high purity.[50,51]

Similarly, G4 was used following the Fmoc/*t*Bu strategy for the synthesis of anti-SARS octapeptide (H–Ala–Val–Leu–Gln–Ser–Gly–Phe–Arg–OH) at the 1 mmol scale. The peptide was obtained in 95% yield and G4 was recovered (90%).[54]

5.4 PolyCarbon-based Tags for LPPS

Hydrophobicity can be fine-tuned by altering the length of the aliphatic chain. These moieties will be soluble in non-polar solvents, whereas they will precipitate out in polar solvents such as ACN and MeOH. This strategy finds application in peptide synthesis during both the coupling and Fmoc removal steps. Figure 5.5 depicts the step-by-step protocol for the synthesis of peptides using PolyCarbon protocols.

The most common PolyCarbon tags (P) used as C-terminal PGs during peptide synthesis are shown in Figure 5.6. Tamiaki and co-workers used P1 to demonstrate long aliphatic-substituted benzyl alcohol for peptide synthesis.[55,56] This approach was further extrapolated by Chiba and co-workers to confer distinct properties to the attached peptide chains, making the isolation process feasible (P1–P5).[57–78] Jitsubo and Bachem when using P2

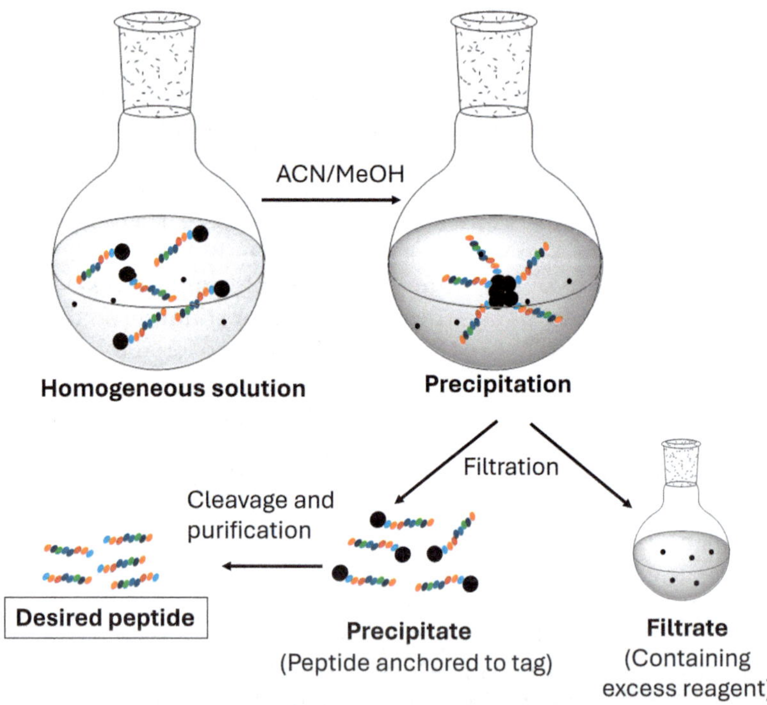

Figure 5.5 Schematic representation of the PolyCarbon strategy. Adapted from ref. 25 with permission from American Chemical Society, Copyright 2022.

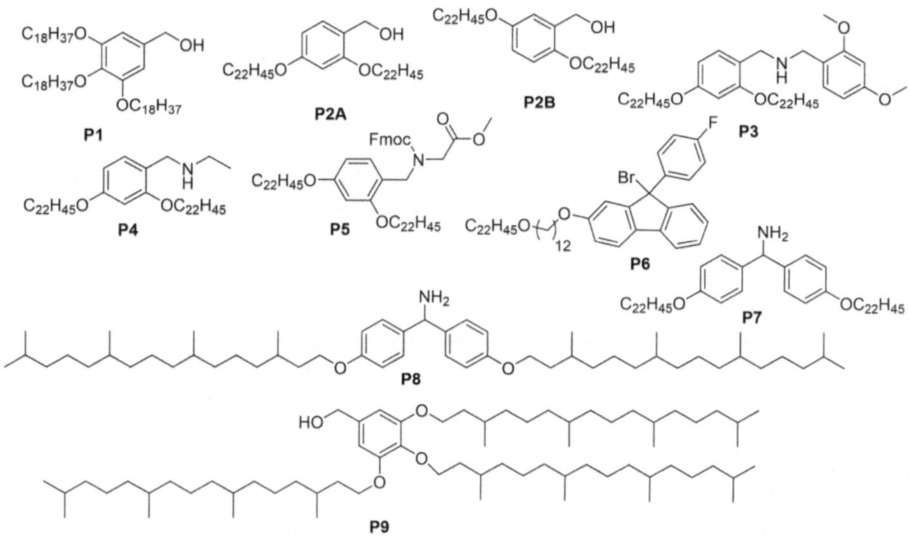

Figure 5.6 Representative structures of PolyCarbon tags. Adapted from ref. 25 with permission from American Chemical Society, Copyright 2022.

referred to this technology as "molecular hiving" for the synthesis of C-terminal acid peptides.[16,71–79] In parallel, Ajinomoto developed the Ajiphase protocol[80–88] for the preparation of both C-terminal acid and amide peptides utilizing P6 and P7, respectively.[89–93] A further improvement was made to the Ajiphase technology by using branched PolyCarbons (P8 and P9).[18]

The first demonstration of P1 for peptide synthesis was reported by Tamiaki and co-workers.[55,56] The first amino acid is attached to P1 *via* an ester bond. The stability of the ester is dependent on the phenyl substitution. An electron-donating group enhances the acid lability whereas an electron-withdrawing group increases the stability towards acid treatment. The tag P1 bearing large alkoxy substituents was synthesized[55,56] but it showed a lower precipitation ability and hence was unsuitable for application in large-scale peptide synthesis.[60,61]

P2A was used for the synthesis of linear and cyclic peptides by applying the Fmoc/*t*Bu strategy. Bivalirudin and h-ghrelin were synthesized at the 2 and 10 mmol scale, respectively, by employing a convergent synthetic protocol.[64] Two cyclic peptides containing a disulfide bridge, H–c(Cys–Ala–Cys)–OH and somatostatin [H–c(Ala–Gly–Cys–Lys–Asn–Phe–Phe–Trp–Lys–Thr–Phe–Thr–Ser–Cys)–NH_2], were also synthesized. Coupling was performed using azabenzotriazoletetramethyluronium hexafluorophosphate–1-hydroxy-7-azabenzotriazole (HATU–HOAt) (or HBTU–HOBt–DIEA) in DCM and Fmoc removal was achieved using 1% DBU–PIP (1:1) in THF. After each reaction, the reaction mixture was treated with excess ACN, which precipitated the required product. After filtration, the residue was used in the next step directly. This process of coupling and Fmoc removal was repeated until the final desired peptide was assembled on the tag. The tagged peptides were subjected to global deprotection and precipitated using diisopropyl ether (DIPE). The yields obtained for bivalirudin and h-ghrelin were 64 and 68%, respectively.[64]

In the case of cyclic peptides reported by Kitada *et al.*, the disulfide bridge was formed using I_2.[61] The cyclic tagged peptide was cleaved and isolated as explained earlier. Somatostatin was obtained in 95% yield. In another report, a disulfide bridge in the tagged peptide (peptide P2A) was formed using electrochemical reactions, followed by global deprotection and precipitation as explained above.[94]

A hydrophobic elastin peptide [H–(Pro–Gly–Val–Gly–Val)$_2$–OH] was synthesized using P2A as a tag under high pressure and also using a magnetic nanoparticles (MNPs) strategy.[65] During the synthesis, hydrophobic elastin peptides are prone to gelation and aggregation. The main problem in elastin peptide synthesis is the increase in viscosity during synthesis, which leads to slower kinetics. The first coupling was performed using Fmoc–Val–OH and P2A where DIC–4-dimethylaminopyridine (DMAP) was used as coupling reagent in THF. After coupling, excess ACN was added to precipitate Fmoc–Val–P2A. Fmoc was removed as explained earlier followed by precipitation using excess ACN. The next coupling was performed with 1-{[1-(cyano-2-ethoxy-2-oxoethylideneaminooxy)dimethylaminomorpholino]} uronium hexafluorophosphate (COMU)–DIEA in cyclohexane–ACN to avoid gelation problems. Owing to aggregation, the reaction mixture turned into a viscous gel, thereby disturbing the fluidity. To resolve this issue, a gas-tight syringe was used under pressure. The plunger was pushed manually to maintain a pressure of around 0.10–0.20 MPa, which increased the fluidity and aided successful coupling using COMU–DIEA as a coupling reagent This approach was iterated until Fmoc–Pro–Gly–Val–Gly–Val–O–P2A was synthesized. The crude product was divided into two parts. Fmoc was removed from one part, and the second part was subjected to global deprotection using TFA–trifluoroethanol (TFE)–DCM to obtain H–Pro–Gly–Val–Gly–Val–O–P2A and Fmoc–Pro–Gly–Val–Gly–Val–OH, respectively. These two independent parts were then coupled using COMU–DIEA as coupling reagent in cyclohexane–acetone as solvent to obtain [Fmoc–(Pro–Gly–Val–Gly–Val)$_2$–O–P2A]. Fmoc was then removed to obtain H–(Pro–Gly–Val–Gly–Val)$_2$–O–P2A, which was treated with TFA in TFE–DCM (1 : 5). The reaction mixture was diluted with MeOH to obtain a precipitate. The filtrate was then concentrated, followed by acidification with 6 N HCl. The precipitate was further washed with ACN to afford crude elastin peptide in 88% yield.

Owing to the cumbersome procedure, another approach for the synthesis of peptides prone to gelation and aggregation was adopted using MNPs for purification after each step.[68] The key highlight of this technique is the use of Fe_3O_4 MNPs, which are tagged to long aliphatic chains and are referred to as MNPs. This aids in capturing aggregated MNPs and tagged peptides with the help of a magnet. Simple decantation helps in removing excess reagents and by-products. Several washings using ACN provide the desired tagged peptide. Elastin peptide (H–Pro–Gly–Val–Gly–Val–OH) was synthesized using P2A at the 0.8 mmol scale and Fe_3O_4 with octadecyl isocyanate (as MNPs). Esterification was performed as explained earlier.[65,68] Coupling and Fmoc removal were performed using the conditions mentioned earlier. Sonication was applied during the reaction to disaggregate the MNPs and the product of interest (tagged peptide). The desired peptide was obtained by cleaving it from the tag and following the above procedure for work-up. In this one-pot synthetic protocol, elastin peptide was obtained in 87% yield, which is comparable to that using cumbersome high-pressure LPPS.

Another approach to obtain the desired product from the viscous reaction mixture was adopted by Ma and co-workers. This process was referred to as a "capture and release approach" and was used for the synthesis of semaglutide and TP4 (or Oreoch-2) along with its analogues following a convergent strategy.[95–97] To avoid gelation, a small amount of diatomite was added to the suspension. This addition efficiently precipitated the diatomite-labelled tagged peptide, which could be separated from impurities by simple filtration. This was then subjected to ultrasonication to separate the tagged peptide from the diatomite. The filtrate containing the tagged peptide was used directly in the next step.[95–97] This process of isolation of the desired product is referred to as "diatomite mixing wash". For peptide synthesis, esterification was performed using the above procedure. Fmoc–AA–O–P2A was isolated by using diatomic mixing wash as explained above. Fmoc was removed following the above protocol. After diatomite mixing wash, H–AA–O–P2A was obtained. Coupling was then carried out using HBTU–HOBt–DIEA in THF–DMF (9 : 1), again followed by diatomite mixing wash. This cycle of coupling and deprotection was repeated until all the fragments were successfully synthesized. TFA–TFE–DCM (1 : 10 : 90) was used to remove the tags from the fragment and the desired product was obtained as a precipitate upon treatment with DIPE. The fragments were then coupled in the appropriate order to obtain the desired protected peptides. Global deprotection afforded crude peptides, which upon purification by RP-HPLC gave pure semaglutide and TP4.[95,96] Six distinct analogues of TP4 were synthesized similarly.[97]

Chiba and co-workers further developed P2B, with several distinct properties. This tag takes a distinct green colour upon acid treatment, making it suitable for quantitative acidic colorimetric analysis.[69] P2B is compatible with Boc chemistry owing to its high tolerance towards TFA and it was used for angiotensin III (H–Arg–Val–Tyr–Ile–His–Pro–Ile–OH) synthesis at the 1 mmol scale. The reaction was sequentially monitored using acidic colorimetric analysis along with photo-triggered fluorometric assay. After global deprotection and precipitation with DIPE, angiotensin III was obtained in 90% yield.

P2B was recently reported for the synthesis of Leu-enkephalin using a green coupling reagent [cyclic propylphosphonic anhydride (T3P)] at the 0.66 mmol scale.[98] Esterification was performed using DIC–DMAP in DCM. After completion, the product was precipitated using excess ACN and filtered. Fmoc removal was achieved using standard Fmoc removal conditions employing piperidine (16 equiv.) in DCM. The product was obtained as a precipitate by addition of excess ACN to the reaction mixture. Coupling was performed in DCM using T3P (1.5 equiv.; 50% in DCM) and DIEA (4 equiv.). The reaction was completed in 5 min. Excess ACN was added to precipitate the desired product. This cycle of coupling and deprotection was continued until the peptide was assembled. Global deprotection was performed and DIPE was added, then the reaction mixture was cooled to 0 °C and stirred for 30 min. Leu-enkephalin was obtained in >99% purity as a white precipitate.[98]

A Rink-amide linker-based tag P3 was used for the synthesis of α-conotoxin MII {H–Gly–c[Cys^1–c(Cys^2–Ser–Asn–Pro–Val–Cys^1)–His–Leu–Glu–His–Ser–Asn–Leu–Cys^2]–NH_2} at the 1 mmol scale.[63] Two different Cys PGs, Fmoc–Cys(Acm)–OH (for Cys at positions 2 and 8) and Fmoc–Cys(Mmt)–OH (for Cys at positions 3 and 16) were used. Linear peptide was assembled onto P3 following the above protocol. The tagged peptide was treated with TFA–TIS–DCM (1 : 5 : 94) to selectively eliminate the Mmt (monomethoxytrityl) protecting group followed by neutralization by adding DIEA. Extraction was effected using DCM and the organic layer upon concentration was subjected to oxidation using diethyl azodicarboxylate (DEAD) in THF. After complete cyclization, the reaction mixture was concentrated, followed by treatment with I_2 in THF. The reaction was quenched by addition of ascorbic acid and extraction was performed as explained above. Excess ACN was added to the organic layer to precipitate the product. Global deprotection afforded the crude product, which after passing through a hydrophilic PTFE filter was treated with DIPE. α-Conotoxin MII was obtained as a precipitate in 71% yield.[63]

The C-terminal amidated peptide ABT-510 was synthesized using the tag P4, which aids in incorporation of an ethylenamide linker.[66] This tag eliminates diketopiperazine (DKP) formation and epimerization. For the coupling step, HATU–HOAt–DIEA (1 : 1 : 2; 1.2 equiv.) in THF (with 1% DMF) as solvent was used. When the reaction was complete, excess ACN was added to precipitate the desired product. Fmoc removal was performed in 1% DBU with THF as solvent. Excess of ACN afforded the desired tagged peptide as a precipitate. Repetitive coupling and Fmoc removal were performed until the peptide was assembled onto the P4 tag. Acetic anhydride–pyridine (1 : 1) was then used for acetylation. The tagged peptide was subjected to global deprotection followed by washing with DIPE to afford crude ABT-510 (89%), which was further purified by preparative RP-HPLC.[66]

An electrochemical technique was also used for peptide synthesis using the tag P4.[99] Triphenylphosphine (Ph_3P) generates phosphine radical cation, which aids in activating carboxylic acids, thus acting like a coupling reagent. First coupling was performed using COMU–DIEA in THF. After completion, the reaction mixture was treated with excess ACN to precipitate the desired product. Fmoc was then removed using 1% DBU–PIP (1 : 1) in THF. The triphenylphosphine oxide ($Ph_3P{=}O$) that was formed as a stoichiometric by-product could be isolated selectively from the desired peptide and recycled for further use.[99] To elongate the peptide chain, the tagged amino acid was dissolved in ACN–cyclohexane (c-Hex) solvent mixture followed by addition of

2,4-lutidine (0.6 equiv.), Bu_4NClO_4 (electrolyte, 0.5 equiv.), Ph_3P, and Fmoc–AA–OH. The solution was then electrolysed under argon at 0.67 mA cm^{-2} at room temperature, followed by passing 4.8 F mol^{-1} of electricity. After coupling was complete, an excess of ACN was added to the c-Hex layer to precipitate the desired product. The ACN layer from the reaction mixture was then subjected to activation with charcoal, filtered, and dried. The residual filtrate was dissolved in toluene–*n*-hexane (30 : 1) to precipitate the electrolyte (Bu_4NClO_4), again followed by filtration. The filtrate was treated for Fmoc removal with DBU–PIP (1 : 1), followed by neutralization with 1 N HCl. The organic layer was separated and dried to allow isolation of the by-product $Ph_3P{=}O$. Global deprotection and washing with DIPE gave leuprorelin.[99]

Usually, the head-to-tail cyclization of a peptide is performed on SPPS-synthesized sequences, still protected and loaded on the solid support. The carboxylic moiety may be present in the side-chain, as in aspartic or glutamic acid,[100] or may be the C-terminus when the carboxylic group is part of the linker structure.[101] Head-to-tail peptide cyclization was carried out using P5 tag to synthesize mahafacyclin B at the 0.5 mmol scale.[62] Coupling and Fmoc removal were performed using HATU–HOAt–DIEA (1 : 1 : 2; 1.2 equiv.) and DBU–PIP (1 : 1) in THF, respectively. Excess ACN was added after each step to recover the intermediates as precipitates. The methyl ester was then cleaved under basic conditions using 1 M aqueous LiOH in THF, followed by neutralization with 12 N HCl. Addition of excess ACN afforded the linear tagged peptide as a precipitate. Cyclization was then performed using HATU–HOAt–DIEA with DMF–THF (1 : 9) solvent mixture. Crude tagged peptide was obtained as a precipitate upon addition of excess ACN. Global deprotection and precipitation using DIPE afforded crude mahafacyclin B (71% yield), which was purified using preparative RP-HPLC.

Stellarin E peptide was synthesized using P5 (and also P2) at the 1.8 mmol scale *via* three distinct routes in which the position of the tag was changed: C-terminus, N-terminus, and centre of the sequence.[70] P2 was used as it is highly labile under acidic conditions. The idea behind this approach was to induce the *cis* conformation for better cyclization (see Figure 5.7).

A linear protected peptide using P5 at the C-terminus was prepared following the earlier procedure. P2 was also used in two other cases, displaying

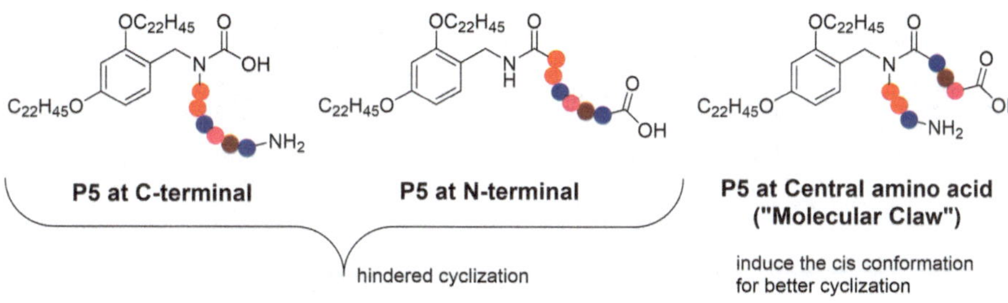

Figure 5.7 Strategies for stellarin E synthesis. Adapted from ref. 25 with permission from American Chemical Society, Copyright 2022.

different C-terminal amino acids. However, in both cases a Gly residue was tagged using P5. Therefore, in both cases, a double tag strategy was used. After the desired peptide had been assembled, P2 was selectively removed with TFA–TFE (0.5 : 8) in DCM. This afforded the protected peptide with a tag either at the N-terminus or at the central position. After neutralization with TEA, excess ACN was added to obtain a product with a Gly "N-tag". Cyclization was then achieved using COMU–DIEA in $CHCl_3$ under high dilution conditions. Excess ACN was added to precipitate the tagged cyclic peptide. After filtration, global deprotection was performed followed by washing with DIPE to afford crude stellarin E. Among the three approaches, the best yield was obtained by using the central tag (see Figure 5.8). In this case, the pure peptide was purified by RP-HPLC and a yield of 57% was obtained.[70]

The central tag strategy, also known as the "molecular claw" approach, presents several advantages. This strategy involves the *cis* conformation effect in the middle of the peptide sequence, thereby reducing the hindrance around the point of cyclization.

Ajiphase technology was developed by Ajinomoto for the synthesis of C-terminal acid/amidated peptides.[80–88] Initially, P6 and P7 were developed with linkers, one for C-terminal acid peptides and the other for C-terminal amide peptides, respectively. P6, being a halo-based hindered tag, was designed to mimic 2-chlorotrityl chloride (CTC) resin to avoid DKP formation.[102,103] Intermediates were isolated as precipitates by addition of excess MeOH–ACN, and were used in the next step without further purification.

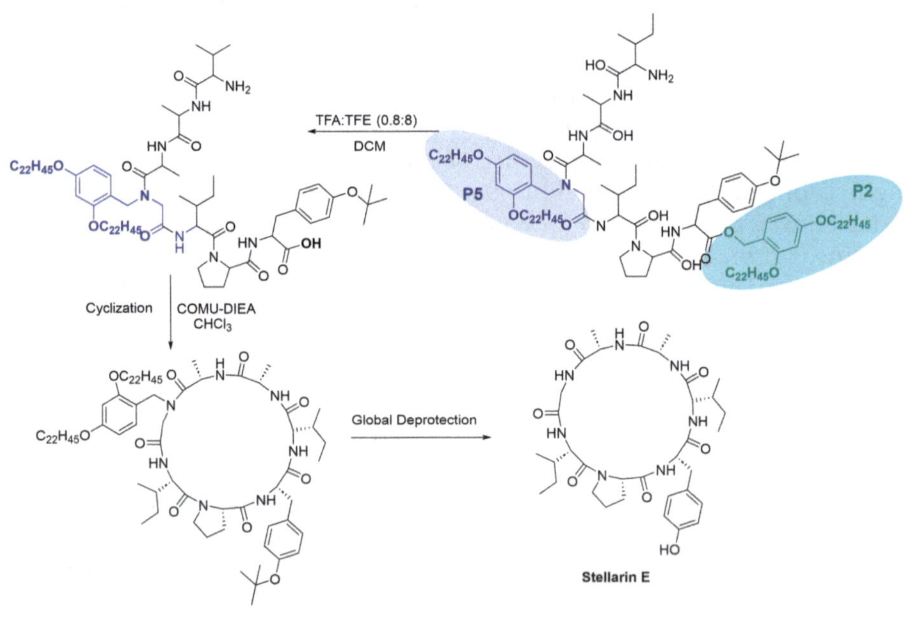

Figure 5.8 "Molecular claw" strategy for stellarin E synthesis. Adapted from ref. 25 with permission from American Chemical Society, Copyright 2022.

However, P6 and P7 are soluble in either THF or halogenated solvents, which increases the viscosity of the reaction mixture, making the extraction process inefficient. Therefore, P8 and P9 were developed for one-pot peptide elongation.[18] The presence of ramified phytol chains in both anchors improves the lipophilicity, thereby enhancing the removal of by-products in one-pot extraction, hence reducing the consumption of solvents. In both cases, washing after coupling was avoided and the reaction mixture was used directly in the next step. P8 was used to synthesize C-terminal amidated degarelix at the 2.38 mmol scale and P9 was used to synthesize the C-terminal acid peptide bivalirudin at the 10 mmol scale. Esterification was performed using EDC–DMAP and coupling using EDC–HOBt. The reaction mixture was then washed using 20% NaCl. Fmoc removal was achieved using thiomalic acid–DBU (3 : 8) for 2 h at room temperature. This was followed by treatment with methanesulfuric acid and washing with aqueous $NaHCO_3$ and DMF. Thiomalic acid, along with methanesulfuric acid, was used as a scavenger to remove dibenzofulvene from the reaction mixture. Without any further treatment, the organic layer was used for the next coupling as no impurities derived from Fmoc removal conditions or fulvene were observed. Coupling and Fmoc removal were iterated, as explained above, to obtain the desired tagged peptides. Acetylation was performed on the degarelix-tagged peptide before final deprotection. Then two reaction mixtures containing the tagged peptides were concentrated and precipitated by addition of excess ACN. Global deprotection was performed to obtain the pure peptides. Degarelix was obtained in 85% yield with 89% purity, whereas bivalirudin was obtained in 79% yield with 84% purity.[18]

5.5 Silicon-based Tags for LPPS

Silyl groups are known to reduce viscosity and increase hydrophobicity. A series of silicon-based hydrophobic tags were designed and evaluated for LPPS.[104,105] Si-based tags confer higher hydrophobicity than PolyCarbon tags, which allows better solubility in non-halogenated solvents, hence making peptide synthesis an environmentally friendly and sustainable process. In addition to Si-based tags, super silyl tags (containing Si–Si bonds) have also been evaluated in an LPPS strategy.[106] Figure 5.9 illustrates the silylated tags used for peptide synthesis. These tags have been successfully demonstrated to be compatible with Boc and Cbz (carboxybenzyl) chemistry in addition to Fmoc chemistry, providing a wide range of possibilities for peptide synthesis and avoiding side reactions such as DKP formation or racemization. S1 was used to synthesize difficult sequences containing *N*-methylated amino acids.[105] S2 and S4 are compatible with Fmoc and Cbz chemistry whereas S3 and S5 are compatible with Boc chemistry. Being compatible with Boc and Cbz chemistry, carbamate esters of silyl tags have been used for N-terminal protection and C→N chain elongation has been attempted (compared with regular N→C peptide synthesis) by avoiding DKP formation and racemization.[106,107]

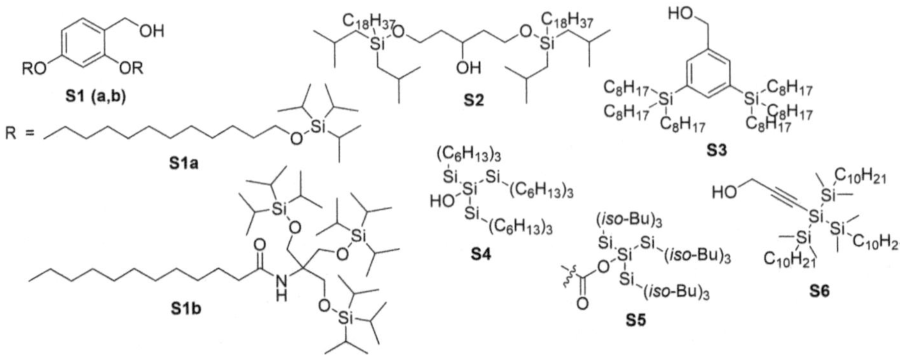

Figure 5.9 Representative structures of silicon-based tags.

The higher hydrophobicity and lower viscosity of the silylated chemical tag S1 were exploited for the synthesis of peptides containing *N*-methylated amino acids.[105] The first amino acid was anchored using EDC–DMAP in cyclopentyl methyl ether (CPME). Upon coupling, the excess of active ester was quenched by adding 2-aminoethoxyethanol (AEE). Fmoc was removed using DBU. Sodium 3-mercaptopropanesulfonate (MPSNa) in DMSO was added to remove dibenzofulvene (DBF). The reaction mixture was neutralized using HCl. Extraction was performed to remove excess reagents and by-products in the aqueous layer. The CPME layer was used directly in the next coupling using COMU–DIEA. This cycle of coupling and deprotection was carried out until the peptide was built onto the tag S1. Global deprotection was carried out with TFA–TIS–3,6-dioxa-1,8-octanedithiol–H_2O (92.5 : 2.5 : 2.5 : 2.5). Pure peptides (H–Tyr–Ala–Ala–Phe–Leu–OH, H–Tyr–Me–Leu–Me–Leu–Phe–Leu–OH, and H–Tyr–Me–Phe–Me–Phe–Phe–Leu–OH) were obtained as precipitates by addition of *tert*-butyl methyl ether (TBME).[105]

S2 was used to synthesize a 14-mer DRGN-1 antimicrobial peptide by a convergent approach (0.006 mmol). The synthesis was performed for two fragments (10-mer and 4-mer). The first fragment was synthesized using S2 whereas the second fragment was synthesized by a classical solution-phase approach. Esterification was performed using S2 (1 equiv.) and Fmoc–Ala–OH, DCC, and DMAP (2.5 : 2.5 : 1.2) in DCM. The reaction mixture was stirred at room temperature overnight, then filtered and washed with DCM. The filtrate was concentrated and purified to obtain pure Fmoc–Ala–O–S2. Fmoc was removed using DBU (1 equiv.) in $CHCl_3$. The reaction mixture was stirred at room temperature for 3 h, then concentrated and purified using silica gel column chromatography. H–Ala–O–S2 was then coupling to Fmoc–Val–SNp (1.3 equiv.) in $CHCl_3$. The reaction mixture was stirred at room temperature for 36 h, then diluted with DCM and washed with saturated Na_2CO_3 solution. The aqueous layer was separated and washed with DCM. The combined organic layer was dried using Na_2SO_4, filtered and concentrated. The residue was purified using silica gel column chromatography. This cycle of coupling and Fmoc removal was repeated until fragment 1–S2 was obtained. Fragment

2 was prepared as per classical solution peptide synthesis. Fragment 1–S2 and fragment 2 (1 : 1.3) were dissolved in $CHCl_3$ and EDC–HOBt (1.4 : 1.2) was added. The reaction mixture was stirred at room temperature for 1 h. After completion, the reaction mixture was diluted with $CHCl_3$ and extracted using water. Work-up was performed as explained above followed by column chromatography to obtain the protected DRGN-1–O–S2 peptide. Deprotection was performed using DRGN-1–O–S2 (1 equiv.) and NaOH (3.2 equiv.) in a mixture of solvents ($CHCl_3$–THF–MeOH–H_2O). The reaction mixture was stirred at room temperature for 3.5 days, then after concentration it was diluted with $CHCl_3$ and HCl was added. Again, the reaction mixture was concentrated and the residue was washed with *n*-hexane (S2 was recovered in 81% yield). The remaining residue was diluted with $CHCl_3$ and filtered. The filtrate was concentrated to afford protected DRGN-1 peptide in 75% yield.[107]

S3 was used to synthesize H–Ala–Ala–Ala–Ala–Ala–Ala–Ala–OH. H–Ala–O–S3 was prepared as mentioned above. Coupling was then performed using H–Ala–O–S3 and Fmoc–AA–OPfp where Pfp = pentafluorophenyl (1 : 1.2) in $CHCl_3$. The reaction mixture was stirred for 19 h at room temperature. 2-Aminoethanol (1.65 equiv.) was added to the reaction mixture, which was again stirred for 20 min at room temperature (to remove excess Fmoc–A–OPfp). Saturated Na_2CO_3 was added and the layers were separated. Work-up was performed as explained earlier (eliminating the column chromatography for purification). Fmoc was then removed from Fmoc–Ala–Ala–O–S3 (1 equiv.) using DBU (2 equiv.) in $CHCl_3$ for 30 min at room temperature. HCl (1 equiv.) was added while maintaining the pH at 6–7 to remove DBU. The reaction mixture was extracted with water and the organic layer was washed with Na_2CO_3 followed by drying using Na_2SO_4, filtered, and concentrated to afford H–Ala–Ala–O–S3. The coupling and deprotection were continued until the peptide was synthesized. H–Ala–Ala–Ala–Ala–Ala–Ala–Ala–O–S3 (1 equiv.) was dissolved in $CHCl_3$–MeOH–H_2O and LiOH (2 equiv.) was added. The reaction mixture was stirred for 43 h at room temperature then concentrated, and the residue was dissolved in $CHCl_3$. HCl (2 equiv.) was added and the mixture was filtered. The residue was washed with H_2O and $CHCl_3$, then dissolved in TFA. The suspension was filtered and the filtrate was concentrated to afford H–Ala–Ala–Ala–Ala–Ala–Ala–Ala–OH in >99% yield with 77% purity.[107]

Wu and Yamamoto used S4 and S5 to synthesize 9-mer nelipepimut-S (a peptide cancer vaccine) at the 0.024 mmol scale using the convergent synthesis approach.[106] S4 was used as a C-terminal protecting group and S5 was used as an N-terminal protecting group to synthesize the 7-mer and 2-mer, respectively. Dipeptide (Fmoc–Phe–Leu–OH) was synthesized prior to the attachment to the tag. S4 (1.5 equiv.) was added to triflic acid (1 equiv.) in DCM and the reaction mixture was stirred for 2 h at room temperature, then Fmoc–Phe–Leu–OH (1 equiv.) and 1-methylimidazole (2.2 equiv.) were added. The reaction mixture was stirred for 16 h at room temperature, then concentrated and purified using silica gel column chromatography to afford Fmoc–Phe–Leu–O–S4. Fmoc was removed using excess diethylamine (Et_2NH) in DCM. The reaction mixture was stirred for 1.5 h at room temperature, then water was added and the organic layer was collected, dried over Na_2SO_4, filtered, and concentrated. The residue was used for coupling without any further purification. Coupling was performed using Fmoc–AA–OPfp as explained above. This coupling and deprotection were performed until the entire peptide were assembled onto S4 to afford fragment 1–S4. In parallel, fragment 2 was synthesized wherein S5 was used as a carbamate ester for N-terminal protection. H–Lys(Boc)–OBn (1.3 equiv.) and TEA (1.5 equiv.) were dissolved in DCM and the reaction mixture was cooled to −78 °C. Separately, tag (1 equiv.) was added to a stirred solution of triflic acid (1.35 equiv.) in DCM and the reaction mixture was stirred for 1.2 h at room temperature. This solution was added to the above mixture at −78 °C and the reaction mixture was stirred at room temperature for 16 h. After completion, the reaction mixture was concentrated and purified using silica gel column chromatography to afford S5–Lys(Boc)–OBn. Catalytic hydrogenation was performed using Pd/C with H_2 at 50 °C for 4 h, to remove the Bn group. The reaction mixture was filtered over Celite and the filtrate was concentrated to afford S5–Lys(Boc)–OH, which was used in the next step without purification. S5–Lys(Boc)–OH was dissolved in DCM and $HSi[OCH(CF_3)_2]_3$ (1.5 equiv.), H–Ile–OBn (1.5 equiv.), and PMBN-$HSi[OCH(CF_3)_2]_3$ (0.05 equiv.) were added. The reaction mixture was stirred for 12 h at 40 °C. After completion, the reaction mixture was directly purified using silica gel column chromatography to afford pure S5–Lys(Boc)–Ile–OBn. The Bn group was removed as explained earlier to afford S5–Lys(Boc)–Ile–OH (fragment 2). Fragment 1–S4 and fragment 2 were then coupled following a similar procedure as in the case of DRGN-1 peptide. The N- and C-terminal-protected peptide was then treated with TFA–TIS–H_2O (95 : 2.5 : 2.5) and the reaction mixture was stirred for 17 h at room temperature, then concentrated, and the residue was washed with diethyl ether. The residue was again dissolved in TFA and filtered to remove impurities. The filtrate was concentrated to afford the peptide in >99% yield with 73% purity.[106]

In the same study, S6 was also used to synthesize nelipepimut-S (a peptide cancer vaccine) at the 0.15 mmol scale.[106] Esterification was performed using S6 (1 equiv.) and Boc–Leu–OH–DCC–DMAP (2 : 2 : 1.2) in DCM. The reaction mixture was stirred for 3 h at room temperature. After completion, the reaction mixture was filtered, washed with DCM, and concentrated. The product Boc–Leu–O–S6 was purified by silica gel column chromatography. After the pure Boc–Leu–O–S6 was obtained, Boc was removed by dissolving the compound in DCM and adding 4 M HCl in dioxane. The reaction mixture was stirred for 1–2 h until completion. Saturated Na_2CO_3 was added and the layers were separated. The aqueous phase was washed with DCM three times. The combined DCM layers were dried using Na_2SO_4, filtered, and concentrated. H–Leu–O–S6 was then coupled with Boc–AA–OH–EDC·HCl–HOBt (1.1 : 1.2 : 0.3) in DCM as solvent. The reaction mixture was stirred at room temperature until completion. Work-up was performed as explained above. This cycle of deprotection and coupling was repeated until the entire sequence was assembled. S5 was removed by treating the tagged peptide with LiOH (2.5 equiv.) in DCM as solvent. The reaction mixture was stirred for 24 h at room temperature. After completion, the reaction mixture was concentrated, then DCM was added followed by HCl and the layers were separated. The organic layer was concentrated and washed with hexanes to obtain the protected peptide in 93% yield with 83% purity.[106]

5.6 Miscellaneous Tags for LPPS

Other tags that have been synthesized for LPPS include fluorous-, ionic liquid (IL)-, and hydrophobic polymer (HP)-based tags.[25] Fluorous tags mainly depend on the F content present. If the F content is >60% the compound can be extracted using fluorous FC72 (perfluorohexane). Fluorous solvents are immiscible in both organic and aqueous layers. Being recyclable and bearing this unique solubility property, these tags are elegant anchors for peptide synthesis. However, these tags have found suitable application only for smaller sequences. It is difficult to synthesize longer peptides, as the F content needs to be maintained at >60%, which is a serious limitation of this strategy. In addition, the cost involved in the synthesis of small peptides is quite high owing to the use of expensive fluorine, which makes this approach impractical for industrial use. Finally, the use of fluorine tags cannot be considered a green strategy.

Compared with fluorous tags, ILs offer a greener alternate. ILs provides a wide range of solubility in polar and non-polar solvents. However, this strategy is not scalable and hence does not find application at the production level in industry. Soluble polymeric tags (polynorbornene) are prepared by ring metathesis (controlled polymerization). This approach allows modulation of the characteristics of the tags by varying the composition of the polymer for branching, thereby modulating the loading and solubility. Controlled polymerization and increased loading make them the ideal choice for peptide synthesis at a large scale. However, production-scale synthesis using this technology has not been reported.

5.7 Conclusion

SPPS is highly efficient and hence is strategically adopted for the production-level synthesis of peptide-based active pharmaceutical ingredients (APIs). However, SPPS is subject to several drawbacks concerning the use of hazardous chemicals and huge volumes of solvents during washing steps, the latter leading to large amounts of waste compared with small organic molecule-based API production. The high solvent consumption issue can be significantly reduced in LPPS protocols. The key step in peptide synthesis *via* either SPPS or LPPS is to reduce the solvent consumption, which can be achieved by eliminating the washing/purification steps. This aids in sustainable and economical synthesis. In this regard, the LPPS strategy, irrespective of the protocol (MEPS, molecular hiving, GAP, Ajiphase, *etc.*) aims to lower the solvent consumption significantly, thereby reducing the PMI and cEF. However, the road to ideal green peptide synthesis still seems to have innumerable gaps. With the current status and considering all the knowledge achieved so far, sustainable peptide synthesis seems to be an achievable goal.

Acknowledgements

The authors wish to thank colleagues working in the field for fruitful discussions during the preparation of this chapter. This work was funded by the National Research Foundation (NRF) (Blue Sky's Research Programme #120386).

References

1. B. G. de la Torre and F. Albericio, *Molecules*, 2023, **28**, 1038.
2. D. J. Drucker, *Nat. Rev. Drug Discovery*, 2020, **19**, 277–289.
3. F. Albericio and H. G. Kruger, *Future Med. Chem.*, 2012, **4**, 1527.
4. O. Al Musaimi, D. Al Shaer, F. Albericio and B. G. de la Torre, *Pharmaceuticals*, 2023, **16**, 336.
5. A. Henninot, J. C. Collins and J. M. Nuss, *J. Med. Chem.*, 2018, **61**, 1382.
6. S. Jadhav, W. Seufert, C. Lechner and R. Schönleber, *Chimia*, 2021, **75**, 476.
7. M. W. Pennington, B. Zell and C. J. Bai, *Med. Drug Discovery*, 2021, **9**, 1.
8. J. H. Rasmussen, *Bioorg. Med. Chem.*, 2018, **26**, 2914.
9. A. A. Zompra, A. S. Galanis, O. Werbitzky and F. Albericio, *Future Med. Chem.*, 2009, **1**, 361.
10. L. Wang, N. Wang, W. Zhang, X. Cheng, Z. Yan, G. Shao, X. Wang, R. Wang and C. Fu, *Signal Transduction Targeted Ther.*, 2022, **7**, 48.
11. W. Cabri, P. Cantelmi, D. Corbisiero, T. Fantoni, L. Ferrazzano, G. Martelli, A. Mattellone and A. Tolomelli, *Front. Mol. Biosci.*, 2021, **8**, 697586.
12. P. Lloyd-Williams, F. Albericio and E. Giralt, *Chemical Approaches to the Synthesis of Peptides and Proteins*, CRC Press, Boca Raton (FL, USA), 1997, vol. 10, p. 304.
13. I. F. Eggen, F. T. Bakelaar, A. Petersen, P. B. W. Ten Kortenaar, N. H. S. Ankone, H. E. J. M. Bijsterveld, G. H. L. Bours, F. E. L. Bellaj, M. J. Hartsuiker, G. J. Kuiper and E. J. M. J. Ter Voert, *Pept. Sci.*, 2005, **11**, 633.
14. R. B. Merrifield, *J. Am. Chem. Soc.*, 1963, **85**, 2149.
15. O. Al Musaimi, B. G. de la Torre and F. Albericio, *Green Chem.*, 2020, **22**, 996.
16. L. Ferrazzano, M. Catani, A. Cavazzini, G. Martelli, D. Corbisiero, P. Cantelmi, T. Fantoni, A. Mattellone, C. De Luca, S. Felletti, W. Cabri and A. Tolomelli, *Green Chem.*, 2022, **24**, 975.
17. R. A. Sheldon, *ACS Sustainable Chem. Eng.*, 2018, **6**, 32.
18. D. Takahashi, T. Inomata and T. Fukui, *Angew. Chem., Int. Ed.*, 2017, **56**, 7803.
19. A. Isidro-Llobet, M. N. Kenworthy, S. Mukherjee, M. E. Kopach, K. Wegner, F. Gallou, A. G. Smith and F. J. Roschangar, *J. Org. Chem.*, 2019, **84**, 4615.
20. M. Mutter, H. Hagenmaier and E. Bayer, *Angew. Chem., Int. Ed. Engl.*, 1971, **10**, 811.

21. E. Bayer and M. Mutter, *Nature*, 1972, **237**, 512.
22. E. Bayer, M. Mutter, R. Uhmann, J. Polster and H. Mauser, *J. Am. Chem. Soc.*, 1974, **96**, 7333.
23. M. Mutter and E. Bayer, *Angew. Chem., Int. Ed. Engl.*, 1974, **13**, 88.
24. E. Bayer and M. Mutter, *Chem. Ber.*, 1974, **107**, 1344.
25. A. Sharma, A. Kumar, B. G. de la Torre and F. Albericio, *Chem. Rev.*, 2022, **122**, 13516.
26. P. Marchetti, M. F. Jimenez Solomon, G. Szekely and A. G. Livingston, *Chem. Rev.*, 2014, **114**, 10735.
27. A. G. Livingston, P. Marchetti, P. Gaffney and R. Liu, Process for the production of solvent stable polymeric membranes, *WO pat.*, WO2017212246, 2017.
28. A. G. Livingston, L. G. Peeva and S. So, Degradable supports for tide synthesis and use of membrane filtration to cleave tides, *WO pat.*, WO2010055343, 2010.
29. A. G. Livingston, L. G. Peeva, S. W. J. So, R. Campos Vasconceles, R. J. Leatherbarrow, E. W. Tate and P. R. J. Gaffney, Solvent resistant diafiltration of peptides, PNA or oligonucleotides, *US pat.*, US8664357, 2014.
30. C. Noti, M. Cristau, J. Riegler, V. d.l.P. d.M. Castro Pinzon, H. M. Rodriguez Cabrera, F. Albericio Palomera, W. Chen and A. G. Livingston, Preparation of branched soluble polymeric supports for membrane enhanced peptide synthesis, *WO pat.*, WO2016188835, 2016.
31. S. So, L. G. Peeva, E. W. Tate, R. J. Leatherbarrow and A. G. Livingston, *Chem. Commun.*, 2010, **46**, 2808.
32. S. So, L. G. Peeva, E. W. Tate, R. J. Leatherbarrow and A. G. Livingston, *Org. Process Res. Dev.*, 2010, **14**, 1313.
33. V. Castro, C. Noti, W. Chen, M. Cristau, A. Livignston, H. Rodríguez and F. Albericio, *Macromolecules*, 2017, **50**, 1626.
34. J. F. Kim, A. M. Freitas da Silva, I. B. Valtcheva and A. G. Livingston, *Sep. Purif. Technol.*, 2013, **116**, 277.
35. J. Yeo, L. Peeva, S. Chung, P. Gaffney, D. Kim, C. Luciani, S. Tsukanov, K. Seibert, M. Kopach, F. Albericio and A. Livingston, *Angew. Chem., Int. Ed.*, 2021, **60**, 7786–7795.
36. M. E. Kopach, E. S. Murzinski, V. Rustagi and H. A. F. I. Salim, Compounds and methods for liquid phase synthesis, *WO pat.*, WO2023028466A1, 2023.
37. J. Wu, G. An, S. Lin, J. Xie, W. Zhou, H. Sun, Y. Pan and G. Li, *Chem. Commun.*, 2014, **50**, 1259.
38. G. An, C. Seifert and H. Sun, *Heterocycles*, 2015, **90**, 344.
39. G. An, W. Zhou and X. Xu, *Heterocycles*, 2015, **90**, 1405.
40. G. An, C. Seifert and G. Li, *Org. Biomol. Chem.*, 2015, **13**, 1600–1617.
41. C. W. Seifert, A. Paniagua, G. A. White, L. Cai and G. Li, *Eur. J. Org Chem.*, 2016, **2016**, 1714.
42. S. Qiao, J. Wu, J. Mo, P. T. Spigener, B. N. Zhao, B. Jiang and G. Li, *Synlett*, 2017, **28**, 2483.

43. H. Zhang, Z. Yang, B. N. Zhao and G. Li, *J. Org. Chem.*, 2018, **83**, 644.
44. P. Kaur, S. Pindi, W. Wever, T. Rajale and G. Li, *J. Org. Chem.*, 2010, **75**, 5144.
45. G. Li and C. Seifert, System and method for solution phase GAP peptide synthesis through the design of a benzyl-type group-assisted purification protecting group and its application in the Fmoc/tBu synthesis of thymopentin free from polymers, chromatography and recrystallization, *WO pat.*, WO2017112809, 2017.
46. C. Seifert, Method for solution-phase peptide synthesis and protecting strategies therefore, *WO pat.*, WO2019231760, 2019.
47. C. Seifert, Method for solution-phase peptide synthesis, *WO pat.*, WO2019217116, 2019.
48. C. Seifert, Synthesis strategy for gap protecting group, *WO pat.*, WO2020159837, 2020.
49. C. Seifert, Compositions and methods for solution phase group-assisted purification peptide synthesis, *US pat.*, US20210079036, 2021.
50. H. Li, J. Chao, J. Hasan, G. Tian, Y. Jin, Z. Zhang and C. Qin, *J. Org. Chem.*, 2020, **85**, 6271.
51. H. Li, J. Chao, G. Tian, J. Hasan, Y. Jin, Z. Zhang and C. Qin, *Org. Chem. Front.*, 2020, **7**, 689.
52. H. Li, J. Chao, Z. Zhang, G. Tian, J. Li, N. Chang and C. Qin, *Org. Lett.*, 2020, **22**, 3323.
53. H. Li, J. Li, J. Chao, Z. Zhang and C. Qin, *Org. Chem. Front.*, 2022, **9**, 946.
54. H. Li, J. Ren, J. Li, Z. Zhang, N. Chang and C. Qin, *Org. Biomol. Chem.*, 2020, **18**, 8433.
55. H. Tamiaki, T. Obata and K. Toma, *Pept. Sci.*, 1998, **1998**, 125.
56. H. Tamiaki, T. Obata, Y. Azefu and K. Toma, *Bull. Chem. Soc. Jpn.*, 2001, **74**, 733.
57. K. Chiba, Y. Kono, S. Kim, K. Nishimoto, Y. Kitano and M. Tada, *Chem. Commun.*, 2002, **2002**, 1766.
58. K. Chiba, M. Sugihara, K. Yoshida, Y. Mikami and S. Kim, *Tetrahedron*, 2009, **65**, 8014.
59. M. Iijima, Y. Mikami, T. Yoshioka, S. Kim, H. Kamiya and K. Chiba, *Langmuir*, 2009, **25**, 11043.
60. G. Tana, S. Kitada, S. Fujita, Y. Okada, S. Kim and K. Chiba, *Chem. Commun.*, 2010, **46**, 8219.
61. S. Kitada, S. Fujita, Y. Okada, S. Kim and K. Chiba, *Bioorg. Med. Chem. Lett.*, 2011, **21**, 4476–4479.
62. Y. Fujita, S. Fujita, Y. Okada and K. Chiba, *Org. Lett.*, 2013, **15**, 1155.
63. S. Kitada, S. Fujita, Y. Okada, S. Kim and K. Chiba, *Tetrahedron*, 2013, **69**, 2555.
64. Y. Okada, H. Suzuki, T. Nakae, S. Fujita, H. Abe, K. Nagano, T. Yamada, N. Ebata, S. Kim and K. Chiba, *J. Org. Chem.*, 2013, **78**, 320.
65. Y. Okada, S. Hosoya, H. Suzuki and K. Chiba, *Org. Lett.*, 2014, **16**, 6448.
66. E. Matsumoto, Y. Fujita, Y. Okada, E. I. Kauppinen, H. Kamiya and K. Chiba, *J. Pept. Sci.*, 2015, **21**, 691.

67. Y. Okada, H. Wakamatsu, M. Sugai, E. I. Kauppinen and K. Chiba, *Org. Lett.*, 2015, **17**, 4264.
68. Y. Okada, H. Asama, H. Wakamatsu, K. Chiba and H. Kamiya, *Eur. J. Org Chem.*, 2017, 5961.
69. H. Wakamatsu, Y. Okada, M. Sugai, S. R. Hussaini and K. Chiba, *Asian J. Org. Chem.*, 2017, **6**, 1584.
70. S. Yamagami, Y. Okada, Y. Kitano and K. Chiba, *Eur. J. Org Chem.*, 2021, 3133.
71. K. Chiba and T. Marui, Electrochemically cleavable liquid-phase peptide synthesis supports and the synthesis using them, *JP pat.*, JP2004067555, 2004.
72. K. Chiba and T. Marui, Amino acid reagents for liquid-phase synthesis of peptides, *JP pat.*, JP2004059509, 2004.
73. K. Chiba and T. Marui, Apparatus for liquid-phase peptide synthesis, *JP pat.*, JP2004035521, 2004.
74. J. Nakano, M. Horiuchi, K. Chiba, S.-H. Kim, H. Tanaka and M. Ono, Automated continuous liquid-phase peptide synthesis apparatus, *JP pat.*, JP2005206525, 2005.
75. K. Chiba, Preparation of support for crystallization separation and method of separating compounds such as amino acids or peptides, *WO pat.*, WO2006104166, 2006.
76. K. Chiba, S. Kim and Y. Kono, Preparation of 2,4-bis(hydrocarbyloxy) benzyl alcohol as carrier for separation, method for separation of compound, and method for synthesis of peptide using the carrier, *WO pat.*, WO2007034812, 2007.
77. K. Chiba, T. Higashiguchi, K. Abe and A. Tsuruyama, The reactor for liquid phase peptide synthesis and the liquid phase peptide synthesis method, *JP pat.*, JP2009114096, 2009.
78. K. Chiba, Y. Okada and M. Takahashi, Method for synthesis of cysteine-containing peptides with disulfide bond, *JP pat.*, JP2015027957, 2015.
79. H. Abe, H. Suzuki, S. Fujita, K. Nagano, T. Yamada, N. Ebata and K. Chiba, *Pept. Sci.*, 2012, **48**, 131.
80. D. Takahashi, Preparation of diphenylmethane compounds as protecting reagents and anchors for liquid-phase peptide synthesis, *US pat.*, US20100249374, 2010.
81. D. Takahashi, Preparation of 2-alkoxy-9-halo-9-phenylfluorene compounds as protecting reagents and anchors for liquid-phase peptide synthesis, *WO pat.*, WO2010104169, 2010.
82. D. Takahashi, Preparation of benzyl compounds for use as protecting reagents for amino acid and peptide synthesis, *WO pat.*, WO2011078295, 2011.
83. D. Takahashi, Method for producing peptide by anchor method, *WO pat.*, WO2012165545, 2012.
84. D. Takahashi, Method for producing peptides by anchor method using nonnucleophilic base for deprotecting Fmoc group, *WO pat.*, WO2012165546, 2012.

85. D. Takahashi, Method for removing Fmoc groups in the solution-phase synthesis of peptides, *WO pat.*, WO2013089241, 2013.
86. D. Takahashi, M. Bert, K. Kawajiri and R. Nakaya, Method for producing peptide continuously, *WO pat.*, WO2020218497, 2020.
87. D. Takahashi and T. Inomata, Cyclic peptide production method, *WO pat.*, WO2020262590, 2020.
88. D. Takahashi, T. Inomata and Y. Shinohara, Method for producing cyclized peptide having intramolecular S–S bond, *WO pat.*, WO2020101032, 2020.
89. D. Takahashi and T. Yamamoto, *Tetrahedron Lett.*, 2012, **53**, 1936.
90. Y. Hayashi, W. Fukasawa, T. Hirose, M. Iwatsuki, R. Hokari, A. Ishiyama, M. Kanaida, K. Nonaka, A. Také, K. Otoguro, S. Ōmura, K. Shiomi and T. Sunazuka, *Org. Lett.*, 2019, **21**, 2180.
91. D. Takahashi, T. Yano and T. Fukui, *Org. Lett.*, 2012, **14**, 4514.
92. K. Aihara, C. Komiya, A. Shigenaga, T. Inokuma, D. Takahashi and A. Otaka, *Org. Lett.*, 2015, **17**, 696.
93. T. A. Sanguliya, A. O. Antipova, S. V. Shkavrov and N. B. Épshtein, *Pharm. Chem. J.*, 2019, **53**, 462.
94. S. Kitada, M. Takahashi, Y. Yamaguchi, Y. Okada and K. Chiba, *Org. Lett.*, 2012, **14**, 5960.
95. X. Liu, N. Zhang, X. Gu, Y. Qin, D. Song, L. Zhang and S. Ma, *ACS Comb. Sci.*, 2020, **22**, 821.
96. N. Zhang, X. Gu, D. Song, P. Zhang, N. Zhang, L. Zhang and S. Ma, *Tetrahedron Lett.*, 2021, **79**, 1.
97. N. Zhang, X. Gu, D. Song, P. Zhang, N. Zhang, W. Chen, S. Ji, Y. Qi and S. Ma, *Bioorg. Chem.*, 2022, **119**, 1.
98. A. Mattellone, D. Corbisiero, P. Cantelmi, G. Martelli, C. Palladino, A. Tolomelli, W. Cabri and L. Ferrazzano, *Molecules*, 2023, **28**, 7183.
99. S. Nagahara, Y. Okada, Y. Kitano and K. Chiba, *Chem. Sci.*, 2021, **12**, 12911.
100. S. A. Kates, N. A. Solé, C. R. Johnson, D. Hudson, G. Barany and F. Albericio, *Tetrahedron Lett.*, 1993, **34**, 1549.
101. K. J. Jensen, J. Alsina, M. F. Songster, J. Vágner, F. Albericio and G. Barany, *J. Am. Chem. Soc.*, 1998, **120**, 5441.
102. K. Barlos, D. Gatos, J. Kallitsis, G. Papaphotiu, P. Sotiriu, Y. Wenqing and W. Schäfer, *Tetrahedron Lett.*, 1989, **30**, 3943.
103. K. Barlos, D. Gatos, S. Kapolos, G. Papaphotiu, W. Schäfer and Y. Wenqing, *Tetrahedron Lett.*, 1989, **30**, 3947.
104. A. Nagaya, S. Murase, Y. Mimori, K. Wakui, M. Yoshino, A. Matsuda, Y. Kobayashi, H. Kurasaki, D. R. Cary, K. Masuya, M. Handa and N. Nishizawa, *Org. Process Res. Dev.*, 2021, **25**, 2029.
105. S. Yano, T. Mori and H. Kubota, *Molecules*, 2021, **26**, 3497.
106. A. Wu and H. Yamamoto, *Chem. Sci.*, 2023, **14**, 5051.
107. A. Wu, I. Ramakrishna, T. Hattori and H. Yamamoto, *Org. Biomol. Chem.*, 2022, **20**, 8685.

CHAPTER 6

Sustainable Approaches to Chemo-enzymatic Peptide Synthesis (CEPS)

A. TOPLAK[a*], A. RICCI[b*] AND W. CABRI[b]

[a]Fresenius Kabi, via Roma 108, 20060 Cassina de Pecchi (MI), Italy; [b]P4i Laboratory Department of Chemistry "Giacomo Ciamician" Alma Mater Studiorum – University of Bologna, via P. Gobetti 85, 40129 Bologna, Italy
*E-mail: ana.toplak@fresenius-kabi.com, antonio.ricci@fresenius-kabi.com

6.1 Introduction to Peptide Therapeutics

A better understanding of biological systems has expanded the opportunities for exciting new therapeutic targets that can be modulated by types of molecules other than the traditional small-molecule drugs. These scientific advances have turned the pharmaceutical industry's focus towards larger molecules such as peptides and protein-based therapeutics. This is especially visible in the substantial peptide market size, exceeding US$ 40 billion in 2022, and representing 5% of the global pharma market with >60 marketed peptide drugs and generously supported by a pipeline that counts more than 150 peptides in clinical development and an estimated >200 molecules in various preclinical phases.[1–4]

Although therapeutic indications are versatile and mostly related to endocrinology, metabolic diseases, and oncology treatments, the market is strongly dominated by diabetic drugs, such as insulin and its analogs,

Green Chemistry Series No. 85
Sustainability in TIDES Chemistry: Green Approaches to Oligonucleotides and Oligopeptides Synthesis
Edited by Alessandra Tolomelli, Lucia Ferrazzano and Walter Cabri

Published by the Royal Society of Chemistry, www.rsc.org

followed by glucagon-like peptide 1 receptor agonists (GLP-1RA) for the treatment of diabetes and obesity (dulaglutide, liraglutide, and more recently semaglutide and the dual receptor agonist tirzepatide).[2,4] With the unprecedented results of clinical trials in treating obesity[5] coupled with cardiovascular benefits,[6] some of these molecules are on their way to becoming blockbusters, meaning that the increase in demand will also drive the need for their efficient and sustainable production.

In the early days, peptide therapeutics were limited to endogenous hormones, but now include a diverse range of molecules. In addition to predominant native peptides and their analogs, bioactive peptides are emerging that are unrelated to native sequences and are identified by synthetic library screenings or (phage) displays.[7,8] This notable expansion in variety is primarily driven by the medicinal chemistry approaches to minimize the shortcomings of peptide therapeutics and maximize their unique advantages.[9] It is crucial for all peptide therapeutics, regardless of their source, to be produced in a cost-effective, reliable, and environmentally sustainable manner.

6.2 Synthetic Methods for Peptide Therapeutics

The choice of production methodology largely depends on the peptide sequence, length and the properties of the peptide. Two commonly recognized methods are fermentation and synthetic methods.[10] Biocatalytic methods involving enzymes can be combined with both methodologies either to create the backbone peptide bond or to modify the side-chain residues. To date, there is no single best methodology for producing peptides. A combination of technologies, typically predetermined by the primary sequence, is often utilized to achieve the desired results.

6.2.1 Fermentation

It is attractive to employ the cell's natural ability to synthesize peptides owing to the high chemo- and regioselectivity, lack of racemization, mild process conditions, and low environmental impact.[11] Although the initial development costs are high, and developing such a process for each specific peptide takes time, fermentation is more cost-efficient and greener at a large scale. This could be attributed to the lower cost of fermentation starting materials and the possibility of using renewable sources. It is estimated that only 6% of the active pharmaceutical ingredient (API) cost is attributable to fermentation,[12] whereas for chemical routes, raw materials and solvents account for up to 60% of the cost.[10]

Ribosome-dependent and -independent syntheses are two methods for the production of peptides by fermentation. The ribosomal route, also called the recombinant method, is based on the transcription and translation of DNA coding sequences. The first fully recombinant peptide drug to reach the market was the 51 amino acid (AA)-long hormone insulin in 1982.[13] Since then, many analogs of insulin with different properties (fast-acting or long-lasting) have been developed and produced using genetic engineering[14,15] combined with some post-recombinant modification steps (chemical or enzymatic modification). Although many total synthetic routes[16] are now available for insulin and its analogs, they are still produced recombinantly. Further, such approach was successfully extended to semisynthetic liraglutide and semaglutide. However, the recombinant method has the major limitation that it cannot easily incorporate unnatural amino acids and modifications that are often necessary to enhance the drug-like properties of a peptide. For the introduction of a plethora of non-canonical amino acids (ncAAs), genetic code expansion technology was developed.[17] The current options allow the incorporation of two or three different ncAAs into the peptide sequence *in vivo*.[18] However, ncAA needs to be fed into the organism, and a better option would be to introduce pathways for ncAA production in the host itself. ncAAs can be further functionalized by classical reactions and newly developed bioorthogonal chemistries compatible with living systems.[19] With these tools, many new scaffolds and *in vivo* modulations of peptides will become feasible.

Another route to produce (cyclic) peptides in whole cells is to employ non-ribosomal peptide synthetases (NRPSs),[20] as in the case of vancomycin (antibiotic)[21] and cyclosporin A[22] (immunosuppressive drug), which are exclusively produced in this way. NRPS is often described as a specialized peptide assembly line consisting of modules, each performing a defined task based on covalent linkage and at the expense of ATP. The particular sequential order of these modules determines the primary sequence of the peptide. However, the current inability to easily tailor and redesign these large dedicated multimeric enzyme complexes with high turnover rates has limited their widespread use in peptide synthesis.

6.2.2 Synthetic Methods

Although some short peptides (up to 10 AAs) are made very efficiently using liquid-phase peptide synthesis (LPPS),[23] the most straightforward and widely used method to prepare peptides synthetically nowadays is solid-phase peptide synthesis (SPPS). The ease of use, modularity, and automation of steps have made this technology popular among peptide scientists since its introduction in 1963 by Bruce Merrifield. It is usually the method of choice to synthesize peptides on a laboratory scale and a widely chosen method to produce them on a large scale. Nowadays, fluorenylmethyloxycarbonyl (Fmoc)-based SPPS prevails over *tert*-butoxycarbonyl (Boc)-based SPPS owing to the milder reaction conditions and the more facile scalability.[24] Using SPPS, incorporation of multiple unnatural amino acids and modified amino acids [poly(ethylene glycol) (PEG), fatty acid, azido] and C-terminal (*e.g.* amidation), and N-terminal modification (*e.g.* acetylation) are achievable without drastic changes to the SPPS protocol.[25] Despite numerous improvements over the years (discussed in Chapter 4), the process mass intensity (PMI) of the SPPS still persists as one of its main challenges to be overcome.[26] Furthermore, the synthesis of long peptides shows a non-linear decrease in yield with increase in the peptide length, caused by the limited access to the terminal amino function and possible aggregation of the growing chain. With each step, impurities accumulate, leading to the need for tedious purification of the final long peptide. A way to overcome this challenge for longer peptides is to use SPPS for shorter (up to 20 AAs) peptide fragments produced in good yields and subsequently assemble/ligate these fragments into the final target product. In this way, the full potential of SPPS sequence space freedom can be harnessed, yields can be improved, and the manufacturing process can be derisked. Furthermore, the synthesis of smaller sequences opens up the possibility of exploiting recent approaches, such as molecular hiving,[27] which allow a reduction of the solvent volumes and the replacement of the environmentally detrimental dimethylformamide (DMF). Several chemical and enzymatic strategies have been developed for the subsequent assembly/ligation of short peptide fragments and are discussed in the following sections.

6.3 Chemical Methods for Peptide Ligation and Cyclization

6.3.1 Peptide Fragment Condensation (Hybrid Approach)

Peptide fragments synthesized through SPPS can be joined together in a solution through the classical solution peptide synthesis (CSPS) or through liquid-phase peptide synthesis (LPPS) that utilizes soluble tags or anchors. This coupling method is often referred to as a hybrid or fragment condensation approach, which reduces the use of excess reagents and solvents required to assemble the desired product.[28]

Enfuvirtide[29] is a 36-AA-long antiviral peptide that is used for the treatment of HIV infection. It is a linear peptide composed of all L-amino acids, with an acetylated N-terminus and an amide at its C-terminus. The synthesis of enfuvirtide by a hybrid approach from three SPPS-produced fragments at an unprecedented scale of 3 tonnes per year marked a significant advance in synthetic peptide production methodology and revolutionized the commercial synthetic peptide production landscape. Although as a drug enfuvirtide did not reach the status of a blockbuster, the synthesis of enfuvirtide is in many aspects exemplary. Its preparation represents the most complex synthesis of a pharmaceutical peptide to date, involving 106 synthetic steps and a 4-month stringent controlled process with solvent recycling loops, resulting in a 30% overall yield. Recently, a similar approach was applied to the synthesis of the 39-mer tirzepatide that contains two ncAAs and a lysine side-chain modification. A 46% overall yield and 70% crude purity were obtained when the continuous manufacturing approach and real-time analytics were bundled with nanofiltration for intermediate purification.[30] Still, this approach to peptide synthesis is hampered by the costs of SPPS and additional intrinsic limitations such as the (in)solubility of the protected peptide fragment and the availability of coupling positions in the sequence that minimize racemization. To overcome some of these issues, several chemoselective ligation methods have been developed, such as native chemical ligation (NCL), α-keto acid–hydroxylamine (KAHA) ligation,[31] traceless Staudinger ligation,[32] and serine–threonine ligation (STL).[33]

6.3.2 Native Chemical Ligation (NCL) of Peptides and Proteins

The most notable example of chemoselective ligation is NCL, which is performed in an aqueous solution with unprotected peptide fragments (as found in proteins). This has opened the door to semisynthesis and total chemical synthesis of several hundred peptides and proteins. Notably, a 352-residue-long polymerase was made by NCL in 2017.[34] The term 'native chemical ligation' was originally coined in the laboratory of Kent and co-workers in 1994,[35] describing the assembly of a 72-mer protein human interleukin IL-8 from two segments using a thioester bond-forming chemical reaction in aqueous solution at pH 7. NCL exploits the unique properties of peptide thioesters (thiophenol thioesters) that undergo selective intramolecular aminolysis in the presence of cysteine as a terminal amino function under mild conditions. The NCL reaction of coupling an activated acyl donor peptide in the form of a thioester with a nucleophile (a peptide with an N-terminal cysteine) proceeds in two steps (see Scheme 6.1).

The first step is a capture step, in this case a reversible transthioesterification or thiol–thioester exchange between an electrophilic (aryl) thioester at the C-terminus of the N-peptide and the nucleophilic thiol of a Cys residue located at the N-terminus of the C-peptide. The second step is rearrangement, a spontaneous S→N acyl shift that generates a native peptide bond at the ligation junction. All of these features allow chemoselectivity as only N-terminal Cys is involved in assembly while the remaining Cys in the fragment sequence remain unaffected. NCLs are typically carried out in an aqueous phosphate buffer at near-neutral pH ranging from 6.8 to 7.2 to avoid thioester hydrolysis at higher pH. A denaturing agent, guanidine hydrochloride (GdnHCl), is used to improve the solubility of fragments, and an exogenous aromatic thiol (*e.g.* thiophenol) is added to allow transesterification and facilitate the attack of the cysteine thiol. The efficiency of the reaction is influenced by various factors such as the steric properties of the reacting partners, the reactivity of exogenous thiol, and the type of thioester used.[33] For example, different amino acids such as C-terminal thioesters exhibited different reactivities, with Gly, Cys, and His reacting the fastest (2 h), whereas β-branched amino acids (Leu, Thr, Val, Ile) or Pro were not converted even

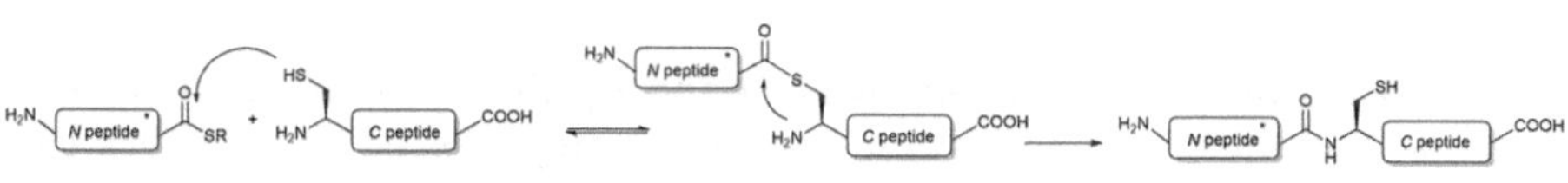

Scheme 6.1 Mechanism of NCL.

after 48 h.[36] The type of ester used is also an important factor affecting the ligation efficiency. Aryl derivatives tend to be more reactive than alkyl counterparts. Several exogenous thiols have been reported for use in NCL, with (4-carboxymethyl)thiophenol [4-mercaptophenylacetic acid (MPAA)] being the most common owing to its water solubility, ability to exchange with thioesters quickly, and effectiveness as a leaving group.

Traditionally made by Boc chemistry, the synthesis of peptide thioesters is not a facile, straightforward process, and many creative methods have been developed to make the synthesis compatible with Fmoc chemistry and NCL.[37] When assembling more than two peptide fragments in the C–N direction, it is essential to protect the N-terminal Cys of the fragment that carries a thioester group to prevent its cyclization. For the assembly of more than two sequences in the N–C direction, the use of surrogate thioesters prepared by Fmoc chemistry is particularly useful.[33] These latent thioesters can either undergo exchange to give an aryl thioester *in situ* during the NCL reaction (hydrazide method), or they can remain inert under the NCL conditions until they are activated [bis(2-sulfanylethyl)amino (SEA) linker]. Others, such as pivonylguanidine groups, can distinguish between different exogenous thiols. Recent NCL protocols have been summarized by Dawson and co-workers,[38] and a detailed overview of extended NCL methods, including extension to selenocysteine, can be found in a review by Melnyk and co-workers.[37]

The requirement for the cysteine in the coupling position is often regarded as the main limitation of NCL. To circumvent the need for a cysteine found only in 1.8% of natural proteins, several methodologies have been developed. Desulfurization of cysteine to alanine by selective desulfurization opened up the possibility of using non-natural amino acids, extending the ligation position to Phe, Leu, Val, Lys, Gln, Pro, and Thr. A milder, metal-free method for desulfurization was developed using tris(2-carboxyethyl)phosphine (TCEP), a water-soluble radical initiator, and a thiol. However, for the above-described methods, all other cysteines in the fragment need to be suitably protected.[33]

6.3.2.1 *Extension of NCL to Recombinant and Enzymatic Methods (Greener Fragments)*

A greener method to obtain very long peptides/protein thioesters is expressed protein ligation (EPL), where the protein of interest is recombinantly expressed as N-terminal fusion with mini-intein carrying an affinity tag. Those constructs can be purified from cell lysates by affinity chromatography and cleaved on the resin by a thiol to yield a C-terminal thioester.[39] Thiol-mediated intein cleavage is a prolonged process dependent on the nature of the C-terminal AA preceding the intein. Using split inteins can improve the scope of thiol cleavage somewhat. Nevertheless, the EPL method is better suited for proteins than peptides as the significantly lower molecular weight of peptides results in a smaller mass of peptides compared with the equimolar large fusion protein. In addition to the low yield limitation (ng–mg L^{-1}), incorporating non-canonical amino acids is challenging, as with all recombinant methods. Apart from direct cleavage of peptide/protein–intein fusion with thiols to yield C-terminal thioesters, ammonia and hydrazine cleavage can yield amides and hydrazides, respectively.[40,41] EPL-produced thioesters can be used as substrates for NCL or peptiligase/subtiligase chemo-enzymatic peptide synthesis (CEPS) in a process called expressed enzyme ligation (EEL),[42] where even cysteine-free peptides can be ligated to the corresponding thioester. C-terminal hydrazides can be functionalized *in situ* to thioesters and used sequentially in NCL/CEPS. A hydrazide tag can be added *via* a peptide fragment by EPL or NCL, or enzymatically using sortase A (scar present) or traceless by peptide amidase.[43] The C-peptide/protein containing N-terminal cysteine can also be produced recombinantly, typically as a fusion construct incorporating a specific proteolytic site.

NCL has evolved significantly since its introduction, with many extended and refined methodologies emerging to minimize its limitations. However, despite these developments, NCL is not widely used on an industrial scale, primarily because of the challenges associated with the production of thioesters at scale (synthetic, recombinant) and their inherent chemical reactivity. Furthermore, the low substrate concentration and lengthy reaction times required for NCL lead to thioester hydrolysis and unwanted epimerization, making NCL unsuitable for industrial-scale use.

6.4 Enzymatic Methods for Peptide Ligation and Cyclization

The use of biocatalysts in the field of peptides/macromolecules is still underexplored. Nevertheless, enzymatic ligation has gained popularity owing to its mild conditions and excellent regio- and chemoselectivity.[44,45] This section is focused exclusively on enzymes forming native peptide bonds for linear peptide synthesis and macrocyclization. This set of enzymes includes engineered proteases (*e.g.* subtilisin/trypsin type), transpeptidases (*e.g.* sortases, asparaginyl endopeptidases) and macrocylases. To date, only peptiligases have been utilized for the industrial manufacture of peptides. A recently published protocol collection should remove the virtual barrier to enzymatic technologies and further stimulate their use.[46]

6.4.1 Engineered Proteases

Application of proteases in the reverse direction is a logical approach for peptide synthesis.[47] Since the 1980s this field has been stimulated by the applications of enzymes in neat organic solvents to promote the coupling reaction and eliminate hydrolysis by water. Subtilisin, a serine protease, and a few thermostable subtilisin-like enzymes have been used for peptide fragment ligations in water-free systems.[48,49] Here, in the kinetic approach, activated acyl donor [peptide (thio)ester] forms an acyl–enzyme intermediate that can be attacked by a nucleophile (acyl acceptor) to create the final product in high yield. However, the low solubility of long protected peptide fragments in neat organic solvents, poor enzyme activity, and the need to strictly control the water activity to prevent the hydrolysis of ester and product rendered this approach unsuitable for large-scale applications.

A more promising approach was ligation in a water system by an enzyme, a thiol-subtilisin originally generated by the chemical conversion of active-site serine into cysteine BPN′ protease from *Bacillus amyloliquefaciens*. The mechanism involves the formation of an acyl–enzyme thioester, which prefers intermolecular aminolysis to hydrolysis.[50] Seleno-subtilisin is another variant with improved aminolysis over hydrolysis.[51] These variants had poor activity, and only with the introduction of recombinant methods was a better-performing thiol-subtilisin (subtiligase) obtained. Wells and co-workers changed the active site S221C and a nearby residue P225A to reduce the steric crowding.[52] These mutations improved the ligation activity, but subtiligase efficiency was limited owing to the high excess (10×) of the nucleophile fragment required to achieve yields and minimize the unwanted hydrolysis. Stabiligase, a (thermo)stable variant, was prepared[53] and mutant libraries have been constructed to modify subtiligase specificity. Synthetic, phage display, or proteome-based substrates were used in the screenings.[53–55] Results of PILS (proteomic identification of ligation sites) identified Y217 as the primary determinant of P1′ specificity and F189 for P2′ specificity. Subtiligase and its variants have been used in a wide range of applications, such as protein semisynthesis (ribonuclease A), peptide cyclization, synthesis of thioesters/acids, and N-terminal labeling of peptides/proteins.[56] The main limitations of subtiligase are still poor reagent efficiency, calcium dependence, and the requirement for protease-assisted folding during its production in the *Bacillus* host.

A novel thiol-subtilisin termed peptiligase was developed by EnzyPep BV to overcome some of these limitations. Peptiligase, a stable and calcium-independent thiol-subtilisin, was created by incorporating active-site mutations (S221C, P225A) into stabilized, Ca^{2+}-independent subtilisin BS149.[57] A peptide ester fragment is preferably carboxamide methyl (OCam), which readily forms a thioester complex with the active site of the enzyme and can be efficiently coupled to an amine nucleophile. Hydrolysis by water nucleophile is minimal, leading to a high synthesis/hydrolysis ratio (S/H) (see Figure 6.1). Peptiligase has improved efficiency, requiring a stoichiometric amount or a slightly excess of nucleophile to drive reactions to completion with minimal or no hydrolysis. Furthermore, the equivalency ratio can be tuned according to the cost and complexity of the fragments that have to be coupled, choosing the most critical one as the rate-limiting reagent. Ligations are usually performed with unprotected peptide fragments in buffered solutions (pH 7.5–8.5) at room temperature for a short time. For ligations with poorly soluble fragments, reactions with peptiligase can be performed at higher temperatures (peptiligase T_m = 66 °C) and in the presence of cosolvents such as DMF and dimethyl sulfoxide (DMSO) (up to 50% v/v) and denaturants (2 M urea or GdnHCl). Just like BPN′ protease, peptiligase has six recognition pockets:

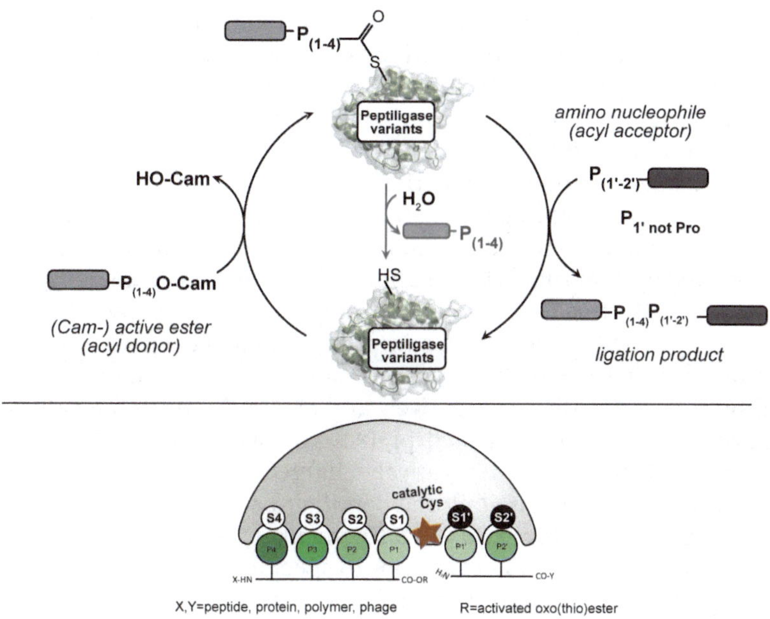

Peptiligase variants	Substrate scope
Peptiligase	**narrow P1'(A,S,G)**;medium P3-P1/P2'; P4 hydrophobic
Thymoligase	**specific P1 (K,R)/P1'(D,E)**;P4 hydrophobic
Omniligase-1	**broad P3/P1'/P2'**;medium P2/P1; P4 hydrophobic

Figure 6.1 Mechanism of peptiligase-mediated CEPS. Reproduced from ref. 44, https://doi.org/10.3389/fchem.2019.00829, under the terms of the CC BY 4.0 license, https://creativecommons.org/licenses/by/4.0/.

four on the acyl donor side (S1–S4) and two that recognize the acyl acceptor (nucleophile) (S1′ and S2′). It was found that the remaining limitation of peptiligase was its narrow acyl acceptor scope. Only small amino acids, Gly, Ala, and Ser, were efficiently ligated, thus limiting its practical applications. Using structure-guided site-evaluation libraries at the S1′ (M222, L217, P225) and S2′ pocket (F189), the scope of peptiligase could be drastically broadened and modulated.[58] On the other hand, the acyl donor scope of peptiligase is similar to that of the parent protease BS149. It is relatively broad, with a slight preference for large hydrophobic and aromatic residues in S4 and hydrophobic residues in S1. Ligation rate was further improved by S4 pocket mutation (I107). Engineering efforts to design an efficient and broad substrate-scope peptiligase led to the discovery of pentamutant omniligase-1.[59]

Applications of this broad enzyme range from efficient peptide to peptide ligation, peptide to polymer, and peptide to protein (human serum albumin). Omniligase-1 can also efficiently cyclize peptides longer than 12 AA in a head-to-tail manner, tolerating the presence of isopeptide bonds, D-AA, or

PEG moieties. Cyclotides (McoTI–II) and disulfide-rich peptides (kalata B1) could also be produced and folded on the gram scale using omniligase-1.[60,61] Recently, the cyclization by omniligase-1 was combined with several other technologies such as oxime ligation, click chemistry, and/or rigidification using tris(bromomethyl)benzene to access tri- and tetracyclic bioactive and bispecific peptides in a one-pot reaction.[62]

However, if a broad peptiligase is applied for fragment ligations, the N-terminus of the acyl donor should be protected by a suitable protecting group to avoid unwanted cyclizations. A better approach is to use a selective enzyme that solely recognizes the free N-terminus of the acyl acceptor (C peptide) as a nucleophile. For this reason, engineering efforts were applied to create a set of selective peptiligases that could distinguish between small and large side-chains and hydrophobic and polar or negatively and positively charged residues. One such enzyme is thymoligase, which prefers positively charged Lys/Arg in P1 (mutations S166D/S156N) and negatively charged Asp/Glu in P1′ (mutations M222G/L217R). This substrate-tailored peptiligase was used to synthesize thymosin α_1 drug (28-mer) from two SPPS fragments (14-mers) in high yield and purity.[63] Selective enzymes open the door to specific labeling of heterodimeric proteins (*e.g.* antibodies, insulin)[64] and, in the peptide sequential ligation reactions, eliminate the need for N-terminal protection of the acyl donor.

As a general approach to peptiligase CEPS, a set of hydrolase versions of peptiligases (where the catalytic C221 in the active site is back-mutated to serine) are used to identify potential coupling site(s) on a selected peptide API. Then, the best coupling site is chosen taking into consideration the enzyme efficiency and the fragment manufacturing complexity.[65] Following this, iterative peptiligase tailoring cycles are performed to increase the efficiency and selectivity for the desired coupling site, until the enzyme is finally fine-tuned to meet the desired reaction process conditions. In this way, the technology was further developed in the Fresenius Kabi R&D laboratories, producing new mutants for the synthesis of several therapeutic peptides, namely dasiglucagon, glepaglutide, elsiglutide, teriparatide, salmon calcitonin, bivalirudin, and the blockbusters liraglutide and semaglutide.[66]

Another advantage of peptiligases is the possibility of using the *Bacillus* host in the fermentation process, which allows its extracellular production at a concentration of grams per liter, simplifying the downstream process to purify and recover the desired enzyme. A potential drawback of all thiol-subtilisins is the need for oxo- or thioesters as acyl donors, which can be easily obtained for peptides but, for large macromolecules, could represent a hurdle.

6.4.1.1 *Biocatalytic Manufacture by Peptiligase CEPS*

Omniligase-1 and selective peptiligase have evolved as highly efficient biocatalysts for chemo-enzymatic peptide synthesis, as demonstrated on a gram scale. In the next step, omniligase-1 was evaluated as an industrial biocatalyst in the synthesis of the antidiabetic drug GLP1-RA exenatide. Exenatide is a 39-mer peptide amide containing only natural AAs. The CEPS strategy was designed to ligate purified fragments H-1–21-OCamLeuNH$_2$ and H-22–39-NH$_2$ using omniligase-1. The scale-up from the laboratory protocol encountered several difficulties; in particular, ester fragment synthesis required diketopiperazine control, and eventually multi-kilogram scale OCamLeu ester was replaced with the more stable hydroxymethylbenzoic acid (HMBA)–Lys ester. In addition to purified fragments, more economical crude SPPS fragments were also tested. A change of ester type required an increase in the amount of biocatalyst. Optimal ligation conditions were found to be a crude ester to crude amine equivalents ratio of 1 : 1.2, 0.2 M potassium phosphate (KPi) buffer (pH 8.0) supplemented with 10% v/v acetonitrile, and omniligase-1 and TCEP solution at final pH 7.5 and 26 °C for 4 h. The yield was 88% and the S : H ratio was 98 : 2. After purification, the API yield was 60%, and the overall yield was 35% compared with the original laboratory-scale process yield of 44% (see Figure 6.2).[67]

Omniligase-1 can accommodate process-required changes of ester type and use of crude SPPS substrates in the presence of a cosolvent, thus demonstrating that it is an efficient catalyst with a high turnover number for CEPS on a large scale. The quality of the API was demonstrated and could even be improved further by opting for pure instead of crude SPPS fragment ligation. For successful CEPS manufacture, substrate engineering, biocatalyst engineering, and reaction engineering are required.

The authors compared full SPPS and final CEPS processes for exenatide synthesis and found that compared with full SPPS, the CEPS yield was 60% higher, total costs were 3.5 times lower, the E-factor was seven times lower, and a 10-fold reduction in CO_2 kg^{-1} API was achieved (see Table 6.1).

The full potential of the CEPS can be exploited exponentially by combining the use of alternative, more sustainable and green approaches to manufacture the peptide fragment. Apart from the molecular hiving approach mentioned in the previous chapter, alternative solvents can be applied to the standard SPPS protocol.

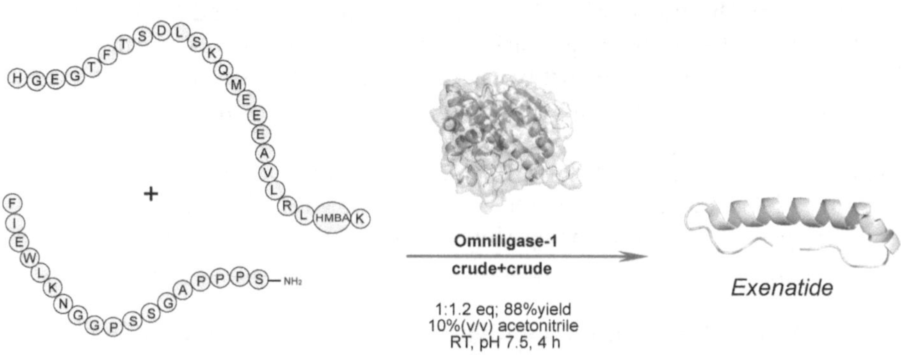

Figure 6.2 CEPS approach for exenatide synthesis. Reproduced from ref. 67 with permission from the Royal Society of Chemistry.

Table 6.1 Comparison of process characteristics of full SPPS and CEPS processes for exenatide API using omniligase-1. Reproduced from ref. 67 with permission from the Royal Society of Chemistry.

Process characteristic	Full SPPS	Optimized CEPS
Overall yield/%	22	35
Total cost/k€ per kg^{-1} API	147.9	41.8
Calculated E-factor	33 613	4793
Carbon intensity	117.7	12.1

6.4.2 Transpeptidases

6.4.2.1 Sortases (Srts)

Naturally occurring sortase A (SrtA) was identified in the pathogenic bacterium *Staphylococcus aureus*, which catalyzes covalent anchoring of the host cell surface proteins to its cell wall envelope.[68] These protease-catalyzed transpeptidations are a common feature in many Gram-positive bacteria. Owing to their virulence, various sortases have been identified and targeted for treating infectious diseases. The mechanism of SrtA involves recognition of a peptide motif LPXTG (where X = any amino acid, usually E) and amide bond cleavage between threonine and glycine that proceeds *via* an acyl–enzyme (thioester) intermediate followed by coupling to the pentaglycine nucleophile of a cell wall envelope. In chemical biology applications, the nucleophile can be diglycine carrying various payloads (proteins, peptides, conjugation handles/reactive groups, toxins, lipids, sugars, nucleotides, solid supports, and cells) (see Figure 6.3). In addition to peptide/protein conjugation, SrtA can mediate head-to-tail cyclizations. Consequently, the term *sortagging*[69] was coined due to the widespread use of sortase A in biotechnology.

SrtA is a robust enzyme, readily produced recombinantly in *Escherichia coli* host (*ca.* 40 mg L^{-1}) and is commercially available (see Table 6.2). The crystal structure and the mechanism have been elucidated. The addition of Ca^{2+} is required for activity, although variants exist that do not rely on the calcium ion for the activity. Native sortase A exhibits low catalytic activity; in particular, the K_m of the acyl donor is very high, resulting from its biological role where it operates at high local concentrations of both substrates bound to the bacterial cell wall. Although poor kinetics can be circumvented through proximity-based sortase-mediated ligation (PBSML), where the SpyTag–SpyCatcher system is used to bring the reactants in proximity to permit >95% ligation efficiencies, a better approach is to improve the activity by enzyme engineering.[70]

Active SrtA-5M (calcium-dependent) and SrtA-7M (calcium-independent) variants were identified by yeast display/fluorescence-activated cell sorting (FACS)[71] in combination with rational design[72] and have become the starting point for any further engineering efforts. Two notable variants, namely

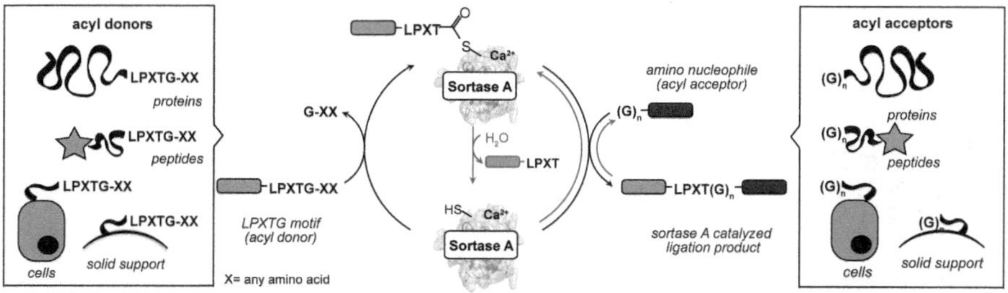

Figure 6.3 Mechanism of sortase A-mediated CEPS and possible applications. Reproduced from ref. 46 with permission from Elsevier, Copyright 2017.

SrtA-5M + Y187E/E189R and SrtA-5M + D124G,[73] have been developed for labeling antibodies. It is essential to understand that transpeptidation reactions are reversible, and the products of ligation can act as substrates. In the laboratory, the most common approach to prevent the reverse reaction is the addition of diglycine nucleophile in large amounts to drive the reaction in the desired direction. However, this can be costly and impractical on a large scale. Other engineering approaches include flow setup[74] and substrate engineering. Substrates may carry activated bonds placed in the substrate tag, self-inactivating leaving groups (both chemically and sterically), or a hexahistidine tag that can be removed in a complex with Ni^{2+}.[75,76] In summary, poor sortase-mediated ligation (SML) kinetics can be improved to some extent, and SML ligation can be used as an adjunct to recombinant methods, SPPS methods, and NCL (hydrazide tag addition). Careful selection of sortase variants for specific applications may minimize unwanted side reactions (hydrolysis of acyl–thioester intermediate, reversal of reaction).[77] However, the main limitation is the requirement for a recognition motif that must be introduced in the acyl donor and, therefore, present in the final product. Engineering studies in which large mutant libraries were screened in a high-throughput fashion have identified variants with altered specificity. However, it is often observed that an increase in selectivity is accompanied by a decrease in activity. Therefore, starting engineering campaigns with a variant that already has improved kinetic properties is advantageous.[78] Such specific sortases can be used in orthogonal iterative reactions to create a wide range of modified peptides/proteins.[79] Enzyme stability is just as important as its specificity and activity. Enzymes resistant to unfolding or denaturation caused by various factors such as temperature, pH, or cosolvent are more suitable for industrial applications. These enzymes are easier to handle and store and have a prolonged shelf-life. Several stability-improved variants of SrtA have been developed, mainly by rational design or directed evolution. Some of those SrtA variants do not rely on calcium for the activity[68] and can tolerate up to 45% DMSO,[80] and others can tolerate denaturants[69] or elevated temperatures.[81]

Table 6.2 Overview of sortase A variants.

Selected sortase A sources	Name/mutations/PDB	Recognition motif	Nucleophile
Staphylococcus aureus	Sa-SrtA, SrtA (PDB: 1T2P, 1T2W)	LPXTG (X = any AA), LPETG	H-GlyGly
Streptococcus pyogenes[a]	Sp-SrtA (PDB: 3FN6)	LPXLG	H-AlaAla, H-AlaGly, H-GlyGly, H-SerGly
Corynebacterium diphtheriae[a]	Cd-SrtA (PDB: 5K9A)	LPLTG	Internal Lys
Lactobacillus plantarum	Lp-SrtA[102]	PQTSEQ	Primary amines
Variants of Sa-SrtA with altered specificity			
Selective variant with reduced activity	SrtA-F40;[103] SrtA T164Q/V168M/L169H/D170L/E171A/Q172E	XPKTG (X = A, D, S), APATG	H-GlyGly
Selective variant with reduced activity	SrtA-F1-21;[104] SrtA V161Y/K162W/P163A/T164N/D165E/V166R/G167I/V168F/L169H/D170V/E171L	APXTG/FPXTG (X = any AA)	H-GlyGly
Selective variant with enhanced activity, selectivity orthogonal to SrtA and SrtA-5M	eSrtA(4S-9);[78] SrtA-5M + N98D/S102C/A104V/A118T/F122A/K134R/F144L/I182V/E189F	LPEXG (X = A, C, S)	H-GlyGly
Selective variant with enhanced activity, selectivity orthogonal to SrtA and SrtA-5M	eSrtA(2A-9)[78] SrtA-5M + 102C/A104H/E105D/K138P/K152I/N160K/K162H/T164N/K173E/I182V/T196S	LXETG (X = A, S, R)	H-GlyGly
Evolved SrtA that recognizes amyloid-β protein	SrtAβ;[105] SrtA-4S.6 variant + I76L/S102C/E105D/N107E/S118I/I123L/D124L/N127H/G134R/K138L/G139D/M141I/K145T/K152R/M155I/R159C/K162R/Q172H/K73E/K177R/V182A/V189Y/T196S/R197S/K206R	LMVGG	H-GlyGly
SrtA-F40 variant that acts on histones	SrtA-W4;[106] SrtA-F40 + E105K/E108Q/D124G/Y187L/E189R/K190E/K138M/D160N	HPDTG	H-GlyGly

Variants of Sa-SrtA with enhanced activity			
120× higher catalytic efficiency than SrtA	SrtA-5M;[71] SrtA + P94R/D160N/D165A/K196T	LPXTG	H-GlyGly
Efficient for ligations on antibodies	SrtA-5M + Y187E/E189R; SrtA-5M + D124G; SrtA-5M + D124G/Y187L/E189R[73]	LPXTG	H-GlyGly
Active and Ca^{2+}-independent SrtA, high hydrolytic side reaction	SrtA-7M;[67] SrtA-P94R/E105K/E108A/D160N/D165A/K190E/K196T	LPXTG	H-GlyGly
Variants of Sa-SrtA with improved stability			
Ca^{2+}-free variant, with lowered activity	SrtA E105K/E108A[68]	LPXTG	H-GlyGly
DMSO resistant, enhanced activity, but low thermal stability	SaSrtA-4M; Sa-SrtA + P94S/D160N/D165A/K196T[80]	LPXTG	H-GlyGly
Variant with increased thermostability/activity of 4M	SrtA-6M; SrtA-4M + R159N/K162P[81]	LPXTG	H-GlyGly
Thermostable and active variant with moderate tolerance to cosolvents and denaturants	CyM6 cyclic SrtA-6M[81]	LPXTG	H-GlyGly
Temperature- and denaturant-tolerant SrtA	Bicyclic SrtxS11;[69] Srt5M/D124G/Y187L/E189R variant	LPXTG	H-GlyGly

[a]Calcium independent.

6.4.2.2 Asparaginyl Endopeptidases (AEPs)

Members of this cysteine protease family include recently discovered and synthetically useful plant enzymes termed Aspx-ligases or peptidylasparaginyl ligases (PALs) that include representative butelase-1 (PDB: 6DHI) from *Clitoria ternatea*[82] and others such as OaAEP1 (PDB: 5HOI) from *Oldenlandia affinis*,[83] VyPAL2 (PDB: 7F5Q) from *Viola yedoensis*,[84] and engineered VcAEP (PDB: 5ZBI)[85] from *Viola canadensis*. Compared with SrtA, PALs are more efficient transpeptidases and require 100–1000-fold less enzyme than SrtA-catalyzed reactions.[45] In addition, they recognize a smaller sorting motif, namely N/D-HV, for prototypic butelase-1. The mechanism is similar to that for SrtA; butelase-1 cleaves after the Asn/Asp residue of the recognition motif and forms an acyl–enzyme thioester that is subsequently attacked by various nucleophiles. At neutral pH, butelase-1 can accept nucleophiles with an N-terminal NH_2–X1X2 sequence (X1 = not proline, X2 = Leu, Ile, Cys, or Val), resulting in a minimal footprint in the final peptide product (see Figure 6.4). Since cyclization reactions are preferred with butelase-1, for linear ligations the N-peptide should carry the C-terminal recognition tag (NHV), but its N-terminus should be masked for butelase-1 recognition (no Ile, Val, Leu/Cys at the P2 position). Reaction of such an N-peptide with C-peptide starting with GI will produce a tag NGI not recognized by butelase-1 in the final product, thus eliminating reverse transpeptidation reactions. Similar selectivity can be obtained with the OaAEP1 variant that recognizes NGL but not the NGV motif.[86]

This tag can be easily incorporated by recombinant or SPPS techniques. In a transpeptidation reaction, butelase-1 can further functionalize peptides/proteins as hydrazides (precursors of thioesters for NCL) or as glycine thioesters. The comprehensive applications range from cyclizations of various (non-)peptidic molecules that carry a butelase tag to creating dendritic structures and bioconjugate antibodies.[87–89] Difficulties in recombinant production of butelase-1 have turned the research focus to OaAEP1,

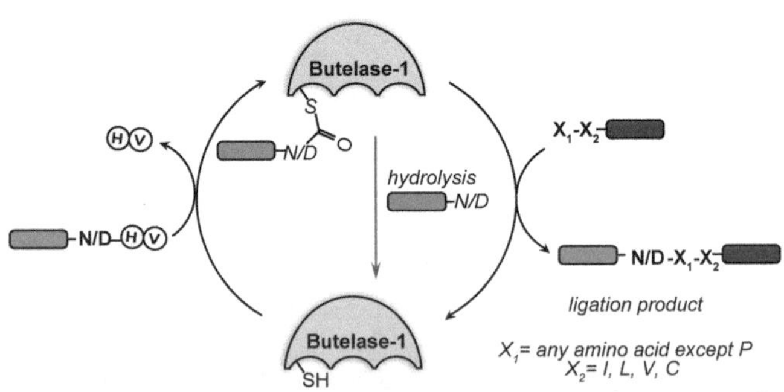

Figure 6.4 Mechanism of butelase-1-mediated CEPS. Reproduced from ref. 44, https://doi.org/10.3389/fchem.2019.00829, under the terms of the CC BY 4.0 license, https://creativecommons.org/licenses/by/4.0/.

which could be produced recombinantly in *E. coli*, albeit with low yields (2 mg L^{-1}). Recently, VyPAL2 and VcAEP, identified by genome mining, were characterized and recombinantly produced. Interestingly, to obtain an active enzyme form, they require N-terminal solubility tags followed by an autoactivation step at low pH that mimics maturation in plants. The ability to produce PALs recombinantly permits further improvements. In addition, understanding which residues are critical for ligase activity [ligase activity-determining (LAD) residues] may improve the efficiency and allow the re-engineering of some proteolytic AEPs into efficient transpeptidase. Residues found in S2 (gate-keeper position, or LAD1) and S1′ substrate-binding pockets (LAD2) have been reported as ligase activity modulators. This hypothesis is supported by variant OaAEP1 C247A (LAD1 region), resulting in increased activity,[90] and by dual-functional AEP variant VcAEP–V238A/Y168A that exhibits >90% cyclization compared with 5% activity of the unmodified variant.[84] For sequential ligation reactions or dual-labeling reactions, orthogonal enzymes with complementary specificities (*e.g.* butelase-1 and VyPAL2; NHV and NGL) can be used, or a pH-controlled reaction that can drive specificity (DSL at pH 4.5 *versus* NGL at neutral pH).[91] Driven by their scarcity, butelase-1 and VyPAL2 were successfully immobilized and reused for up to 100 cycles. Immobilized versions increased the effective enzyme concentrations and promoted reactions in a one-pot or a continuous-flow reactor system.[92] With recent success in obtaining fusion butelase-1 recombinantly in *E. coli* with a titer of 100 mg L^{-1},[93] the door to its industrial applications is wide open. In addition to advances in enzymatic applications of PALs *in vitro*, they can also be used successfully *in planta* to produce therapeutic cyclotides such as cyclized [T20K]kalata B1.[94]

6.4.2.3 Trypsiligase

An engineered trypsin variant with improved ligation activity reported by Bordusa and co-workers, termed trypsiligase (trypsin + K60E/N143H/Y151H/D189K; PDB: 4NIV),[95] is an enzyme with applications in N- and C-terminal labeling of peptides/proteins. The enzyme is in a disordered zymogen-like form and is only activated by the presence of a substrate that contains the recognition Y-RH motif and Zn^{2+} ions. Efficient labeling of the N-terminus was reported for nucleophiles having RH in P1′P2′ and activated acyl donors in the form of Ogp esters (peptidyl-4-guanidinophenyl esters) similar to endopeptidases. C-terminal modifications occur *via* a transpeptidation reaction similar to that for sortases.[96] Trypsiligase recognizes the Y-RH motif and cleaves after tyrosine to form an acyl–enzyme complex that can be attacked by an RH-X nucleophile (X = peptide, tag).[97] As the product carries a C-terminal YRH scar, the reaction is reversible, leading to hydrolysis (see Figure 6.5). Trypsiligase recognizes the very rare motif YRH found in only 0.5% of all sequences, which limits its traceless application in protein and peptide synthesis.

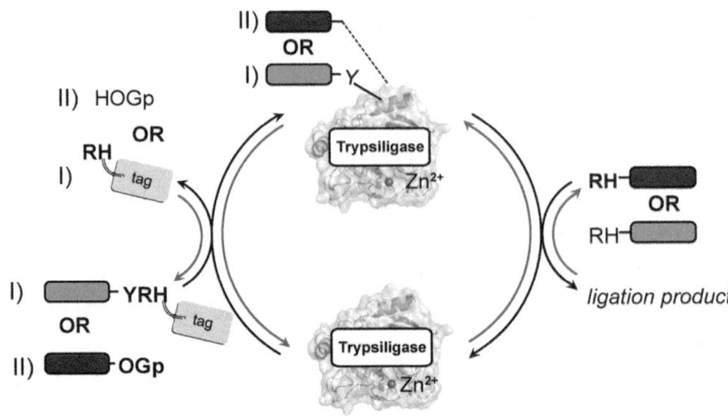

Figure 6.5 Mechanism of trypsiligase-mediated CEPS. Reproduced from ref. 44, https://doi.org/10.3389/fchem.2019.00829, under the terms of the CC BY 4.0 license, https://creativecommons.org/licenses/by/4.0/.

6.4.3 Macrocyclases

These enzymes are part of the biosynthetic pathways of bioactive cyclic compounds and have only recently been characterized as stand-alone biocatalysts, highlighting the importance of harnessing the natural diversity of enzymes. Prokaryotic PatG (PDB: 4AKT) recognizes XP-AYDGE (X = peptide) with obligatory P1 proline or ncAA that can form a *cis* peptide bond between P1 and P2. In the second step, an acyl–enzyme oxo ester is cyclized by nucleophilic attack of the free N-terminus of the bound peptide X. This enzyme contains a crucial helix–loop–helix insertion that shields the acyl–enzyme complex from water nucleophile and promotes cyclization.[98,99] In addition to PatG, the TE domain from NRPS tyrocidine synthetase TycTE can efficiently catalyze the cyclization of decapeptide thioesters to form various bioactive cyclic peptides.[100,101] These enzymes (and improved versions) can potentially become valuable tools for peptide chemists in the future.

In the growing market of protein therapeutics, we envision that the application of engineered proteases (*e.g.* omniligase-1, selective peptiligase, or selective subtiligases) could be extended from industrial peptide synthesis to efficient N-terminal labeling of large macromolecules such as antibody–drug conjugates.[76] In addition, enzymes such as efficient/engineered transpeptidases are of particular use for the C-terminal labeling of large macromolecules, where the introduction of the activated group (*e.g.* ester) is not straightforward.

6.5 Conclusion

Owing to their unique position between small-molecule drugs and biologics, peptide therapeutics are constantly evolving. Whether it is a change of therapeutic source (natural sources, synthetic, recombinant) or diversity introduced by the drug development process (native, analogs, heterologs, and conjugates), the peptide field is constantly driven by innovations across various disciplines. Simultaneously, sustainable synthetic approaches are emerging and will eventually drive the field towards highly designed biological systems. Currently, enzyme-driven processes are a significant step forward towards green peptide production processes. Alongside enzyme specificity (yielding site-specific and well-defined products), efficiency (time and reagent efficiency), robustness (thermostability and cosolvent stability), the biocatalyst availability also plays a role in selecting and applying a biocatalyst for the manufacture of peptides. In this scenario, peptiligases play a primary role owing to their availability and their versatility in engineering with respect to the desired peptide sequence.

Immobilization and recycling of the biocatalyst in the flow reactor setup will further widen the industrial application possibilities. Many of these properties can be evolved by protein engineering methods, starting either from naturally occurring enzymes or from scratch (*de novo* method). A toolbox of enzymes with orthogonal specificities or on–off switches is highly desirable for synthetic assemblies.

Until (microbial) cell factories become available for the manufacture of peptide/protein therapeutics, enzymatic methods based on improved efficient variants can serve as a powerful and sustainable alternative approach to the existing chemical and fermentation methodologies.

List of Abbreviations

AA	Amino acid
AEP	Asparaginyl endopeptidase
API	Active pharmaceutical ingredient
Boc	*tert*-Butoxycarbonyl
CEPS	Chemo-enzymatic peptide synthesis
CSPS	Classical solution peptide synthesis
DMF	Dimethylformamide
DMSO	Dimethyl sulfoxide
EEL	Expressed enzyme ligation
EPL	Expressed protein ligation
FACS	Fluorescence-activated cell sorting
Fmoc	Fluorenylmethyloxycarbonyl
GdnHCl	Guanidine hydrochloride
GLP-1RA	Glucagon-like peptide 1 receptor agonist
HMBA	Hydroxymethylbenzoic acid
KAHA	α-Keto acid–hydroxylamine
LAD	Ligase activity-determining
LPPS	Liquid-phase peptide synthesis
MPAA	4-Mercaptophenylacetic acid
ncAA	Non-canonical amino acid
NCL	Native chemical ligation

NRPS	Non-ribosomal peptide synthetases
PAL	Peptidylasparaginyl ligase
PBSML	Proximity-based sortase-mediated ligation
PEG	Poly(ethylene glycol)
PILS	Proteomic identification of ligation sites
PMI	Process mass intensity
SEA	Bis(2-sulfanylethyl)amino
SML	Sortase-mediated ligation
SPPS	Solid-phase peptide synthesis
SrtA	Sortase A
STL	Serine–threonine ligation
TCEP	Tris(2-carboxyethyl)phosphine reducing agent

Acknowledgements

The authors would like to acknowledge the whole team of EnzyPep BV and Fresenius Kabi for their continuous efforts in improving the peptiligase chemo-enzymatic peptide synthesis approach.

References

1. G. Rossino, E. Marchese, G. Galli, F. Verde, M. Finizio, M. Serra, P. Linciano and S. Colina, *Molecules*, 2023, **23**, 7165.
2. M. Muttenthaler, G. F. King, D. J. Adams and P. F. Alewood, *Nat. Rev. Drug Discovery*, 2021, **20**, 309.
3. A. Henninot, J. C. Collins and J. M. Nuss, *J. Med. Chem.*, 2018, **61**, 1382.
4. L. Ferrazzano, M. Catani, A. Cavazzini, G. Martelli, D. Corbisiero, P. Cantelmi, T. Fantoni, A. Mattellone, C. De Luca, S. Felletti, W. Cabri and A. Tolomelli, *Green Chem.*, 2022, **24**, 975.
5. J. P. H. Wilding, R. L. Batterham, S. Calanna, M. Davies, L. F. Van Gaal, I. Lingvay, B. M. McGowan, J. Rosenstock, M. T. D. Tran, T. A. Wadden, S. Wharton, K. Yokote, N. Zeuthen and R. F. Kushner, *N. Engl. J. Med.*, 2021, **384**, 989.
6. I. Fernández-Ruiz, *Nat. Rev. Cardiol.*, 2013, **20**, 719.
7. C. Heinis, T. Rutherford, S. Freund and G. Winter, *Nat. Chem. Biol.*, 2009, **5**, 502.
8. Y. Huang, M. M. Wiedmann and H. Suga, *Chem. Rev.*, 2019, **119**, 10360.
9. L. Di, *AAPS J.*, 2015, **17**, 134.
10. J. H. Rasmussen, *Bioorg. Med. Chem.*, 2018, **26**, 2914.
11. H. P. Meyer and O. Werbitzky, in *Biocatalysis for Green Chemistry and Chemical Process Development*, John Wiley & Sons, London, 2011, ch. 2.
12. A. Johnson, *CRS Report No. IF11026*, 2018.
13. I. S. Johnson, *Science*, 1982, **219**, 632.
14. A. N. Zaykov, J. P. Mayer and R. D. Dimarchi, *Nat. Rev. Drug Discovery*, 2016, **15**, 425.

15. G. Walsh, *Appl. Microbiol. Biotechnol.*, 2005, **67**, 151.
16. J. A. Karas, J. D. Wade and M. A. Hossain, *Chem. Rev.*, 2021, **121**, 4531.
17. Y. Ryu and P. G. Schultz, *Nat. Methods*, 2006, **3**, 263.
18. J. S. Italia, P. S. Addy, S. B. Erickson and J. C. Peeler, *J. Am. Chem. Soc.*, 2019, **141**, 6204.
19. C. S. Diercks, D. A. Dik and P. G. Schultz, *Chemistry*, 2021, **7**, 2883.
20. R. Finking and M. A. Marahiel, *Annu. Rev. Microbiol.*, 2004, **58**, 453.
21. V. Le and W. Siriwatwechakul, *Bioresour. Technol. Rep.*, 2021, **15**, 100805.
22. S. A. Survase, L. D. Kagliwal, U. S. Annapure and R. S. Singhal, *Biotechnol. Adv.*, 2011, **29**, 418.
23. M. Verlander, *Int. J. Pept. Res. Ther.*, 2007, **13**, 75.
24. M. Amblard, J. Fehrentz, J. Martinez and G. Subra, *Mol. Biotechnol.*, 2006, **33**, 239.
25. K. J. Jensen, in *Peptide Synthesis and Applications*, Humana Totowa, New York, 2013, vol. 1047, ch. 1.
26. A. Isidro-Llobet, M. N. Kenworthy, S. Mukherjee, M. E. Kopach, K. Wegner, F. Gallou, A. G. Smith and F. Roschangar, *J. Org. Chem.*, 2019, **84**, 4615.
27. Y. Okada, R. Takasawa, D. Kubo, N. Iwanaga, S. Fujita, K. Suzuki, H. Suzuki, H. Kamiya and K. Chiba, *Org. Process Res. Dev.*, 2019, **23**, 2576.
28. A. Sharma, A. Kumar, B. G. De La Torre and F. Albericio, *Chem. Rev.*, 2022, **122**, 13516.
29. B. L. Bray, *Nat. Rev. Drug Discovery*, 2003, **2**, 5.
30. M. O. Frederick, R. A. Boyse, T. M. Braden, J. R. Calvin, B. M. Campbell, S. M. Changi, S. R. Coffin, C. Condon, O. Gowran, J. McClary Groh, S. R. Groskreutz, Z. D. Harms, A. A. Humenik, N. J. Kallman, N. D. Klitzing, M. E. Kopach, J. K. Kretsinger, G. R. Lambertus, J. T. Lampert, L. M. Maguire, H. A. Moynihan, N. S. Mullane, J. D. Murphy, M. E. O'Mahony, R. N. Richey, K. D. Seibert, R. D. Spencer, M. A. Strege, N. Tandogan, F. L. Torres Torres, S. V. Tsukanov and H. Xia, *Org. Process Res. Dev.*, 2021, **25**, 1628.
31. J. W. Bode, *Acc. Chem. Res.*, 2017, **50**, 2104.
32. M. Köhn and R. Breinbauer, *Angew. Chem., Int. Ed.*, 2004, **43**, 3106.
33. J.-C. M. Monbaliu and A. R. Katritzky, *Chem. Commun.*, 2012, **48**, 11601.
34. W. Xu, W. Jiang, J. Wang, L. Yu, J. Chen, X. Liu, L. Liu and T. F. Zhu, *Cell Discovery*, 2017, **3**, 7008.
35. P. E. Dawson, T. W. Muir and S. B. H. Kent, *Science*, 1994, **266**, 27.
36. T. M. Hackeng, J. H. Griffin and P. E. Dawson, *Proc. Natl. Acad. Sci. U. S. A.*, 1999, **96**, 10068.
37. V. Agouridas, E. Mahdi, V. Diemer, M. Cargoe, J. M. Monbaliu and O. Melnyk, *Chem. Rev.*, 2019, **119**, 7328.
38. P. A. Cistrone, M. J. Bird, D. T. Flood, A. P. Silvestri, C. J. Jordi, D. A. Thompson and P. E. Dawson, *Curr. Protoc. Chem. Biol.*, 2019, **11**, 1.
39. T. W. Muir, *Annu. Rev. Biochem.*, 2003, **72**, 249.
40. J. Liu, O. Ekanayaje, D. Santoleri, W. Kelsi and S. Rozovsky, *ChemBioChem*, 2020, **21**, 346.

41. I. R. Cottingham, A. Millar, E. Emslie, A. Colman, A. E. Schnieke and C. McKee, *Nat. Biotechnol.*, 2001, **19**, 974.
42. S. H. Henager, N. Chu, Z. Chen, D. Bolduc, D. R. Dempsey, Y. Hwang, J. Wells and P. A. Cole, *Nat. Methods*, 2016, **13**, 925.
43. R. Li, M. Schmidt, T. Zhu, X. Yang, J. Feng, Y. Tian, Y. Cui, T. Nuijens and B. Wu, *Natl. Sci. Rev.*, 2021, **9**, 158.
44. T. Nuijens, A. Toplak, M. Schmidt, A. Ricci and W. Cabri, *Front. Chem.*, 2019, **7**, 1.
45. H. E. Morgan, W. B. Turnbull and M. E. Webb, *Chem. Soc. Rev.*, 2022, **51**, 4121.
46. T. Nuijens and M. Schmidt, Enzyme-mediated ligation methods, in *Springer Methods in Molecular Biology*, Humana, New York, 2019, vol. 2012.
47. H.-D. Jakubke, *Angew. Chem., Int. Ed. Engl.*, 1995, **34**, 175.
48. T. Nuijens, A. H. M. Schepers, C. Cusan, J. a. W. Kruijtzer, D. T. S. Rijkers, R. M. J. Liskamp and P. J. L. M. Quaedflieg, *Adv. Synth. Catal.*, 2013, **355**, 287.
49. A. Toplak, T. Nuijens, P. J. L. M. Quaedflieg, B. Wu and D. B. Janssen, *Enzyme Microb. Technol.*, 2015, **73–74**, 20.
50. L. Polgar and M. L. Bender, *Biochemistry*, 1967, **6**, 610.
51. Z. Wu and D. Hilvert, *J. Am. Chem. Soc.*, 1989, 4513.
52. L. Abrahamsén, J. Tom, J. Burnier, K. A. Butcher, A. Kossiakoff and J. A. Wells, *Biochemistry*, 1991, **30**, 4151.
53. T. K. Chang, D. Y. Jackson, J. P. Burniert and J. A. Wells, *Proc. Natl. Acad. Sci. U. S. A.*, 1994, **91**, 12544.
54. S. Atwell and J. A. Wellst, *Proc. Natl. Acad. Sci. U. S. A.*, 1999, **96**, 9497.
55. A. M. Weeks and J. A. Wells, *Nat. Chem. Biol.*, 2018, **14**, 50.
56. A. M. Weeks and J. A. Wells, *Chem. Rev.*, 2020, **120**, 3127.
57. A. Toplak, T. Nuijens, P. J. L. M. Quaedflieg, B. Wu and D. B. Janssen, *Adv. Synth. Catal.*, 2016, **358**, 2140.
58. T. Nuijens, A. Toplak, P. J. L. M. Quaedflieg, J. Drenth, B. Wu and D. B. Janssen, *Adv. Synth. Catal.*, 2016, **358**, 4041.
59. A. Toplak, E. F. Teixeira de Oliveira, M. Schmidt, H. J. Rozeboom, H. J. Wijma, L. K. M. Meekels, R. de Visser, D. B. Janssen and T. Nuijens, *Comput. Struct. Biotechnol. J.*, 2021, **19**, 1277.
60. M. Schmidt, A. Toplak, P. J. L. M. Quaedflieg, H. Ippel, G. J. J. Richelle, T. M. Hackeng, J. H. van Maarseveen and T. Nuijens, *Adv. Synth. Catal.*, 2017, **359**, 2050.
61. M. Schmidt, Y. H. Huang, E. F. Texeira de Oliveira, A. Toplak, H. J. Wijma, D. B. Janssen, J. H. van Maarseveen, D. J. Craik and T. Nuijens, *ChemBioChem*, 2019, **20**, 1524.
62. D. E. Streefkerk, M. Schmidt, J. H. Ippel, T. M. Hackeng, T. Nuijens, P. Timmerman and J. H. Van Maarseveen, *Org. Lett.*, 2019, **21**, 2095.
63. M. Schmidt, A. Toplak, H. J. Rozeboom, H. J. Wijma, P. J. L. M. Quaedflieg, J. H. van Maarseveen, D. B. Janssen and T. Nuijens, *Org. Biomol. Chem.*, 2018, **16**, 609.

64. T. Nuijens, A. Toplak, M. B. A. C. van de Meulenreek, M. Schmidt, M. Goldbach, D. B. Janssen and P. J. L. M. Quaedflieg, *Chem. Today*, 2016, **34**, 16.
65. T. Nuijens, P. Quaedflieg and J. M. Leonard, *US pat.*, 10752931, 2020.
66. T. Nuijens, J. Mateman, R. De Visser, W. Cabri, K. Wallraven, E. Wijker, A. Ricci and A. Toplak, *EP pat.*, 4291647, 2023.
67. J. Pawlas, T. Nuijens, J. Persson, T. Svensson, M. Schmidt, A. Toplak, M. Nilsson and J. H. Rasmussen, *Green Chem.*, 2019, **21**, 6451.
68. S. K. Mazmanian, G. Liu, H. Ton-That and O. Schneewind, *Science*, 1999, **285**, 760.
69. M. Ritzefeld, *Chemistry*, 2014, **20**, 8516.
70. C. Freund and D. Schwarzer, *ChemBioChem*, 2021, **22**, 1347.
71. I. Chen, B. M. Dorr and D. R. Liu, *Proc. Natl. Acad. Sci. U. S. A.*, 2011, **108**, 11399.
72. H. Hirakawa, S. Ishikawa and T. Nagamune, *Biotechnol. Bioeng.*, 2012, **109**, 2955.
73. L. Chen, J. Cohen, X. Song, A. Zhao, Z. Ye, C. J. Feulner, P. Doonan, W. Somers, L. Lin and P. R. Chen, *Sci. Rep.*, 2016, **6**, 1.
74. R. L. Policarpo, H. Kang, X. Liao, A. E. Rabideau, M. D. Simon and B. L. Pentelute, *Angew. Chem., Int. Ed.*, 2014, **53**, 9203.
75. J. M. Antos, M. C. Truttmann and H. L. Ploegh, *Curr. Opin. Struct. Biol.*, 2016, **36**, 111.
76. A. Debon, E. Siirola and R. Snajdrova, *J. Am. Chem. Soc.*, 2023, **3**, 1267.
77. J. Li, Y. Zhang, O. Soubias, D. Khago, F. A. Chao, Y. Li, K. Shaw and R. A. Byrd, *J. Biol. Chem.*, 2020, **295**, 2664.
78. B. M. Dorr, H. O. Ham, C. An, E. L. Chaikof and D. R. Liu, *Proc. Natl. Acad. Sci. U. S. A.*, 2014, **111**, 13343.
79. M. Fottner, M. Weyh, S. Gaussmann, D. Schwarz, M. Sattler and K. Lang, *Nat. Commun.*, 2021, **12**, 1.
80. Z. Zou, H. Alibiglou, D. M. Mate, M. D. Davari, F. Jakob and U. Schwaneberg, *Chem. Commun.*, 2018, **54**, 11467.
81. Z. Zou, D. M. Mate, M. Nöth, F. Jakob and U. Schwaneberg, *Chem. – Eur. J.*, 2020, **26**, 13537.
82. G. K. T. Nguyen, A. Kam, S. Loo, A. E. Jansson, L. X. Pan and J. P. Tam, *J. Am. Chem. Soc.*, 2015, **137**, 15398.
83. K. S. Harris, T. Durek, Q. Kaas, A. G. Poth, E. K. Gilding, B. F. Conlan, I. Saska, N. L. Daly, N. L. Van Der Weerden, D. J. Craik and M. A. Anderson, *Nat. Commun.*, 2015, **6**, 10199.
84. D. Zhang, Z. Wang, S. Hu, J. Lescar, J. P. Tam and C. F. Liu, *Int. J. Mol. Sci.*, 2022, **23**, 458.
85. Y. Chen, D. Zhang, X. Zhang, Z. Wang, C. F. Liu and J. P. Tam, *Front. Chem.*, 2021, **9**, 1.
86. F. B. H. Rehm, T. J. Harmand, K. Yap, T. Durek, D. J. Craik and H. L. Ploegh, *J. Am. Chem. Soc.*, 2019, **141**, 17388.

87. T. J. Harmand, D. Bousbaine, A. Chan, X. Zhang, D. R. Liu, J. P. Tam and H. L. Ploegh, *Bioconjugate Chem.*, 2018, **29**, 3245.
88. Y. Cao, G. K. T. Nguyen, S. Chuah, J. P. Tam and C. F. Liu, *Bioconjugate Chem.*, 2016, 27, 2592.
89. G. K. T. Nguyen, Y. Qiu, Y. Cao, X. Hemu, C.-F. Liu and J. P. Tam, *Nat. Protoc.*, 2016, **11**, 1977.
90. A. M. James, J. Haywood, J. Leroux, K. Ignasiak, A. G. Elliott, J. W. Schmidberger, M. F. Fisher, S. G. Nonis, R. Fenske, C. S. Bond and J. S. Mylne, *Plant J.*, 2019, **98**, 988.
91. D. Zhang, Z. Wang, S. Hu, S. Balamkundu, J. To, X. Zhang, J. Lescar, J. P. Tam and C. F. Liu, *J. Am. Chem. Soc.*, 2021, **143**, 8704.
92. X. Hemu, J. To, X. Zhang and J. P. Tam, *J. Org. Chem.*, 2020, **85**, 1504.
93. J. Zhao, R. Fan, F. Jia, Y. Huang, Z. Huang, Y. Hou and S. Q. Hu, *J. Agric. Food Chem.*, 2021, **69**, 5976.
94. M. A. Jackson, J. Xie, L. T. T. Nguyen, X. Wang, K. Yap, P. J. Harvey, E. K. Gilding and D. J. Craik, *Transgenic Res.*, 2023, **32**, 121.
95. S. Liebscher, M. Schöpfel, T. Aumüller, A. Sharkhuukhen, A. Pech, E. Höss, C. Parthier, G. Jahreis, M. T. Stubbs and F. Bordusa, *Angew. Chem., Int. Ed. Engl.*, 2014, **53**, 3024.
96. C. Meyer, S. Liebscher and F. Bordusa, *Bioconjugate Chem.*, 2016, **27**, 47.
97. S. Liebscher, P. Kornberger, G. Fink, E.-M. Trost-Gross, E. Höss, A. Skerra and F. Bordusa, *ChemBioChem*, 2014, **15**, 1096.
98. J. Koehnke, A. Bent, W. E. Houssen, D. Zollman, F. Morawitz, S. Shirran, J. Vendome, A. F. Nneoyiegbe, L. Trembleau, C. H. Botting, M. C. M. Smith, M. Jaspars and J. H. Naismith, *Nat. Struct. Mol. Biol.*, 2012, **19**, 767.
99. J. A. McIntosh, C. R. Robertson, V. Agarwal, S. K. Nair, G. W. Bulaj and E. W. Schmidt, *J. Am. Chem. Soc.*, 2010, **132**, 15499.
100. E. Yeh, H. Lin, S. L. Clugston, R. M. Kohli and C. T. Walsh, *Chem. Biol.*, 2004, **11**, 1573.
101. M. L. Adrover-Castellano, J. J. Schmidt and D. H. Sherman, *ChemCatChem*, 2021, **13**, 2095.
102. T. Matsumoto, R. Takase, T. Tanaka, H. Fukuda and A. Kondo, *Biotechnol. J.*, 2012, 7, 642.
103. K. Piotukh, B. Geltinger, N. Heinrich, F. Gerth, M. Beyermann, C. Freund and D. Schwarzer, *J. Am. Chem. Soc.*, 2011, **133**, 17536.
104. L. Schmohl, J. Bierlmeier, F. Gerth, C. Freund and D. Schwarzer, *J. Pept. Sci.*, 2017, **23**, 631.
105. C. J. Podracky, C. An, A. DeSousa, B. M. Dorr, D. M. Walsh and D. R. Liu, *Nat. Chem. Biol.*, 2021, **17**, 317.
106. Z. A. Wang, S. D. Whedon, M. Wu, S. Wang, E. A. Brown, A. Anmangandla, L. Regan, K. Lee, J. Du, J. Y. Hong, L. Fairall, T. Kay, H. Lin, Y. Zhao, J. W. R. Schwabe and P. A. Cole, *J. Am. Chem. Soc.*, 2022, **144**, 3360.

CHAPTER 7

Recombinant and Semisynthesis of Peptides

P. FONTE*, A. MALCANGI, M. STUCCHI AND A. OPPEDISANO

Olon SpA, via Solferino, 10036 Settimo Torinese (TO), Italy
*E-mail: pfonte@olonspa.it

7.1 Introduction

7.1.1 Manufacture of Peptide Therapeutics

Peptide therapeutics are emerging as a potent new class of active pharmaceutical ingredients (APIs), even though historically their original usage dates back to the 1920s, when insulin was first used in humans to treat diabetes, and in the last decade more and more peptide drugs have entered the market. This is a rich field of research, as illustrated by the fact that both academic groups and pharma industries are actively researching and investing money. This is a clear sign of a field that is not yet mature, with many possibilities for further improvements and discoveries.

The preparation of peptides is a very deceptive area of work, where the apparently simple chemistry involved can lead to underestimation of the complexity of the system. Indeed, most of the synthetic work revolves around the formation of an amide bond to forge the peptide carbon backbone. However, when applied to real systems, this simple operation can meet unexpected difficulties: the unique chemical and physical properties of peptides,

Green Chemistry Series No. 85
Sustainability in TIDES Chemistry: Green Approaches to Oligonucleotides and Oligopeptides Synthesis
Edited by Alessandra Tolomelli, Lucia Ferrazzano and Walter Cabri

Published by the Royal Society of Chemistry, www.rsc.org

the formation of secondary structures, the low solubilities, the unusual and surprising reactivity of certain amino acid sequences, and so on. All of these properties can lead to many struggles along the path to the final target. In light of the issues, most syntheses rely on the leverage of solid-phase peptide synthesis (SPPS), which is a very well-established technique, with a deep background of knowledge and a vast database of solutions. SPPS is a widely known methodology that is generally economically advantageous and easy to develop in order to prepare a target molecule; moreover, it has some clear strong points on its side: excess reagents/reactants are easy to remove, scale-up is very straightforward, there is ample opportunity to implement automation in the process, and lastly the insertion of non-natural amino acids or non-canonical structures is facile. The other most common methodology employed in the preparation of peptides is liquid-phase peptide synthesis (LPPS), which is very similar to SPPS but performed in solution.

On the other hand, both techniques have clear disadvantages: often peptides and amino acids show very limited solubility in the solvent or system used in the synthesis, which can be a cause, with other factors, of the low yields that are sometimes observed; in addition, both methodologies use hazardous solvents and/or reagents, which makes them not very appealing from a sustainability point of view. Finally, the cost of these techniques increases with each amino acid added to the sequence, and the cost impact for peptides with sequences longer that 20–25 amino acids can be very high, with a negative impact on the cost of the final drug substance and on the profit for the manufacturer.

While peptides with longer sequences (40 or even 50 amino acids long) are making their way to the market, recombinant technologies are now becoming of commercial interest; more specifically, in the field of pharmaceutical peptides there is a growing requirement for more complex structures, which are not simple to access *via* SPPS or LPPS. Generally, the chemical synthesis of peptides longer than 20 amino acids is cumbersome and expensive. As more complex structures and longer amino acid chains are coming through the pipeline, the classical methodologies of SPPS and LPPS for the preparation of APIs are less competitive, and at this point recombinant deoxyribonucleic acid (rDNA) technologies are gaining the edge (see Figure 7.1).

Currently, about 80 peptides are approved for human use, more than 150 are in clinical development, and many more are in preclinical studies.[1] Every year new therapeutic peptides are approved by the US Food and Drug Administration (FDA) and regulatory agencies around the world. Peptide therapeutics have distinct features compared with small molecules: they are generally more potent and more specific and can establish interactions with the target receptors that are just not feasible for small molecules. In addition, peptides can target more types of receptors and, thanks to their nature, peptides are less harmful to the human body. On the other hand, peptides are easily degraded *in vivo*, making them difficult to administer orally, and they do not easily cross the cell membrane.

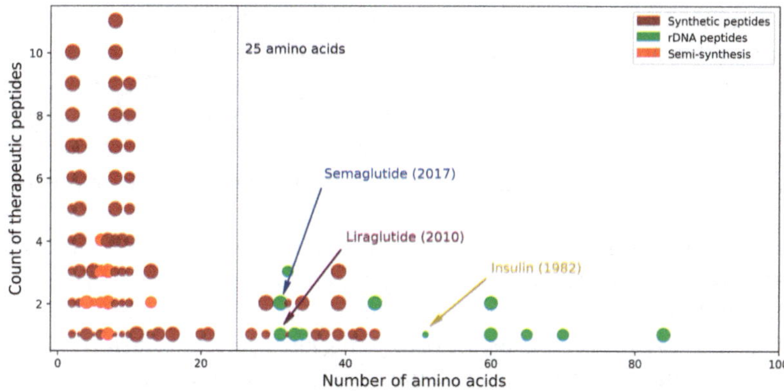

Figure 7.1 Number of therapeutic peptides as a function of the number of amino acids. Dark brown dots represent APIs manufactured *via* synthesis, light brown dots represent semisynthetic APIs and green dots represent APIs manufactured *via* rDNA. The size of the dots is proportional to the year of production (larger = more recent).

In light of these considerations, rDNA routes for the preparation of new peptides are increasing. With this methodology, it is possible to engineer bacteria and yeasts, among others, to naturally express the desired target peptide. As one can imagine, the costs associated with such production methods are minimally dependent on the peptide length, hence the preparation of long peptides is economically advantageous at this point. Overall, rDNA techniques have many appealing aspects: natural organisms are exploited and engineered to deliver therapeutic peptides, carried out in aqueous systems, employing non-hazardous reagents such as molds and syrups. The sustainability of the process in this case goes hand in hand with the economic advantages of the methodology. In addition, the building of the molecule is performed by the natural organism, and this comes with the clear advantage of the natural precision in codifying the correct structure, with the correct amino acid sequence and the required stereo form. Nevertheless, the industrialization of such processes is more difficult to develop and more time is required in the R&D phase to scale up a productive process. Additionally, some restriction is mandatory on the fermented peptides, as natural organisms cannot express non-natural structures or non-canonical amino acids. Finally, when comparing and exploring the various possibilities for producing peptides, one must also consider the differences in the quality of the products: the quality profiles of the same peptides prepared *via* biotechnologies or *via* SPPS are radically different: in one case the most common impurities includes deletion and duplication of amino acids in the sequence, with many process-related impurities being formed during the many synthetic steps; in the other situation, most of the impurities are product related, where degradation pathways and bio-derived residues are the most critical issues.

Taken together, the rDNA and SPPS processes for manufacturing therapeutic peptides are two possibilities that do not overlap but on the contrary cover different areas of needs in the industry panorama. The ever-changing scenario of API manufacturing, with multiple players, such as regulatory agencies, sustainability targets, and market requirements themselves, will naturally set the emphasis on the most suitable technologies from process to process.

Currently, the high impact of APIs prepared *via* biotechnologies is exemplified by the striking success of semaglutide. This is a peptide produced by Novo Nordisk, originally prescribed as an injectable antidiabetic (Ozempic) and subsequently marketed as an orally disposable formulation (Rybelsus), with a large impact on the market. Thereafter, the drug was shown to have an impressive weight-loss effect, and sales skyrocketed. Many countries, such as the USA, saw a huge increase in medical prescriptions of the drug and Novo Nordisk, the company holding the patents for semaglutide, in 2023 registered a market value higher than the gross domestic product (GDP) of Denmark (the country of location of the company headquarters), and it is the second most valuable company in Europe.

7.1.2 Manufacturing Sustainability

The concept of industrial sustainability and the idea of realizing clean industrial processes have gained importance over time. There is a need to reshape industrial process technologies to prevent pollution at the source. Biotechnology represents an opportunity to move from fossil-based to bio-based industry, reducing global warming. Fermentation, as a water-based process, has clear ecological advantages. The raw materials used are renewable resources and wastes may be biodegraded in wastewater treatment plants or recycled to produce clean fuel. Nevertheless, an assessment of the environmental impact of novel developed processes is important to have a clear understanding of the product life cycle, from raw materials sourcing, processing, manufacture, and transport to end-of-life disposal, and to identify more sustainable process options.

The use of specific analytical tools such as life-cycle assessment (LCA) can provide decision support on how to design future technologies. LCA provides a holistic view of a process. The application of this methodology may support the identification of the types of microorganisms to be utilized, can drive the selection of optimal substrates, and can allow the choice of process model and bioreactor technology to minimize energy consumption.[2–4]

The result of the scale-up of a process to the full commercial scale can also be predicted by LCA and negative environmental impact effects can be reduced.[4] The identification of hotspots can, in fact, be used for process optimization. The environmental impact of a bioprocess is primarily due to energy and water consumption and the improvement of the process yield through strain optimization is the most effective strategy to reduce it.[5] Continuing technical innovation and the development of recombinant DNA technology increased the number of industrial biotechnology applications. This technology provides a powerful method for combining diverse genetic capabilities, thus permitting the engineering of organisms with new specific capabilities. The tailor-made synthesis of products by modified microorganisms opens up new possibilities for the production of chemicals for different purposes.

This approach can be useful every time there is a problem in the supply chain of the product, for example if the product is present in low concentration in the natural source or if it is difficult to extract and purify it from the natural source. rDNA technology can be an opportunity in the case of a product that is difficult to produce by chemical synthesis or may be used to design chemicals that would not occur by natural pathways.

An example of an important application of rDNA is represented by human insulin. This hormone cannot be obtained directly from the natural source and extraction from animal tissues has several drawbacks. The possibility of producing recombinant human insulin in microorganisms represented a major success and paved the way to the production of recombinant human hormones and their therapeutic use.

7.2 Human Insulin: The First Hormone Produced Using Recombinant DNA Technology

7.2.1 History of Insulin

Human insulin is comprised of 51 amino acids and has a molecular weight of 5808 Da. It is produced by beta cells of the pancreas and plays a key role in regulating carbohydrate and fat metabolism in the body. Insulin is synthesized as a single polypeptide known as preproinsulin (110 amino acids). The mature form consists of an A-chain with 21 amino acids and a B-chain containing 30 amino acids and the two polypeptides are linked together by two disulfide bonds. Further, the A-chain has an intrachain disulfide bond.[6,7]

Insulin is secreted by beta cells of islets of Langerhans in a glucose-dependent manner and acts to stimulate glucose uptake in the liver, muscle, and adipose tissues. Like other protein hormones, insulin binds with specific receptors on the outer membrane of target cells, which causes metabolic processes to take place within those cells. Insulin was the first human hormone produced using rDNA technology.

Diabetes mellitus was first described over 2000 years ago and represented a fatal medical condition until the early 1920s.[8] The discovery of insulin was a milestone in medical science. The work leading to the discovery of insulin was started in 1921 by the Canadian surgeon Frederick Grant Banting and his assistant Charles Best, who worked at the University of Toronto under the supervision of John Macleod. They isolated insulin from pancreatic animal tissues and carried out the initial studies administering insulin preparations to diabetic patients.[9,10] Banting and Best started a collaboration with Eli Lilly that led in 1922 to the first commercially available insulin.[11] In Europe, two Danish companies, which in the late 1980s merged into Novo Nordisk, started to produce insulin.

Insulin from cows and pigs was used to treat diabetes for many years and saved millions of lives. Nevertheless, the use of animal insulin was problematic because its composition was not identical to that of human insulin and stimulated the production of antidrug antibodies, preventing the effective treatment of diabetic patients or even leading to worsening of their diabetes.[12] In addition, there was the problem of contamination of insulin of animal origin with proteases and the risk of hydrolysis of the product.[11] Major efforts were made in subsequent years to improve purification to achieve safety and stability of the product.

The bovine insulin amino acid sequence was elucidated by the British biochemist Frederick Sanger in 1955. For this discovery, Sanger was awarded the Nobel Prize for Chemistry in 1958.[13] In 1960, the human insulin sequence was elucidated and some comparative studies with bovine and porcine insulin evidenced a three and a single amino acid difference, respectively.[11] In order to obtain a therapeutic product identical in sequence with native human insulin, the total chemical synthesis was considered, but was found to be economically unaffordable.

The enzymatic conversion of porcine into human insulin was developed in the 1970s and found to be commercially feasible.[11,14] This result was achieved by considering that protein semisynthesis could be carried out very conveniently if protein fragments obtained by the action of proteinases could be recombined again by the action of the same proteinases. In order to achieve such an objective, techniques were needed to shift the conditions in such a manner that the enzymes remained catalytically active, and under one set of conditions peptide bond hydrolysis was favored and under the other peptide bond synthesis was favored. The addition of organic cosolvents was used to shift the equilibria in enzyme-catalyzed reactions towards synthesis.[14] The product obtained was termed "semisynthetic human insulin" or "human insulin (emp)" (emp = enzyme-modified porcine).

In the 1970s, rDNA methods for cloning and expressing genes in *Escherichia coli* were under intense development. The invention of DNA cloning by Stanley Cohen and Herbert Boyer led to the creation of various recombinant proteins with medicinal uses. In 1978 and in 1979, the successful expression in *E. coli* of two mammalian hormones, first somatostatin and then human insulin, was achieved by Genentech. Somatostatin represented only a proof of concept to test the feasibility of the method on a 14-amino acid mammalian hormone. The method used for insulin was essentially the same as that for somatostatin.[15] Human insulin, which is comprised of a total of 51 amino acids (21 in the A-chain and 30 in the B-chain), was the first therapeutically useful protein product resulting from rDNA technology.

The first recombinant insulin was approved by the FDA in 1982 for the treatment of diabetic patients. The successful production of human insulin in bacteria provided, for the first time, a practical, scalable source of human insulin.

7.2.2 Insulin Production Process

The use of cows and pigs to obtain the pancreatic hormone had a negative footprint in terms of sustainability and could not support the increasing world demand for insulin. The change from extraction to recombinant insulin production process led to several advantages:

- an improvement in quality;
- a reduction of the risk of immunogenic reaction that could compromise the efficacy of the therapy;
- the possibility of producing insulin on a large scale at an affordable cost;
- an improvement in the sustainability of the production process.

The first production process developed by Genentech used an approach that required the expression of chemically synthesized complementary DNA (cDNA) encoding for the insulin A and B chains separately in *E. coli*. Each chain was fused to a carrier protein and produced independently. After expressing the fusion proteins and releasing the two chains by chemical cleavage, the two free chains were purified and co-incubated under optimum reaction conditions that promoted the generation of intact and bioactive insulin by disulfide bond formation. The first recombinant insulin available on the market was produced using this method.

Another approach consisted in the expression of a single chemically synthesized cDNA encoding for native human proinsulin in *E. coli* as inclusion bodies followed by solubilization and a refolding procedure. After purification, excision of C-peptide was performed with a bioconversion step (proteolytic digestion). This approach was more efficient and convenient than the two-chains combination approach for the large-scale production of therapeutic insulin. The method required a single fermentation and purification process and was economically affordable. Eli Lilly adopted this technology to produce Humulin, which entered the market in 1986.

Recombinant insulin represented a significant improvement in purity and quality consistency over semisynthetic insulin and permitted a worldwide supply.[16]

E. coli is not the only host utilized for insulin production.[17,18] In Europe, *Saccharomyces cerevisiae* was used by Novo Nordisk and currently several recombinant commercial insulin products are expressed in this strain.[7] The advantage of the yeast expression system is the extracellular secretion, which can simplify the purification of the protein. Although *S. cerevisiae* is still the predominant yeast system for insulin production, several alternative yeast hosts have become available in recent years. An example is *Pichia pastoris*, which is a very useful expression host with superior features. The use of integrative vectors allows the genetic stability of the strain to be improved. Its strong and tightly regulated methanol-inducible alcohol oxidase 1 (AOX1) promoter and its capacity to reach very high cell densities, by simple cultivation strategies, contribute to the high-level production of recombinant proteins. *P. pastoris* has proven its applicability for insulin production, and a comparative study revealed a performance equal to or even better than that of *S. cerevisiae*. Biosimilar insulin analogs produced by Mylan (rh-insulin glargine and rh-insulin aspart) were obtained in *P. pastoris* and approved in 2018–2021. Also, Baxter received approval in 2022 for a biosimilar rh-insulin using *P. pastoris* as the expression system.

Except for Novo Nordisk, which is still using *S. cerevisiae*, all other players are using the higher performance expression system in *E. coli* or *P. pastoris* for the recombinant production of insulin and insulin analogs.

7.3 Glucagon and GLP-1 Analogs

The history of insulin is tightly bound to glucagon, another hormone peptide secreted by alpha cells in the pancreas, which has an opposite effect as it counter-regulates insulin and raises the levels of glucose in the blood.[19] Glucagon itself was isolated by Kimball and Murlin in 1922, while they were working on the isolation of insulin.[20] Glucagon is a 29-mer peptide, produced in the human body by proglucagon, which has a wide range of functions in health, and in particular is related to type I and type II diabetes. Like insulin, its amino acid sequence was clarified in the 1950s by Eli Lilly and its functions in the human body became more thoroughly understood in the 1970s. Its preeminent role in today's medicine is exemplified by its inclusion in the WHO's *Model List of Essential Medicines*. Its manufacture by rDNA technology was finally introduced in the early 1990s to supplant the isolation of the API from natural sources, both for ease of availability and to prevent contamination of the drug by infectious residues of animal origins. The first manufacturing route employed an engineered version of *S. cerevisiae*, providing a glucagon perfectly resembling that derived from animals.[21] The production process involves expression of the glucagon in a cultivation broth on agar, fed with yeast and glucose, followed by the recovery phase of the API from the fermentation broth: after completion of the productive phase, the pH of the both is increased to solubilize the target molecule, then the product is recovered by centrifugation to remove the yeast and other solid residues, and finally a purification step by reversed-phase high-performance liquid chromatography (RP-HPLC) is performed to achieve the required quality of the product. An alternative process was developed later, based on treatment on a cation-exchange resin of the supernatant obtained after the centrifugation step, followed by crystallization at the isoelectric point and a final RP-HPLC step to fix the quality of the API. The processes for the isolation of glucagon are complicated by its natural trend to aggregate and form fibrils; this event is apparently reversible in basic environments and therefore the manufacturing route involves steps of basic treatment of the stream to prevent any undesired phenomena.

GLP-1 (glucagon-like peptide 1) is a 37-amino acid peptide involved in glucose homeostasis and in the regulation of insulin and glucagon levels. Owing to its antidiabetic function, it is an important structure target for new APIs;

moreover, compared with other antidiabetes drugs, it has a much safer profile, with lower risks of hypoglycemia, and at the same time can have a body weight-lowering effect and a cardioprotective role.[22] However, the usefulness of GLP-1 is plagued by a very short half-life *in vivo* due to the proteolytic activity of the enzymes dipeptidyl peptidase-4 (DPP-4) and neutral endopeptidase (NEP 24.11); depending on the dosing method, its half-life ranges from 1.5 min to 1.5 h.[23] Interestingly, it seems that the cleavage by DPP-4 yields a GLP-1(9–36) product that acts as an antagonist of GLP-1 itself.

7.4 Liraglutide and Semaglutide

Native peptides are defined as peptides with the same amino acid sequence as in natural peptides, and they can either be produced by isolation from natural sources, prepared synthetically, or obtained by recombinant DNA techniques. They can exhibit distinct advantages over either small molecules or larger biologics: compared with small molecules, peptides exhibit higher target specificity and a better interaction with the receptors, and compared with larger biologics, such as proteins and antibodies, the cost of production is lower and they show less immunogenicity. On the other hand, peptides have weak points as drugs: they cannot easily permeate through the cell membranes and they are easily degraded *in vivo*, resulting in low stability when administered. These drawbacks result in the necessity to target mainly extracellular receptors and the need to adopt subcutaneous injection of the drug, because the oral bioavailability is very low.

This easy degradation of native peptides in the human body is a major hurdle and much research effort has been poured into the improvement of the pharmacokinetic properties of peptides. Selected engineering of the structure of the peptides, with insertion of non-natural amino acids and backbone modifications, is now widely employed to increase stability towards biodegradation and improve physical properties. The purpose is to open up new and novel interactions with target receptors. At the same time, developments in the formulation methodologies are aimed at increasing the solubility and bioavailability of peptides, so that oral and other routes of administration become feasible.[24] These modified peptides are defined as analogs, in contrast to native peptides.

Many efforts were then directed at modification of the backbone of GLP-1 and addition of side-chains to improve the physicochemical properties of peptides. Both liraglutide and semaglutide, among other APIs, were developed starting from GLP-1, with precise structural modification to reduce the metabolic degradation and improve pharmacokinetics.

Liraglutide closely resembles GLP-1, with 97% identity, so the modifications made to the peptide structure are minimal, which should guarantee retention of the activity of GLP-1. In more detail, liraglutide is the 7–37 sub-sequence of the original backbone of GLP-1; there is a single amino acid substitution compared with GLP-1(7–37). During the medicinal chemistry development, many amino acid substitutions and modifications to the backbone were tested, and finally it was concluded that the most valuable ones, in terms of stability *versus* proteolytic degradation and activity *in vivo*, are the following:

- replacement of a lysine at position 34 with an arginine residue;
- acylation of lysine at position 26 with a fatty acid side-chain, linked to the ε-amino group through a glutamic acid spacer.[25]

These apparently simple, but crucial, modifications play a key role in the behavior of the API: the amino acid substitution prevents degradation *in vivo*, and the acylation increases the binding to the receptor. The outcome is that liraglutide has a half-life of about 8–10 h when administered *via* intravenous injection and up to 13–15 h for subcutaneous administration, making it viable for daily injection (see Figure 7.2).

The production of liraglutide has been patented with both SPPS[26] and fermentation processes, even though the manufacture of liraglutide for commercial purposes is carried out only by fermentation, for economic reasons. A caveat is necessary: the engineered organism used in the process can express the amino acid modification, as the arginine inserted is a natural amino acid, but cannot handle the insertion of an acylated lysine. The process therefore entails a hybrid approach: the peptide backbone is expressed by the natural organism, the peptide is isolated and purified as necessary, and at this point the peptide is modified chemically to introduce the modification of interest. In the production process for liraglutide, as patented by Novo Nordisk, this path is described as follows: the expression of the peptide backbone is carried out by properly engineered *S. cerevisiae*, then the peptide is purified and a chemical acylation takes place. More recent reports described the use of either *S. cerevisiae* or *E. coli*, both ending in a similar downstream process.[27]

The original report by Novo Nordisk very simply described a method for the preparation of acylated peptides: the fermentation broth is processed by initial purification by RP-HPLC, followed by precipitation of the compound

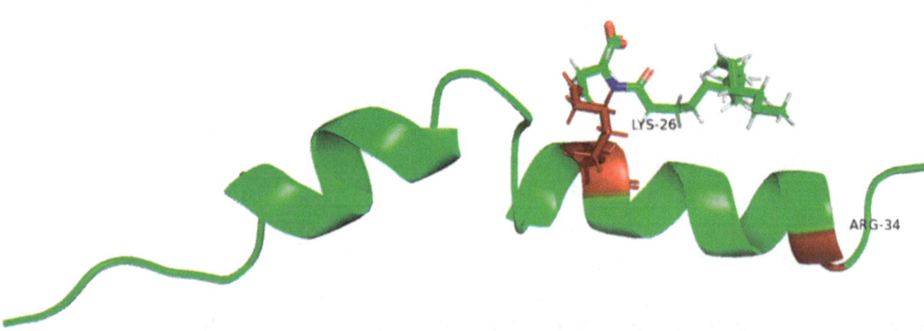

Figure 7.2 Secondary structure of liraglutide. Highlighted in red are the modified portions compared with GLP-1. 3D structure re-elaborated from Protein Data Bank, ref. 64.

at isoelectric pH (~5.4) and its isolation by centrifugation. Here the semisynthesis begins, with the necessity to acylate the lysine residue. Therefore, the isolated compound is at first treated with Et_3N to reach a basic pH of 11.6, then an active ester of the desired lipophilic side-chain carboxylic acid is added (palmityl–Glu–OH in the case of liraglutide preparation).[28] This acylation is the key step in the preparation: Lys26 has to be acylated with the required side-chain, while at the same time leaving unreacted the free amino group of the N-terminal histidine. Key to this selectivity is the pH of the reaction conditions: by operating in an aqueous environment, at a controlled pH of 10.5–11.5, there is a synthetically controlled chemoselectivity towards the primary amino group of the lysine. Indeed, under these conditions the free N-terminal amino group of histidine is less reactive, and the potentially interfering amidine group of arginine, which is protonated under these conditions, does not become acylated.

The final step of the semisynthesis is the removal of the protecting group used in the production process of the side-chain, needed to shield the carboxylic acid functional group on the glutamic acid spacer. Depending on the protecting group employed (usually a methoxy ester or a *tert*-butyl ester), a basic aqueous treatment or an acidic treatment with trifluoroacetic acid (TFA) is applied, and finally the target functionalized peptide is ready for a final purification. The final purification stage using RP-HPLC is crucial to attain a purity that meets the standard required by regulatory agencies; most often to achieve the target a C_{18} stationary phase is used in combination with acidic water and acetonitrile as the mobile phase. As alternatives, C_8- or even C_4-based-resins can be used, and ethanol and formic acid are common modifiers for eluting mixtures. To achieve a higher purity, in some processes even a double RP-HPLC purification has been applied. In cases such as liraglutide, where a semisynthetic approach is followed, the purification needs to address both process-related and product-related impurities. As one can imagine, very fine control of the conditions is required to achieve the target quality; therefore, during the R&D development phase, a crucial point is the determination of the structures of key impurities, whether they are critical from the purification point of view or not, and if they are critical then care must be taken to limit their formation by carefully tuning the process. All these evaluations can take up a considerable amount of the development time of the process and they are part of the control strategy of a process.

Further studies have been carried out to improve the manufacturing steps and achieve better processes, with the introduction of new methodologies. To improve expression and reduce the degradation by host enzymes, it was found that the inclusion in the expressed molecule of a tag or carrier sequence could help. New expression of fusion proteins, made up of the peptide backbone and a tailored tag or carrier, was developed and used in manufacturing processes. The most commonly used tags are polyhistidine tag (His-tag), ubiquitin or ubiquitin-like, and thioredoxin, and clearly the overall process needs to be redefined to account for their implementation. The peptide, now fused to the tag, is recovered from the broth in an appropriate manner, depending on how the target is expressed (intra- or intercellular, related to the producing organism); then a bioconversion step is required to cleave off the tag and free the desired peptide. Subsequently, depending on the type of carrier or tag cleaved, a different purification step is required to upgrade the stream. Generally, this purification can be achieved using affinity chromatography and the tags used are tailored to be easy to purify. Even if more steps are added to the process, the advantages gained in terms of both ease of purification and product recovery offset the negative impact.[29] After purification of the peptide, the steps of the semisynthesis are performed on the new material. Once the complete structure of the API has been constructed, most often a final deprotection step is needed, as the free carboxylic acid of the glutamic acid contained in the side-chain is protected, by either a methyl or a *tert*-butyl ester. In the former case, basic aqueous treatment is sufficient to reveal the carboxylic moiety; in the latter case, an acidic treatment, usually with TFA, is required to reveal the final molecule.

Interesting new developments in the preparation of liraglutide include the following:

- the overexpression of liraglutide by concatamers by an appropriate gene in a host cell (*i.e. E. coli*), the expressed multimer then being digested by proteases and carboxypeptidases to provide the peptide;[30]
- the use of new fusion proteins, which permits faster, cheaper, and more environmentally sustainable processes.[31,32]

Notwithstanding the successful results obtained with liraglutide, new and more efficient APIs were investigated to improve the drug profiles further. The strategic pathway for the modification of the structure of GLP-1 was repurposed by Novo Nordisk. Research was carried out with the objective of increasing the affinity for the receptor, which is positively correlated with the extension of activity efficacy and half-life, while at the same time retaining the positive attributes developed with liraglutide.[33] After initial studies, it became clear that modification of the amino acid sequence, the side-chain, and the spacer of the side-chain itself from the peptidic backbone had the most important influences on the binding and potency of the newly developed compounds. Therefore, the minimum extent of modifications was carried out, so that the profile of the API would remain as similar as possible to that of GLP-1 and liraglutide and minimize the risk of an autoimmune response *in vivo*. What resulted was semaglutide, an antidiabetic peptide analog administered by injection; compared structurally with liraglutide it has a further amino acid substitution where Aib8, a non-natural amino acid, is inserted in the sequence in place of an Ala, a γ-Glu-2xOEG spacer is used in place of a simple glutamic acid, and lastly a C_{18} diacid is used rather than a shorter monoacid side-chain. With these modifications, the half-life of semaglutide in humans is increased to ~165 h, opening up the way for a weekly dose administration (see Figures 7.3 and 7.4).

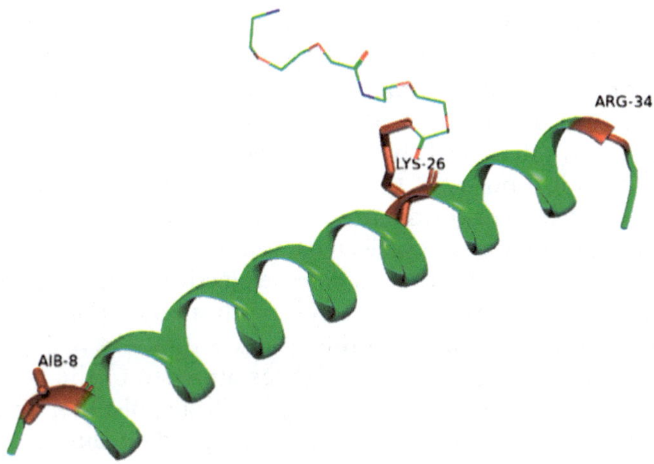

Figure 7.3 Secondary structure of semaglutide. Highlighted in red are the modified portions compared with GLP-1. 3D structure re-elaborated from Protein Data Bank, ref. 65 and 66.

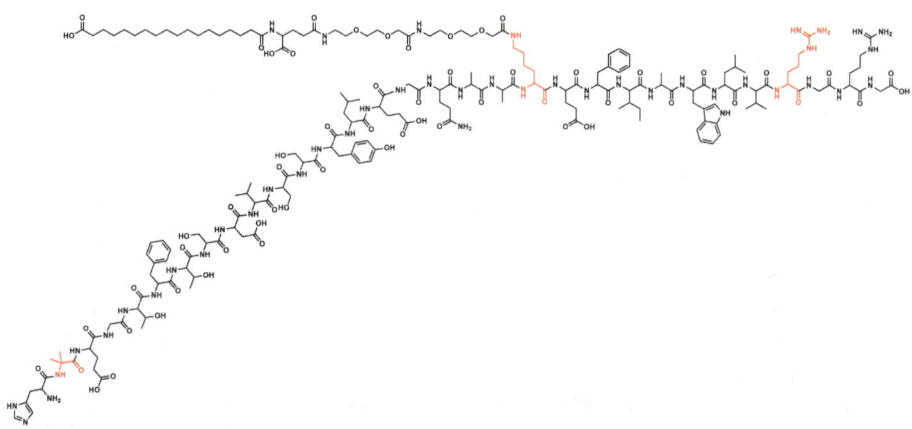

Figure 7.4 Primary structure of semaglutide. Highlighted in red are the modified portions compared with GLP-1.

The manufacturing process of semaglutide *via* semi-recombinant DNA preparation was patented by Novo Nordisk.[34] The process is similar to that for liraglutide, but in this case the insertion of a non-natural amino acid means that the full peptide backbone cannot be obtained *via* fermentation, but instead the semisynthetic approach this time needs to include a step of chain elongation to insert the missing amino acids. The manufacturing process begins with the upstream phase, with propagation of the expression plasmids in *E. coli*, then these plasmids are recovered and analyzed to confirm the correct inclusion of the proper sequence and finally transformed into

S. cerevisiae to produce the target peptide backbone. Two possible excisions of amino acids are reported: either two (His and Aib) or four (His, Aib, Glu, and Gly); the reasoning behind this choice is simple: activation of the C-terminal amino acids in these fragments harbors no risk of racemization of the amino acid involved (both Aib and Gly have no stereocenter in the α-position), thus preserving the optical purity of the target peptide. The expressed molecule is then recovered from the fermentation broth and purified as required; the final product of this phase is the selected shorter peptide, short by either two or four amino acids. At this point, a chemoselective acylation, on Lys26, in the presence of a free amino group at the N-terminus is applied, in a similar manner to that described for liraglutide. What is different with semaglutide is the nature of the side-chain that becomes attached to the lysine: in this case, a longer diacid is employed with a hydrophilic polyether amide spacer used to fuse the two parts of the molecule. The product can now be further processed to add the missing amino acids to the carbon backbone. The bis- or tetrapeptide (depending on the route chosen) is preactivated in the form of an active ester with standard coupling reagents and finally reacted with the acylated peptide. At this point, the synthesis is one step away from the target API; just a deprotection is required, as both the smaller peptide and the side-chain are added in the synthesis in a protected form, to avoid any formation of impurities due to cross-reactivity. Once the target structure has been revealed, a final RP-HPLC purification takes place with precise tuning of the conditions to achieve the required purity.

Further process developments now allow the use of a different culturing organism, either *E. coli* or *P. pastoris*, in addition to *S. cerevisiae*. This is obviously reflected in the requirement to adapt the upstream and downstream processes to isolate a peptide of adequate quality to be employed in the semisynthetic part of the process. The advantages of using fusion proteins or carriers to stabilize the molecule in the fermentation broth and simplify the purification were also introduced in the manufacturing of semaglutide.

Thanks to its very good pharmacokinetics and dynamics, more development work was carried out on the formulation of semaglutide: in addition to the classic injection, a different additive, SNAC {sodium 8-[(2-hydroxybenzoyl)amino]octanoate}, permits semaglutide to be formulated as tablets and be used in oral dosage form. More recently, even a route of administration *via* small rubber patches that adhere to the internal part of the cheek and release the drug has been found to be possible, opening up new frontiers for the delivery of peptides to patients.

As can be easily imagined, semaglutide is attracting considerable attention, and Novo Nordisk announced that, in addition to its antiobesity and diabetes regulation functions, semaglutide has been found to reduce the risk of cardiovascular diseases.[35] In addition, new limited reports have claimed the ability of semaglutide to act against drug and alcohol abuse, reducing dependence in patients under treatment. However, these reports have yet to find confirmation.

All this activity has accelerated the market demands and in turn the annual production of semaglutide is now over the metric ton level, and its manufacture must be able to withstand these rhythms without causing shortages and the production costs must be kept reasonable to allow patients to have access to the required medication. All these parameters need to be considered with the sustainability of the whole package in mind, from plasmid propagation to delivery to the patient, and the rDNA techniques are demonstrating themselves to be able to fulfill this role.

7.5 Recombinant Peptides and Protein Production: The Expression Systems

There is no universal expression platform for the production of recombinant peptides that is optimal for all of them. The choice of the host strain depends principally on protein structure, the presence of post-translational modifications, and the opportunity to have the product inside the cell or in the cultivation medium.

The fermentation yield is an important parameter in the evaluation of process costs and sustainability. The process mass intensity (PMI) depends on the production yield. The higher the productivity, the lower is the consumption of water, raw materials, and energy per kilogram of protein produced. The fermentation output depends on the cell density achieved and on the specific productivity of the strain.

Microorganisms with higher growth rates, such as *E. coli*, and that can reach a high cell density, such as *P. pastoris*, have clear economic and environmental advantages over less efficient expression systems.[36] A high cell density corresponds to high recombinant protein production.

The specific productivity depends on the genetic characteristics of the recombinant strain. Among the genetic characteristics, the presence of a strong inducible promoter in the expression cassette plays an important role. The expression pattern and efficiency rely on the selected promoter. If an inducible promoter is chosen, the growth and the production phase can be decoupled to reduce the metabolic load imposed by recombinant expression, while constitutive high-level expression frequently depletes the host cells.

An increase in yield through strain optimization is probably the most effective action to obtain environmental improvements.

7.5.1 *Escherichia coli*

E. coli, first isolated by Theodor Escherich in 1885, was the first system to be used for recombinant protein production. For some *E. coli* strains the complete genome sequence has been published, making its genetic manipulation easy.

The *E. coli* expression system is the most extensively used bacterial expression system; its transformation is simple and has high efficiency. In addition to its easy manipulation, *E. coli* has the advantage of a high growth rate. The fermentation time is short and the energy consumption is very low. Using *E. coli* it is possible to produce huge amounts of proteins with an economical and low environmental impact process.

An essential component of *E. coli*-based expression systems is the vector used to transfer the genetic information inside the bacterial cell.[37] The vector replicates independently inside the cell and is transferred to daughter cells. Generally, multiple copies of the vector are produced containing the rDNA. The copy number is controlled by the replicon (origin of replication) and varies from one vector to another. The use of a high copy number plasmid, such as the plasmid cloning vector pUC, does not always mean higher efficiency of recombinant protein production because the large amount of plasmid affects the bacterial growth rate. The high metabolic burden increases cellular stress level and can lead to plasmid rejection. The cells, in fact, eliminate the plasmid DNA to decrease the metabolic burden.[38,39] Over the course of cultivation, the number of bacterial cells bearing plasmids can decrease. The plasmid free cells have a higher growth rate because they do not have to replicate the plasmid DNA or produce the recombinant protein and consume the carbon source present in the cultivation medium, resulting in a limitation of carbon availability for the recombinant cells and consequently low recombinant protein production.

In order to avoid plasmid loss, a marker selection gene is usually present in the vector. The marker could be an antibiotic resistance gene or a metabolic selection marker. In the case of the presence on the plasmid of an antibiotic resistance gene, addition of antibiotic to the culture medium is required to maintain a high selectivity pressure. In this case, the recombinant protein purification process has to prove complete elimination of the antibiotic using sensitive analytical methods.

The metabolic selection marker represents a better choice. This system requires the use of a host strain that is lacking an essential gene, which was deleted from the genome and placed on the plasmid. Only the recombinant microorganisms will be able to grow.

The gene encoding the recombinant protein is cloned into the multiple cloning site (MCS) of the vector under the control of a promoter that will regulate the expression of the gene. The promoter controls the binding of RNA polymerase to DNA. The use of a strong promoter, with a high affinity for RNA polymerase, is essential to obtain an increase in the transcription of the heterologous gene. The T7 promoter system present in the pET plasmid cloning vector is the preferred one. This system requires the presence of the T7 RNA polymerase-encoding DE3 phage fragment, integrated in the bacterial chromosome, and under the control of lacUV promoter. The lacUV promoter is regulated by a chemical inducer, isopropyl-β-D-thiogalactopyranoside (IPTG), which activates transcription of T7 RNA polymerase when added to the culture. T7 RNA polymerase is faster than native *E. coli* enzyme and transcribes the recombinant gene under the control of T7 promoter that is present on the plasmid.[38–40]

The use of a strong promoter and a system with a high transcription efficiency is a key factor to achieve a high yield of the fermentation process. The recombinant protein production in the inducible expression system can be controlled and separated from the cell growth phase. Normally there is a high specific growth rate initially to support biomass formation and then there is a decrease in specific growth rate in the production stage to achieve maximum productivity.[41]

The direct expression of small recombinant peptides is very difficult, because they are easily degraded by the proteases present in the host cell. In addition, small peptides are not easily isolated and separated from the host proteins. In order to circumvent these technical difficulties, other essential elements are included in the expression cassette, namely the protein tags. There are different types of tags that have various functions. The protein tag can be formed by a short amino acids tail or by a protein. They can help the purification of the product, protect it from enzymatic degradation, or modify its solubility properties.

The affinity tags, generally constituted by a histidine tail, are used for the purification of the peptide drug by affinity chromatography. His-tagged proteins can be recovered by immobilized metal ion affinity chromatography using nickel or cobalt ion-loaded nitrilotriacetic acid–agarose resins. The host proteins are easily eliminated in the flow through. In other cases, the tag is a carrier protein that can increase or decrease the solubility of the recombinant product. The variation in solubility can give an advantage in the purification of the peptide or protect it from the protease enzymes present in the cytoplasm.

Heterologous proteins generally accumulate in the cytoplasm as inclusion bodies, which comprise insoluble misfolded aggregates of proteins. The use of carrier proteins that enhance the solubility of the recombinant protein can avoid the formation of inclusion bodies. In the case of peptides that are sensitive to the proteolytic action of the cytoplasmic enzymes, the formation of inclusion bodies may represent a form of protection, and the use of a different carrier protein, that reduces solubility, can induce recombinant protein precipitation as inclusion bodies.

If co-expression of a protein tag is introduced, a single chimeric polypeptide chain will be produced from the cell and an enzymatic or chemical cleavage will be necessary to release the protein drug. In the case of enzymatic removal, a protease cleavage site will be introduced in the expression cassette between the protein tag and recombinant drug. The correct removal of the protein tag is a very important step and high-fidelity endoprotease enzymes are used to guarantee the safety and efficacy of the protein therapy. The effect of an imprecise enzymatic cut is a product having one or more additional amino acids or a deletion or even fragmentation of the protein. For this reason, the choice of the endoprotease enzyme has to be carefully assessed for each single recombinant product, comparing the amino acid sequence of the protein expressed and the enzymatic cleavage site sequence, to ensure that only the cleaving site can be recognized by the endoprotease.

The introduction of a specific genetic signal in the expression cassette can lead the recombinant protein into the periplasm region and allow the formation of disulfide bonds. The periplasm provides an oxidizing environment and helps in forming disulfide bonds, which facilitate the proper folding of recombinant proteins. This occurs when the disulfide bond formations between cysteine amino acids are present in the peptide chain. The protein is produced in a soluble form even if at a relatively low concentration.

E. coli is usually preferred because it offers several advantages, including high growth rate, simple media requirement, ease of handling, high yield, and very cost-effective large-scale production. However, there are also some disadvantages, such as improper eukaryotic protein refolding and low expression level, lack of post-translational modification (glycosylation), endotoxin contamination, and proteolytic digestion of the product. Several genetic modifications of *E. coli* strains have been introduced in order to overcome these problems and some examples are presented below.

N-Linked glycosylation is the most common post-translational modification of proteins in eukaryotes. It has been discovered that the bacterium *Campylobacter jejuni* possesses the capability to glycosylate proteins and it was also shown that a functionally active *N*-glycosylation pathway could be transferred to *E. coli*, obtaining protein glycosylation.[42,43]

The codon usage of the heterologous protein also plays a major role in determining the expression level of recombinant proteins. If the codon usage of the heterologous protein differs significantly from the average codon usage of the *E. coli* host, it could result in very low expression. Usually, the frequency of the codon usage reflects the abundance of their corresponding tRNA.

Expression of heterologous proteins in *E. coli* can be improved by replacing codons that are rarely found in highly expressed *E. coli* genes with more favorable major codons. Similarly, co-expression of the genes encoding for a number of the tRNA for rare codons may enhance the expression of heterologous proteins in *E. coli*.

The use of protease-deficient *E. coli* strains, which carry mutations that eliminate the production of proteases, may also improve the yield of recombinant proteins by reducing proteolytic degradation. Undesired proteolysis, in fact, may lower the product yield or affect biological activity, but also complicates downstream processing as the degradation products will have similar physicochemical properties and will be difficult to eliminate.

The recombinant peptides and proteins are produced and accumulated in the cytoplasm of *E. coli* and their recovery requires cell disruption with release of the lipopolysaccharides, generally known as endotoxins, that compose the bacterial outer membrane. These components are a major concern since they are pyrogenic in humans and can induce immune responses and the production of antidrug antibodies that can reduce the therapeutic efficacy of the recombinant peptide. In the biopharmaceuticals produced using this host, additional purification steps are introduced in the downstream process to guarantee their elimination.[44]

7.5.2 Yeast Expression Systems

Yeast expression systems are used when secretion of the recombinant protein is required. In yeast cells, the compartmentalization into endoplasmic reticulum (ER) and Golgi guarantees the correct folding of the protein along the secretory pathway, allowing the secretion of the recombinant protein in soluble form. The presence of the product in soluble form in the medium helps its recovery and purification, improving the economics of the production process. The natural secretory machinery may not be sufficient under the conditions of high recombinant protein production and might lead to protein degradation and reduction of protein secretion. In this case, metabolic engineering of the secretory pathway is recommended in order to obtain high protein yields.[45] Among yeast strains, *S. cerevisiae* and *P. pastoris* are very commonly used for the production of heterologous proteins.

Yeast is a preferred host for the expression of various proteins that require post-translational modifications for their biological activity. Yeast cells have the ability to carry out numerous post-translational modifications such as phosphorylation, *O*-linked glycosylation, *N*-linked glycosylation, acetylation, and acylation. Recombinant proteins are expressed in yeast in properly folded and functionally active form. Production of biopharmaceuticals using a yeast expression system is also very cost-effective.

However, one major concern with producing therapeutic glycoproteins for human applications is that yeast *N*-glycosylation is of the high-mannose type, which confers a short half-life *in vivo* and hyper-immunogenicity and thus renders the therapeutic glycoprotein less effective. The initial glycosylation that occurs in the ER is the same in yeast and humans, but completely different at the Golgi level where, in humans, the mannosylated proteins are converted to galactosylated and sialylated forms. For this reason, various human therapeutic proteins, including therapeutic monoclonal antibodies, are being produced in mammalian cell lines. Recombinant proteins expressed in mammalian cells are properly folded, glycosylated, and generally produced as functionally active proteins. However, the production cost of biopharmaceuticals using a mammalian expression system is very high and other concerns related to viral contamination and oncogene transmission have to be considered.

Recently, using genetic engineering techniques, some mutant yeasts with deletion of the natural glycosylation pathway and having a humanized glycosylation pattern were produced and used for the production of erythropoietin and antibodies.[40,45] The transformation with expression vectors containing the gene for the recombinant protein production has a high efficiency and, depending on the type of plasmid used, can lead to the integration of the gene in the yeast chromosome or to an autonomously replicating vector. The integration of the gene in the chromosome confers stability to the population without selective pressure. *S. cerevisiae* represents a specific platform for recombinant insulin, insulin analogs, and GLP-1 agonists used by Novo Nordisk to develop a Crabtree negative and auxotrophic strain with improved secretory function.[7,46] Constitutive glycolytic promoters can be present in the vector [ADH1 (alcohol dehydrogenase), PGK (phosphoglycerate kinase), GAP (GTPase-accelerating protein)], but inducible promoters are preferred because the strong constitutive expression influences the growth rate negatively. The strongest regulated promoters are induced by galactose (Gal) and strongly repressed by glucose.[44] This type of yeast also has some disadvantages because it cannot reach a high cell density, the secretion of proteins is generally poor compared with other yeasts, the tendency to produce ethanol (Crabtree effect) may reduce the fermentation yield, and its glycosylation produces hyper-mannosylated glycans.[40]

An alternative yeast strain used for the large-scale production of biopharmaceuticals is *Pichia pastoris* (*Komagataella phaffii*). *P. pastoris* has two alcohol oxidase enzymes, AOX1 and AOX2, that are transcripts in the presence of methanol and allow the growth of yeast on methanol as the sole carbon and energy source. There are mutant strains defined as Muts with only AOX2 gene and a lower efficiency of growth on methanol and others (Mut–) without any AOX gene.[47] The first application of *P. pastoris* dates back 50 years to the commercial production of single-cell proteins as animal feed additives.[48] This process, owing to an increase in the price of methanol, was abandoned, but the same fermentation method was subsequently used for recombinant protein production. The genomic sequence of *P. pastoris* was published in 2009 and favored the development of biopharmaceuticals using *P. pastoris* as a host.[49] Using genetic engineering, the strain was modified and several customized improved strains were produced. *P. pastoris* has the ability to reach high cell densities with a very high level of expression of heterologous proteins by its robust and tightly regulated methanol-inducible alcohol oxidase 1 (AOX1) promoter and has the advantage of the glycosylation of secreted proteins because it does not hyper-glycosylate the heterologous proteins, so glycoproteins produced in *P. pastoris* may be more suitable for therapeutic use in humans.[50]

P. pastoris can therefore be an attractive alternative for the large-scale production of recombinant insulin and insulin analogs.[6] Heterologous proteins can be expressed at high levels and efficiently secreted. Purification is relatively straightforward owing to the small amount of host proteins. For this reason, the number of recombinant product applications in *P. pastoris* has been increasing over time.

The AOX1 promoter (derived from *P. pastoris* alcohol oxidase 1 gene) is induced on the addition of methanol. The induction system is very efficient, but it has a drawback in scale-up compared with a standard fermentation plant, because of the need for plant adaptation in order to manage the use of an inflammable solvent. Alternative promoters that do not require methanol for induction are also available, such as GAPDH (glyceraldehyde-3-phosphate dehydrogenase), for constitutive expression of recombinant proteins and other methanol-free expression systems.[51]

7.6 Fermentation Process and Sustainability

7.6.1 Sustainability Assessment

The very first concept of "green chemistry" can be traced back to the Pollution Prevention Act of 1990 in the USA, which introduced antipollution generation as a national policy.[52] The term itself was coined in the early 1990s by the staff of the Office of Pollution Prevention and Toxins at the US Environmental Protection Agency (EPA).[53] Many other milestones in the history of green chemistry contributed to its development and diffusion, but the definition of the "12 principles of green chemistry" by Paul Anastas and John Warner in 1998 is considered to be by far the most important one (see Table 7.1).[54] In fact, they were able to define the boundaries of the green chemistry concept, and their work continues to serve as the foundation of the field today.

To really determine if a product, process, or technology could be considered "green", most or all the "12 principles of green chemistry" need to be intrinsically present. Nonetheless, these principles were designed to be the broadest possible, and not all of them could be applied to a specific topic. In the context of peptides production, the 2nd and 10th principles cannot be fully applied, since they refer, at least partially, to the product and not the process. On the other hand, the 9th and 11th principles are specific to industrial plant and engineering technologies, and are outside the scope of a treatise on a chemical topic such as that presented in this chapter. Therefore, we focused our attention on the other principles to assess if the production of peptides by microbial fermentation could be considered a "green" technology. Moreover, we did not consider purification, since it is covered by other technologies presented in this book (*e.g.* SPPS), and its impact is often related to the specific product and not the technology.

Microbial fermentation is an intrinsically safe technology: usually the only solvent employed is water, which is non-toxic and non-flammable, and the temperatures involved are usually low (<50 °C). Moreover, most of the reagents employed are safe, since all we need to do to obtain the desired peptide or nucleotide is to feed energy in the form of sugars to the microbe.

Table 7.1 The 12 principles of green chemistry.[54]

1	Prevent waste
2	Design safer chemicals and products
3	Design less hazardous chemical syntheses
4	Use renewable feedstocks
5	Use catalysts, not stoichiometric reagents
6	Avoid chemical derivatives
7	Maximize atom economy
8	Use safer solvents and reaction conditions
9	Increase energy efficiency
10	Design chemicals and products to degrade after use
11	Analyze in real time to prevent pollution
12	Minimize the potential for accidents

Therefore, we can conclude that microbial fermentation is fully in compliance with the 2nd, 3rd, 8th, and 12th principles of green chemistry.

Almost no fossil-based materials are employed in a microbial fermentation process. In fact, both the microbe and the reagents present in such a process are completely renewable, and water is basically the only solvent needed during the fermentation. The 4th principle perfectly fits the microbial fermentation technology.

The synthesis of peptides is performed by a microbe through the use of specific enzymes, which are the "catalysts of nature". Even if reagents are fed (*e.g.* sugar) to the microbe, they are not considered stoichiometric. Moreover, no intermediate or protection/deprotection steps are needed, since enzymes are extremely selective catalysts. Therefore, the synthesis performed by the microbe is perfectly in line with both the 5th and 6th principles.

Even though it is not trivial to identify a specific chemical transformation in a microbial fermentation process, since the chemical bonds are formed through catalysts (enzymes) and without protection/deprotection steps, its atom economy should be the highest possible for these transformations. For these reasons, we can conclude that the atom economy is maximized if a microbial fermentation process is employed for the synthesis of peptides and nucleotides (7th principle).

Often, biotransformation processes have been considered environmentally friendly, but in many cases these conclusions are not sustained by real metrics.[55] In particular, aqueous-based processes with a relatively low concentration of product generate large amounts of wastes. Even though the reduction of generated wastes should always be the driver to a greener process, peptide microbial fermentation consists mainly of water and not toxic raw materials, as illustrated in Figure 7.5. After the inactivation of the microbe and degradation of the residual peptide, the waste could be directly released into the environment. Therefore, we could even argue whether such processes actually generate wastes (1st principle).

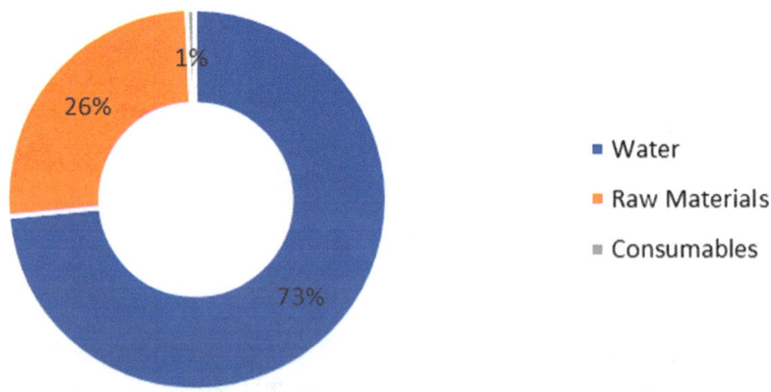

Figure 7.5 Composition of process mass intensity (PMI) from microbial fermentation industrial processes.

Finally, even if the 10th principle mainly refers to the product, most of the materials employed in a microbial fermentation process are biodegradable by definition.

Hence most of the 12 principles of green chemistry are respected in microbial fermentation for the production of peptides and nucleotides, and we can safely conclude that this technology can be considered green.

However, all that glitters is not gold, and much more work needs to be done to improve the greenness of this technology further.

7.6.2 Further Improvements of Sustainability

In order to design a sustainable process, an increase in the amount of output product per unit input of ingredients and energy is mandatory. Mutation of the producing host to create efficient producing strains can lead to a significant improvement in process efficiency and a reduction of environmental impacts. The microorganism is not the only key point of the development activity. Optimization of the cultivation conditions by optimizing the media, temperature, pH, inoculum, and production kinetics is important to achieve the maximum potential of the strain in terms of growth and production of recombinant proteins. The raw material consumed in the largest quantity in fermentation is water, which is used for media preparation, for equipment cleaning, and for cooling during fermentation. The reuse or efficient use of water is of high priority in the scale-up of a fermentation process. The cultivation media are composed of carbon and nitrogen sources and a salt solution containing all the essential elements and some vitamins. The constituents of the media must satisfy the elemental requirements for cell biomass and recombinant protein production giving an adequate supply of energy for biosynthesis and cell maintenance. The selection of appropriate carbon and nitrogen sources and optimization of their concentrations in the media are essential to obtain a high yield and might have a direct impact on process sustainability. Substrates produced using energy- and chemical-intensive processes play a negative role. In contrast, many substrates (starch, proteins, sugar) that are co-produced with other agricultural products and represent a waste of the primary production process have a low environmental impact. For example, glycerol is a cheap carbon source that can be an alternative to glucose.

The carbon source is the energy source of the microorganism. Rapidly metabolized carbon sources lead to a high oxygen demand and high power absorption. Oxygen is a key substrate of the fermentation. If there is a risk of oxygen limitation due to fast metabolism, it may be overcome by decreasing the concentration of the carbon source in the batch medium and adding additional quantities of the same as a continuous or semicontinuous feed during fermentation, or otherwise changing the carbon source with a complex raw material (starch or dextrin) in order to slow the metabolism.

The fermentation kinetics have a great impact on the final product yield and consequently also on process sustainability. Not only the quantity but also the quality of the recombinant protein could vary depending on the fermentation model implemented. There are several fermentation models that can be adopted, outlined below.

Feed-batch fermentation allows good control of the carbon source availability, which is very important to avoid a shift of metabolism and the production of unwanted by-products. This strategy combines the possibility of separating the growth and production stages, ensuring good cellular activity by means of controlled nutrient utilization.

Typically, in inducible recombinant protein expression systems, there is a high specific growth rate phase, initially, to obtain a large amount of biomass, and a production phase in which the growth rate is decreased to drive the carbon into recombinant protein production. The uncoupling of growth and protein expression can be advantageous for both biomass formation and protein yield.

The reduction of the kinetics during the production phase can improve protein folding and facilitate its exportation outside the cell in yeast.[41]

The feeding profile can vary from linear to exponential, depending on the type of strain and promoter utilized and also on the fermentation phase considered. The use of a defined feeding of substrate allows an ideal growth rate to be maintained, preventing both over-nutrition and cell starvation.

Models of continuous feeding of media with intermittent drawing of broth can also be adopted. Novo Nordisk developed a continuous fermentation process for the production of insulin and GLP-1 analogs.[46] Nevertheless, continuous fermentation is generally considered risky at the industrial scale. In fact, contamination, strain mutations, and product instability may reduce the fermentation yield significantly.

High cell density cultures are a preferred strategy for optimizing recombinant protein volumetric productivity, which is a key parameter of bioprocess cost-effectiveness. It is important to consider that protein production may be negatively affected by a high growth rate, when defining the fermentation model. Impaired folding and subsequent degradation of the product might occur. Although the preferred growth temperature of the host is 28–30 °C, lowering the temperature to 20–25 °C during the induction phase could help to reduce the growth rate and to obtain the correct protein folding and decrease the proteolytic activities responsible for recombinant protein degradation.

An improvement in recombinant protein production may be obtained by reducing the metabolic load. The metabolic load corresponds to the proportion of resources devoted to the basic cellular metabolism (cell maintenance) and dedicated to recombinant gene maintenance and expression. The metabolic load increases with increase in the genetic load, gene size or gene copy number, and the expression level. An increase in the gene copy number may not result in increased recombinant protein production if an overload problem has been determined. The effect produced by the metabolic load is a decrease in the growth rate of the host and mistranslation of the recombinant protein. To reduce the metabolic load, integration of the recombinant gene in the chromosome might be helpful, because maintenance of plasmids inside the cell requires a large amount of energy. Optimization of the formula of the medium to guarantee the availability of amino acids for recombinant protein production is also an essential factor.[56]

Contributions to global warming are to a large extent driven by energy consumption.[5] Fermentations require steam sterilization of the equipment and stirring of the media during cultivation, which has a certain energy consumption.[3] Energy consumption is an important aspect of the fermentation process and its minimization is a crucial factor to improve sustainability. Host strains with high growth rates, such as *E. coli*, represent an advantage in terms of energy depletion during cultivation because their fermentation cycle time is short. An *E. coli* fermentation usually lasts 24–48 h, whereas a yeast fermentation normally takes 5–7 days. Even though the peak of energy absorption may be higher, a shortened fermentation time has a positive effect on the total energy consumption.

To guarantee good oxygen transfer in high cell density fermentations, a higher stirring speed is applied and an increase in power absorption is observed. Nevertheless, the increase in the process yield can result in an improvement in PMI and lead to an overall positive life-cycle assessment.

The power consumption of the fermentation is not only related to the fermentation time. A decrease in energy consumption during cultivation can be obtained also by choosing efficient mixing equipment in the bioreactor. There are innovative impeller types that were designed to improve mixing and allow the conduction of the fermentation at a lower stirring speed compared with classical Rushton turbines. The result is a decrease in power absorption.

7.7 Scale-up of a Recombinant Product

To ensure the reliability of the scale-up, it is important to adopt a suitable scale-up criterion. A constant oxygen mass transfer coefficient (k_La), constant specific power input (P/V), constant impeller tip speed, and constant dissolved oxygen concentration are four commonly used scale-up criteria used in the fermentation industry.[57]

One of the major substrates of the fermentation is oxygen, which is provided by aeration. For optimal oxygen transfer, the air flow and agitation are adjusted automatically to maintain a minimum dissolved oxygen set point, which is usually fixed at 20%. Oxygen transfer capacity coupled with excellent cooling facilities is crucial to achieve the maximum potential of fermentation. As the culture scale increases, it becomes more difficult to ensure dissolved oxygen availability throughout the culture and oxygen often becomes the limiting nutrient for the fermentation process. A constant k_La (volumetric oxygen transfer coefficient) is the best option for scale-up.[58] A minimum dissolved oxygen set point can be considered as an alternative to k_La.

The internal asset of the fermenter and the type and dimensions of the impellers and their reciprocal distances are key factors. To guarantee maximum success in scaling up, it is important to define a good scale-down model to be used in the development of the fermentation process. If methanol is used in the fermentation, specific safety plant adaptations have to be introduced to minimize the fire risk.

In general, *E. coli* fermentation can be easily scaled up and the risk of failure is very low. Using this bacterial strain, huge amounts of recombinant protein can be obtained in a short time at a low cost with an environmentally sustainable process. The modulation of the growth rate during production is the main key point. If this aspect is not controlled, the risk of plasmid loss or formation of inclusion bodies would be very high. Genetic stability is higher if yeast is selected as a host, because the recombinant gene is generally integrated in the chromosome. Eukaryotic cells require a fermentation model of greater complexity and consequently the scale-up is more tricky. Oxygen limitation could have a severe effect on the fermentation yield. Ethanol formation in *S. cerevisiae* could completely compromise fermentation.

Strict control of the feed rates and the study of the optimal feed profile are mandatory in yeast fermentations and the fermentation kinetics have to be well defined to guarantee the secretion of a properly folded product. Once the upstream phase is complete, where the purpose is to have the expressing organism to produce the target molecule at the highest possible concentration, the ensuing part is the downstream process. Here the target molecule must be isolated and purified, to meet the required quality criteria for the drug substance. This is achieved in a series of steps, each one increasing the stream quality from the previous step, in a continuous enhancement from the broth to the API. The first goal of the downstream process is to have the peptide in aqueous solution or in the supernatant phase of the broth. This is more easily achieved when yeasts (*i.e. S. cerevisiae* or *P. pastoris*) are employed to express the API, as these organisms generally are able to produce the desired molecule in an extracellular fashion and the target compound is already in the liquid mixture. In contrast, when employing other types of bacteria that express the target peptide in an intramolecular fashion (*i.e. E. coli*), a cell disruption step is required. In any case, most often a volume reduction is useful before any other operation; centrifugation or membrane filtration are common operations. After this, the batch volume is reduced, making the process less expensive and less time consuming, at the same time some of the most dissimilar components of the broth are removed from the broth itself. In the case that a lysis is necessary, two main types of cell disruption are available to the process chemist to perform the operation: chemical lysis and mechanical lysis. Based on several factors, such as stability of the cell, economic considerations, and technological feasibility, many options can be considered. Chemical lysis can be brought about by (among others) osmotic or temperature shocks, enzyme digestions, or lipid membrane dissolution. Mechanical lysis, on the other hand, can be realized by homogenization, grinding, or sonication, all methods that are able to rupture the cell membrane and release the peptide in solution.

Once the target peptide is in solution, the first stage of purification starts, namely the removal of destabilizing or degrading impurities. As one can imagine, the broth as is, or the lysate broth in the case of lysis, contains many undesired or even deleterious contaminants, such as pyrogens, viruses, proteases, and other high molecular weight protein-based materials. Once again, a few process options are available: filtrations are common choices, and also capture and release chromatography techniques. Filtrations are often employed to improve the quality of the mixture under cheap conditions: with proper choice of the filtration media, one can remove solid particles from a complex mixture, or even separate soluble compounds. In addition to conventional filtrations, which are based on flowing the mixture in a perpendicular direction through a filtration medium to build up a cake of filtered material, tangential flow filtration (TFF) (sometimes referred to as membrane filtration) is commonly performed as a bioseparation step. In TFF, the flow of the mixture is parallel to the filtration membrane, running over its surface at a defined flow-rate. By regulating the operating parameters, the process developer can manipulate the system and apply an apparent pressure (defined as transmembrane pressure) that forces the mixture onto the membrane surface and, consequently, the filtered solution flows through it (see Figure 7.6). The main advantage of this technique is the avoidance of the buildup of a cake on the filtration medium, as the continuous flow removes any deposited particles from the surface and recycles them back to the feed tank; the result is a steady filtration flow-rate.[59] Membranes range greatly in size, type of filtration material, porosity, and physical disposition, giving the process developer a vast range of choices to select the most suitable one for the process at hand. In many cases when dealing with complex and inherently inhomogeneous mixture such as broths, a flocculation or sedimentation operation can help to keep the phase efficient.

At this stage of the downstream process, the target is to remove most of the compounds with high molecular weight and improve the purity of the stream. Microfiltration TFF techniques are employed, where filtration

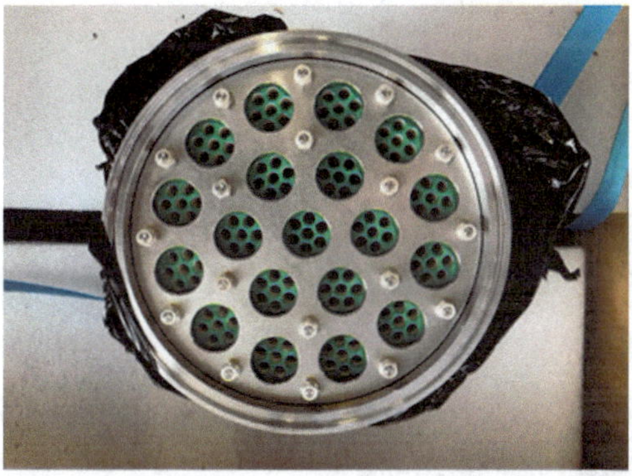

Figure 7.6 Head-on view of the interior of a microfiltration membrane.

media ranging from 0.1 to 0.45 μm allow the efficient filtration of smaller compounds (which generally are the target product) away from the bulk of contaminants. The recovered permeate from this phase is a mostly clear solution, which often needs to be concentrated to a more reasonable volume to achieve process efficiency. Once again TFF techniques are employed in this phase, where smaller pore membranes are used to retain this time the target compound in the retentate solution (see Figure 7.7).

Another technique commonly applied to upgrade the quality of the stream is capture and release chromatography (also called affinity chromatography). In this purification step, it is possible to achieve both a volume reduction of the batch and removal of bulk impurities, such as endotoxins and DNA-related

Figure 7.7 Microfiltration apparatus: at (a) industrial scale, and (b) pilot scale.

residues, among others. The methodology is based on a specific binding between the target molecule and the adsorbent, which is functionalized with specific functionalities. Therefore, the starting solution is loaded into the system (usually a packed column), then the system undergoes a wash phase, in which all the unbound product is physically eluted from the column, and finally an eluent able to displace the bound product from the resin is introduced and the eluates can be selectively collected to obtain a solution with highly improved quality and reduced volumes. Significantly, high yields and material recovery are easily achievable with capture and release purification. There are several possible interactions and bindings that can be exploited to carry out an affinity purification on proteins: enzyme–ligand type of interactions, receptor–agonist systems, and also metal–ligand complexations. The last system is denoted immobilized metal affinity chromatography (IMAC) and exploits the ability of some amino acids, the most utilized being histidine, to chelate transition metals such as nickel or copper (see Figure 7.8).

These methods of stream upgrading can be so efficient that target peptides sometimes are strategically expressed as a fusion protein, including a tail or 'tag' specially tailored to have a strong affinity with the selected adsorbent and allow for a clean capture stage. Obviously, the ease of purification is not the only advantage of expressing a fusion protein, it also comes with

Figure 7.8 Pilot-scale IMAC column.

increased stability of the fusion protein itself in the fermentation broth and in the lysis step. Substantial research has been reported on this topic over the years, and its use is a relatively standard option in the arsenal of a process developer. The down side of this strategy is plainly the need to excise the 'extra' part of the molecule, but, again, the tag is generally engineered in such a manner that the removal is fairly straightforward and does not impair the process. Very defined amino acid sequences are inserted where required, and these sequences are the target for a protease-type enzyme that performs a clean hydrolysis of the peptidic bond, finally releasing the peptide in its correct primary structure and removing the fusion tag.

In most processes, an intermediate purification step is required before moving to the final stages of API manufacture. Often an ion-exchange resin is used, but affinity chromatography or size-exclusion chromatography can also prove useful. This stage of purification is aimed at removing the last remnants of host cell proteins and process-related impurities and to achieve virus clearance. In cases where a bioconversion was performed, the added enzyme and the excised fusion tag also need to be removed.

Finally, the stage is set to perform a fine chromatographic purification (see Figure 7.9), the aim being to remove the product-related impurities and achieve the quality target profile. Broadly, two types of purification are used: low–medium-pressure chromatography and high-pressure chromatography, both of which are based on hydrophobic interactions working in reverse phase. Low-pressure purification is the most appealing from the process point of view, owing to both the easier operation and the lower overall cost

Figure 7.9 Example of an industrial chromatographic system.

of operation. In low-pressure purifications, several resins with various particle sizes and pore distributions can be used, whereas high-pressure purifications generally rely on a silica matrix functionalized with alkyl chains (C_4, C_8, and C_{18} are the most common).

Following the purification step, one can proceed to isolate the API. In bioprocesses the target peptide is a relatively sensitive product, and heat and physical shocks can be critical, hence normal drying in a standard heating apparatus can be deleterious to the quality of the product. Freeze-drying is applied instead: the high-purity eluates obtained from the purification stage are frozen at cryogenic temperatures and then lyophilized. A treatment of the eluted solution is generally necessary before freeze-drying, as organic solvents have to be removed, in addition to pH corrections and eventually addition of stabilizers. When complete, the downstream process results in the isolation of an amorphous powder of pure peptide for commercialization.

The ultimate goal of the development phase is to scale up the process from the laboratory scale to the pilot scale or even directly to the plant scale. When doing this, additional considerations must be taken into account. The product quality is clearly the main factor, but other parameters also come into play: economics and instrumental availability are critical to a successful process definition. In light of this, volume reduction and time efficiency play a crucial role in the process development. Moreover, it is important to match upstream and downstream phases of the same process: the timelines need to align to achieve the required efficiency. A single fermentation broth can sustain more than one downstream process; at the same time, the completion of a downstream process can require less time than an upstream process. To avoid any plant downtime, prior planning is critical.

When developing a new process in R&D, correct downscaling is important: knowledge and experience of the plant are essential and the involvement of all the necessary personnel can help to anticipate many issues. The time scale to consider in plant operations is on a different level compared with the laboratory, and product stability must be correctly assessed to avoid incurring any unexpected surprises. Most of the instrumentation used in plants can be correctly reproduced in the laboratory, and a robust process design considers the scalability of all parameters, such as longer operating times and more complex unit operations. This implies a thorough knowledge of product stability in any phase of the process, both chemical stability, to control the impurity profile, and physical stability, to prevent undesired phenomena such as precipitation, aggregation, or clogging, or even more subtle fibrillation, all of which cause losses of material. Many other parameters can change dramatically with an increase in batch size during the scale-up phase, and all of them must be kept under control, *e.g.* pressure, local concentration inhomogeneity, or pH variations.

A powerful tool in a process chemist's hands is the design of experiment (DoE), a tool highly supported by the "quality by design" (QbD) concept expressed by International Council for Harmonisation of Technical Requirements for Pharmaceuticals for Human Use (ICH) guidelines (ICH Q8).[60] By applying this statistical instrument one can define an area of feasibility of the process with focused experiments; this results in a much safer scalability of the process itself: rather than a working on a defined value per parameter, a wider working area is defined as safe. This in turn allows one to modify, and most importantly retain total control over, any parameter during the scale-up without concerns for the product quality or yield. This is a "win–win" situation where the process guarantees the output quality of the product and eventual issues normally encountered during the scale-up phase can be treated under control.

7.8 Future Directions: Beyond the Canonical 20 Amino Acids

The new frontier in recombinant protein production is represented by the possibility of introducing unnatural amino acids into proteins. A possible strategy used to achieve this aim employs the three stop codons Amber, Ochre, and Opal to specify the unnatural amino acid. The first experiments were carried out with Amber codon in a cell-free system. Subsequently, the translation machinery was introduced in recombinant cells to achieve the direct incorporation of unnatural amino acids into the protein during the protein synthesis.[61]

The Amber codon is the least-used stop codon and its reassignment to unnatural amino acids was expected to have no impact on the synthesis of other proteins. An aminoacyl tRNA synthetase that specifically aminoacylates the suppressor tRNA with the unnatural amino acid was introduced in the recombinant strain genome. Feeding the fermentation with the unnatural amino acid, its incorporation into the protein was achieved.[62] The efficiency of incorporation is still low because the Amber stop codon is also recognized by the cellular release factor, which determines the termination of the protein synthesis.[63] This research, which is in an early phase, will potentially allow the creation of new proteins that harbor one or more unnatural amino acids and reduce the gap between biotechnology and chemical synthesis.

A very promising new tool, with great potential to streamline the development of new projects and scale-up effectiveness, among many other applications, is the coming of age of bioinformatic and chemoinformatic tools. The ability to rationally design new ideas *a priori* and test them *in silico*, and the possibility to exploit the capacity of algorithms to direct the research in a more focused manner, are all areas with great impact. When applied to the field of enzymes, for example, control and prediction of the activity of enzymes become a real possibility, leading to shorter development times, more efficient results, and overall more economical solutions, coupled with less waste. This is already taking place in academic laboratories, where these abilities are being combined with the direct evolution of enzymes with excellent results, opening up the way to improve the robustness and reactivity of enzymes, together with a broader selectivity for a wider range of substrates.

These advances can positively impact the industrial environment with effects on development timelines and material waste.

List of Abbreviations

ADH1	Alcohol dehydrogenase
AOX1, AOX2	Aldehyde oxidase enzymes
API	Active pharmaceutical ingredient
cDNA	Complementary DNA
DoE	Design of experiment
DPP-4	Dipeptidyl peptidase-4
emp	Enzyme-modified porcine
ER	Endoplasmic reticulum
FDA	US Food and Drug Administration
GAP	GTPase-accelerating protein
GAPDH	Glyceraldehyde-3-phosphate dehydrogenase
GDP	Gross domestic product
GLP-1	Glucagon-like peptide 1
ICH	International Council for Harmonisation of Technical Requirements for Pharmaceuticals for Human Use
IMAC	Immobilized metal affinity chromatography
IPTG	Isopropyl-β-D-thiogalactopyranoside
lacUV	Promoter from the *Escherichia coli* lac operon
LCA	Life-cycle assessment
LPPS	Liquid-phase peptide synthesis
MCS	Multiple cloning site
NEP 24.11	Neutral endopeptidase
PGK	Phosphoglycerate kinase
pUC	Plasmid cloning vector
QbD	Quality by design
rDNA	Recombinant DNA
rh-insulin	Recombinant human insulin
RP-HPLC	Reversed-phase high-performance liquid chromatography
SNAC	Sodium 8-[(2-hydroxybenzoyl)amino]octanoate
SPPS	Solid-phase peptide synthesis
TFA	Trifluoroacetic acid
TFF	Tangential flow filtration

References

1. L. Ferrazzano, M. Catani, A. Cavazzini, G. Martelli, D. Corbisiero, P. Cantelmi, T. Fantoni, A. Mattellone, C. De Luca, S. Feletti, W. Cabri and A. Tolomelli, *Green Chem.*, 2022, **24**, 975.
2. K. Wowra, E. Hegel, A. Scharf, A. Gruenberger and K. Rosenthal, *Trends Biotechnol.*, 2023, **41**, 1199.

3. M. Becker, N. Kolheb, S. Hunger, R. Ecshrich, R. Mueller and A. Aurich, *Eng. Life Sci.*, 2020, **20**, 90.
4. S. Feijoo, S. Gonzalz-Garcia, J. M. Lema and M. T. Moreira, *J. Cleaner Prod.*, 2017, **165**, 204.
5. H. Nielsen, K. Oxenboll and H. Wenzel, *Int. J. Life Cycle Assess.*, 2007, **12**, 432.
6. N. A. Baeshen, M. N. Baeshen, A. Sheikh, R. S. Bora, M. M. M. Ahmed, H. A. Ramadan, K. S. Saini and E. M. Redwan, *Microb. Cell Fact.*, 2014, **13**, 141.
7. T. Kjeldsen, P. Balschmidt, I. Diers, M. Hach, N. C. Kaarsholm and S. Ludvigsen, *Biotechnol. Genet. Eng. Rev.*, 2001, **18**, 89.
8. R. E. Chance, N. B. Glazer and K. L. Wisher, in *Biopharmaceuticals An Industrial Perspective*, 1999, p. 149.
9. F. G. Banting, C. H. Best and J. J. R. Macleod, *Am. J. Physiol.*, 1922, **59**, 479.
10. D. Mohajan and H. K. Mohajan, *Stud. Soc. Sci. Humanit.*, 2023, **2**, 8.
11. G. Walsh, *Appl. Microbiol. Biotechnol.*, 2005, **67**, 151.
12. L. Fryklund, *Med. Res. Arch.*, 2023, 11.
13. I. Vecchio, C. Tornali, N. L. Bragazzi and M. Martini, *Front. Endocrinol.*, 2018, **9**, 1.
14. G. A. Homandberg, J. A. Mattis and M. Laskowski, *Biochemistry*, 1978, **17**, 5220.
15. A. Riggs, *Endocr. Rev.*, 2021, **42**, 374.
16. W. Landgraf and J. Sandow, *Diabetes*, 2016, **12**, 12.
17. J. Alyas, A. Rafiq, H. Amir, S. U. Khan, T. Sultana, A. Ali, A. Hameed, I. Ahmad, A. Kazmi, T. Sajid and A. Ahmad, *Biomed. Res. Ther.*, 2021, **8**, 4540.
18. G. Walsh and E. Walsh, *Nat. Biotechnol.*, 2022, **40**, 1722.
19. P. Fonjallaz and E. Loumaye, *J. Biotechnol.*, 2000, **79**, 185.
20. Editorial, *Nat. Rev. Endocrinol.*, 2023, **19**, 311.
21. I. Mullerup, T. Kornfelt, P. Oreskov and H. Tottrup, *Glucagon III*, Springer Berlin, Heidelberg, 1996.
22. B. Manandhar and J. M. Ahn, *J. Med. Chem.*, 2015, **58**, 1020.
23. L. B. Knudsen and J. Lay, *Front. Endocrinol.*, 2019, **10**, 1.
24. J. L. Lau and M. K. Dunn, *Bioorg. Med. Chem.*, 2017, **26**, 2700.
25. I. Guryanov, A. Bondesan, D. Visentini, A. Orlandin, B. Biondi, C. Toniolo, F. Formaggio, A. Ricci, J. Zanon and W. Cabri, *J. Pept. Sci.*, 2016, 471.
26. D. Carbajo Lopez and F. Albericio Palomera, Method for preparation of Liraglutide using BAL linker, *WO pat.*, WO138855, 2017.
27. M. L. Puppala and V. Mudii, Process for preparation of Liraglutide using recombinant *Saccharomyces cerevisiae*, *WO pat.*, WO 082138, 2019.
28. D. L. Dunweber, I. H. Jensen and L. B. Hansen, Method for producing acylated peptides, *WO pat.*, WO029077, 2004.
29. T. Moks, L. Abrahmsen, B. Osterlof, S. Josephson, M. Ostlig, S. O. Enfors, I. Persson, B. Nilsson and e M. Uhlen, *Nature*, 1987, **5**, 379.
30. S. Gupta, S. S. Salunkhe, B. Varshney and R. S. Mody, Expression and large-scale production of peptides, *WO pat.*, WO172921, 2018.

31. N. Cheng, L. Yang, N. Dai, Z. Hu, F. Yang, R. Chen, J. Chen, Y. Zhou, Y. Huang and Z. Su, *Process Biochem.*, 2008, **43**, 1253.
32. S. Ahmadi, M. B. Shahsavani, Z. Tavaf, R. M. Albaghlany, A. Kumar, A. A. Moosavi-Movahedi and R. Yousefi, *PLoS One*, 2022, **17**, 1.
33. J. Lau, P. Bloch, L. Schaffer, I. Petterson, J. Speltzer, J. Kofoed, K. Madsen, L. B. Knudsen, J. McGuire, D. B. Steensgaard, H. M. Strauss, D. X. Gram, S. M. Knudsen, F. S. Nielsen, P. Thygesen, S. Reedtz-Runge and T. Fruse, *J. Med. Chem.*, 2015, **58**, 7370.
34. J. F. Lau, A. S. Andersen, P. Bloch, J. Lau, P. W. Garibay, T. Kruse, I. S. N. Noerby, C. U. Jessen, C. Christensen and J. C. Norrild, Semi-recombinant preparation of GLP-1 analogues, *WO pat.*, WO083549, 2009.
35. M. Lenharo, *Nature*, 2023, **620**, 480.
36. A. M. Vieira Gomes, T. S. Carmo, L. S. Carvalho, F. M. Bahia and N. S. Parachin, *Microorganisms*, 2018, **6**, 1.
37. M. Pouresmaeil and S. Aziz-Dargahlou, *Arch. Microbiol.*, 2023, **205**, 1.
38. T. W. Overton, *Drug Discovery Today*, 2014, **19**, 590.
39. G. L. Rosano and E. A. Ceccarelli, *Front. Microbiol.*, 2013, **40**, 257.
40. A. Berlec and B. Strukelj, *J. Ind. Microbiol. Biotechnol.*, 2013, **40**, 257.
41. A. Madhavan, K. B. Arun, R. Sindhu, J. Krishnamoorthy, R. Reshmy, R. Sirohi, A. Pugazhendi, M. K. Awasthi, G. Szakacs and P. Binod, *Microb. Cell Fact.*, 2021, **20**, 1.
42. M. Wacker, D. Linton, P. G. Hitchen, M. Nita-Lazar, S. M. Haslam, S. J. North, M. Panico, H. R. Morris, A. Dell, B. W. Wren and M. Aebi, *Science*, 2002, **298**, 1790.
43. N. Ferrer-Miralles, J. Domingo-Espin, J. L. Corchero, E. Vazquez and A. Villaverde, *Microb. Cell Fact.*, 2009, **8**, 1.
44. S. Wegmueller and S. Schimd, *Curr. Org. Chem.*, 2014, **18**, 1005.
45. G. Wang, M. Huang and J. Nielsen, *Curr. Opin. Biotechnol.*, 2017, **48**, 77.
46. A. Srivastava, J. Piskur, J. Nielsen and M. Egel-Mitani, Method for the production of heterologous polypeptides in transformed yeast cells, *US pat.*, US6190883, 2001.
47. M. Karbalaei, S. A. Rezaee and H. Farsiani, *J. Cell. Physiol.*, 2020, **235**, 5867.
48. M. Ahmad, M. Hirz, H. Pichler and H. Schwab, *Appl. Microbiol. Biotechnol.*, 2014, **98**, 5301.
49. K. De Schutter, Y. Lin, P. Tiels, A. Van Hecke, S. Glinka, J. Weber Lehmann, P. Rouze, Y. Van de Peer and N. Callewaert, *Nat. Biotechnol.*, 2009, **27**, 561.
50. N. K. Tripathi and A. Shrivastava, *Front. Bioeng. Biotechnol.*, 2019, **7**, 1.
51. S. Takagi, N. Tsutsumi, Y. Terui, X. King, H. Yurimoto and Y. Sakai, *FEMS Yeast Res.*, 2019, **19**, 1.
52. Summary of the Pollution Prevention Act: 42 U.S.C. §13101 et seq, 1990.
53. Green Chemistry History, available at: https://www.acs.org/greenchemistry/what-is-green-chemistry/history-of-green-chemistry.html.
54. P. T. Anastas and J. C. Warner, *Green Chemistry: Theory and Practice*, Oxford University Press, New York, 1998.

55. P. D. De Maria, *Curr. Opin. Green Sustainable Chem.*, 2021, **31**, 100514.
56. M. Vandermies and P. Frickers, *Microorganisms*, 2019, **7**, 1.
57. M. Sun, A. X. Gao, X. Liu, Y. Yang, R. Ledesma-Amaro and Z. Bai, *Microb. Cell Fact.*, 2023, **22**, 1.
58. A. H. M. Gameil, F. Yusuf, A. S. Azmi and N. I. M. Puad, *Biol. Nat. Res. Eng. J.*, 2021, **5**, 37.
59. P. Agrawal, K. Wilkstein, E. Giunn, M. S.-M. C. I. Mason and J. Sayle, *Org. Process Res. Dev.*, 2023, **27**, 571.
60. ICH Q8 (R2), Pharmaceutical development – Scientific guideline, available at: https://www.ema.europa.eu/en/ich-q8-r2-pharmaceutical-development-scientific-guideline.
61. T. S. Young and P. G. Schultz, *J. Biol. Chem.*, 2010, **285**, 11039.
62. D. De la Torre and J. W. Chin, *Nat. Rev. Genet.*, 2021, **22**, 169.
63. S. Hans, N. Kumar, N. Gohil, K. Khambhati, G. Bhattacharjee, S. S. Deb, R. Maurya, V. Kumar, S. M. S. Reshamwala and V. Singh, *Microb. Cell Factories*, 2022, **21**, 1.
64. S. Ludvigsen, D. B. Steensgaard, J. K. Thomsen, H. Strauss and M. Normann, Liraglutide, 2012, DOI: 10.2210/pdb4apd/pdb.
65. X. Zhang, M. J. Belousoff, R. Danev, P. M. Sexton and D. Wootten, Semaglutide-bound Glucagon-Like Peptide-1 (GLP-1) Receptor in Complex with Gs protein, 2020, DOI: 10.2210/pdb7ki0/pdb.
66. X. Zhang, M. J. Belousoff, R. Danev, P. M. Sexton and D. Wootten, Structure and dynamics of semaglutide- and taspoglutide-bound GLP-1R-Gs complexes, *Cell Rep.*, 2021, **36**, 109374.

CHAPTER 8

Oligopeptide Regulatory Affairs and Challenges

A. J. WOLF[a*] AND A. RICCI[b*]

[a]Eli Lilly and Company, 1200 Kentucky Avenue, Indianapolis, IN 46221, USA; [b]Fresenius Kabi, via Roma 108, 20060 Cassina de Pecchi (MI), Italy
*E-mail: wolf_allison@lilly.com, antonio.ricci@fresenius-kabi.com

8.1 Introduction

8.1.1 Drugs *Versus* Biologics

In the USA, drugs and biologics are regulated under different regulatory pathways. Drugs are approved under the Food and Drug and Cosmetic Act and the Food and Drug Modernization Act *via* a New Drug Application (NDA)[1] procedure or *via* an abbreviated New Drug Application (ANDA)[2] procedure in the case of generics. Biologics are licensed under Section 351 of the Public Health Service (PHS) Act *via* a Biologics License Application (BLA) procedure.[3] Historically, therapeutic peptides have been classified as either drugs or biologics depending on the number of amino acids in the peptide and the method of manufacture (*e.g.* synthetic, recombinant). The Further Consolidated Appropriations Act, enacted on December 20th, 2019, further amended the definition of a "biological product" in the PHS Act to remove the parenthetical "(except any chemically synthesized polypeptide)" from the statutory category of "protein", making the number of amino acids the sole distinction between a protein/biologic and a drug.[4] Under this Act, a

Green Chemistry Series No. 85
Sustainability in TIDES Chemistry: Green Approaches to Oligonucleotides and Oligopeptides Synthesis
Edited by Alessandra Tolomelli, Lucia Ferrazzano and Walter Cabri

Published by the Royal Society of Chemistry, www.rsc.org

protein is any α-amino acid polymer with a specific defined sequence that is greater than 40 amino acids in length and is therefore considered a biologic and requires a BLA filing for commercialization.

In countries outside the USA, categorization of a drug *versus* a biologic is based on the method of manufacture. In these countries, any protein manufactured using chemical synthesis is regulated as a chemical drug regardless of the number of amino acids in the peptide. This difference in definitions could result in different categorizations of the same molecule across different geographies, which is the case with rDNA-derived teriparatide. This 34-amino acid oligopeptide is approved *via* an NDA in the USA, but is considered a biologic in all other geographies.

It is important to understand the regulatory pathway for a peptide early in clinical development. US regulations and expectations differ for biologics *versus* drugs in terms of potency requirements and process validation requirements. Polypeptide molecules containing more than 40 amino acids are considered proteins, and therefore need a potency determination based on the bioactivity regardless of the method of manufacture (*e.g.* synthetic, rDNA microbial expression) for initial phase I clinical trials run under an investigational new drug application (IND). This requirement is typically not flexible and could result in a clinical hold in the USA if an acceptable potency assay (*i.e.* bioassay) is not included in the specifications and stability studies.

As clinical development proceeds and planning begins for commercial applications, it is important to understand if the product will be filed through an NDA or BLA mechanism in the USA. If the product will be licensed through a BLA filing, process validation results will need to be included in the initial application for the drug substance, drug product, and device (if applicable). For European Union (EU), Japanese, and rest of the world submissions, synthetic peptides will not require drug substance process validation in the initial application unless the drug substance is sterile. However, for a parenteral drug product that has been sterilized by membrane filtration followed by aseptic filling, process validation data are required in the initial EU application because aseptic processing is considered non-standard in the EU Annex II. Language from Section 5.1 of the November 2016 *Guideline on Process Validation for Finished Products – Information and Data to be Provided in Regulatory Submissions* is as follows:[5]

> In certain cases, however, it is considered necessary to provide production-scale validation data in the marketing authorization dossier at the time of regulatory submission, for example when the product is a biological/biotech product or where the applicant is proposing a non-standard method of manufacture (see section 8 and Annex II). In these cases, data should be provided in the dossier on a number of consecutive batches at production scale prior to approval.

This guidance does provide additional context on non-standard processing and provides a possibility of providing justification that the process can be considered standard for a particular manufacturer and site taking into account the risk to the patient of failure of the product or process. It may be in the sponsor's best interest to negotiate with the EU regulatory agencies on this requirement as a delay to the timing of process performance qualification could allow for these batches to be used to support initial commercial supplies. Given the environmental footprint inherent to the manufacture of commercial batches of synthetic peptides, the EU regulatory agencies may be willing to partner with the sponsor to devise creative regulatory strategies that will allow for the first commercial batches to be able to be used within their shelf life. One approach that may be acceptable would be to delay the manufacture of process performance qualification batches such that the batch data would be available at the time of the issuance of the first assessment report on day 121 of the procedure for a standard marketing authorization application.

8.1.2 Regulatory Guidance and General Monographs

Novel synthetic peptides do not currently have dedicated guidance documents from the International Council for Harmonisation of Technical Requirements for Pharmaceuticals for Human Use (ICH), US Food and Drug Administration (FDA), or European Medicines Agency (EMA). Key ICH documents (*e.g.* Q3A, Q3B, Q6A, Q6B, and M7) exclude synthetic peptides from their scope. Synthetic peptides are included in the general chapters of some compendia, including *Substances for Pharmaceutical Use* (Ph. Eur. 2034), which establishes impurity reporting, identification, and qualification thresholds for peptide-related impurities (Table 8.1).[6] Although these thresholds are acceptable for EU submissions, some FDA reviewers have stated in public forums that these thresholds are not broadly accepted and would require product-specific justification by the sponsor. On the other side, the FDA issued guidance dedicated to the acceptance criteria for the ANDA filing of a generic peptide manufactured with a synthetic approach where the originator adopted a fermentation process.[7] This guidance originated as a consequence of a citizen petition issued by Novo Nordisk on the generic version of liraglutide, claiming that a different manufacturing strategy will generate a new impurity profile with unknown and potentially risky toxicological concerns.

The core of the guidance, drafted in October 2017 and approved in May 2021, set the clear expectation of not more than 0.10% for the unknown related substances and not more than 0.5% for the identified substances that were not present in the reference listed drug (RLD), whereas

Table 8.1 Thresholds for organic impurities in synthetic peptides.

Source	Reporting limit/%	Unknown limit/%	Identification limit/%	Qualification limit/%
Ph. Eur. 2034	>0.1	—	>0.5	>1.0
FDA Guidance 2021	—	<0.10	<0.5	>0.5

identification comprises not only their chemical properties (including aggregation) but also their potential immunogenicity response. Albeit this guidance is dedicated to the ANDA application of five specific peptides (glucagon, liraglutide, nesiritide, teriparatide, and teduglutide), it was considered as a reference starting point in the characterization of peptides. As a demonstration of this, the EMA Draft *Guideline on the Development and Manufacture of Synthetic Peptides*[8] introduced the control of the immunogenic potential and fibrillation (aggregation) of the peptide and their related impurities.

In the case of the peptide-related impurity specification limit, the EMA recommends the identification of the impurities below the 0.5% for generic synthetic peptides, even keeping valid the original specification limit:[8]

> In general, these synthetic peptides have to comply with the requirements of the Ph. Eur. General Monograph *Substances for Pharmaceutical Use*. This monograph allows an identification threshold of 0.5%. However, for comparability purposes a full evaluation of the impurity profile also covering impurities lower than 0.5% is expected.

Members of the USP BIO1 Expert Committee published a three-part series of papers covering different aspects of the control strategy for synthetic therapeutic peptides. The papers covered Analytical Considerations, Raw Material Considerations, and Manufacturing Process Considerations. This Expert Committee has also recently taken the principles from these publications and issued draft monographs USP ⟨1503⟩ *Quality Attributes of Synthetic Peptide Drug Substances* and USP ⟨1504⟩ *Quality Attributes of Starting Materials for the Chemical Synthesis of Therapeutic Peptides* for public comment. Although none of these articles are legally binding, they provide a generally accepted standard adopted by many synthetic peptide manufacturers.

The lack of guidance and standards has been acknowledged at external conferences by various regulatory agency representatives. In September 2022, the EMA issued a draft concept paper outlining topics for inclusion in a synthetic peptide guidance document, which was issued in draft form for comment in October 2023. Topics covered in the EMA draft guideline[8] include the following:

- integrated control strategy development;
- requirements for the 3.2.S.2 sections of the Common Technical Document (CTD) for solid-phase peptide synthesis (SPPS) manufacturing including manufacturing process descriptions and starting material expectations;
- characterization approaches including impurity profile;
- purity/impurity control strategy.

Although consensus regulatory guidance is lacking for synthetic peptides, externally available information and publications provide strong precedents that can guide the development and commercialization of therapeutic peptides. Control of impurities is a primary focus of any regulatory review of a synthetic peptide therapeutic. Later sections in this chapter will focus on the different aspects of impurity control for a synthetic peptide, as this control is foundational to ensuring the safety and efficacy of a therapeutic peptide.

8.2 Regulatory Considerations for Manufacturing Oligopeptide Therapeutics

All regulatory dossiers related to therapeutic products including synthetic peptides are required to provide detailed information on the manufacture of the drug substance and the drug product. This information is essential for regulatory agencies to assess the capability of the manufacturing process to produce consistent material that is of a quality comparable to that of the products tested during pivotal clinical trials. Throughout the manufacture, there are quality controls built into the process to ensure that the product meets its quality standards for use as a therapeutic product. The primary objective of the manufacturing process and control strategy is to provide acceptable control of identity, purity, strength, and quality, while also ensuring adequate supply of the drug substance and drug product to support commercial demand.

8.2.1 Control of Drug Substance Manufacturing Processes

Manufacturing protocols for the production of therapeutic peptides typically consist of two main processes: the upstream part, where the crude peptide primary sequence is assembled, and the downstream part, where the crude peptide is purified and isolated.[9–12] Synthesis of the crude peptide can be achieved through recombinant cDNA or chemical approaches such as liquid-phase peptide synthesis (LPPS)[13] or solid-phase peptide synthesis (SPPS),[14,15] and hybrid recombinant–chemical strategies are also possible.[16] For the purposes of this chapter, SPPS and LPPS protocols also include the utilization of peptide fragments (as opposed to being restricted to only single amino acids) and LPPS comprises classical solution peptide synthesis (CSPS), the novel liquid-phase approaches as classified by Albericio and co-workers[17] (see Figure 8.1), and the chemo-enzymatic peptide synthesis (CEPS) as described in Chapter 6.

The current peptide design trend focuses on the use of unnatural residues as a means to enhance the physicochemical and pharmacological properties of the resulting peptide.[18,19] An excellent example is the incorporation of a lysine derivative bearing a mixed poly(ethylene glycol) (PEG)–fatty acid moiety on its ε-amino group into a glucagon-like peptide 1 (GLP-1) derivative to reduce its aggregative tendency and enhance its binding to albumin, thus increasing its bioavailability.[20,21]

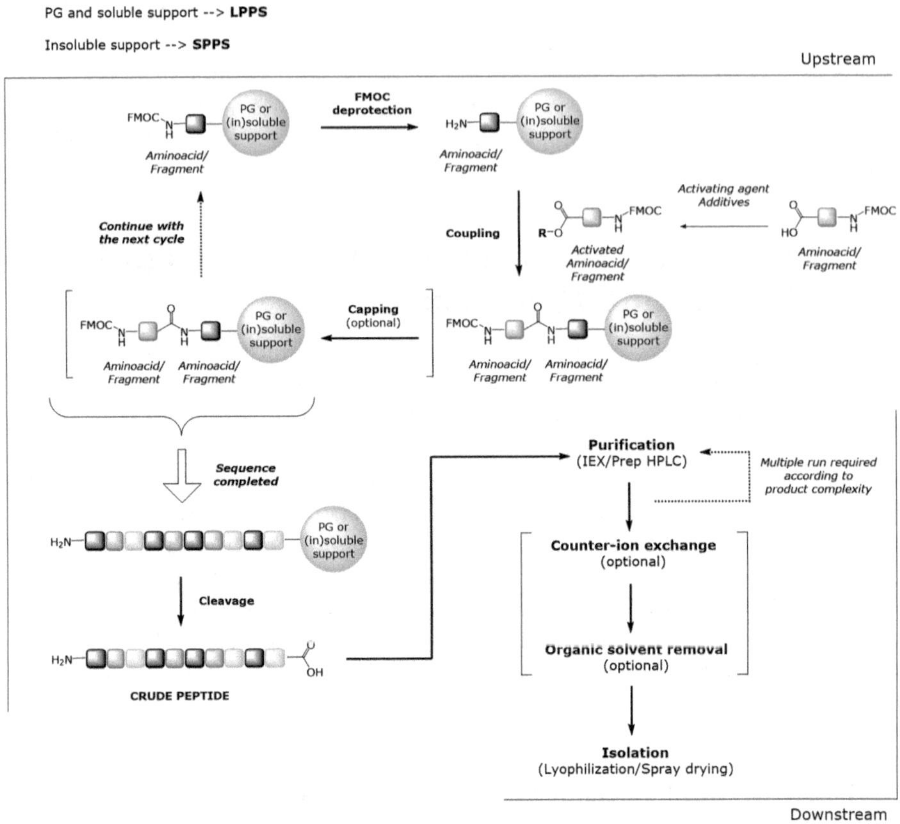

Figure 8.1 Schematic overview of the upstream (*via* SPPS/LPPS) and downstream processes in peptide manufacture.

This trend has led to the classification of peptides into three main categories:[22]

- *Native*, where the structure fully mimics the natural product (*e.g.* glucagon).
- *Analog*, where only a few modifications are made to enhance the biological properties (*e.g.* liraglutide).
- *Heterolog*, where the primary structure contains several unnatural residues and is significantly different from the parent native peptide (*e.g.* degarelix). These are typically short peptides of approximately 10 residues.

In this context, chemical approaches are generally preferred owing to the ease with which unnatural amino acids can be inserted into the primary sequence.[23] Furthermore, impressive improvements in the synthetic protocols over the past several decades have caused these approaches to become preferred over recombinant strategies even for native peptides. Within the realm of chemical approaches, evaluation of the various manufacturing options for synthetic peptides requires consideration of numerous factors, including technical aspects. Some methods may be more favorable in situations where the anticipated commercial demand is high, whereas others may be preferred for a low-volume therapeutic peptide. However, each approach is associated with distinct regulatory risks and will require different considerations and supporting data to provide the greatest probability of regulatory approval.

In the case of analogs and heterologs, SPPS can be considered the method of choice, and it is a method with which regulatory agencies have become familiar. The SPPS process entails repetitive cycles of deprotection, activation, and amide bond formation to incorporate each amino acid in the sequence, as depicted in Figure 8.1. Since the cornerstone achievement of the hybrid SPPS–LPPS protocol developed for enfuvirtide (brand name: Fuzeon) by Merck,[24] the fluorenylmethyloxycarbonyl (Fmoc) moiety represents the most common protective group of the N-terminus owing to its ability to be removed under mildly basic conditions.[15,25] Despite the high degree of optimization that has been attained for the SPPS approach, which typically allows for an amino acid coupling efficiency of >99%, its complexity is significantly greater than that of traditional small-molecule manufacturing processes. In particular, SPPS requires a large number of starting materials and in-process controls and may generate numerous impurities. During the review of a marketing authorization application, the regulatory reviewer will assess the adequacy of the described upstream process and associated controls to ensure the consistent management of impurities and other critical quality attributes. Sufficient manufacturing detail combined with a demonstrated understanding of the synthetic protocol must be provided in the application to allow for this assessment.

Following peptide assembly, purification and isolation steps are needed irrespective of the method of synthesis selected. Even the best controlled process has the potential to generate undesirable closely related compounds, which must be removed or reduced to yield a drug substance with acceptable purity. Column chromatography steps are often employed to provide high-purity oligopeptides.[11,26] Regulatory agencies will expect to see a comprehensive description of the purification process, including details of the stationary phase(s), mobile phase(s), and criteria used to evaluate the purity of the process streams collected during the chromatography steps. Equipment size constraints often necessitate the splitting or pooling of intermediate batches during the purification process. It is expected that the manufacturer will demonstrate control and traceability of the batch genealogy, including the criteria applied for combining intermediate batches during manufacture.

8.2.1.1 Control of Impurities in the Drug Substance

The impurity profile undoubtedly plays a pivotal role in the quality attributes of a therapeutic peptide. As with all products obtained through a (bio)chemical manufacturing process, the impurities present in a peptide product can be divided into peptide-related impurities, which are impurities associated with the peptide structure (*e.g.* diastereomers), and non-peptide impurities, such as residual solvents, ions, or other contaminants originating from the starting material or generated during the manufacturing process (*e.g.* nitrosamines).[8] Peptide-related impurities are primarily formed during the manufacturing process and display a characteristic pattern depending on the synthetic method adopted. Furthermore, critical peptide-related impurities may be carried over from the protected amino acid starting materials used during the synthesis. Finally, impurities resulting from peptide degradation must be considered. Table 8.2 presents a list of the most frequently encountered peptide-related impurities, which have been the topic of several excellent reviews and books.[27,28] This section will review only the most critical and common mechanisms associated with the impurity profile of isolated drug substances (or drug products), skipping those that affect other parameters (*e.g.* yield in the case of diketopiperazine formation).[29]

As can be appreciated from Table 8.2, several peptide-related impurities may originate from multiple sources, including the starting material, the manufacturing process, or the occurrence of degradation. For this reason, the first key step is to ensure adequate control of the protected amino acids (or dipeptides) used during peptide assembly; this aspect is further discussed in USP chapter ⟨1504⟩, which is dedicated entirely to this topic.[30] Specifically, any possible amino acid-related impurities, such as unwanted enantiomers, unprotected residues (particularly those lacking the Fmoc protective group or completely unprotected, which may lead to double insertion of the desired amino acid in the peptide sequence), dimers, and β-Ala–amino acid adducts should be considered because they may result in peptide-related impurities that are structurally very similar to the intended parent peptide. The satisfactory monitoring of such impurities may necessitate tailored analytical methods, such as the use of chiral stationary phases for enantiomers or alternative detectors for free amino acids that lack UV absorption. The final section of the aforementioned USP chapter is dedicated to non-peptide impurities carried over from the starting materials and outlines the need to evaluate their fate in the isolated peptide drug substance. One of these impurities that warrants special mention is ethyl acetate, a solvent that is often used for the crystallization of certain protected amino acids. Ethyl acetate can hydrolyze over time to liberate free acetic acid, an unwanted side product that is activated during the coupling step and then acts as a capping agent to generate a truncated peptide sequence. For this reason, whenever ethyl acetate is used, a negligible acetic acid content must be confirmed so as to suppress such side reactions. This is even more critical when ethyl acetate is used as a solvent, such as in combination with dimethyl sulfoxide in the case of a recently reported novel SPPS protocol.[31]

The most common peptide-related impurities formed during the manufacturing process are *endo*/des amino acid sequences, which resemble the desired primary sequence except for one amino acid having been coupled twice (in the case of *endo* sequences) or not at all (in the case of

Table 8.2 Overview of critical peptide-related impurities in SPPS/LPPS processes.[a]

Impurity	Description	Origin	Control strategy
endo-Amino acid	$N+1$ amino acid	Process (coupling or starting material)	Adjustment of process parameters or of starting material impurity profile
Des-amino acid	$N-1$ amino acid	Process (coupling, Fmoc deprotection)	Adjustment of coupling reaction condition or Fmoc deprotection
+ Protecting group/linker	Incomplete deprotection	Process (cleavage)	Adjustment of cleavage reaction conditions, selection of alternative linker
+ *endo*-β-Ala	N + *endo*-β-Ala residue	Process (starting material)	Control of *endo*-β-Ala impurity in the starting material
Truncated sequence	Capped sequence	Process (coupling)	Adjustment of coupling reaction conditions
Epimerization	Inversion of one amino acid stereocenter	Process (coupling, Fmoc deprotection, downstream or starting material); degradation	Adjustment of coupling/Fmoc deprotection conditions, tuning of basic conditions in downstream, control of starting material purity
Aspartimide	Aspartic acid to maleimide	Process (Fmoc deprotection, cleavage)	Use of Tmb-AA; adjustment of the process parameters
Deamidation	Hydrolysis of asparagine and glutamine side-chains	Process (upstream/cleavage, downstream or starting material); degradation	Water contamination in SPPS, Asp or Glu in the starting materials, residual moisture control
Oxidation	Oxidation of Met and Cys functions	Process (cleavage, downstream); degradation	Use of reducing agents, packaging under inert atmosphere
Disulfide bridge oligomerization	Monomer dimerization/polymerization through Cys–Cys bridge	Process (bridge formation, downstream; degradation in RTU FDF)	Diluted solution or portion-wise addition during cyclization, diluted FDF

[a]Tmb-AA, *N*-Fmoc-*N*-(2,4,6-trimethoxybenzyl)glycine; RTU, ready-to-use; FDF, finished dosage form.

des sequences). Truncated sequences originate from the missing insertion of one amino acid (as in the des-amino acid case) in an SPPS protocol where a capping step is performed after the coupling step. In this case, however, the primary sequence differs significantly from that of the parent peptide, unless the incomplete coupling occurred in the final stages of peptide assembly. *Endo* sequences may also be caused by the presence of amino acid dimer impurities or Fmoc-deprotected or free amino acids in the starting material, whereas des sequences can originate only during the manufacturing process.

Endo impurities formed during the manufacturing process are usually associated with the presence of a base in the coupling reaction medium, such as residual piperidine (or any other base) from the Fmoc deprotection.[32] The selection of mildly acidic conditions for the amino acid coupling, such as using diisopropylcarbodiimide (DIC) and OxymaPure, together with careful washing after Fmoc deprotection, can help reduce the formation of such impurities. A special case also worth mentioning is the double coupling of the first amino acid in the chain, particularly when trityl resins are used; in this case, the underlying mechanism has been identified as the activation of the penultimate amino acid, as demonstrated by the Ferring R&D team.[33]

The formation of des-amino acid impurities is dependent on the primary sequence and the manufacturing approach selected, because these impurities are associated with complex coupling positions that hamper the insertion of the next amino acid or the removal of the Fmoc protective group (which may be removed in the subsequent Fmoc deprotection step, allowing continuation of the sequence). Incomplete coupling can be mitigated by optimization of the reaction conditions (*e.g.* amino acid equivalents, reaction temperature, reaction time, choice of coupling reagent), repetition of the coupling step, or modification of the synthetic strategy. Several tools are currently available to avoid the problems arising from these "difficult sequences", such as pseudoproline or depsipeptide building blocks, and the insertion of such residues in critical positions may reduce the complexity of the coupling.[34–36] Similar strategies have also been successfully applied in the case of incomplete Fmoc deprotection. Finally, innovative approaches may assist in controlling these impurities. In the case of chemo-enzymatic methods, the synthesis of smaller fragments considerably reduces the likelihood of generating difficult sequences and thus des-amino acid impurities.[37,38]

Other purely process-related impurities are those originating from incomplete deprotection of the side-chain protective groups or incomplete removal of the solid support linker. In both cases, the impurity formation occurs during the cleavage step, where the completed crude peptide is detached from the solid support. Careful selection of the scavenger cocktail used in combination with the acidic medium [usually trifluoroacetic acid (TFA)], together with fine-tuning of the reaction conditions, should be performed at this stage to ensure total removal of the protective groups and complete trapping of the generated reactive species, thus avoiding their subsequent reattachment to the crude peptide. An interesting trend that has recently emerged is the use of amino acids featuring unprotected side-chains as starting materials in SPPS.[39] Seminal studies have demonstrated that several side-chain functionalities can be tolerated in the SPPS protocol without affecting the coupling/deprotection cycle efficiency, such as hydroxy (phenoxy), amide, imidazole, and guanidinium moieties. This approach drastically decreases the number of protective groups required, thus simplifying the cleavage procedure and mitigating the formation of protected impurities. Solid-phase linker removal follows a similar chemistry to that used for lateral chain protective group deprotection, where the carbocation liberated during the cleavage step can irreversibly reattack the crude peptide;[40–42] stable linkers such as the Wang linker exhibit a strong tendency for this owing to the high reactivity of the generated carbocation.[43] Also in this case, the use of unprotected amino acids will allow the adoption of acid-labile linkers (*e.g.* trityl-like linker) that generate more stable carbocations with a lower tendency to reattack the peptide chain during cleavage.

Peptide epimers are among the most common impurities in peptides because the racemization of individual amino acids can occur at practically every step of the manufacturing process, without even considering the possible presence of enantiomers in the protected amino acid starting materials.[44] The probability of diastereomeric impurities (where more than one amino acid stereocenter is inverted) is statistically far lower than that of epimeric impurities. The degree of epimerization is strongly dependent on the acidity of the α-proton of the amino acid and the reaction conditions employed. Side-chain-mediated epimerization has also been reported for histidine and cysteine residues.[45] The oxazolone side reaction that occurs during carboxylic acid activation is the most recognized pathway for epimerization during coupling.[46] The basic conditions used for Fmoc deprotection may also induce the epimerization of sensitive amino acids, particularly at the N-terminal residues of the growing chain. Downstream processes can further contribute to epimerization if basic mobile phases are required for purification, such as during the counterion-exchange steps. Lastly, epimerization can take place during storage, particularly at the level of the drug product if the finished dosage form (FDF) is a ready-to-use solution formulated at basic pH, as in the case of the GLP-1 analogs liraglutide (brand names: Victoza and Saxenda) and semaglutide (brand names: Ozempic and Wegovy).

Given these problems, methods to suppress epimerization have been extensively studied by chemists since the development of the SPPS approach and a plethora of strategies are currently available to mitigate this side reaction. For the amino acid activation step, an impressive number of coupling reagents have been developed and claimed to suppress epimerization even for complex products.[47] In general, the use of mildly acidic conditions in conjunction with a carbodiimide (*e.g.* DIC) and an additive such as OxymaPure allows the insertion of the desired Fmoc-protected amino acid with high yield and minimal epimerization.[47] Nonetheless, epimerization can still occur under these conditions if the coupling step requires a prolonged reaction time, as in the case of peptide fragments, where careful selection of the coupling reagent cocktail is essential. One example of this is the synthetic route to tirzepatide (brand name: Mounjaro) developed by Eli Lilly,[48] where the optimal coupling conditions between each of the four fragments were independently studied to reduce the C-terminal epimerization to the maximum extent possible. Greener approaches, such as CEPS as discussed in Chapter 6, represent a far more efficient alternative to chemical coupling reagents for fragment condensation because they can accomplish the coupling step without epimerization. The Fmoc deprotection conditions should also be tuned (*i.e.* reducing the amount of base and reaction time) to minimize the epimerization of sensitive amino acids. Finally, the downstream steps should be designed to avoid extremely basic pH values, which may be detrimental to peptide stability.

Moving to amino acid-related impurities, aspartimides may be formed during both synthetic processes and storage. These compounds are generated by nucleophilic attack of the amide backbone on either the protected side-chains of aspartic acid residues (during synthesis) or the side-chains of asparagine residues (mainly during storage). The succinimide ring formed can subsequently reopen, leading to several side products, such as isoaspartic acid and piperazine amide derivatives.[49] These impurities primarily originate under basic conditions (*e.g.* during Fmoc deprotection), although strongly acidic conditions can trigger the ring closure (*e.g.* during the cleavage step).[49] The severity of aspartimide formation depends strongly on the sequence, with the Gly–Asx sequence exhibiting the highest rate of formation. An example of this problem can be found in the case of teduglutide (brand name: Gattex), which contains the aforementioned sequence in the early stage of the SPPS assembly.[50] Several strategies for avoiding aspartimide formation have been developed over the years, including the utilization of bulky side-chain protective groups,[51,52] the use of N-backbone protective groups,[53] the inclusion of additives during Fmoc deprotection,[54] and masking the carboxylic acid functionality as an unreactive derivative.[45] Lastly, a systematic study of the influence of sustainable solvents indicated a lower tendency for aspartimide formation in these solvents compared with dimethylformamide (DMF), although this was not definitively confirmed.[31]

Hydrolysis of the side-chain amide moieties of asparagine and glutamine can also result in amino acid-related impurities. Aside from the contamination of the respective starting materials with aspartic acid or glutamic acid, the formation of these impurities during SPPS is usually associated with the aforementioned aspartimide mechanism followed by hydrolysis of the resulting succinimide.[55] Asparagine/glutamine side-chain hydrolysis can also occur at the cleavage stage when water is used as a scavenger, albeit to a low extent. In contrast, the downstream steps play a crucial role in the generation of these impurities because peptides are routinely dissolved in acidic or basic aqueous buffers. Hence careful control of the chromatographic conditions and robust stability assessment of the pooled fractions are crucial for minimizing the formation of these impurities. Finally, these impurities can be formed during storage of the isolated drug substance owing to the unavoidable presence of residual water. Storage at a controlled low temperature (*e.g.* −20 °C) typically overcomes this problem. A similar degradation pathway also has to be expected for drug products, particularly in the case of ready-to-use solutions. Careful selection of the formulation pH should therefore be undertaken.[56]

Sulfur-containing amino acids are susceptible to oxidation (particularly in the case of methionine) and oligomerization (in the case of cysteine) owing to the peculiar reactivities of these residues. Methionine oxidation can be suppressed by incorporating reducing agents in the cleavage step[57] and working under an inert atmosphere. Oligomerization occurs during the cyclization step of peptides containing one or more disulfide bridges, where intermolecular S–S bond formation competes with the desired intramolecular process to afford parallel or antiparallel dimers.[58] The complexity increases exponentially when the primary structure contains more than one disulfide bridge and attaining control over the regiochemistry has to be adequately addressed.[59] In general, the use of dilute reaction conditions or careful portion-wise addition of the monomer to the oxidative reaction mixture favors the intramolecular cyclization and suppresses oligomerization. Whereas the latter approach is preferred during the manufacture of peptide drug substances on account of the much lower volume required, the former strategy is the only viable option for ready-to-use FDFs such as vasopressin (brand names: Pitressin and Vasostrict).[60]

Finally, we consider peptide analogs and heterologs, where the range of potential peptide-related impurities extends beyond the classical patterns observed for native peptides, including completely new mechanisms that require tailored synthetic strategies to control. A representative example of such exotic impurities is the hydantoin isomer of degarelix (brand name: Firmagon), which originates from isomerization of the dihydroorotic acid residue bound to the phenylalanine at position 5 (see Figure 8.2) under the basic aqueous conditions commonly encountered during the piperidine-mediated Fmoc deprotection step of SPPS.[61] In this case, although piperidine promotes the isomerization to only a low extent, the best control strategy would be to completely avoid exposing the dihydroorotic acid moiety to any base. This was accomplished by replacing the substituted phenylalanine residue at position 5 with nitrophenylalanine, which was subsequently reduced and functionalized after the SPPS assembly of the sequence.[62]

Non-peptide impurities are treated in a similar manner to small-molecule drugs, where the same guidelines are applied with respect to residual solvents,[63] elemental impurities,[64] and nitrosamines.[65]

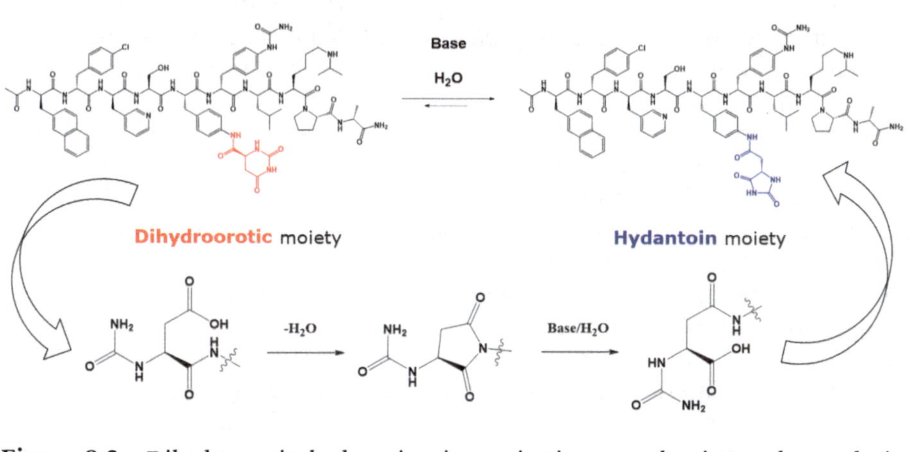

Figure 8.2 Dihydroorotic–hydantoin isomerization mechanism observed in degarelix.

The residual solvents that are routinely monitored as a critical quality parameter for isolated peptide drug substances are those used during the downstream processes. Accordingly, the acetonitrile commonly used as organic modifier for the mobile phase preparation must satisfy a residual level of not more than 410 ppm.[63] Other solvents potentially used in downstream steps must also be routinely quantified and remain within the specification limits defined in the ICH guidelines. Solvents used in upstream processes are typically completely removed during the downstream steps. Nonetheless, this should be adequately demonstrated during the validation procedures.

Similarly to residual solvents, all ions used during manufacture, particularly those utilized in the downstream steps, have to be monitored or justification for not doing so must be provided. In this case, specification limits are not clearly defined and should be evaluated on the basis of toxicological parameters such as permitted daily exposure (PDE). In the case of peptides, acetic acid and TFA warrant special mention. Acetic acid is employed as the counterion of choice for peptides, and TFA is routinely used as a solvent/reagent during cleavage of the crude peptide from the solid support in Fmoc-SPPS.[15] In both cases, their amounts in the isolated peptide drug substance must be carefully defined and routinely monitored as part of the release specification. The key role of these two species in the critical quality attributes of peptide drug substances is underlined by the presence of two USP general chapters dedicated to their measurement.[66,67] The same discussion applies when other counterions are selected, as reflected in the recommendation of the EMA draft guidelines.[8]

8.2.1.2 Analytical Considerations in the Regulation of Therapeutic Peptides

The need for accurate identification of peptide-related impurities very close in structure to the corresponding parent peptides has motivated the development of new analytical approaches that exploit the most recent advances in the field. Similarly to small molecules, the method of choice for monitoring such impurities is usually high-performance liquid chromatography (HPLC) with UV detection. This technique is especially useful for peptides considering that they often contain multiple chromophores, such as the backbone amide moieties and the aromatic side-chains of phenylalanine, tyrosine, tryptophan, and histidine, which ensure adequate UV absorption. However, although the current reported methods in the Pharmacopoeia peptide monographs are largely based on HPLC, this approach often does not allow satisfactory separation of structurally isobaric impurities such as isomers (*e.g.* epimers or regioisomers), which frequently exhibit the same retention time as the parent product. To overcome this problem, researchers combined the use of new technologies such as ultra-performance liquid chromatography (UPLC) with innovative approaches to the chemistry underlying the interactions between the stationary phase, mobile phase, and analytes to maximize the chromatographic separation. Prominent examples of this approach include the application of dynamic electrostatic repulsion reversed-phase (dERRP) chromatography and static electrostatic repulsion reversed-phase (sERRP) chromatography to characterize the glucagon impurity profile[68] and to separate the highly challenging L-Arg1 isomer of icatibant (brand name: Firazyr), where the latter originated from the terminal sterilization step during manufacture of the drug product and was not included in the RLD despite being present at levels of up to 1.5%.[69]

In addition to the appropriate quantification of the product-related impurity profile, adequate peptide characterization requires the extensive use of other analytical methods to properly evaluate the physicochemical properties and biological activity. An overview of the characterization landscape with the key underlying technologies is presented in Figure 8.3 (the analysis of non-peptide-related impurities is not considered here).

Among all of these parameters, the aggregation behavior and immunogenicity response profile play crucial roles, as emphasized by both the EMA and the FDA. These parameters are highly dependent on the peptide primary structure, where any modifications of one or more amino acids (*e.g.* insertion/deletion impurities or epimers) can consistently modify their aggregative or immunogenic properties.

Peptide aggregates may originate from multiple pathways involving either non-covalent or covalent interactions of the monomers, which ultimately lead to the reversible or irreversible formation of complex polymeric structures. Depending on the nature of the aggregate(s) formed, they can compromise the potency of the peptide or lead to toxic side effects such as an immunogenic response. The role of primary sequence modification in aggregation propensity has been reviewed.[70] However, the most common analytical techniques used to monitor aggregation fail to comprehensively evaluate the aggregative tendencies of designed peptides, mainly because they are performed under non-native conditions that may induce the disruption of the aggregate. A clear example is size-exclusion chromatography (SEC), where a stationary phase and an organic mobile phase must be used. This problem motivated attempts to develop technologies that are able to monitor aggregation under native-like conditions. An excellent example of this research

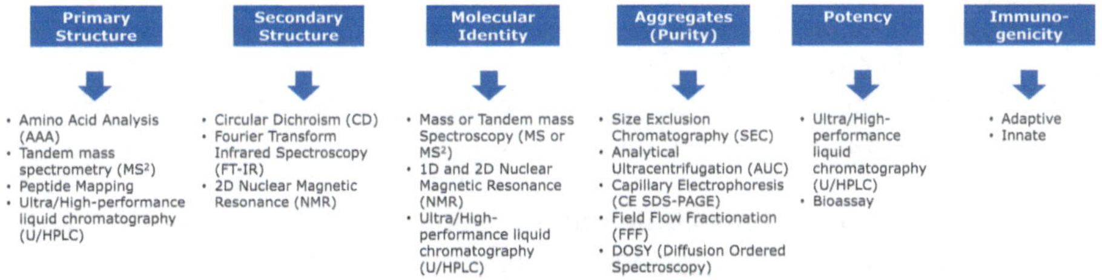

Figure 8.3 Overview of the characterization landscape in therapeutic peptides.

is the use of asymmetric flow field-flow fractionation (AF4) to evaluate the aggregative behavior of the GLP-1 analog liraglutide (brand names: Victoza and Saxenda).[71] When the FDF is a suspension, other techniques such as diffusion ordered spectroscopy (DOSY) and atomic force microscopy (AFM) can be applied to successfully characterize the supramolecular structure, as in the case of lanreotide (brand name: Somatuline).[72] Such techniques will play a pivotal role in the comparison of generic APIs with the reference products already on the market, which is a key aspect for the submission of an ANDA or an EU marketing authorization application.[71]

Similarly, adequate evaluation of the potential immune responses to peptide-related impurities is crucial for guaranteeing the safety profiles of manufactured peptide APIs. In fact, any new peptide drug candidate requires anti-drug antibody (ADA) measurements as part of the clinical trial protocols to assess its immunogenic potential. Of course, the greater the similarity of the peptide sequence to its endogenous peptide target (*e.g.* for native peptides), the less immunogenic potential is expected, even if problems have been observed also in the case of similar sequences (*e.g.* taspoglutide[73]). Evaluation of the potential immunogenic response has also become critical for generic versions of marketed peptides submitted through the ANDA process, because in this case no clinical trial to assess safety and efficacy is performed. In this case, a different peptide-related impurity profile with respect to the marketed product can lead to the presence of one or more sequences able to trigger an immunogenic response (also known as epitopes).[74] Immunogenic responses can be divided into two main classes: innate and adaptive responses.[75] Innate responses can be triggered by innate immune response modulating impurities (IIRMIs) and may be associated with non-peptide-related species introduced during the manufacturing process.[76] Innate immunogenic tests are typically performed at the drug product level to compare the innate response with that of the RLD. Several tests are available to assess innate immunogenicity, such as cytokine secretion from human donor peripheral blood mononuclear cells (PBMCs).[76]

Adaptive responses are associated with the presence of one or more epitopes in the peptide primary structure able to interact with both T-cell receptors and human leukocyte antigen (HLA) receptors on mature antigen-presenting cells (APCs), triggering T-cell activation and initiating the immunogenic response. The peptide epitopes can generally be considered as 9-mer sequences that usually contain five residues participating in the interaction with the HLA site and four residues involved in the interaction with the T-cell receptor (see Figure 8.4).[77] The nature of the sequence is fundamental for the proper interaction between the two proteins, and any minimal modification in the primary structure can drastically alter the immunogenic potential of the sequence itself. In this regard, any of the peptide modifications described in the previous section can greatly impact the immunogenic response (see Figure 8.4). In particular, the over-insertion or deletion of amino acid residues can have dramatic effects because these modifications greatly alter the primary structure of the parent product. Deamidation-related impurities act in a similar manner,

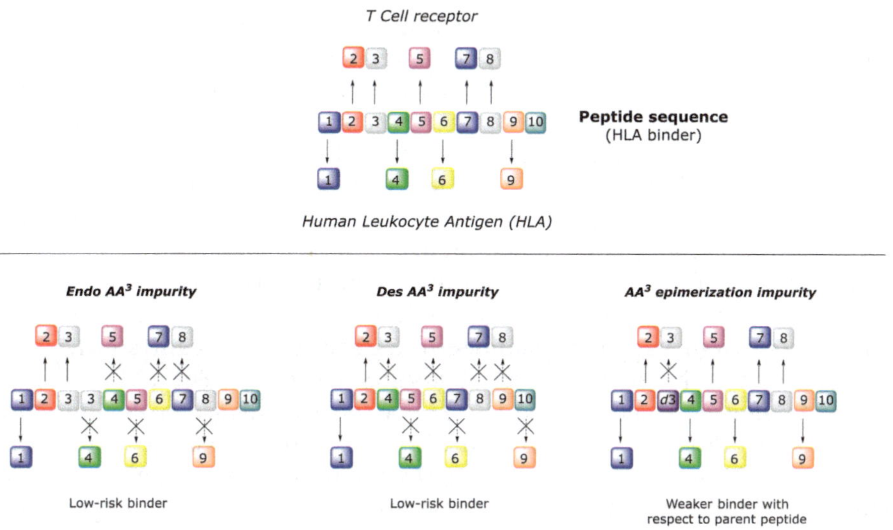

Figure 8.4 Impact of structural modification on the primary sequence of peptides towards immunogenic response.

although they are expected to be present also in the reference products (*e.g.* RLD) owing to their degradative origin. The other peptide-related impurities discussed (*e.g.* epimers, aspartimides, and impurities generated from unnatural amino acids) lead to primary sequences that are not proteinogenic in nature and thus generate non-predictable immunogenic responses.

The risk of an adaptive immunogenic response can be assessed through *in vitro* tests to evaluate the affinity of the target peptide for the HLA binding domain or its ability to activate T-cells (T-cell assays).[78] *In silico* approaches are also available, in which the primary peptide structure of interest is split into 9-mer sequences for comparison against a set of known epitopes to provide a potential immunogenic ranking score. This strategy proved useful for performing a preliminary evaluation of the immunogenic potential of the selected peptide,[79] even if regulatory agencies do not currently consider *in silico* results alone to be sufficient for an adequate immunogenic evaluation.[8] All of these tests are conducted at the drug substance level, in which the drug substance itself and each of the peptide-related impurities not present in the RLD are examined separately to independently evaluate their adaptive immunogenic potential.

8.2.2 Regulatory Considerations for Drug Product Manufacture

Therapeutic peptides are typically formulated as injectable drug products as the amide bonds are not able to withstand the acidic conditions associated with oral dosing. Manufacture of a parenteral peptide drug product typically involves the dissolution of the drug substance in an appropriate matrix suitable for parenteral administration. The solution drug product has contact with manufacturing equipment and container closure components throughout the product life cycle. Key knowledge and understanding of the compatibility of the drug product during manufacture, packaging, and storage are needed to ensure safety of the product. During a review of a marketing authorization application, a regulatory reviewer will assess the data package to support the compatibility with excipients, manufacturing equipment, and container closure components.

A primary requirement in drug product manufacture is to provide a sterile drug product that can be safely injected into patients. Typically, small-molecule drug substances can withstand terminal moist heat (steam) sterilization, which provides the greatest level of sterility assurance. The EMA has published a *Guideline on the Sterilization of the Medicinal Product, Active Substance, Excipient and Primary Container* (EMA/CHMP/CVMP/QWP/850374/2015) that includes a decision tree for selecting the appropriate sterilization process. However, synthetic peptides can sometimes show significant degradation under the terminal sterilization conditions and must use alternative sterilization procedures. A representative example is the icatibant case described in the previous section, where most of the impurities present in the drug product are generated from the terminal sterilization performed on the drug substance. In this situation, the common approach is to use membrane filtration as a means of sterilizing the drug product solution, which is then aseptically filled into sterilized container closure components. This type of approach has a higher risk than terminal sterilization and must therefore be thoroughly justified in the marketing authorization application by showing the degradation observed under the conditions outlined in the decision tree. Additionally, as aseptic processing is considered non-standard in EU regulations, results from process performance qualification studies need to be provided in the marketing authorization application to support approval.

The heat sensitivity of synthetic peptide drug products brings additional concern associated with the ability of the drug product to retain its quality during shipping and storage. Many of these products require refrigerated storage conditions (*e.g.* 2–8 °C) to ensure that the product quality will remain within specification during the shelf life of the product.

For products with this amount of sensitivity to environmental conditions, regulatory agencies will expect to see studies that simulate the shipping of the finished product, storage of product, and use of the product out of refrigeration during dosing. These studies will further support the ability of the product to withstand the conditions of manufacture, distribution, and patient use while remaining within the product specifications.

8.3 Regulatory Support for Innovation

Many of the major global health authorities have initiatives to support and enable continuous improvements in the development and manufacture of therapeutic medicinal products. There are also initiatives geared towards the development of sustainable processes that minimize the environmental impact inherent in the commercial manufacture of active substances. Synthetic peptides are known to have a significant environmental footprint when traditional SPPS manufacturing processes are employed, with process mass intensity (PMI) values that go far above 1000 for a 30-mer peptide.[12,80] Even worse, a relevant percentage of this PMI is due to *N,N*-dimethylformamide (DMF), a reprotoxic solvent that, since December 12th, 2023, has been regulated by the European Chemicals Agency (ECHA) to be used so as to ensure a worker inhalation exposure limit of not more than 6 mg m^{-3} (2 ppm).[81]

These are the drivers behind the extensive research performed by academia and industry in recent years to seek alternative solvents in SPPS or for innovative manufacturing strategies, such as those described in the previous chapters. All of these new approaches will generate process changes at the regulatory level (particularly for already filed DMF) that usually are evaluated by regulatory agencies with a major impact [Type II in Europe or Prior Approval Supplement (PAS) in the USA]. However, in such a critical context, it is likely that a health authority would welcome a manufacturing innovation that seeks to reduce the environmental impact, evaluating a faster approval route. Regulatory frameworks are well established in the USA and EU to seek advice during the course of development of a new product, but more recently additional mechanisms have been established to aid industry in the implementation of new technologies in manufacturing.

8.3.1 European Union (EMA)

In 2007, the European Commission established regulations that restrict the use of chemicals of very high concern to the environment. This REACH Regulation (EC 1907/2006) also encourages the replacement of materials of high concern (*e.g.* reagents, solvents) with less dangerous substances where possible. A commercial peptide manufacturer and/or drug developer working on a sustainable peptide synthesis platform could leverage this regulation to justify the switch to a new sustainable manufacturing platform and could expect some support from the EMA to implement the change.

Marketing authorization holders and sponsors of clinical trials can seek scientific advice from the Committee for Medicinal Products for Human Use (CHMP) at any stage of development to seek alignment with the data requirements to support a marketing authorization application (or a variation to an approved marketing authorization). Detailed procedures and timelines are published by the EMA for scientific advice that can help sponsors through this process. The sponsor is tasked with preparing a briefing document to outline the development plan, including any changes being made and a proposed testing plan to support the change(s). Early alignment with the health authorities on the study design and supporting data is essential to avoiding major objections or delays in approval of the marketing authorization application or variation application. The scientific advice procedure is best used in situations where there is no clear regulatory guidance or precedent, which is often the case for synthetic peptide therapeutics.

Another potential mechanism to obtain development feedback and support from the EMA is through the EMA Innovation Task Force (ITF) that was created in August 2014. This pathway is meant to guide development in scenarios where there is no established EMA scientific, legal, and regulatory experience. It can be useful for scenarios where an early dialog may be needed. A completely novel manufacturing technology may benefit from this type of procedure if specialized expertise is needed from the regulatory authorities or if there is an area that requires some action by the regulatory authorities to update or create guidance and/or legislation.

8.3.2 United States (FDA)

In the USA, clinical trials are conducted under an investigational new drug application (IND). IND sponsors have multiple options for interacting with the FDA during the course of development to seek advice and input on key topics through formal sponsor meetings. The different types of formal sponsor meetings allowed in this framework for novel drugs and biologics are summarized in Table 8.3.

Table 8.3 Formal meetings between the FDA and sponsors under the Prescription Drug User Fee Act (PDUFA).

Meeting type	Description	Timing	Examples
Type A	Typically, these are meetings necessary to discuss a stalled program or application	A meeting scheduled 30 days after the meeting is requested	A meeting to discuss a clinical hold or a complete response letter
Type B	Typically, these are meetings at key milestones where alignment is beneficial to plan further development	A meeting scheduled 60 or 70 days after the meeting is requested	A pre-IND meeting, end of phase II meeting, pre-NDA meeting, pre-BLA meeting
Type C	Any other meeting to discuss the development program	A meeting scheduled 75 days after the meeting is requested	A meeting to discuss a significant process change and the supporting comparability plan
Type D	A meeting focused on a narrow set of issues at a key decision point	A meeting scheduled 50 days after the meeting is requested	A meeting to discuss transition to a new analytical method for purity testing
INTERACT	A meeting to discuss novel products with unique challenges very early in development	A meeting scheduled 75 days after the meeting is requested	A meeting to discuss a novel modality prior to starting first in humans enabling non-clinical studies

The FDA acknowledged that sponsors looking to use innovative new approaches in the development or manufacture of a therapeutic product would benefit from additional opportunities to seek feedback from experts at the FDA and established the Emerging Technology Team (ETT) in 2014. Sponsors can submit a written request to apply to participate in the Emerging Technology Program (ETP) if:

- their technology has the potential to improve product safety, identity, strength, quality, and purity and
- the technology is an innovative or novel approach in terms of product technology, manufacturing processes, or control strategy.

If accepted into the ETP, sponsors are then able to have interactions with ETT members to discuss the technology and align with regulatory submission requirements to facilitate implementation of the new technology. The ETT can identify potential policy barriers and work towards resolution to enable registration of the new technology. Synthetic peptide manufacturers could leverage the expertise in FDA's ETT to allow for new green chemistry approaches to peptide synthesis if the technology meets the criteria for inclusion in the ETP. Through this program, the sponsor would be allowed interactions with the ETT and review staff at the agency to align with requirements and expectations. Manufacturing site visits can be a part of these interactions, which would allow FDA experts the opportunity to see the technology in action and gain a better understanding of the manufacturing process and control strategies. These types of interactions can facilitate more rapid implementation of these technologies through a shared understanding of the science.

8.4 Conclusion

Regulatory expectations for therapeutic oligopeptides encompass some concepts that are well established for both small-molecule chemical drugs and large-molecule biologics. Detailed regulatory guidance specific to therapeutic peptides is not widely available, but some guidance documents are starting to emerge that outline expectations for manufacturers, especially in the key topic of impurity control. The regulatory agency's review will largely focus on the ability of the process to produce a consistent product with control of impurities. Applications of novel technologies such as green chemistry approaches are likely to be supported by the regulatory agencies provided that the control of impurities is not negatively impacted. Chemical approaches for peptide manufacture have reached an impressive level of knowledge, providing relatively easy access to the production of complex peptide sequences of >20 residues with no distinction between native and unnatural amino acids. Despite this, the level of expectation raised by the regulatory authorities on the control of the (non)peptide-related impurities further challenges the limit of the current chemical and analytical technologies, particularly for generic registrations. Furthermore, the SPPS approach suffers from very high PMI values due to the large amount of DMF used for the washing of the solid support, a reprotoxic solvent that is currently under the spotlight of the ECHA.

These are the critical aspects that a novel sustainable approach must consider to be successfully implemented, aiming at the consistent reduction of the PMI, the application of alternative solvents to DMF, and comparable, even if not better, control of the impurities with respect to the existing chemical approaches.

References

1. FDA New Drug Application, https://www.fda.gov/drugs/types-applications/new-drug-application-nda, accessed on January 2024.
2. FDA Abbreviated New Drug Application, https://www.fda.gov/drugs/types-applications/abbreviated-new-drug-application-anda, accessed on January 2024.
3. FDA Biological License Application Process, https://www.fda.gov/vaccines-blood-biologics/development-approval-process-cber/biologics-license-applications-bla-process-cber, accessed on January 2024.
4. FDA website, https://www.fda.gov/drugs/guidance-compliance-regulatory-information/deemed-be-license-provision-bpci-act#:~:text=The%20Further%20Consolidated%20Appropriations%20Act%2C%202020%2C%20enacted%20on,%E2%80%9Cbiological%20product%E2%80%9D%20and%20subject%20to%20the%20statutory%20transition, accessed on January 2024.
5. EMA Guideline, Guideline on process validation for finished products – information and data to be provided in regulatory submissions, https://www.ema.europa.eu/en/documents/scientific-guideline/guideline-process-validation-finished-products-information-and-data-be-provided-regulatory-submissions-revision-1_en.pdf, 21 November 2016.
6. European Pharmacopoeia general monograph 2034, *Substances for Pharmaceutical Use*, 2024, Table 2034.2.
7. FDA guidance, *ANDAs for Certain Highly Purified Synthetic Peptide Drug Products That Refer to Listed Drugs of rDNA Origin Guidance for Industry*, May 2021.
8. EMA, *Guideline on the Development and Manufacture of Synthetic Peptide*, draft, October 2023.
9. S. Kota, in *Peptide Therapeutics: Strategy and Tactics for Chemistry, Manufacturing and Controls*, Royal Society of Chemistry, Cambridge, 2021, ch. 5, p. 111.
10. M. W. Pennington, B. Zell and C. J. Bai, *Med. Drug Discovery*, 2021, **9**, 100071.
11. C. De Luca, C. G. Lievore, D. Bozza, A. Buratti, A. Cavazzini, A. Ricci, M. Macis, W. Cabri, S. Felletti and M. Catani, *Molecules*, 2021, **26**, 4688.

12. L. Ferrazzano, M. Catani, A. Cavazzini, G. Martelli, D. Corbisiero, P. Cantelmi, T. Fantoni, A. Mattellone, C. De Luca, S. Felletti, W. Cabri and A. Tolomelli, *Green Chem.*, 2022, **24**, 975.
13. A. Sharma, A. Kumar, B. G. de la Torre and F. Albericio, *Chem. Rev.*, 2022, **122**, 13516.
14. M. Verlander, *Int. J. Pept. Res. Ther.*, 2007, **13**, 75.
15. W. Chan and P. White, *Fmoc Solid Phase Peptide Synthesis: A Practical Approach*, Oxford University Press, Oxford, 1999.
16. A. Staby, D. B. Steensgaard, K. F. Haselmann, J. S. Marino, C. Bartholdy, N. Videbæk, O. Schelde, H. Bosch-Traberg, L. T. Spang and D. J. Asgreen, *Pharm. Res.*, 2020, **37**, 120.
17. A. Sharma, A. Kumar, B. G. de la Torre and F. Albericio, *Chem. Rev.*, 2022, **122**, 13516.
18. L. Wang, N. Wang, W. Zhang, X. Cheng, Z. Yan, G. Shao, X. Wang, R. Wang and C. Fu, *Signal Transduction Targeted Ther.*, 2022, **7**, 48.
19. T. G. Castro, M. Melle-Franco, C. E. A. Sousa, A. Cavaco-Paulo and J. C. Marcos, *Biomolecules*, 2023, **13**, 981.
20. L. B. Knudsen and J. Lau, *Front. Endocrinol.*, 2019, **10**, 155.
21. L. V. Jacobsen, A. Flint, A. K. Olsen and S. H. Ingwersen, *Clin. Pharmacokinet.*, 2016, **55**, 657.
22. J. L. Lau and M. K. Dunn, *Bioorg. Med. Chem.*, 2018, **26**, 2700.
23. A. M. Thayer, *Chem. Eng. News*, 2011, **89**, 21.
24. B. L. Bray, *Nat. Rev. Drug Discovery*, 2003, **2**, 587.
25. L. A. Carpino and G. Han, *J. Org. Chem.*, 1970, **37**, 3404.
26. O. Ludemann-Hombourger, P. Möller and F. Kuster, *Chem. Today*, 2021, **39**, 50.
27. Y. Yang, *Side Reactions in Peptide Synthesis*, Academic Press, New York, 2015.
28. M. D'Hondt, N. Bracke, L. Taevernier, B. Gevaert, F. Verbeke, E. Wynendaele and B. De Spiegeleer, *J. Pharm. Biomed. Anal.*, 2014, **101**, 2.
29. Y. Yang and L. Hansen, *ACS Omega*, 2022, **7**, 12015.
30. United States Pharmacopeia (USP) General Chapter ⟨1504⟩, *Quality Attributes of Starting Materials for the Chemical Synthesis of Therapeutic Peptides*, USP-NF, Rockville, MD, 2023.
31. S. Jadhav, V. Martin, P. H. G. Egelund, H. J. Castro, T. Krüger, F. Richner, S. T. Le Quement, F. Albericio, F. Dettner, C. Lechner, R. Schönleber and D. S. Pedersen, *Green Chem.*, 2021, **23**, 3312.
32. M. D'Hondt, N. Bracke, L. Taevernier, B. Gevaert, F. Verbeke, E. Wynendaele and B. De Spiegeleer, *J. Pharm. Biomed. Anal.*, 2014, **101**, 2.
33. Y. Yang, L. Hansen, A. Baldi and F. Badalassi, *Org. Process Res. Dev.*, 2021, **25**, 250.
34. D. A. Senko, N. D. Timofeev, I. E. Kasheverov and I. A. Ivanov, *Amino Acids*, 2021, **53**, 665.
35. I. Coin, M. Beyermann and M. Bienert, *Nat. Protoc.*, 2007, **2**, 3247.
36. S. R. Manne, K. Rustler, T. Bruckdorfer, B. G. de la Torre and F. Albericio, *Tetrahedron Lett.*, 2023, **115**, 154301.

37. J. Pawlas, T. Nuijens, J. Persson, T. Svensson, M. Schmidt, A. Toplak, M. Nilsson and J. H. Rasmussen, *Green Chem.*, 2019, **21**, 6451.
38. T. Nuijens, A. Toplak, M. Schmidt, A. Ricci and W. Cabri, *Front. Chem.*, 2019, **7**, 829.
39. Y. Yang, L. Hansen and P. Ryberg, *Org. Process Res. Dev.*, 2022, **26**, 1520.
40. M. Giraud, F. Cavelier and J. Martinez, *J. Pept. Sci.*, 1999, **5**, 457.
41. P. Stathopoulos, S. Papas and V. Tsikaris, *J. Pept. Sci.*, 2006, **12**, 227.
42. F. Yraola, R. Ventura, M. Vendrell, A. Colombo, J. Fernàndez, N. de la Figuera, D. Fernández-Forner, M. Royo, P. Forns and F. Albericio, *QSAR Comb. Sci.*, 2004, **23**, 145.
43. V. Castro, H. Rodriguez and F. Albericio, *Org. Lett.*, 2013, **15**, 246.
44. D. S. Kemp, in *The Peptides: Analysis, Synthesis, Biology*, Academic Press, 1979, vol. 7, p. 315.
45. C. A. Arbour, L. G. Mendoza and J. L. Stockdill, *Org. Biomol. Chem.*, 2020, **18**, 7253.
46. M. Goodman and W. J. McGahren, *Tetrahedron*, 1967, **23**, 2031.
47. F. Albericio and A. El-Faham, *Org. Process Res. Dev.*, 2018, **22**, 760.
48. M. O. Frederick, R. A. Boyse, T. M. Braden, J. R. Calvin, B. M. Campbell, S. M. Changi, S. R. Coffin, C. Condon, O. Gowran, J. McClary Groh, S. R. Groskreutz, Z. D. Harms, A. A. Humenik, N. J. Kallman, N. D. Klitzing, M. E. Kopach, J. K. Kretsinger, G. R. Lambertus, J. T. Lampert, L. M. Maguire, H. A. Moynihan, N. S. Mullane, J. D. Murphy, M. E. O'Mahony, R. N. Richey, K. D. Seibert, R. D. Spencer, M. A. Strege, N. Tandogan, F. L. Torres Torres, S. V. Tsukanov and H. Xia, *Org. Process Res. Dev.*, 2021, **25**, 1628.
49. R. Subirós-Funosas, A. El-Faham and F. Albericio, *Tetrahedron*, 2011, **67**, 8595.
50. K. Neumann, J. Farnung, S. Baldauf and J. W. Bode, *Nat. Commun.*, 2020, **11**, 982.
51. A. Karlström and A. Undén, *Tetrahedron Lett.*, 1996, **37**, 4243.
52. R. Behrendt, S. Huber, R. Martí and P. White, *J. Pept. Sci.*, 2015, **21**, 680.
53. A. M. Abdel-Aal, G. Papageorgiou, R. Raz, M. Quibell, F. Burlina and J. Offer, *J. Pept. Sci.*, 2016, **22**, 360.
54. T. Michels, R. Dölling, U. Haberkorn and W. Mier, *Org. Lett.*, 2012, **14**, 5218.
55. S. Catak, G. Monard, V. Aviyente and M. F. Ruiz-López, *J. Phys. Chem. A*, 2009, **113**, 1111.
56. S. Patel, V. K. Vyas and P. J. Mehta, *Int. J. Pept. Res. Ther.*, 2023, **29**, 22.
57. H. Huang and D. L. Rabenstein, *J. Pept. Res.*, 1999, **53**, 548.
58. G. A. Grant, *Synthetic Peptides: A User's Guide*, W. H. Freeman and Company, New York, 1992.
59. T. M. Postma and F. Albericio, *Eur. J. Org Chem.*, 2014, **17**, 3519.
60. Vasostrict® leaflet, https://www.accessdata.fda.gov/drugsatfda_docs/label/2016/204485s003lbl.pdf, accessed on January 2024.
61. I. Guryanov, A. Orlandin, A. Viola, B. Biondi, D. Badocco, F. Formaggio, A. Ricci and W. Cabri, *Org. Process Res. Dev.*, 2019, **23**, 2746.

62. I. Guryanov, A. Orlandin, A. Viola, B. Biondi, F. Formaggio, A. Ricci and W. Cabri, *Org. Process Res. Dev.*, 2020, **24**, 274.
63. ICH guideline, *Q3C (R8) on Impurities: Guideline for Residual Solvents*, November 2022.
64. ICH guideline, *Q3D (R2) on Elemental Impurities*, May 2022.
65. EMA guideline, *Nitrosamine Impurities in Human Medicinal Product*, EMA/369136/2020, June 2020.
66. United States Pharmacopeia general chapter ⟨503⟩, *Acetic Acid in Peptides*, USP-NF, Rockville, MD, 2023.
67. United States Pharmacopeia general chapter ⟨503⟩, *⟨503.1⟩ Trifluoroacetic Acid (TFA) in Peptides*, USP-NF, Rockville, MD, 2023.
68. G. Mazzoccanti, S. Manetto, M. Bassan, A. Foschini, A. Orlandin, A. Ricci, W. Cabri, O. H. Ismail, M. Catani, A. Cavazzini and F. Gasparrini, *RSC Adv.*, 2020, **10**, 12604.
69. G. Mazzoccanti, S. Manetto, M. Bassan, M. Macis, W. Cabri, A. Ciogli, A. Ricci and F. Gasparrini, *Eur. J. Pharm. Sci.*, 2024, **193**, 106682.
70. K. L. Zapadka, F. J. Becher, A. L. Gomes dos Santos and S. E. Jackson, *Interface Focus*, 2017, **7**, 20170030.
71. V. Marassi, M. Macis, S. Giordani, L. Ferrazzano, A. Tolomelli, B. Roda, A. Zattoni, A. Ricci, P. Reschiglian and W. Cabri, *Molecules*, 2022, **27**, 5485.
72. M. Grimaldi, A. Santoro, M. Buonocore, C. Crivaro, N. Funicello, M. Sublimi Saponetti, C. Ripoli, M. Rodriquez, S. De Pasquale, F. Bobba, L. Ferrazzano, W. Cabri, A. M. D'Ursi and A. Ricci, *Pharmaceutics*, 2022, **14**, 681.
73. J. Rosenstock, B. Balas, B. Charbonnel, G. B. Bolli, M. Boldrin, R. Ratner and R. Balena, *Diabetes Care*, 2013, **36**, 498.
74. C. A. Weber, P. J. Mehta, M. Ardito, L. Moise, B. Martin and A. S. De Groot, *Adv. Drug Delivery Rev.*, 2009, **61**, 965.
75. V. Jawaa, L. P. Cousens, M. Awwad, E. Wakshull, H. Kropshofer and A. S. De Groot, *Clin. Immunol.*, 2013, **149**, 534.
76. C. K. Holley, E. Cedrone, D. Donohue, B. W. Neun, D. Verthelyi, E. S. Pang and M. A. Dobrovolskaia, *Molecules*, 2021, **26**, 7461.
77. A. S. De Groot, B. J. Roberts, A. Mattei, S. Lelias, C. Boyle and W. D. Martin, *Drug Discovery Today*, 2023, **28**, 103714.
78. S. Cohen and S. Chung, *Bioanalysis*, 2021, **13**, 1071.
79. A. E. Mattei, A. H. Gutierrez, M. D. William, T. E. Frances, B. J. Roberts, A. S. Rosenberg and A. S. De Groot, *Front. Drug Discovery*, 2022, **2**, 952326.
80. A. Isidro-Llobet, M. N. Kenworthy, S. Mukherjee, M. E. Kopach, K. Wegner, F. Gallou, A. G. Smith and F. Roschangar, *J. Org. Chem.*, 2019, **84**, 4615.
81. The European Commission, *EUR-Lex*, Document 32021R2030, 2021.

CHAPTER 9

Oligonucleotide Synthesis and State-of-the-Art Drug Discovery

T. OSAWA AND S. OBIKA*

Graduate School of Pharmaceutical Sciences, Osaka University, 1–6 Yamadaoka, Suita, Osaka 565-0871, Japan
*E-mail: obika@phs.osaka-u.ac.jp

9.1 Introduction

Oligonucleotide therapeutics[1,2] target RNA, which is upstream of protein synthesis *in vivo*, and control its function; this is different from conventional medicines that target proteins. Until a few decades ago, oligonucleotide therapeutics had difficulties such as easy degradation *in vivo*. Currently, the clinical development of several highly effective oligonucleotide therapeutics is ongoing because of rapid advances in the chemical modification of oligonucleotide therapeutics and drug delivery system (DDS) technology. The term "oligonucleotide therapeutics" as used in this chapter refers to a drug that is composed of an oligonucleotide in which approximately 20 nucleotides are linked and is produced by chemical synthesis (see Table 9.1). mRNA vaccines[3,4] and gene therapy drugs[5,6] are composed of the same oligonucleotides as oligonucleotide therapeutics and have also attracted attention in recent years. mRNA vaccines exert pharmacological effects on proteins that are translated and synthesised *in vivo,* and are produced enzymatically in a cell-free system. Similarly to mRNA vaccines, gene therapy drugs exert pharmacological effects on proteins synthesised through gene expression. They

Green Chemistry Series No. 85
Sustainability in TIDES Chemistry: Green Approaches to Oligonucleotides and Oligopeptides Synthesis
Edited by Alessandra Tolomelli, Lucia Ferrazzano and Walter Cabri

Published by the Royal Society of Chemistry, www.rsc.org

Table 9.1 Main characteristics of oligonucleotide therapeutics and gene therapy drugs.

Characteristic	Oligonucleotide therapeutics	mRNA vaccines	Gene therapy drugs
Structure	15–25-mer single-stranded DNA and double-stranded RNA	Single-stranded RNA consisting of ≥3000 nucleotides	Viral vectors and plasmids
Chemical modification	Introduction of artificial nucleic acids and ligand conjugates	None	None
Manufacturing method	Chemical synthesis	Biological manufacturing method	Biological manufacturing method
Mechanism	Oligonucleotides act directly on target mRNA	Proteins produced by gene expression stimulate the immune system	Proteins produced by gene expression exhibit pharmacological effects

are manufactured by introducing a gene encoding a protein into a plasmid or viral vector using genetic recombination technology. Therefore, oligonucleotide therapeutics can be distinguished from mRNA vaccines and gene therapeutics based on these aspects.

Recently, the depletion of drug discovery targets has become a problem. Therefore, significant efforts have been made to develop new drug modalities that exert pharmacological effects through novel mechanisms. Among these, oligonucleotide therapeutics are highly regarded as a new modality, following conventional small-molecule and antibody drugs. Because small-molecule drugs are manufactured through chemical synthesis, standardisation and quality control are easy; however, drug discovery targets are depleting, as mentioned above. Although antibody drugs have excellent target specificity, drug discovery targets for antibody drugs are generally limited to membranes and secreted proteins, and mass production is difficult. In addition to their high target specificity, oligonucleotide therapeutics are manufactured by chemical synthesis, making standardisation and quality control easier than with antibody-based drugs. Therefore, research into the development of oligonucleotide therapeutics has become increasingly active in recent years. This chapter is focused on oligonucleotide therapeutics, to which TIDES chemistry has contributed significantly, and an overview is provided of recent advances in oligonucleotide design and pharmaceutical profile improvements related to structurally innovative motifs.

9.2 Classification of Oligonucleotide Therapeutics

Oligonucleotide therapeutics have recently attracted attention as a novel drug discovery modality. However, it is difficult to use natural DNA and RNA in their original forms for such technologies because the stability of the duplexes formed by DNA and RNA is insufficient for pharmaceutical applications, and DNA and RNA are rapidly degraded by nucleolytic enzymes that exist inside and outside cells. To overcome these difficulties, chemically modified artificial nucleic acids are essential for oligonucleotide therapeutics, and they have been developed worldwide.[7–10] Since 2023, 20 oligonucleotide therapeutics have been approved, and all of them have been chemically modified (see Table 9.2); the structures and properties of chemically modified nucleic acids will be described later. Oligonucleotide therapeutics are classified into aptamers,[11,12] antisense oligonucleotides (ASOs),[13–15] small interfering RNAs (siRNAs),[16,17] and decoy[18] based on their structure and mechanism of pharmacological action (see Figure 9.1). This chapter describes ASOs and siRNAs, which have a particularly high number of approved cases.

ASOs are 15–25-mer single-stranded oligonucleotides with molecular weights of *ca.* 5000–10 000, whereas siRNAs are generally 20-mer double-stranded RNAs with molecular weights of *ca.* 15 000. The molecular

Table 9.2 Oligonucleotide therapeutics developed to date.

Brand name	Common name	Type	Chemical modification[a]
Vitravene®	Fomivirsen	ASO	Phosphorothiate (PS)
Macugen®	Pegaptanib	Aptamer	FNA, 2′-OMe, PEG conjugate
Kynamro®	Mipomersen	ASO	MOE, PS
Exondys 51®	Eteplirsen	ASO	Phosphorodiamidate morpholino oligomer (PMO)
Spinraza®	Nusinersen	ASO	MOE, PS
HEPLISAV-B®		CpG oligonucleotide	PS
Tegsedi®	Inotersen	ASO	MOE, PS
Onpattro®	Patisiran	siRNA	2′-OMe
Waylivra®	Volanesorsen	ASO	MOE, PS
Givlaari®	Givosiran	siRNA	FNA, 2′-OMe, GalNAc conjugate
Vyondys 53®	Golodirsen	ASO	PMO
Viltepso®	Viltolarsen	ASO	PMO
Oxlumo®	Lumasiran	siRNA	FNA, 2′-OMe, PS, GalNAc conjugate
Leqvio®	Inclisiran	siRNA	FNA, 2′-OMe, PS, GalNAc conjugate
Amondys 45®	Casimersen	ASO	PMO
Amvuttra®	Vutrisiran	siRNA	FNA, 2′-OMe, PS, GalNAc conjugate
Qalosdy®	Tofersen	ASO	2′-OMe, PS
Izevay®	Avacincaptad pegol	Aptamer	2′-OMe, FNA, PEG conjugate
Cyfendus®		CpG oligonucleotide	PS
Rivfloza®	Nedosiran	siRNA	FNA, 2′-OMe, PS, GalNAc conjugate

[a]FNA, 2′-fluoro-RNA; PEG, poly(ethylene glycol); MOE, 2′-*O*-(2-methoxyethyl)-RNA.

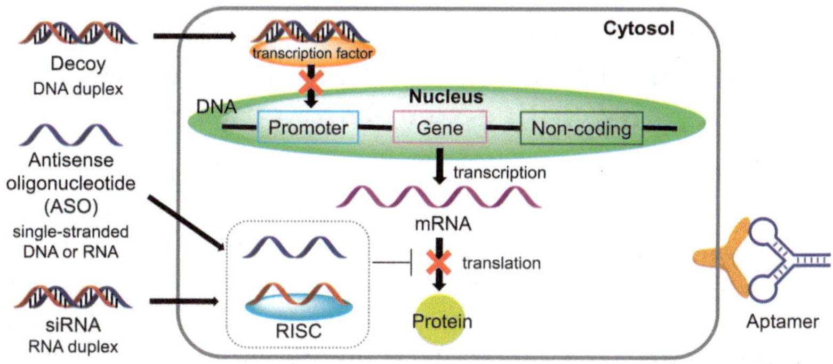

Figure 9.1 Oligonucleotide therapeutics classified by structure and mechanism.

weights of conventional low molecular weight drugs are generally <500, and those of antibody drugs are 100 000–200 000. Therefore, oligonucleotide therapeutics are intermediate in molecular size between these drugs and are also referred to as "middle-molecule drugs". In addition, oligonucleotides are highly water soluble because the phosphodiester bonds that connect the nucleic acid monomers are negatively charged. Oligonucleotide therapeutics have high molecular weights and are readily soluble in water, making it difficult for them to penetrate the biological membranes composed of lipids. Therefore, intracellular delivery of oligonucleotide therapeutics is achieved by (1) chemical modification of oligonucleotides, (2) conjugation of oligonucleotides and ligands, and (3) use of drug delivery system (DDS) carriers such as lipid nanoparticles (LNPs).[19,20] A typical example of the chemical modification of oligonucleotides is phosphorothioate (PS), in which the oxygen atom of a phosphodiester is replaced with a sulfur atom. Additionally, PS modifications have been utilised in numerous oligonucleotide therapeutics, especially ASOs (see Table 9.2). As PS-modified oligonucleotides have improved affinity for proteins and increased fat solubility, PS modification can improve the cell permeability of oligonucleotides.[21–23] Moreover, PS-modified oligonucleotides also have superior resistance to nucleases.[24] Owing to these factors, PS-modified oligonucleotides can be delivered into cells, even in the absence of DDS carriers; however, their intracellular delivery efficiency is extremely low. As will be discussed later, chemical modification techniques for nucleic acids, including PS modification, can markedly improve the stability and effectiveness of siRNAs and ASOs. This chapter focuses on ASOs, which account for half of the oligonucleotide therapeutics approved to date; the pharmacological mechanism of ASOs, notable points in designing ASO sequences, and the improvement of the pharmacological profile of ASOs using chemically modified nucleic acids are discussed. The knowledge gained through the development of chemically modified nucleic acids and their application in ASOs is also discussed.

9.3 Classification of Antisense Oligonucleotides (ASOs) by the Pharmacological Mechanism: Gapmer and Splice-switching Oligonucleotides (SSOs)

The mechanisms of ASOs can be divided into two types: one type depends on RNase H, an enzyme that recognises DNA/RNA heteroduplexes and cleaves the RNA strand, and the other involves RNase H-independent suppression or control of mRNA function. RNase H-dependent ASOs bind to target mRNA through the gapmer structure described below, and intracellular RNase H cleaves the mRNA and suppresses protein synthesis. In addition, RNase H-independent ASOs regulate and inhibit translation into proteins through a steric blocking mechanism by tightly binding to target mRNA, but not by degrading mRNA. A typical example of this mechanism of action is the control of pre-mRNA splicing by ASOs that regulate splicing. These oligonucleotides are referred to as splice-switching oligonucleotides (SSOs), which have undergone the most research and development (R&D) among steric block-type ASOs. Next, RNase H-dependent gapmer-type ASOs and RNase H-independent SSOs are discussed in more detail.

9.3.1 Gapmers

RNase H, which is the key to the pharmacological effects of gapmer-type ASOs, is an endonuclease that exists mainly in the cell nucleus, recognises DNA/RNA heteroduplexes, and selectively cleaves the RNA strand. From a biological perspective, RNase H plays a role in removing primer RNA bound to the Okazaki fragment.[25,26] Based on the crystal structure of the complex of the catalytic domain of RNase H[27] and the DNA/RNA heteroduplex, and the results of RNase H cleavage experiments,[28–30] a region containing at least three to four consecutive natural DNAs is essential for DNA/RNA heteroduplex recognition by RNase H (see Figure 9.2).

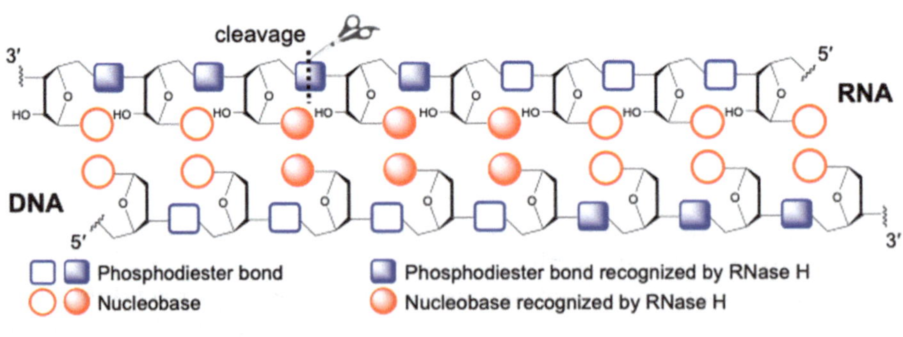

Figure 9.2 Recognition of DNA/RNA heteroduplex by RNase H.

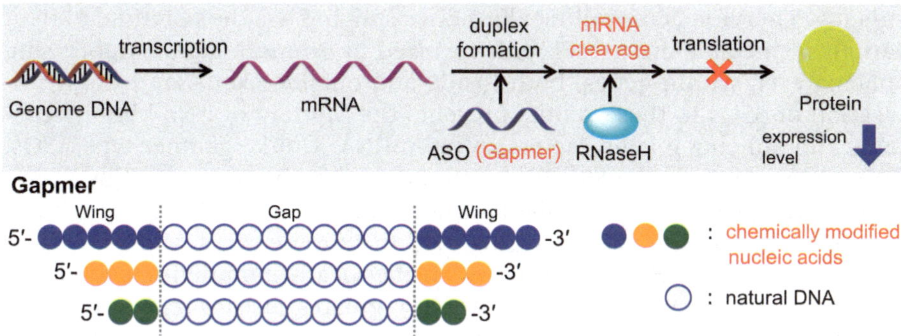

Figure 9.3 Structure and mechanism of a gapmer-type ASO.

Gapmer-type ASOs specifically form a duplex with a sequence complementary to the target mRNA, followed by the recognition of the DNA/RNA heteroduplex structure by RNase H in the cell and cleavage of the mRNA strand (see Figure 9.3). Through this mechanism, gapmer-type ASOs inhibit the translation of target mRNA into proteins and exhibit therapeutic effects. In addition, the cleaved mRNA dissociates from the gapmer-type ASO and a single-stranded gapmer-type ASO is released from RNase H. Through the properties of RNase H, gapmer-type ASOs can catalytically cleave mRNA, resulting in gapmer-type ASOs capable of exhibiting high target mRNA gene expression silencing. Gapmer-type ASOs are generally 15–20-mer single-stranded DNAs with PS modification, and several residues at the 3′- and 5′-ends, referred to as wing regions, are replaced with artificial nucleic acids that have a high mRNA binding ability such as 2′-*O*-(2-methoxyethyl)-RNA (MOE) and 2′,4′-BNA/LNA (bicyclic nucleic acid/locked nucleic acid), introduced in Section 9.5. In contrast, natural DNAs are used in the central 8–10 residues of the ASO sequence, referred to as the gap region, to enable recognition of the ASO/RNA heteroduplex by RNase H. Artificial nucleic acids are not introduced into the gap region because most chemically modified nucleic acids with excellent RNA-binding ability inhibit recognition by RNase H. Therefore, if chemically modified nucleic acids with both a high binding ability to mRNA and RNase H recognition ability can be developed, the clinical development of gapmer-type ASOs is expected to accelerate.

9.3.2 SSOs

Splicing is a reaction that removes "introns" unnecessary for protein synthesis contained in pre-mRNA transcribed from DNA and connects "exons" necessary for protein synthesis. Splicing is highly conserved from yeast to humans and is an important reaction for proper gene function. In addition, splice variant proteins are synthesised *in vivo* through a phenomenon referred to as "alternative splicing", in which multiple mature mRNAs are generated from one pre-mRNA due to differences in the exons and introns removed by splicing. Therefore, controlling alternative splicing has the potential to treat various intractable diseases. Motifs involved in promoting and suppressing splicing exist within genes, both inside and outside exons, and the hybridisation of SSOs to these motifs prevents the binding of RNA-binding proteins and splicing-related factors to pre-mRNA. Unlike gapmer-type ASOs, SSOs do not require the SSO/RNA heteroduplex to be recognised by RNase H. Therefore, chemically modified nucleic acids can be randomly introduced into the oligonucleotides (see Figure 9.4). In extreme cases, all nucleic acid monomers in SSOs can be replaced with chemically modified nucleic acids. SSOs regulate alternative splicing, which regulates disease-related proteins by promoting the splicing of specific exons (exon skipping) or suppressing the splicing of specific exons (exon inclusion). In the following sections, exon skipping and exon inclusion are explained using examples of approved oligonucleotide therapeutics.

Muscular dystrophy is a general term for genetic diseases characterised by skeletal muscles,[31] and Duchenne muscular dystrophy (DMD) is a typical example. In numerous cases of DMD, the dystrophin gene, which consists of 79 exons, is deleted in units of exons, and dystrophin protein, which is essential for maintaining muscle cells, is not produced. Figure 9.5 shows the splicing pattern of the pre-mRNA when exon 50 is deleted. Eteplirsen (see Table 9.2) targets exon 51 and splices exon 51 (exon skipping), converting it into translatable mRNA and leading to the synthesis of a short but functional dystrophin protein, which improves the pathology of DMD.[32]

Spinal muscular atrophy (SMA) is caused by the loss of full-length SMS protein due to mutations in the survival of the motor neuron 1 (*SMN1*) gene. *SMN2*, which has a sequence that is almost homologous to *SMN1*, exists on the same chromosome as *SMN1*.[33] However, *SMN2* is often incomplete, with exon 7 spliced out because of a single nucleotide polymorphism in exon 7. Therefore, the full-length SMN protein produced from the *SMN2* gene is approximately 10% of that of *SMN1*. Nusinersen (see Table 9.2), a therapeutic drug for SMA, inhibited exon 7 skipping by hybridising to *SMN2* pre-mRNA (see Figure 9.6). Full-length SMN protein is synthesised by inclusion of exon 7, that is, exon 7 is incorporated into mature *SMN2* mRNA.[34,35] This mechanism

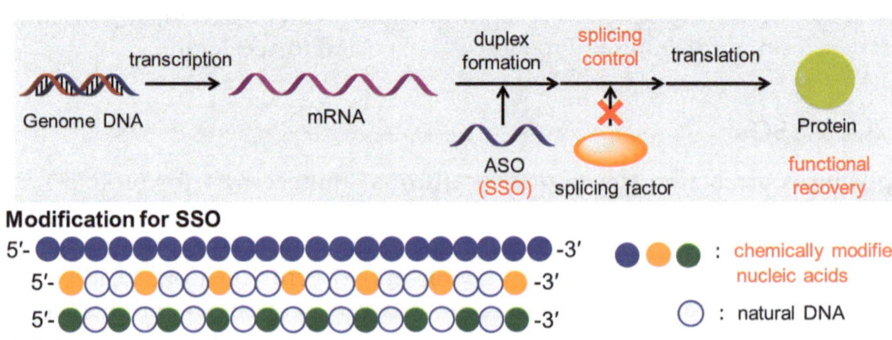

Figure 9.4 Structure and mechanism of SSOs.

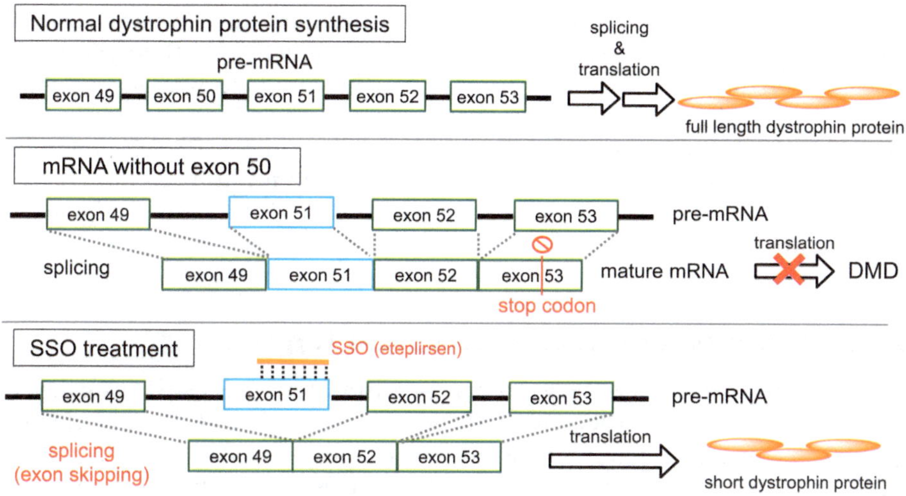

Figure 9.5 Exon skipping in DMD treatment with eteplirsen administration.

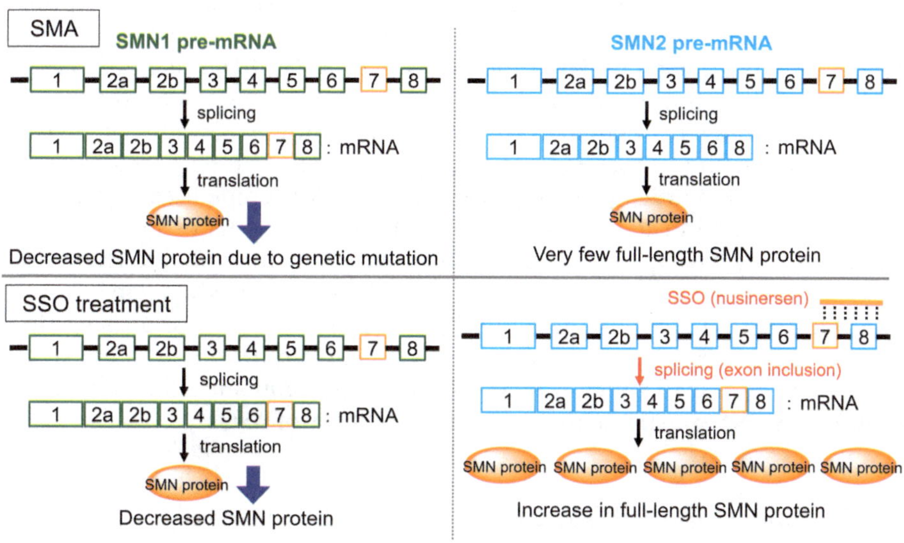

Figure 9.6 Exon inclusion in SMA treatment with nusinersen administration.

of action increases the amount of full-length SMN protein synthesised, providing treatment for the pathological condition that causes SMA.

9.4 ASO Design Strategy

As explained in the previous section, ASOs such as gapmers and SSOs exert medicinal effects by hybridising with the therapeutic target mRNA. Therefore, unlike conventional small-molecule and antibody drugs, oligonucleotide therapeutics directly target genes that cause diseases. One of the most important processes in the R&D of oligonucleotide therapeutics is the selection of target diseases and genes. Furthermore, designing candidate sequences for oligonucleotide therapeutics based on the sequence information of disease-causing genes is a characteristic step in the development of oligonucleotide therapeutics. In this section, the main points in the selection of target diseases and target genes for ASOs, and the design sequences of ASOs, are discussed.

9.4.1 Selection of Target Diseases and Target Genes

The selection of drug discovery targets is an important process in drug development, particularly in the case of ASOs, as the selection of therapeutic target diseases and target genes determines the success or failure of ASO R&D. Oligonucleotide therapeutics such as ASOs recognise the mRNA sequence, hybridise the target specifically, and suppress gene expression. However, because the level of gene expression depends on the type of target gene, ASOs may not be able to knock down the target mRNA completely. If only a small amount of the target gene expression remains, the efficacy of ASOs may not be maximised. Before proceeding with ASO R&D, it is necessary to investigate the extent to which the target gene should be knocked down to demonstrate fully the medicinal efficacy of ASOs. However, when controlling pre-mRNA splicing to increase the expression level of target genes, such as with DMD therapeutics (*e.g.* eteplirsen) and SMA therapeutics (*e.g.* nusinersen) explained in the previous section, the level of gene expression does not necessarily have to be the same as that of healthy individuals. Therefore, in some cases, ASOs exhibit sufficient medicinal efficacy when gene expression levels exceed a certain threshold. In summary, the appropriate selection of therapeutic targets and target genes for ASOs requires careful consideration of the extent to which the target gene expression should be improved or suppressed to maximise the therapeutic efficacy of ASOs.

9.4.2 Selection of DDSs to Treat Target Diseases

Another important factor in selecting a target disease is the DDSs of the oligonucleotides. The success of nusinersen is expected to lead to the development of oligonucleotide therapeutics that target the central nervous system. However, it is difficult to deliver oligonucleotide therapeutics into the central nervous system with existing DDS technology. Hence the development of a DDS suitable for oligonucleotide therapeutics has been a challenge; however, in recent years, LNPs and *N*-acetylgalactosamine (GalNAc) conjugates have been put to practical use as DDSs for siRNA (see Table 9.2). In particular, GalNAc conjugates are used in most approved siRNAs.[36] GalNAc is a ligand for the asialoglycoprotein receptor expressed on the surface of hepatic parenchymal cells, enabling efficient delivery of siRNA to the liver. Additionally,

GalNAc conjugates can be used for the delivery of siRNA and also ASOs to the liver.[37] However, the delivery of ASOs to organs other than the liver remains a significant challenge. Recently, research on DDS ligands other than GalNAc has been actively conducted. For example, palmitic acid,[38,39] which increases the ability to migrate into muscle tissue, and eGLP-1 peptide,[40,41] a ligand for the GLP-1 receptor highly expressed in the pancreas, have recently been discovered as DDS tools for ASOs. The latest DDS technologies are expected to expand the range of diseases that can be treated using ASOs.

9.4.3 Design of the ASO Sequence

After determining the therapeutic targets and target genes for ASOs, the ASO sequence was designed. The length of the mRNA is hundreds to thousands of bases, which is notably longer than the ASOs. Therefore, the efficacy of ASOs greatly depends on the region of mRNA targeted by the ASOs. It is difficult to select an optimal ASO sequence from a large number of candidate sequences. Currently, ASO sequence design is still in the trial-and-error stages, but the points to consider for sequence design can be summarised as follows: (1) secondary structure of mRNA, (2) homology between animal species, (3) secondary structure of the ASO, (4) toxic motifs, (5) off-target effects, and (6) length of the ASO.

9.4.3.1 Secondary Structure of mRNA

In general, mRNAs are rarely found in long single-stranded RNA, and complementary regions in the mRNA form base pairs, leading to the formation of various higher order structures (see Figure 9.7). Compared with ASOs that target double-stranded stem regions, ASOs that target single-stranded loop regions are more easily accessible to mRNA. Therefore, predicting the secondary structure of mRNA and selecting an appropriate region on the mRNA as a target for an ASO are important for obtaining highly active ASOs. Software such as mfold and RNAfold can be used to predict the secondary structure of mRNA. Although the predicted secondary structure can be easily obtained by inputting the sequence of mRNA, it is notable that the secondary structure obtained may change depending on the parameters on the

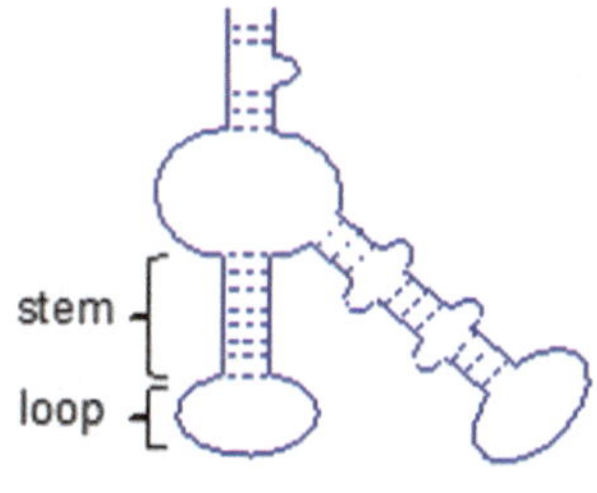

Figure 9.7 Secondary structure of RNA.

software. In addition, intracellular mRNA likely interacts with proteins; however, the effects of such RNA-binding proteins on the secondary structure of mRNA cannot be analysed by secondary structure prediction.

9.4.3.2 Homology Between Animal Species

An important point to consider when selecting the mRNA region to target ASOs is the difference in gene sequences between the animal species. In the R&D of ASOs, *in vitro* and *in vivo* experiments and human clinical trials are required after designing an ASO sequence. However, selecting ASOs with sequences having low homology between animal species makes it difficult to evaluate their efficacy and safety through *in vivo* experiments. Although *in vivo* safety tests using animals cannot evaluate the toxicity caused by the knockdown of genes other than the target mRNA, it is desirable to evaluate the safety margin of ASOs using *in vivo* experiments. From this perspective, when selecting an mRNA region to target using ASOs, homology searches for target genes should be performed considering *in vivo* experiments.

9.4.3.3 Secondary Structure of ASOs

The points discussed above regarding the secondary structures of mRNA are also important for ASO sequence design. As the length of an ASO is approximately 20 mers, the secondary structure of ASOs is not as complex as that of mRNA. However, because chemically modified nucleic acids are generally incorporated into ASOs to form a stable duplex with mRNA, ASOs can form a hairpin or dimeric structure (see Figure 9.8) if the ASO sequence contains only a few base pairs of self-complementary regions. For example, we comprehensively investigated the introduction pattern of a 2′,4′-BNA/LNA, with notably high mRNA-binding ability, into SSOs. This suggests that the introduction of 2′,4′-BNA/LNA stabilises the secondary structure of SSOs, thereby decreasing their activity.[42] Collectively, secondary structures of ASOs can significantly reduce their drug efficacy; therefore, they must be carefully considered in sequence design.

9.4.3.4 Toxic Motif

The toxicity derived from the binding of ASOs to proteins that recognise nucleobases is dependent on short sequence motifs. A typical example is recognition by Toll-like receptors (TLRs). For example, DNA containing CpG

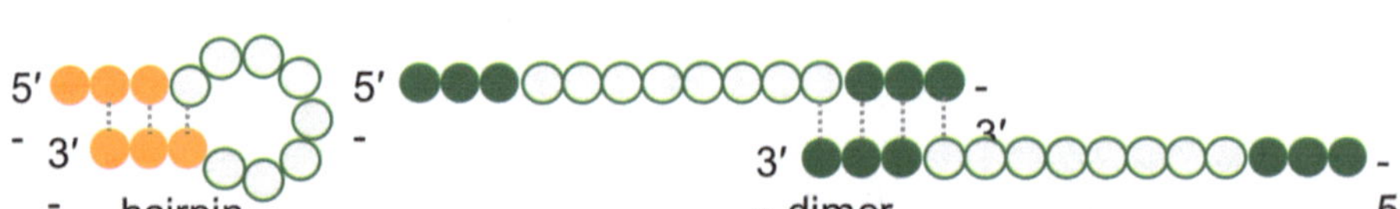

Figure 9.8 Secondary structures that ASOs can form.

dimers is recognised by TLR9 and induces a strong innate immune response.[43] TLRs are also considered to be involved in the sequence-specific thrombocytopenia of ASOs.[44] Additionally, TCC and TGC trimer trinucleotide motifs are known to be strongly associated with hepatotoxicity in mice, and it has been shown that proteins that interact with these motifs may be responsible for the expression of ASO toxicity.[45] Interestingly, in recent years, increasing reports of acute central nervous system toxicity, which is a potential concern in ASOs targeting the central nervous system, have been published. Hagedorn *et al.* used 148 types of ASOs to score acute tolerability behaviour 1 h after intracerebroventricular administration in mice and reported that moderate to severe symptoms were observed in *ca.* 40% of ASOs.[46] According to their report, toxic ASOs tended to decrease calcium oscillations, whereas safe ASOs tended to increase them. As AMPA receptors are known as receptors related to calcium oscillations, the above results suggest that the significant changes in calcium oscillation patterns caused by toxic ASOs may be related to α-amino-3-hydroxy-5-methyl-4-isoxazolepropionic acid (AMPA) receptors. Yoshioka and co-workers reported results that strongly suggest the possibility of AMPA receptor-dependent neurotoxicity of ASOs through experiments to elucidate the mechanism of toxicity using AMPA receptor agonists and antagonists.[47] Although several of the mechanisms of ASO toxicity are currently unclear in detail, predicting and avoiding toxicity when designing ASO sequences will become increasingly important in the future.

9.4.3.5 *Off-target Effect*

Oligonucleotide therapeutics such as ASOs can form duplexes with mRNAs that differ from the target mRNA and suppress their gene expression (off-target effects that depend on duplex formation). Therefore, safe and highly effective oligonucleotide therapeutics, such as ASOs, should specifically hybridise to target genes and not bind to other genes. These off-target effects can be avoided to some extent by designing ASO sequences using existing RNA databases. RefSeq and GENCODE are well-known databases that manage the accuracy of RNA sequence data and annotations. When searching for off-target genes with which the designed ASO can hybridise, the possibility of duplex formation, not only for completely complementary sequences but also for sequences containing several base mismatches, are investigated. Furthermore, because ASOs exhibit medicinal efficacy by binding to pre-mRNA or mature mRNA, the off-target effects of ASOs should be predicted for both pre-mRNA and mature mRNA.

9.4.3.6 *Length of ASOs*

Around 10 years ago, *ca.* 20-mer ASOs (such as 21-mer fomivirsen and 20-mer mipomersen) were often used in practice. In more recent years, chemically modified nucleic acids that can form stable duplexes with RNA have been used for ASOs, and ASOs with <20 mers have been developed. From the viewpoint of manufacturing costs, it is desirable for the ASO length to be shorter. However, considering the off-target effects that depend on duplex formation mentioned above, the shorter the length of the ASO, the higher the risk of interacting with mRNA rather than the target gene.[48] Furthermore, the binding affinity with the target RNA also strongly depends on the length of the ASO, and the longer the ASO, the greater the stability of the ASO/RNA duplex. We and other research teams have shown that the melting temperature (T_m) (an indicator of duplex stability) of the duplex formed by a gapmer-type ASO with complementary RNA (cRNA) does not necessarily correlate with the gene expression suppression effect of ASOs.[49,50] Therefore, although there is a positive correlation between the binding affinity for target RNA and ASO activity up to a certain threshold T_m value, when the RNA-binding affinity of the ASO is too high, it conversely leads to a decrease in ASO activity. To obtain an ASO with high activity, it is necessary to design an ASO sequence with optimal RNA-binding ability and, in addition to the sequence length, the number of chemically modified nucleic acids introduced to increase the RNA-binding ability of the ASO must also be considered.

9.5 Chemically Modified Nucleosides for Improvement of Pharmaceutical Profiles of Oligonucleotide Therapeutics

In the previous sections, we explained the mechanism of action of ASOs, selection of therapeutic target genes for ASOs, and sequence design of ASOs. However, it is difficult to use natural DNA and RNA in medicine. Issues with natural oligonucleotides include the facts that the stability of the double strands formed by DNA and RNA is insufficient for pharmaceutical applications and that DNA and RNA are rapidly degraded by nucleolytic enzymes that exist inside and outside cells. To overcome these issues, chemically modified artificial nucleic acids are essential in nucleic acid medicine. The importance of chemical modification techniques for nucleic acids has increased in recent years. All oligonucleotide therapeutics approved to date in Japan, the USA, and Europe have been chemically modified, as described in Section 9.2 (see Table 9.2). The structure of oligonucleotides is shown in Figure 9.9. Oligonucleotides are composed of three parts: (1) nucleobases, (2) phosphodiesters, and (3) furanoses. Chemically modified nucleic acids are generally introduced into oligonucleotides by solid- or liquid-phase synthesis methods based on phosphoramidite chemistry (see Figure 9.10A),[51–56] as detailed in the following sections. Additionally, a different synthetic approach than the phosphoramidite method is used to synthesise oligonucleotide-like polymers, such as phosphorodiamidate morpholino oligomer (PMO), the phosphate backbone structure of which differs significantly from that of natural DNA or RNA (see Figure 9.10B).[57] Through the chemical modification of these three parts, the physical properties, functions, and drug efficacy of oligonucleotides can be improved. This section provides an overview of the

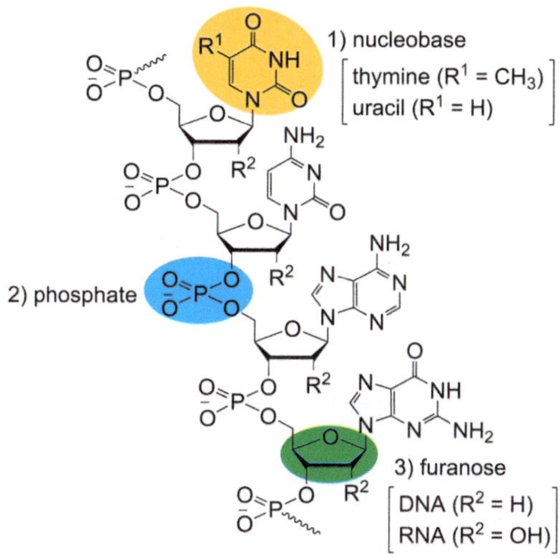

Figure 9.9 Structure of oligonucleotides.

chemical modifications of nucleobases, phosphate backbones, and sugar moieties from the perspective of the pharmaceutical applications of ASOs.

9.5.1 Nucleobase Modification

Chemical modifications of the nucleobases underlying Watson–Crick base pairing can significantly affect the ability of ASOs to hybridise with target mRNAs. Despite intensive research on nucleobase modification, there are few examples of successful applications to ASOs. Nucleobase modifications may affect the base-pairing properties of ASOs, thereby reducing their binding affinity and specificity for target mRNAs. However, the naturally occurring 5-methylcytosine nucleobase has been widely used in oligonucleotide medicinal chemistry to alleviate the immunostimulatory properties of ASOs because of the recognition of CpG motifs by TLR9.[58] Interestingly, 5-methylcytosine represents a chemical modification at the 5-position of pyrimidine nucleobases, and has inspired the design of new heterocycle modifications for use in nucleic acid medicinal chemistry (see Figure 9.11). Appropriate chemical modification at the 5-position of the pyrimidine can stabilise the duplex, which is likely due to efficient stacking between the planar nucleobase and the modified substituent in the major groove. For example, the cytosine nucleobases can be replaced with tricyclic phenoxazine rings. Phenoxazine extends the planar surface area of base pairs and improves stacking while retaining Watson–Crick base-pairing ability with the guanine nucleobase.[59–61] In addition, G-clamp with the addition of an aminopropyloxy side-chain to the phenoxazine nucleobase dramatically increases the thermal stability of the duplex by additional hydrogen bonding between the terminal amino group on the side-chain and the 4-carbonyl group of guanine.[62,63]

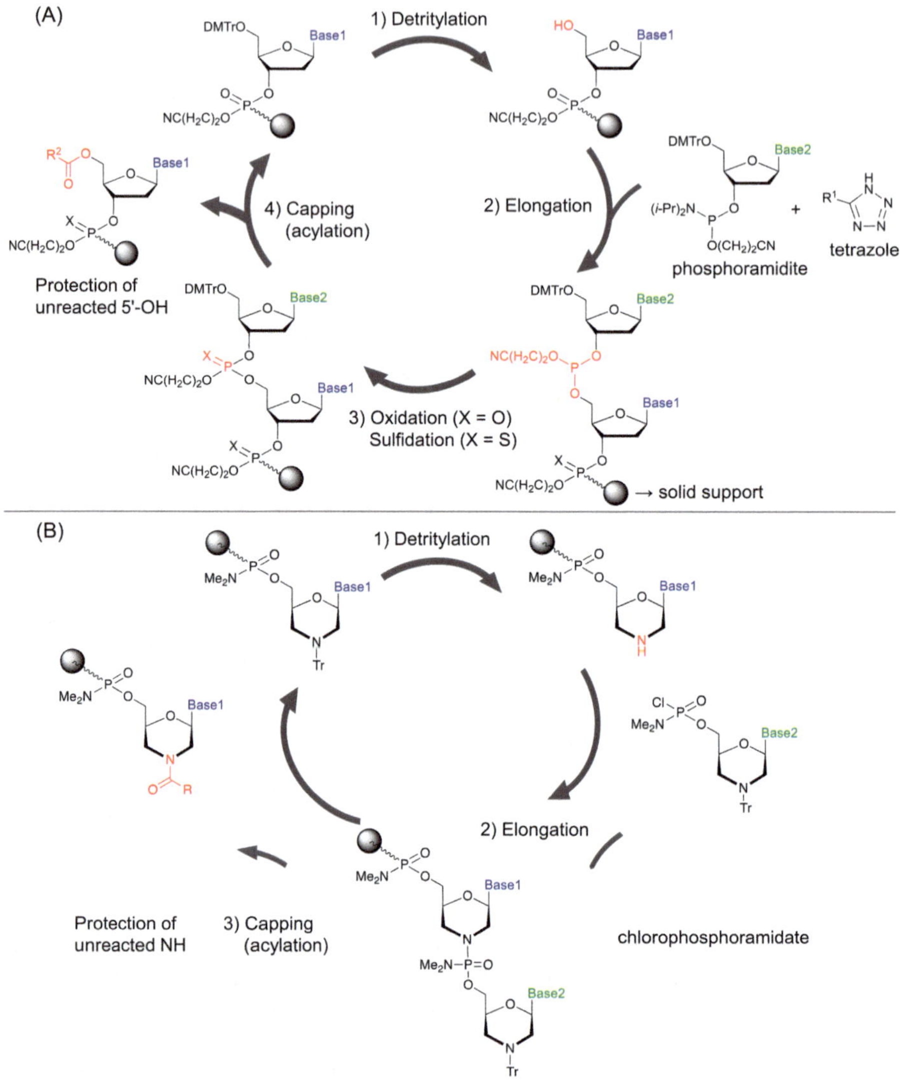

Figure 9.10 Chemical synthesis of oligonucleotides: (A) phosphoramidite method and (B) PMO synthesis.

A few medical applications of G-clamp have shown that G-clamp modification of ASOs improves antisense activity *in vitro* compared with unmodified ASOs.[64,65]

The T–G mismatch base pair is a representative example of a relatively stable wobble base pair. Interestingly, 2-thiothymine[66] maintains the stability of the T–A Watson–Crick base pair, but destabilises the T–G wobble base pair owing to the low hydrogen bond acceptor ability of the sulfur atom.[67,68] Owing to such properties of 2-thiothymine, incorporation of 2-thiothymidine

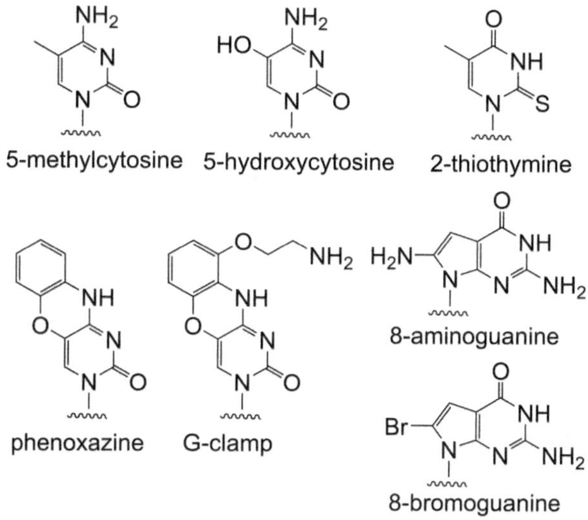

Figure 9.11 Examples of nucleobase modifications for ASOs.

into ASOs allows the recognition of single nucleotide polymorphisms in genes and modulation of allele selectivity.[69–71] In addition, we recently introduced single or multiple combinations of chemically modified nucleobases into hepatotoxic gapmer-type ASOs with 2′,4′-BNA/LNA modification and investigated their effects on hepatotoxicity.[72] As a result, 5-hydroxycytosine, 8-aminoguanine, and 8-bromoguanine were particularly identified as chemically modified nucleobases that exhibited strong toxicity-reducing effects on gapmer-type ASOs. The results of this study demonstrate the effectiveness of chemical modification of the nucleobase moiety of ASOs in avoiding toxicity, and also the potential for chemical modification of the nucleobase moiety of ASOs, which has had few successful applications in ASOs thus far.

9.5.2 Phosphate Backbone Modification

The phosphodiester backbones of natural DNA and RNA are rapidly hydrolysed by intracellular and extracellular nucleases, limiting their usefulness in therapeutic applications. Significant efforts have been made to identify alternative backbones that can increase the resistance to nucleases while maintaining or enhancing the affinity of oligonucleotides for target mRNAs based on chemical modifications. The most common chemical modification of the phosphate backbone involves replacing one of the non-bridging oxygen atoms of the phosphodiester bond with a sulfur or carbon atom (see Figure 9.12). Phosphorothioate (PS) linkages are among the most widely used modifications in oligonucleotide therapeutics. All the ASOs approved to date have been modified using PS (see Table 9.2). PS linkages have chemical properties similar to those of natural phosphodiester linkages and can

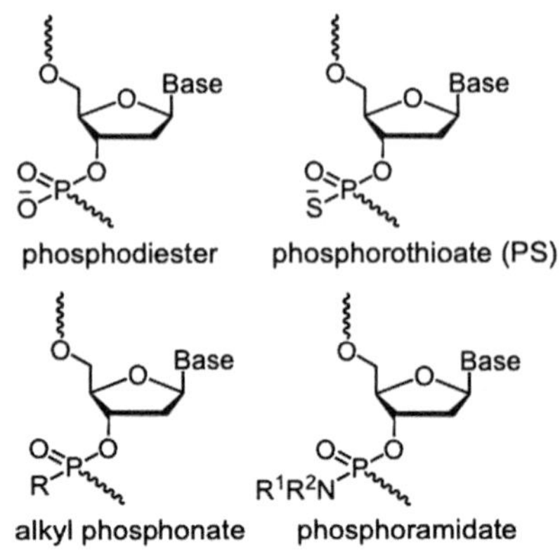

Figure 9.12 Representative examples of backbone modification.

be synthesised under general conditions similar to those of natural DNA and RNA with a phosphodiester backbone. PS modifications can dramatically improve the *in vivo* stability of oligonucleotides,[24] but only slightly decrease their affinity for cRNA: T_m values decrease by *ca.* 0.5 °C per PS modification.[73] PS modifications also provide pharmacokinetic advantages by increasing the binding to plasma proteins, which prevents rapid renal excretion and promotes binding to cell surface proteins, facilitating cellular uptake of ASOs.[21–23] Conversely, PS-modified ASOs can bind to immune receptors and cause non-specific pro-inflammatory toxicity.[74] A minor negative feature of PS modifications is that the incorporated sulfur atom introduces chirality to the phosphorus atom, generating numerous diastereomers of oligonucleotides that are challenging to isolate and purify. The correlation between the stereochemistry of PS and the physical properties of PS-modified oligonucleotides has been studied in detail. For example, the *R*-isomer has a higher cRNA-binding ability than the corresponding *S*-isomer.[75,76] In contrast, the *S*-isomer is considerably more resistant towards some nucleases than the *R*-isomer.[77] Moreover, the synthetic methodology developed by Wada and co-workers based on oxazaphospholidine chemistry permits the preparation of sterically pure PS-modified ASOs[78–81] (see Chapter 11). More recently, drug discovery research on ASOs with stereochemistry-controlled PS through the use of their technology revealed that stereopure ASOs enhance target RNA cleavage by RNase H1 *in vitro* and produce more durable responses in mice than chemically stereorandom ASOs.[82] In contrast, in research by another group, controlling the chirality in each PS linkage in the gap region of gapmer-type ASOs did not provide any appreciable advantage in cell or animal experiments compared with stereorandom ASOs.[83] These studies suggest that the functional improvement of ASOs through stereochemical control of PS may depend on the sequence of the ASO and its therapeutic target. It is necessary to continue to trace trends in research related to the stereochemical control of PS-modified ASOs and their therapeutic applications.

Methylphosphonate modification,[84] in which the oxygen atom of a phosphodiester is replaced with a methyl group, is a type of alkyl phosphonate backbone modification, and one of the earliest reported modifications without a phosphate anion. Although methylphosphonate bonds are stable against nucleases, they are susceptible to cleavage under the basic conditions required for the deprotection of oligonucleotides after chemical synthesis.[85] Because methylphosphonate has no charge, oligonucleotides with a high methylphosphonate content have poor water solubility, leading to a reduced affinity for cRNA. Similarly to the PS linkage, the methylphosphonate linkage also has a chiral centre, and the *R*-isomer exhibits a higher duplex-forming ability than the *S*-isomer.[86] Although alkylphosphonate modification has been considered to have low utility for oligonucleotide therapeutics due to such properties, it has recently been reported that modification of the electroneutral phosphate backbone, including alkylphosphonate modification, can reduce the toxicity of ASOs.[87] Based on this finding, the usefulness of alkylphosphonate-modified ASOs in medicinal chemistry will require further study in the future.

Replacing the 3′-oxygen atom of the phosphodiester with a nitrogen atom results in N3′-phosphoramidate modification (see Figure 9.13). N3′-phosphoramidate-modified oligonucleotides exhibit a high binding affinity with cRNA and high stability towards enzymatic degradation conditions.[88,89] However, N3′-phosphoramidate modifications have lower affinity for intracellular and extracellular proteins and are not stable in acidic conditions compared with their phosphodiester congeners.[90,91] To address these issues, N3′-thiophosphoramidate modifications were developed. Oligonucleotides incorporating thiophosphoramidates were found to have increased acid stability[92] and showed excellent antisense activity for telomerase inhibition.[93] However,

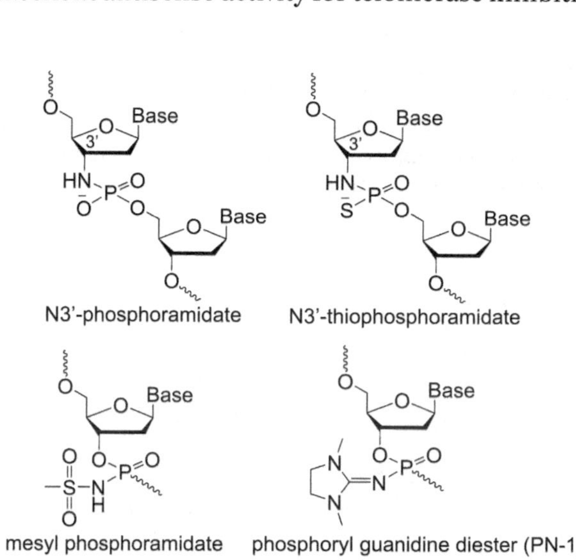

Figure 9.13 Backbone modification with a P–N bond.

almost no research on the development of phosphate backbone modifications with P–N bonds had been performed until *ca.* 10 years ago, and basic drug discovery research on phosphoramidate-modified ASOs has progressed rapidly over the past 10 years. A mesyl phosphoramidate modification with a sulfonamide structure is an example. The P–N bond in mesyl phosphoramidate can be formed using sulfonyl azide during the oxidation of trivalent phosphorus. Phosphoryl guanidine diester, which has a cationic guanidino group and can be synthesised using a method similar to that for mesyl phosphoramidate, has also attracted attention in recent years. Research on oligonucleotide therapeutics with mesyl phophoramidate modification[94–98] or phosphoryl guanidine diester modification has been actively reported.[99–101] Stereochemistry-controlled phosphoryl guanidine diester-modified ASOs exhibit significantly greater activity in cultured neurones under free uptake conditions than similarly modified stereopure PS- and phosphodiester-based ASOs.[101] Furthermore, the same study showed that low doses of PN-1 exhibit profound silencing effects throughout the mouse brain and spinal cord, and both efficacy and persistence in difficult-to-deliver brain tissues.

Several oligonucleotide-like polymers with backbone skeletons that differ significantly from the phosphate backbones of natural DNA and RNA have also been applied to ASOs. A typical example is a chemical modification in which the phosphate backbone is completely replaced with an amide bond (see Figure 9.14, amide linkage). Oligonucleotides with amide linkages have a slightly increased binding affinity for target RNA compared with natural oligonucleotides,[102] although this may depend on the oligonucleotide sequence. Recently, several groups have been conducting research on the pharmaceutical applications of ASOs and siRNAs modified with amide linkages.[103–109] PMO[110] has a structure in which a furanose ring in the sugar moiety is replaced with a morpholine ring. In addition, PMO is linked by a phosphorodiamidate linkage, in which two nitrogen atoms are bonded to a pentavalent phosphorus atom. As the backbone skeleton of PMO is significantly different from that of natural DNA and RNA, it exhibits excellent *in vivo* stability. PMO also has the advantage of low toxicity and it does not contain anionic charges derived from phosphodiesters. However, because its

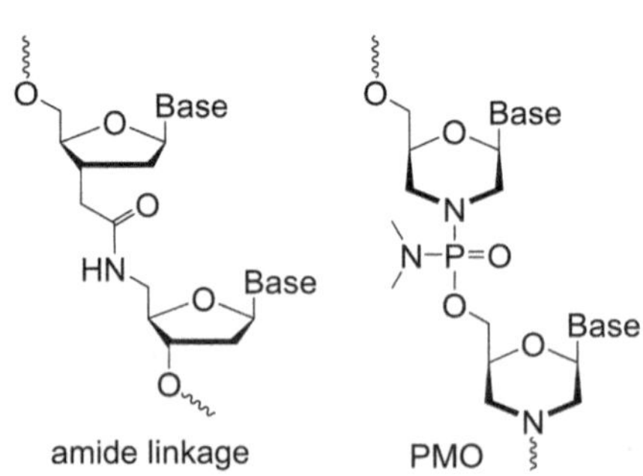

Figure 9.14 Amide linkage and PMO.

backbone structure is significantly different from that of natural oligonucleotides, RNase H does not recognise the PMO/RNA duplex, which is essential for the pharmacological activity of gapmer-type ASOs. Therefore, PMO is often used for exon skipping and exon inclusion-induced ASOs (such as eteplirsen and golodirsen; see Table 9.2).[111–113]

9.5.3 Sugar Modification

The structure of the furanose ring in natural duplexes has an N-type conformation represented by the C2′-*exo*–C3′-*endo* structure in RNA duplexes, and an S-type conformation represented by the C3′-*exo*–C2′-*endo* structure in DNA duplexes (see Figure 9.15). These sugar conformations are determined by the presence or absence and balance of stereoelectronic effects: (1) anomeric effect favouring an N-type conformation, (2) *gauche* effect of O4′–C1′–C2′–O2′ favouring an N-type conformation, and (3) *gauche* effect of O4′–C4′–C3′–O3′ favouring an S-type conformation. In contrast, the conformation of sugar moieties in single-stranded oligonucleotides is mainly in an equilibrium state between N- and S-type conformations, which are structurally free compared with double-stranded DNA and RNA. This indicates that duplex formation limits the fluctuation of the sugar structure and causes entropy loss. Therefore, if the sugar moiety can have a suitable conformation to form duplexes, entropy loss during duplex formation can be reduced and the binding affinity with complementary single strands can be improved.

In addition, nucleic acids exhibit structural fluctuations not only in the sugar moiety but also in the phosphate backbone, which is defined from the torsion angle α to the torsion angle ζ, as shown in Figure 9.16. The phosphate backbone adopts a certain degree of conformation in both RNA and DNA duplexes. Therefore, if the conformation of the phosphate moiety is fixed in a form suitable for duplex formation through chemical modification, the duplex can be thermally stabilised.

Nucleic acids modified at the 2′-position of the sugar moiety, such as 2′-*O*-methyl-RNA (2′-OMe, see Figure 9.17), which is a type of natural chemical modification of nucleotides, can form stable duplexes with complementary

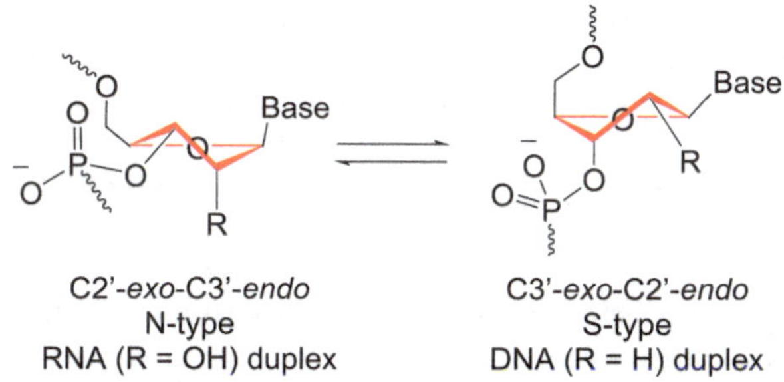

Figure 9.15 Conformational equilibrium of furanose ring.

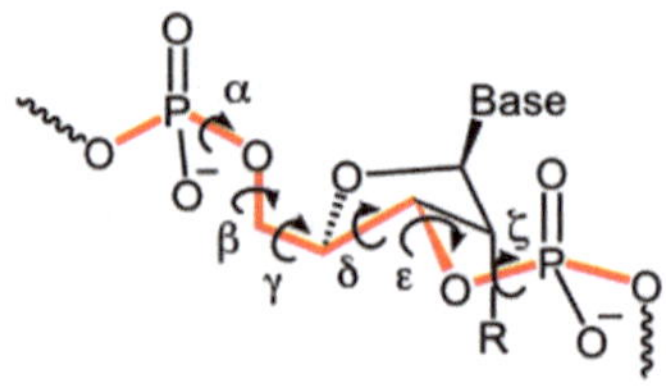

Figure 9.16 Torsion angle of phosphorus backbone.

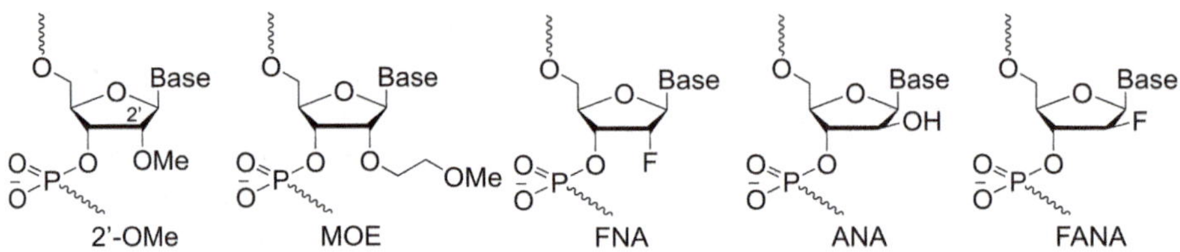

Figure 9.17 Representative 2′-*O*-modified nucleic acids for chemical modification of ASOs.

single-stranded RNA.[114–116] The reason why incorporation of 2′-*O*-modified nucleic acids into oligonucleotides can improve the RNA-binding ability can be explained by the steric electronic effect mentioned above. The sugar conformation of nucleic acids, which has an electronegative functional group at the 2′-position, is predominantly an N-type (see Figure 9.15) due to the *gauche* effect of O4′–C1′–C2′–O2′. Therefore, the thermal stability of the duplex formed with cRNA is improved by 2′-modification. Various artificial nucleic acids have been developed in which the 2′-hydroxyl group is modified through 2′-OMe; however, as the length of the modified substituent increases, the stability of the duplex generally tends to decrease.[116,117] In contrast, MOE shows an excellent RNA binding affinity.[118] The 2-methoxyethyl group of MOE tends to adopt a *gauche* configuration due to the *gauche* effect. As shown in Figure 9.18, the oxygen atom forms a hydrogen bond with the water molecule hydrated in the duplex, thereby stabilising the duplex.[119,120] Furthermore, as the 2-methoxyethyl group moiety is located near the 3′-phosphate, MOE modification can improve the enzymatic stability of oligonucleotides, likely due to its steric hindrance.[120] Because of these characteristics, MOE modification is used in the approved oligonucleotide therapeutics mipomersen, nusinersen, inotersen, and volanesorsen (see Table 9.2).

2′-Fluoro-RNA (FNA) with a fluoro group at the 2′-position is a well-known analogue of RNA.[121] As the fluorine atom has the highest electronegativity among all atoms, the *gauche* effect of O4′–C1′–C2′–F2′ is strengthened by the fluorine atom at the 2′-position. Therefore, the conformational equilibrium of the sugar conformation of FNA was dominated by the N-type, even in a single-stranded oligonucleotide. Additionally, 2′-fluorine modification enhances both Watson–Crick base pairing and π–π stacking.[122,123] Owing to these characteristics of FNA, it can form a more stable duplex with cRNA than

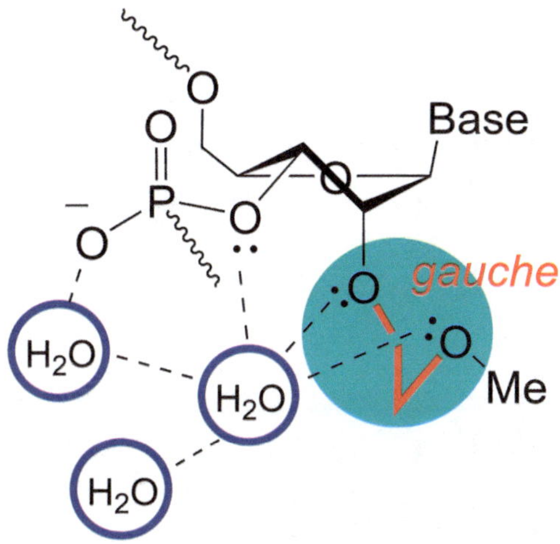

Figure 9.18 Hydration of the 2-methoxyethyl oxygen atom in the duplex.

with 2′-OMe and MOE. In addition, several arabinose-type nucleic acid analogues, such as arabinonucleic acid (ANA) and 2′-fluoroarabinonucleic acid (FANA), are known.[124] In particular, FANA can stabilise duplexes with cRNA, although its sugar structure adopts an O4′-*endo* structure,[125] which differs from the N-type conformation commonly seen in RNA duplexes. Furthermore, ANA- and FANA-modified ASOs exhibit RNase H-mediated RNA strand cleavage activity.[124,126–128] Most existing chemically modified nucleic acids inhibit the recognition of DNA/RNA duplexes by RNase H, making them useful materials for ASOs.

From the research on 2′-modified nucleic acid derivatives described above, if the sugar conformation can be constrained in the same N-type as in the RNA duplex in a more natural manner through the bicyclic structure rather than being controlled by stereoelectronic effects, further improvement in the binding affinity for cRNA can be expected. Based on this concept, we[129,130] and Wengel and co-workers[131,132] independently developed 2′,4′-BNA (bicyclic nucleic acid, also referred to as LNA, locked nucleic acid), which has a methylene bridge between the 2′-hydroxyl group and the 4′-carbon atom (see Figure 9.19). Oligonucleotides incorporating 2′,4′-BNA/LNA formed highly stable duplexes with cRNA, and an increase in T_m of 6–7 °C per modification was observed.[133] This result exceeds that for other sugar-modified nucleic acids (2′-OMe, MOE, and FNA). Furthermore, the 2′,4′-BNA/LNA-modified oligonucleotides also exhibit excellent resistance to nucleolytic enzymes, likely due to the steric effect of the bridged ring moieties. Because of such excellent properties, 2′,4′-BNA/LNA is being actively applied in the clinic to oligonucleotide therapeutics such as ASOs for the treatment of various diseases.[134] Interestingly, it has been reported that the risk of hepatotoxicity of ASOs is significantly decreased when the T_m value of the duplex formed between RNA and an LNA-modified ASO is ≤55 °C.[49] This report shows the importance of

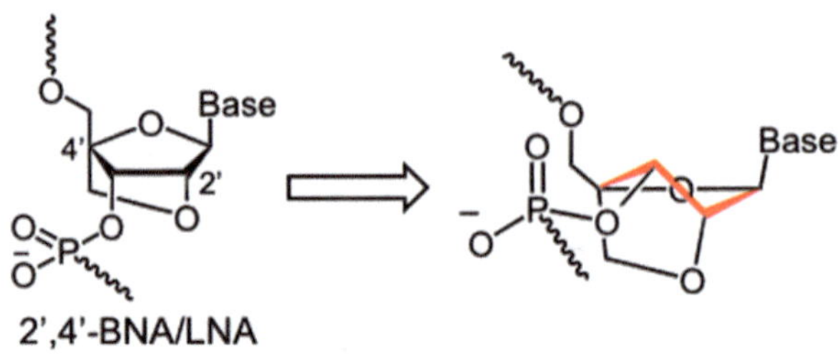

Figure 9.19 2′,4′-BNA/LNA.

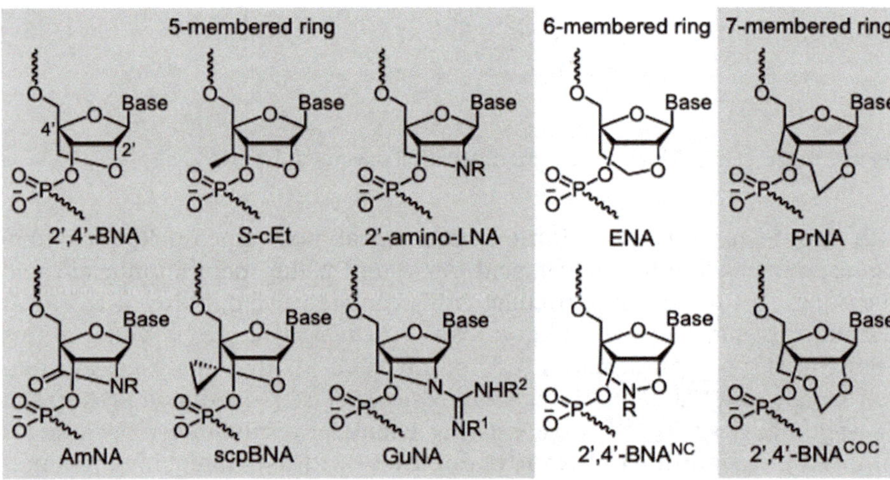

Figure 9.20 Examples of 2′,4′-BNA/LNA derivatives developed to date.

regulating the physical properties of ASOs through chemical modification of sugar moieties in the R&D of ASOs that have both excellent efficacy and safety.

To date, various 2′,4′-BNA/LNA analogues have been synthesised by several research groups, including ours,[9,10,135,136] and some examples are shown in Figure 9.20. Considering the correlation between their bridge structures and physical properties, the larger the ring size, the greater the nucleolytic enzyme resistance, likely because of the greater steric hindrance effect. In terms of cRNA binding ability, 2′,4′-BNA/LNA analogues with smaller ring sizes tend to be better in general. In the 2′,4′-BNA/LNA analogues that we developed by replacing LNA in ASOs with an amido-bridged nucleic acid (AmNA),[137] the hepatotoxicity of ASOs can be reduced while maintaining their antisense efficacy.[138] Our study, which comprehensively examined the introduction of position patterns of LNA-, AmNA-, and guanidine-bridged nucleic acids (GuNAs)[139] into SSOs, revealed that SSOs modified with an appropriate number of AmNA or GuNA moieties were more effective *in vitro* than the corresponding LNA-modified SSOs.[140] In addition, R&D of ASOs incorporating *S*-cEt (cEt = 2′,4′-constrained ethyl),[141,142] led by IONIS Pharmaceuticals, is actively progressing. Particularly interesting findings from their research indicate that *S*-cEt-modified ASOs have improved potency compared with MOE-modified gapmer-type ASOs and improved toxicity profiles compared with 2′,4′-BNA/LNA-modified congeners in animal models and human studies. Furthermore, among the LNAs, cEt, and nucleic acids with 2′-modification such as MOE, PS-modified oligonucleotides incorporating highly lipophilic LNA, *S*-cEt, and FNA interact with a greater number of proteins than the corresponding 2′-OMe- and MOE-modified oligonucleotides.[143] Protein binding with ASOs due to the lipophilicity of artificial nucleic acids may also be a contributing factor to the hepatotoxicity observed with LNA-modified ASOs.[144] Additionally, the same group succeeded in reducing protein binding with ASOs and hepatotoxicity by introducing a 2′-OMe modification with low lipid solubility into the second position from the 5′-end of the gap region of gapmer-type ASOs.[145] These findings strongly support that regulation of protein binding by chemical modification of sugar moieties of nucleic acids is important in reducing the toxicity risk of ASOs.

In addition to restricting the conformation of sugar moieties, restricting the torsion angle of the phosphate backbone is an effective approach for enhancing the duplex-forming ability of oligonucleotides. For example, Leumann and co-workers developed bicyclo-DNA (see Figure 9.21) to constrain the structures of the sugar moiety and phosphate backbone in the DNA duplex.[146,147] However, from structural analysis of the bicyclo-DNA monomer, the sugar conformation of bicyclo-DNA was determined to be a C1′-*exo* type, and the torsion angle γ between the 3′-carbon and the 5′-oxygen was *ca.* 150°, which is significantly different from that in the natural DNA duplex ($\gamma \approx 60°$).[146] Possibly because the structure of bicyclo-DNA differs significantly from that of natural DNA, bicyclo-DNA modification does not improve the thermal stability of the duplex.[148] Therefore, tricyclo-DNA was synthesized by introducing a cyclopropane ring into bicyclo-DNA to adjust the torsion angle γ.[149] As expected, tricyclo-DNA exhibited high affinity with complementary single-stranded oligonucleotides, particularly with cRNA.[150–152] However, molecular modelling of tricyclo-DNA showed that the torsion angle γ was *ca.* 150°, which was not significantly different from that of bicyclo-DNA; instead, the torsion angle β changed significantly.[153] The sugar conformation of

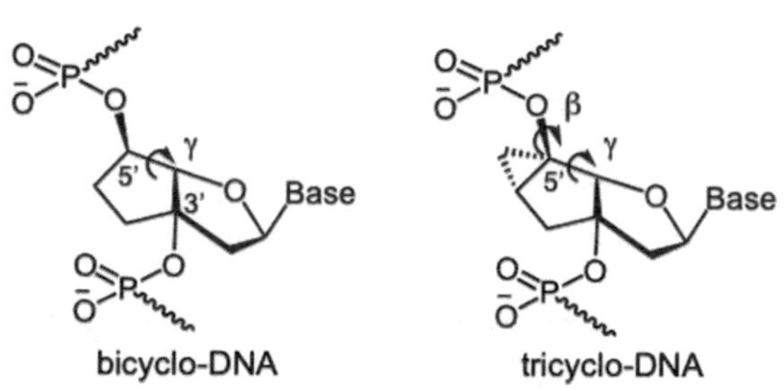

Figure 9.21 Bicyclo-DNA and tricyclo-DNA.

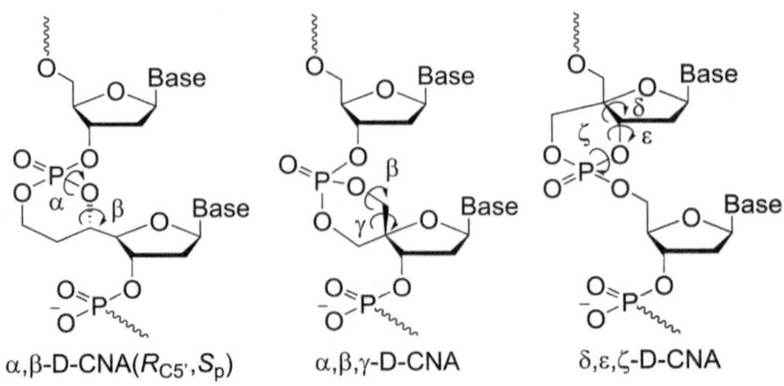

Figure 9.22 CNA analogues.

tricyclo-DNA is of C2′-*endo* type, which is different from that of bicyclo-DNA. It is inferred that tricyclo-DNA succeeded in acquiring a high duplex-stabilising ability owing to a combination of these factors. Furthermore, tricyclo-DNA-modified gapmer-type ASOs have shown potent antisense effects in reducing target mRNAs in the liver, kidneys, diaphragm, quadriceps, and heart of mice.[154] Tricyclo-DNA-modified SSOs have also shown excellent splicing regulation activity in animal models of DMD.[155]

Constrained nucleic acids (CNAs) are another type of artificial nucleic acids with constrained phosphate backbones. α,β-D-CNA[156,157] with 5′-carbon in an *S*-configuration (see Figure 9.22) can increase the binding affinity of oligonucleotides with cDNA and cRNA by constraining torsion angles α and β in the phosphate backbone in the canonical ranges observed in the DNA duplex.[158] In contrast, a diastereomer of α,β-D-CNA in which the 5′-carbon is in the *R*-configuration significantly reduces the thermal stability of the duplex. Modification of the 5′*S*-isomer of α,β-D-CNA to the gap region of gapmer-type ASOs targeting single nucleotide polymorphism (SNP) associated with mutant huntingtin transcripts exhibits the ability to modulate allele-selective silencing.[159] In addition to α,β-D-CNA, CNA analogues such as α,β,γ-D-CNA and δ,ε,ζ-D-CNA have been developed.[160–165] However, research on their application to ASOs has not progressed significantly, perhaps because they cannot form stable duplexes with cRNA.

9.6 Conclusion

This chapter has provided an overview of recent advances in oligonucleotide design and pharmaceutical profile improvements through the chemical modification of ASOs. Owing to space limitations, it is not possible to survey comprehensively chemically modified nucleic acids that can be used in oligonucleotide therapeutics. With this in mind, it is hoped that readers will benefit from this chapter and learn more about the medicinal chemistry of oligonucleotide therapeutics. Additionally, although it was not possible to cover siRNAs in this chapter, they are a modality of oligonucleotide therapeutics that is comparable to ASOs. Similarly to ASOs, the drug discovery chemistry of siRNAs has also been extensively investigated. In short, the chemical modification of nucleic acids is an essential technology for increasing the target mRNA binding affinity and *in vivo* stability of oligonucleotide therapeutics, avoiding toxicity, and accelerating the practical application of oligonucleotide therapeutics. In the future, it is hoped that further progress will be made in nucleic acid chemistry and oligonucleotide medicinal chemistry and that the physical properties and functions of oligonucleotide therapeutics will become more controllable through the combination of chemically modified nucleic acids.

References

1. M. Egli and M. Manoharan, *Nucleic Acids Res.*, 2023, **51**, 2529.
2. L. Moumné, A.-C. Marie and N. Crouvezier, *Pharmaceuticals*, 2022, **14**, 260.
3. A. J. Barbier, A. Y. Jiang, P. Zhang, R. Wooster and D. G. Anderson, *Nat. Biotechnol.*, 2022, **40**, 840.
4. N. Chaudhary, D. Weissman and K. A. Whitehead, *Nat. Rev.*, 2021, **20**, 817.
5. F. Arabi, V. Mansouri and N. Ahmadbeigi, *Biomed. Pharmacother.*, 2022, **153**, 113324.
6. J. R. Mendell, S. A. Al-Zaidy, L. R. Rodino-Klapac, K. Goodspeed, S. J. Gray, C. N. Kay, S. L. Boye, S. E. Boye, L. A. George, S. Salabarria, M. Corti, B. J. Byrne and J. P. Tremblay, *Mol. Ther.*, 2019, **29**, 464.
7. L. K. McKenzie, R. El-Khoury, J. D. Thorpe, M. J. Damha and M. Hollenstein, *Chem. Soc. Rev.*, 2021, **50**, 5126.
8. A. Khvorova and J. K. Watts, *Nat. Biotechnol.*, 2017, **35**, 238.
9. W. Wan and P. P. Seth, *J. Med. Chem.*, 2016, **59**, 9645.
10. S. Obika, S. M. A. Rahman, A. Fujisaka, Y. Kawada, T. Baba and T. Imanishi, *Heterocycles*, 2010, **81**, 1347.
11. J. Yuhan, L. Zhu, L. Zhu, K. Huang, X. He and W. Xu, *J. Controlled Release*, 2022, **346**, 405.
12. S. Ni, Z. Zhuo, Y. Pan, Y. Yu, F. Li, J. Liu, L. Wang, X. Wu, D. Li, Y. Wan, L. Zhang, Z. Yang, B.-T. Zhang, A. Lu and G. Zhang, *ACS Appl. Mater. Interfaces*, 2021, **13**, 9500.
13. J. A. Kulkarni, D. Witzigmann, S. B. Thomson, S. Chen, B. R. Leavitt, P. R. Cullis and R. van der Meel, *Nat. Nanotechnol.*, 2021, **16**, 630.
14. S. T. Crooke, B. F. Baker, R. M. Crooke and X. Liang, *Nat. Rev. Drug Discovery*, 2021, **20**, 427.
15. C. F. Bennett, *Annu. Rev. Med.*, 2019, **70**, 307.
16. M. M. Zhang, R. Bahal, T. P. Rasmussen, J. E. Manautou and X. Zhong, *Biochem. Pharmacol.*, 2021, **189**, 114432.
17. B. Hu, L. Zhong, Y. Weng, L. Peng, Y. Huang, Y. Zhao and X. Liang, *Signal Transduction Targeted Ther.*, 2020, **5**, 101.
18. M. Hecker and A. H. Wagner, *Biochem. Pharmacol.*, 2017, **144**, 29.
19. J. P. Bost, H. Barriga, M. N. Holme, A. Gallud, M. Maugeri, D. Gupta, T. Lehto, H. Valadi, E. K. Esbjörner, M. M. Stevens and S. El-Andaloussi, *ACS Nano*, 2021, **15**, 13992.

20. T. C. Roberts, R. Langer and M. J. A. Wood, *Nat. Rev. Drug Discovery*, 2020, **19**, 673.
21. M. K. Bijsterbosch, M. Manoharan, E. T. Rump, R. L. A. De Vrueh, R. van Veghel, K. L. Tivel, E. A. L. Biessen, C. F. Bennett, P. D. Cook and T. J. C. van Berkel, *Nucleic Acids Res.*, 1997, **25**, 3290.
22. K. Ezzat, Y. Aoki, T. Koo, G. McClorey, L. Benner, A. Corenen-Stass, L. O'Donovan, T. Lehto, A. Garcia-Guerra, J. Nordin, A. F. Saleh, M. Behlke, J. Morris, A. Goyenvalle, B. Dugovic, C. Leumann, S. Gordon, M. J. Gait, S. El-Andaloussi and M. J. A. Wood, *Nano Lett.*, 2015, **15**, 4364.
23. C. M. Miller, A. J. Donner, E. E. Blank, A. W. Egger, B. M. Kellar, M. E. Østergaard, P. P. Seth and E. N. Harris, *Nucleic Acids Res.*, 2016, **44**, 2782.
24. F. Eckstein, *Nucleic Acid Ther.*, 2014, **24**, 374.
25. T. Ogawa and T. Okazaki, *Mol. Gen. Genet.*, 1984, **193**, 231.
26. S. M. Cerritelli and R. J. Crouch, *FEBS J.*, 2009, **276**, 1494.
27. M. Nowothy, S. A. Gaidamakov, R. Ghirlando, S. M. Cerritelli, R. J. Crouch and W. Yang, *Mol. Cell*, 2007, **28**, 264.
28. S. Shibahara, S. Mukai, T. Nishihara, H. Inoue, E. Ohtsuka and H. Morisawa, *Nucleic Acids Res.*, 1987, **15**, 4403.
29. H. Inoue, Y. Hayase, S. Iwai and H. Ohtsuka, *FEBS Lett.*, 1987, **215**, 327–330.
30. H. Donis-Keller, *Nucleic Acids Res.*, 1979, **7**, 179.
31. E. Mercuri, C. G. Bönnermann and F. Muntoni, *Lancet*, 2019, **394**, 2025.
32. C. L. Balden, D. Salgado, S. Monges, M. E. Foncuberta, K. Kekou, K. Kosma, H. Dawkins, L. Lamont, A. J. Roy, T. Chamova, V. Guergueltcheva, S. Chan, L. Korngut, C. Campbell, Y. Dai, J. Wang, N. Barisic, P. Brabec, J. Lahdetie, M. C. Walter, O. Schreiber-Katz, V. Karcagi, M. Garami, V. Viswanathan, F. Bayat, F. Buccella, E. Kimura, Z. Koeks, J. C. van den Bergen, M. Rodrigues, R. Roxburgh, A. Lusakowska, A. Kostera-Pruszczyk, J. Zimowski, R. Santos, E. Neagu, S. Artemieva, V. M. Rasic, D. Vojinovic, M. Posada, C. Bloetzer, P.-Y. Jeannet, F. F. Joncourt, J. Diáz-Manera, E. Gallardo, A. A. Karaduman, H. Topaloglu, R. El Sherif, A. Stringer, A. V. Shatillo, A. S. Martin, H. L. Peay, M. I. Bellgard, J. Kirschner, K. M. Flanigan, V. Straub, K. Bushby, J. Verschuuren, A. Aartsma-Rus, C. Béroud and H. Lochmüller, *Hum. Mutat.*, 2015, **36**, 395.
33. S. Lefebrve, L. Burglen, S. Reboullet, O. Clermont, P. Burlet, L. Viollet, B. Benichou, C. Cruaud, P. Millasseau, M. Zeviani, D. Le Paslier, J. Frézal, D. Cohen, J. Weissenbach, A. Munnic and J. Melki, *Cell*, 1995, **80**, 155.
34. Y. Hua, T. A. Vickers, H. L. Okunola, C. F. Bennett and A. R. Krainer, *Am. J. Hum. Genet.*, 2008, **82**, 834.
35. Y. Hua, K. Sahashi, G. Hung, F. Rigo, M. A. Passini, C. F. Bennett and A. R. Krainer, *Genes Dev.*, 2010, **24**, 1634.
36. J. K. Nair, J. L. S. Wiloughby, A. Chan, K. Charisse, Md. R. Alam, Q. Wang, M. Hoekstra, P. Kandasamy, A. V. Kel'in, S. Milstein, N. Taneja, J. O'Shea,

S. Shaikh, L. Zhang, R. J. van der Sluis, M. E. Jung, A. Akinc, R. Hutabarat, S. Kuchimanchi, K. Fitzgerald, T. Zimmermann, T. J. C. van Berkel, M. A. Maier, K. G. Rajeev and M. Manoharan, *J. Am. Chem. Soc.*, 2014, **136**, 16958.

37. T. P. Prakash, M. J. Graham, J. Yu, R. Carty, A. Low, A. Chappell, K. Schmidt, C. Zhao, M. Aghajan, H. F. Murray, S. Riney, S. L. Booten, S. F. Murray, H. Gaus, J. Crosby, W. F. Lima, S. Guo, B. P. Monia, E. E. Swayze and P. P. Seth, *Nucleic Acids Res.*, 2014, **42**, 8796.
38. T. P. Prakash, A. E. Nullick, R. G. Lee, J. Yu, S. T. Yeh, A. Low, A. E. Chappell, M. E. Østergaard, S. Murray, H. J. Gaus, E. E. Swayze and P. P. Seth, *Nucleic Acids Res.*, 2019, **47**, 6029.
39. M. E. Østergaard, M. Jackson, A. Low, A. E. Chappell, R. G. Lee, R. Q. Peralta, J. Yu, G. A. Kinberger, A. Dan, R. Carty, M. Tanowitz, P. Anderson, T.-W. Kim, L. Fardkin, A. E. Mullick, S. Murray, F. Rigo, T. P. Prakash, C. F. Bennett, E. E. Swayze, H. J. Gaus and P. P. Seth, *Nucleic Acids Res.*, 2019, **47**, 6045.
40. C. Ämmälä, W. J. Drury III, L. Knerr, I. Ahlstedt, P. Stillemark-Billton, C. Wennberg-Huldt, E.-M. Andersson, E. Valeur, R. Jansson-Löfmark, D. Janzén, L. Sundström, J. Meuller, B. P. Monia and S. Andersson, *Sci. Adv.*, 2018, **4**, eaat3386.
41. L. Knerr, T. P. Prakash, R. Lee, W. J. Drury III, M. Nikan, W. Fu, E. Pirie, L. De Maria, E. Valeur, A. Hayen, M. Ölwegård-Halvarsson, J. Broddefalk, C. Ämmälä, M. E. Østergaard, J. Meuler, L. Sundström, P. Andersson, D. Janzén, R. Jansson-Lömark, P. P. Seth and S. Andersson, *J. Am. Chem. Soc.*, 2021, **143**, 3416.
42. T. Shimo, K. Tachibana, Y. Kawasaki, Y. Watahiki, T. Ishigaki, Y. Nakatsuji, T. Hara, J. Kawakami and S. Obika, *Chem. Commun.*, 2019, **55**, 6850.
43. A. M. Krieg, A.-K. Yi, S. Matson, T. J. Waldschmidt, G. A. Bishop, R. Teasgale, G. A. Koretzky and D. M. Klinman, *Nature*, 1995, **374**, 546.
44. P. Narayanan, L. Shen, B. R. Curtis, M. A. Bourdon, J. P. Nolan, S. Gupta, C. Hoffmaster, F. Zhou, B. Christian, J. L. Schaubhut, S. Greenlee, S. A. Burel, J. L. Witztum, J. A. Engelhardt and S. P. Henry, *Toxicol. Sci.*, 2018, **164**, 613.
45. A. D. Burdick, S. Sciabola, S. R. Mantera, B. D. Hollingshead, R. Stanton, J. A. Warneke, M. Zeng, E. Martsen, A. Medvedov, S. S. Makarov, L. A. Reed, J. W. Davis II and L. O. Whiteley, *Nucleic Acids Res.*, 2014, **42**, 4882.
46. P. H. Hagedorn, J. M. Brown, A. Easton, M. Pierdomenico, K. Jones, R. E. Olson, S. E. Mercer, D. Li, J. Loy, A. M. Høg, M. L. Jensen, M. Gill and A. M. Cacace, *Nucleic Acid Ther.*, 2022, **32**, 151.
47. C. Jia, S. S. Lei Mon, Y. Yang, M. Katsuyama, K. Yoshida-Tanaka, T. Nagata, K. Yoshioka and T. Yokota, *Mol. Ther.–Nucleic Acids*, 2023, **31**, 182.
48. H. Yasuhara, T. Yoshida, K. Sasaki, S. Obika and T. Inoue, *Mol. Diagn. Ther.*, 2022, **26**, 117.

49. T. Yamamoto, H. Yasuhara, F. Wada, M. Harada-Shiba, T. Imanishi and S. Obika, *J. Nucleic Acids*, 2012, **2012**, 707323.
50. N. Papargyri, M. Pontoppidan, M. R. Anderson, T. Koch and P. H. Hagedorn, *Mol. Ther.–Nucleic Acids*, 2020, **19**, 706.
51. S. L. Beaucage and M. H. Caruthers, *Tetrahedron Lett.*, 1981, **22**, 1859.
52. M. D. Matteucci and M. H. Caruthers, *J. Am. Chem. Soc.*, 1981, **103**, 3185.
53. L. J. McBride and M. H. Caruthers, *Tetrahedron Lett.*, 1983, **24**, 245.
54. K. K. Ogilvie, N. Usman, K. Nicoghosian and R. J. Cedergren, *Proc. Natl. Acad. Sci. U. S. A.*, 1988, **85**, 5764.
55. K. K. Ogilvie, N. Theriault and K. L. Sadana, *J. Am. Chem. Soc.*, 1977, **99**, 7741.
56. W. J. Stec, G. Zon and W. Egan, *J. Am. Chem. Soc.*, 1984, **106**, 6077.
57. J. Bhadra, S. Pattanayak and S. Sinha, *Curr. Protoc. Nucleic Acid Chem.*, 2015, **62**, 4.
58. M. Jurk, A. Kritzler, H. Debelak, J. Vollmer, A. M. Krieg and E. Uhlmann, *ChemMedChem*, 2006, **1**, 1007.
59. K.-Y. Lin, R. J. Jones and M. Matteucci, *J. Am. Chem. Soc.*, 1995, **117**, 3873–3874.
60. W. M. Flanagan, R. W. Wagner, D. Grant, K.-Y. Lin and M. Matteucci, *Nat. Biotechnol.*, 1999, **17**, 48.
61. P. Sandin, K. Börjesson, H. Li, J. Mårtensson, T. Brown, L. M. Wilhelmsson and B. Alninsson, *Nucleic Acids Res.*, 2008, **36**, 157.
62. K.-Y. Lin and M. Matteucci, *J. Am. Chem. Soc.*, 1998, **120**, 8531.
63. M. A. Maier, J. M. Leeds, G. Balow, R. H. Springer, R. Bharadwaj and M. Manoharan, *Biochemistry*, 2002, **41**, 1323.
64. W. M. Flanagan, J. J. Wolf, P. Olson, D. Grant, K.-Y. Lin, R. W. Wagner and M. D. Matteucci, *Proc. Natl. Acad. Sci. U. S. A.*, 1999, **96**, 3513.
65. R. Stanton, S. Sciabola, C. Salattom, Y. Weng, D. Moshisky, J. Little, E. Walters, J. Kreeger, D. DiMattia, T. Chen, T. Clark, M. Liu, J. Qian, M. Roy and R. Dullea, *Nucleic Acids Ther.*, 2012, **22**, 344.
66. I. Okamoto, K. Shohda, K. Seio and M. Sekine, *J. Org. Chem.*, 2003, **68**, 9971.
67. S. M. Testa, M. D. Disney, D. H. Turner and R. Kierzek, *Biochemistry*, 1999, **38**, 16655.
68. K. Shohda, I. Okamoto, T. Wada, K. Seio and M. Sekine, *Bioorg. Med. Chem. Lett.*, 2000, **10**, 1795.
69. M. E. Østergaard, A. L. Southwell, H. Kordasiewicz, A. T. Watt, N. H. Skotte, C. N. Doty, K. Vaid, E. B. Villanueva, E. E. Swayze, C. F. Bennett, M. R. Hayden and P. P. Seth, *Nucleic Acids Res.*, 2013, **41**, 9634.
70. W. F. Lima, J. G. Nichols, H. Wu, T. P. Prakash, M. T. Migawa, T. K. Wyrzykiewicz, B. Bhat and S. T. Crooke, *J. Biol. Chem.*, 2004, **279**, 36317.
71. M. E. Østergaard, P. Kumar, J. Nichols, A. Watt, P. K. Sharma, P. Nielsen and P. P. Seth, *Nucleic Acid Ther.*, 2015, **25**, 266.

72. T. Yoshida, K. Morihiro, Y. Naito, A. Mikami, Y. Kasahara, T. Inoue and S. Obika, *Nucleic Acids Res.*, 2022, **50**, 7224.
73. S. M. Freier and K. H. Altmann, *Nucleic Acids Res.*, 1997, **25**, 4429.
74. H. Yanai, S. Chiba, T. Ban, Y. Nakaima, T. Onoe, K. Honda, H. Ohdan and T. Taniguchi, *Proc. Natl. Acad. Sci. U. S. A.*, 2011, **108**, 11542.
75. K. Kanamori, Y. Tamura, T. Wada, M. Nishi, H. Kanehara, T. Morii, K. Tajima and K. Makino, *Biochemistry*, 1999, **38**, 16058.
76. M. Boczkowska, P. Guga and E. J. Stec, *Biochemistry*, 2002, **41**, 12483.
77. M. Koziolkiewicz, M. Wojcik, A. Kobylansk, B. Karwowski, B. Rebowska, P. Guga and W. J. Stec, *Antisense Nucleic Acid Drug Dev.*, 1997, **7**, 43.
78. N. Oka, T. Wada and K. Saigo, *J. Am. Chem. Soc.*, 2003, **125**, 8307.
79. N. Oka, T. Kondo, S. Fujiwara, Y. Maizuru and T. Wada, *Org. Lett.*, 2009, **11**, 967.
80. N. Iwamoto, N. Oka, T. Sato and T. Wada, *Angew. Chem., Int. Ed.*, 2009, **48**, 496.
81. N. Oka, M. Yamamoto, T. Sato and T. Wada, *J. Am. Chem. Soc.*, 2008, **130**, 16031.
82. N. Iwamoto, D. C. D. Butler, N. Svrzikapa, S. Mohapatra, I. Zlatev, D. E. Y. Sah, M. S. M. Standley, G. Liu, L. H. Apponi, M. Frank-Kamenetsky, J. J. Zhang, C. Vargeese and G. L. Verdine, *Nat. Biotechnol.*, 2017, **35**, 845.
83. W. B. Wan, M. T. Migawa, G. Vasquez, H. M. Murray, J. G. Nichols, H. Gaus, A. Berdeja, S. Lee, C. E. Hart, W. F. Lima, E. E. Swayze and P. P. Seth, *Nucleic Acids Res.*, 2014, **42**, 13456.
84. P. S. Miller, J. Yano, E. Yano, C. Carroll, K. Jayaraman and P. O. P. Ts'o, *Biochemistry*, 1979, **18**, 5134.
85. R. I. Hogrefe, M. M. Vaghefi, M. A. Reynolds, K. M. Young and L. J. Arnold Jr., *Nucleic Acids Res.*, 1993, **21**, 2031.
86. M. A. Reynolds, R. I. Hogrefe, J. A. Jaeger, D. A. Schwartz, T. A. Riley, W. B. Marvin, W. J. Daily, M. M. Vaghefi, T. A. Beck, S. K. Knowles, R. E. Klem and L. J. Arnold Jr., *Nucleic Acids Res.*, 1996, **24**, 4584.
87. G. Vasquez, M. T. Migawa, W. B. Wan, A. Low, M. Tanowitz, E. E. Swayze and P. P. Seth, *Nucleic Acid Ther.*, 2022, **32**, 40.
88. S. Gryaznov and J.-K. Chen, *J. Am. Chem. Soc.*, 1994, **116**, 3143.
89. S. Gryaznov, T. Skorski, C. Cucco, M. Nieborowska-Skorska, C. Y. Chiu, D. Lloyd, J.-K. Chen, M. Koziolkiewicz and B. Calabretta, *Nucleic Acids Res.*, 1996, **24**, 1508.
90. T. Skorski, D. Perrotti, M. Nieborowska-Skorska, S. Gryaznov and B. Calabretta, *Proc. Natl. Acad. Sci. U. S. A.*, 1997, **94**, 3966.
91. K. Pongracz and S. Gryaznov, *Tetrahedron Lett.*, 1999, **40**, 7661.
92. M. Faria, D. G. Spiller, C. Dubertret, J. S. Nelson, M. R. H. White, D. Scherman, C. Helene and C. Giovannangeli, *Nat. Biotechnol.*, 2001, **19**, 40.
93. B. S. Herbert, G. C. Gellert, A. Hochreiter, K. Pongracz, W. E. Wright, D. Zielinska, A. C. Chin, C. B. Harley, J. W. Shay and S. M. Gryaznov, *Oncogene*, 2005, **24**, 5262.

94. B. P. Chelobanov, E. A. Burakova, D. V. Prokhorova, A. A. Fokina and D. A. Stetsenko, *Russ. J. Bioorg. Chem.*, 2017, **43**, 664.
95. S. K. Miroshnichenko, O. A. Patutina, E. A. Burakova and D. A. Stetsenko, *Proc. Natl. Acad. Sci. U. S. A.*, 2019, **116**, 1229.
96. O. A. Patutina, S. K. Gaponova, A. V. Sen'kova and M. A. Zenkova, *Proc. Natl. Acad. Sci. U. S. A.*, 2020, **117**, 32370.
97. S. M. Hammond, O. V. Sergeeva, P. A. Melnikov, L. Goli, J. Stoodley, T. S. Zatsepin and D. A. Stetsenko, *Nucleic Acid Ther.*, 2021, **31**, 190.
98. L. Zhang, X.-H. Liang, C. L. De Hoyos, M. Migawa, J. G. Nichols, G. Freestone, J. Tian, P. P. Seth and S. T. Crooke, *Nucleic Acid Ther.*, 2022, **32**, 401.
99. E. S. Dyudeeva, M. S. Kupryushkin, A. A. Lomzov, A. A. Pyshnaya and D. V. Pyshnyi, *Russ. J. Bioorg. Chem.*, 2019, **45**, 709.
100. S. A. Zhukov, D. V. Pyshnyi and M. S. Kupryushkin, *Russ. J. Bioorg. Chem.*, 2021, **47**, 380.
101. P. Kandasamy, Y. Liu, V. Aduda, S. Akare, R. Alam, A. Andreucci, D. Boulay, K. Bowman, M. Byrne, M. Cannon, O. Chivatakarn, J. D. Shelke, N. Iwamoto, T. Kawamoto, J. Kumarasamy, S. Lamore, M. Lemaitre, X. Lin, K. Longo, R. Looby, S. Marappan, J. Metterville, S. Mohapatra, B. Newman, I.-H. Paik, S. Patil, E. Purcell-Estabrook, M. Shimizu, P. Shum, S. Standley, K. Taborn, S. Tripathi, H. Yang, Y. Yin, X. Zhao, E. Dale and C. Vargeese, *Nucleic Acids Res.*, 2022, **50**, 5401.
102. E. Rozners, D. Katkevica, E. Bizdena and R. Strömberg, *J. Am. Chem. Soc.*, 2003, **125**, 12125.
103. D. Mutisya, C. Selvam, B. D. Lunstad, P. S. Pallan, A. Haas, D. Leake, M. Egli and E. Rozners, *Nucleic Acids Res.*, 2014, **42**, 6542.
104. D. Mutisya, T. Hardcastle, S. K. Cheruiyot, P. S. Pallan, S. D. Kennedy, M. Egli, M. L. Kelley, A. Smith and E. Rozners, *Nucleic Acids Res.*, 2017, **45**, 8142.
105. S. Epple, C. Thorpe, Y. R. Baker, A. H. El-Sagheer and T. Brown, *Chem. Commun.*, 2020, **56**, 5496.
106. V. Kotikam, J. A. Viel and E. Rozners, *Chem. – Eur. J.*, 2020, **26**, 685.
107. V. Kotikam and E. Rozners, *Acc. Chem. Res.*, 2020, **53**, 1782.
108. Y. R. Baker, C. Thorpe, J. Chen, L. M. Poller, L. Cox, P. Kumar, W. F. Lim, L. Lie1, G. McClorey, S. Epple, D. Singleton, M. A. McDonough, J. S. Hardwick, K. E. Christensen, M. J. A. Wood, J. P. Hall, A. H. El-Sagheer and T. Brown, *Nat. Commun.*, 2022, **13**, 1.
109. M. Ritcher, J. A. Viel, V. Kotikam, P. K. Gajula, L. Coyle, C. Pal and E. Rozners, *Acc. Chem. Res.*, 2023, **18**, 7.
110. J. Summerton and D. Weller, *Antisense Nucleic Acid Drug Dev.*, 1997, **7**, 187.
111. J. Alter, F. Lou, A. Rabinowitz, H. Yin, J. Rosenfeld, S. D. Wilton, T. A. Partridge and Q. L. Lu, *Nat. Med.*, 2006, **12**, 175.
112. P. Sazani, F. Gemignani, S.-H. Kang, M. A. Maier, M. Manoharan, M. Persmark, D. Bortner and R. Kole, *Nat. Biotechnol.*, 2002, **20**, 1228.

113. B. Wu, H. M. Moulton, P. L. Iversen, J. Jiang, J. Li, C. F. Spurney, A. Sali, A. D. Guerron, K. Nagaraju, T. Doran, P. Lu, X. Xiao and Q. L. Lu, *Proc. Natl. Acad. Sci. U. S. A.*, 2008, **105**, 14814.
114. H. Inoue, Y. Hayase, A. Imura, S. Iwai, K. Miura and E. Ohtsuka, *Nucleic Acids Res.*, 1987, **15**, 6131.
115. B. S. Sproat, A. I. Lamond, B. Beijer, P. Neuner and U. Ryder, *Nucleic Acids Res.*, 1989, **17**, 3373.
116. E. A. Lesnik, C. J. Guinosso, A. M. Kawasaki, H. Sasmor, M. Zounes, L. L. Cummins, D. J. Ecker, P. Dan Cock and S. M. Freier, *Biochemistry*, 1993, **32**, 7832.
117. L. L. Cummins, S. R. Owens, L. M. Risen, E. A. Lesnik, S. M. Freier, D. McGee, C. J. Guinosso and P. Dan Cock, *Nucleic Acids Res.*, 1995, **23**, 2019.
118. P. Martin, *Helv. Chim. Acta*, 1995, **78**, 486.
119. V. Tereshko, S. Portmann, E. C. Tay, P. Martin, F. Nat, K.-H. Altmann and M. Egli, *Biochemistry*, 1998, **37**, 10626.
120. M. Teplova, G. Minasov, V. Tereshko, G. B. Inamati, P. Dan Cook, M. Manoharan and M. Egli, *Nat. Struct. Biol.*, 1999, **6**, 535.
121. J. F. Codington, I. Doerr, D. van Praag, A. Bendich and J. J. Fox, *J. Am. Chem. Soc.*, 1961, **83**, 5030.
122. P. S. Pallan, E. M. Greene, P. A. Jicman, R. K. Pandey, M. Manoharan, E. Rozners and M. Egli, *Nucleic Acids Res.*, 2011, **39**, 3482.
123. A. Patra, M. Paolillo, K. Charisse, M. Manoharan, E. Rozners and M. Egli, *Angew. Chem., Int. Ed.*, 2012, **51**, 11863.
124. M. J. Damha, C. J. Wilds, A. Noronha, I. Brukner, G. Borkow, D. Arion and M. A. Parniak, *J. Am. Chem. Soc.*, 1998, **120**, 12976.
125. I. Berger, V. Tereshko, H. Ikeda, V. E. Marquez and M. Egli, *Nucleic Acids Res.*, 1998, **26**, 2473.
126. J.-F. Trempe, C. J. Wilds, A. Y. Denisov, R. T. Pon, M. J. Damha and K. Gehring, *J. Am. Chem. Soc.*, 2001, **123**, 4896.
127. A. Y. Denisov, A. M. Noronha, C. J. Wilds, J.-F. Trempe, R. T. Pon, K. Gehring and M. J. Damha, *Nucleic Acids Res.*, 2001, **29**, 4284.
128. F. Li, S. Sarkhel, C. J. Wilds, Z. Wawrzak, T. P. Prakash, M. Manoharan and M. Egli, *Biochemistry*, 2006, **45**, 4141.
129. S. Obika, D. Nanbu, Y. Hari, K. Morio, T. Ishida and T. Imanishi, *Tetrahedron Lett.*, 1997, **38**, 8735.
130. S. Obika, D. Nanbu, Y. Hari, J. Andoh, K. Morio, T. Doi and T. Imanishi, *Tetrahedron Lett.*, 1998, **39**, 5401.
131. S. K. Singh, P. Koshkin, J. Wengel and P. Nielsen, *Chem. Commun.*, 1998, **4**, 455.
132. A. A. Koshkin, S. K. Singh, P. Nielsen, V. K. Rajwanshi, R. Kumar, M. Meldgaard, C. E. Olsen and J. Wengel, *Tetrahedron*, 1998, **54**, 3607.
133. K. Morihiro, T. Kodama, Kentefu, Y. Moai, R. N. Veedu and S. Obika, *Angew. Chem., Int. Ed.*, 2013, **52**, 5074.

134. P. H. Hagedorn, R. Persson, E. D. Funder, N. Albæk, S. L. Diemer, D. J. Hansen, M. R. Møller, N. Papargyri, H. Christiansen, B. R. Hansen, H. F. Hansen, M. A. Jensen and T. Koch, *Drug Discovery Today*, 2018, **23**, 101.
135. C. Zhou and J. Chattopadhyaya, *Chem. Rev.*, 2012, **112**, 3808.
136. Md. A. Islam and S. Obika, in *Handbook of Chemical Biology of Nucleic Acids*, ed. N. Sugimoto, Springer, Berlin, 2023, p. 497.
137. A. Yahara, A. R. Shrestha, T. Yamamoto, Y. Hari, T. Osawa, M. Yamaguchi, M. Nishida, T. Kodama and S. Obika, *ChemBioChem*, 2012, **13**, 2513.
138. K. Setoguchi, L. Cui, N. Hachisuka, S. Obchoei, K. Shinkai, F. Hyodo, K. Kato, F. Wada, T. Yamamoto, M. Harada-Shiba, S. Obika and K. Nakano, *Mol. Ther.–Nucleic Acids*, 2017, **9**, 170.
139. A. R. Shrestha, Y. Kotobuki, Y. Hari and S. Obika, *Chem. Commun.*, 2014, **50**, 575.
140. T. Shimo, Y. Nakatsuji, K. Tachibana and S. Obika, *Int. J. Mol. Sci.*, 2021, **22**, 3526.
141. P. P. Seth, A. Siwkowski, C. R. Allerson, G. Vasquez, S. Lee, T. P. Prakash, E. V. Wancewicz, D. Witchell and E. E. Swayze, *J. Med. Chem.*, 2009, **52**, 10.
142. P. P. Seth, G. Vasquez, C. A. Allerson, A. Berdeja, H. Gaus, G. A. Kinberger, T. P. Prakash, M. T. Migawa, B. Bhat and E. E. Swayze, *J. Org. Chem.*, 2010, **75**, 1569.
143. X.-H. Liang, W. Shen, H. Sun, G. A. Kinberger, T. P. Prakash, J. G. Nichols and S. T. Crooke, *Nucleic Acids Res.*, 2016, **44**, 3892.
144. E. E. Swayze, A. M. Siwkowski, E. V. Wancewicz, M. T. Migawa, T. K. Wyrzykiewicz, G. Hung, B. P. Monia and C. F. Bennett, *Nucleic Acids Res.*, 2007, **35**, 687.
145. W. Shen, C. L. De Hoyos, M. T. Migawa, T. A. Vickers, H. Sun, A. Low, T. A. Bell III, M. Rahdar, S. Mukhopadhyay, C. E. Hart, M. Bell, S. Riney, S. F. Murray, S. Greenlee, R. M. Crooke, X. Liang, P. P. Seth and S. T. Crooke, *Nat. Biotechnol.*, 2019, **37**, 640.
146. M. Tarköy, M. Bolli, B. Schweizer and C. Leumann, *Helv. Chim. Acta*, 1993, **76**, 481.
147. M. Tarköy, M. Bolli and C. Leumann, *Helv. Chim. Acta*, 1994, **77**, 716.
148. M. Bolli, H. U. Trafelet and C. Leumann, *Nucleic Acids Res.*, 1996, **24**, 4660.
149. R. Sreffens and C. Leumann, *Helv. Chim. Acta*, 1997, **80**, 2426.
150. R. Steffens and C. J. Leumann, *J. Am. Chem. Soc.*, 1997, **119**, 11548.
151. R. Steffens and C. J. Leumann, *J. Am. Chem. Soc.*, 1999, **121**, 3249.
152. D. Renneberg and C. J. Leumann, *J. Am. Chem. Soc.*, 2002, **124**, 5993.
153. P. S. Pallan, D. Ittig, A. Héroux, Z. Wawrzak, C. J. Leumann and M. Egli, *Chem. Commun.*, 2008, **7**, 883.
154. S. Murray, D. Ittig, E. Koller, A. Berdeja, A. Chappell, T. P. Prakash, M. Norrbom, E. E. Swayze, C. J. Leumann and P. P. Seth, *Nucleic Acids Res.*, 2012, **40**, 6135.
155. A. Goyenvalle, G. Griffith, A. Babbs, S. E. Andaloussi, K. Ezzat, A. Avril, B. Dugovic, R. Chaussenot, A. Ferry, T. Voit, H. Amthor, C. Buhr,

S. Schurch, M. J. A. Wood, K. E. Davies, C. Vaillend, C. Leumann and L. Garcia, *Nat. Med.*, 2015, **21**, 270.
156. I. L. Clézio, J. M. Escudier and A. Vigroux, *Org. Lett.*, 2005, **5**, 161.
157. C. Dupouy, I. L. Clézio, P. Lavedan, H. Gornitzka, J. M. Escudier and A. Vigroux, *Eur. J. Org Chem.*, 2006, **24**, 5515.
158. C. Dupouy, N. Iche-Tarrat, M. P. Durrieu, F. Rodriguez, J. M. Escudier and A. Vigroux, *Angew. Chem., Int. Ed.*, 2006, **45**, 3623.
159. M. E. Østergaard, B. Gerland, J. M. Escudier, E. E. Swayze and P. P. Seth, *ACS Chem. Biol.*, 2014, **9**, 1975.
160. I. L. Clézio, H. Gornitzka, J. M. Escudier and A. Vigroux, *J. Org. Chem.*, 2005, **70**, 1620.
161. C. Dupouy, P. Lavedan and J. M. Escudier, *Tetrahedron*, 2007, **63**, 11235.
162. I. L. Clézio, A. Vigroux and J. M. Escudier, *Eur. J. Org Chem.*, 2007, **63**, 1935.
163. C. Dupouy, P. Lavedan and J. M. Escudier, *Eur. J. Org Chem.*, 2008, **2008**, 1285.
164. D. A. Catana, M. Maturano, C. Payrastre, P. Lavedan, N. Tarrat and J. M. Escudier, *Eur. J. Org Chem.*, 2011, **2011**, 6857.
165. M. Maturano, D. A. Catana, P. Lavedan, N. Tarrat, N. Saffon, C. Payrastre and J. M. Escudier, *Eur. J. Org Chem.*, 2012, **2012**, 721.

CHAPTER 10

Sustainable Approaches in Solid-phase Oligonucleotide Synthesis: Current Status and Future Directions

Y. S. SANGHVI[a*], L. FERRAZZANO[b*], W. CABRI[b] AND A. TOLOMELLI[b]

[a]Rasayan Inc., 2802 Crystal Ridge Road, Encinitas, CA 92024-6615, USA; [b]P4i Laboratory, Department of Chemistry "Giacomo Ciamician", Alma Mater Studiorum – University of Bologna, via P. Gobetti 85, 40129 Bologna, Italy
*E-mail: ysanghvi@rasayan.us, lucia.ferrazzano4@unibo.it

10.1 Introduction

Solid-phase oligonucleotide synthesis (SPOS) using nucleoside phosphoramidites has been practiced since the early 1980s.[1] This protocol is exceedingly efficient in the assembly of monomeric amidite units in a sequential manner without the need for the isolation of intermediates after each step. Remarkably, the reproducibility, robustness, and high yields make the SPOS protocol a top choice for the production of all therapeutic oligonucleotides on the market and in clinical trials. In 2001, we summarized the applications of green chemistry in the manufacture of oligonucleotide drugs,[2] and during the past two decades we have witnessed increasing efforts to develop sustainable processes to produce these medicines. In this chapter, enabling

Green Chemistry Series No. 85
Sustainability in TIDES Chemistry: Green Approaches to Oligonucleotides and Oligopeptides Synthesis
Edited by Alessandra Tolomelli, Lucia Ferrazzano and Walter Cabri

Published by the Royal Society of Chemistry, www.rsc.org

technologies are discussed that will positively influence the process mass intensity (PMI) during oligonucleotide production. Although SPOS is a highly sought-after platform technology, several barriers remain to fulfilling the principles of green chemistry (Table 10.1).[3]

An excellent review on this topic was written by experts from industry further endorsing that much work is required to address the remaining challenges.[4] For example, the calculated PMI for 1 kg of 20-mer therapeutic oligonucleotide production was ~4300 kg of various materials.[5] In 2023, a head-to-head comparison of solid-phase peptide synthesis (SPPS) and SPOS focused on sustainability was published.[6] These reports do not account for the PMIs for manufacturing of various amidites, reagents, and solid supports essential for the SPOS of these molecules. Therefore, the overall environmental footprint caused by the production of these compounds is expected to be rather large, calling for more help from process chemists across the industry.

Solid-phase syntheses, the topic of this chapter, share several common features. The main attributes are the easy separation of the growing molecule from reagents and removal of by-products by simple washings, the equipment is fairly compact, and the level of automation is very high. On the other hand, reagent excess is always required to guarantee high conversions, despite careful analysis of the washings. Chain assembly analysis is still based on small-sample cleavage/high-performance liquid chromatographic (HPLC) analysis that is unnecessary taking into consideration how fast these single steps are.

The general production scheme is outlined in Figure 10.1. The following sections are mainly focused on the upstream process performed using

Table 10.1 Oligonucleotide synthesis and the principles of green chemistry.[4]

Principle	Assessment
Prevention of waste	Large amount of organic waste from upstream and downstream (*e.g.* PMI of ~4300)
Atom economy	Poor atom economy (*e.g.* 36% for phosphoramidites for oligonucleotides)
Less hazardous chemical synthesis	Some hazardous reagents [*e.g.* tetrazole, dichloroacetic acid (DCA), iodine, triethylamine trihydrofluoride (TEA·3HF), ammonium hydroxide]
Safer solvents and auxiliaries	Some issues with acetonitrile and toluene
Design for energy efficiency	Reactions are generally performed at room temperature, lyophilization is energy intensive
Use of renewable feedstocks	Many starting materials and reagents from non-renewable sources
Reduce derivatives	Protecting groups are extensively used
Catalysis	No catalytic process used
Design for degradation	Unmodified oligonucleotides generally break down into naturally occurring degradation products
Real-time analysis for pollution prevention	Some real-time data are generated but not used for pollution prevention
Inherently safer chemicals for accident prevention	Cleavage and deprotection involve heating under pressure in aqueous ammonium hydroxide

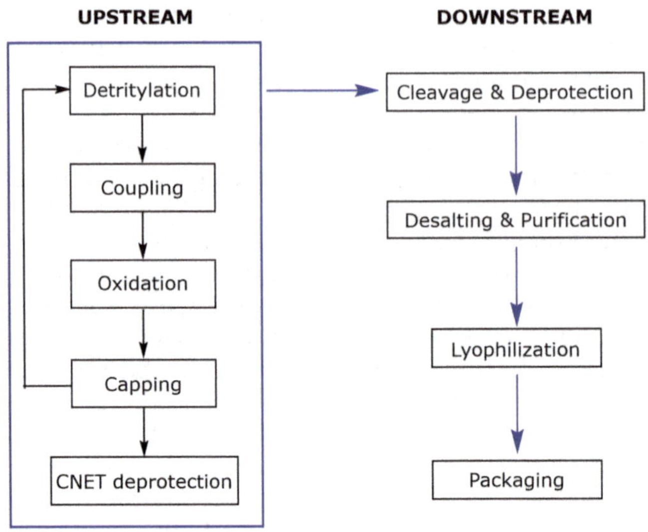

Figure 10.1 Main production steps in SPOS.

SPOS.[7–9] Both the upstream and downstream processes are highly automated on both laboratory and industrial scales.[10]

10.2 Starting Materials

The starting materials such as the solid support, the phosphoramidites, and the solvents are critical in determining the overall sustainability of the process. Particularly, the loading of the solid support, the reaction time, and the excess usage of phosphoramidites contribute to the overall PMI of oligonucleotide synthesis.

10.2.1 Solid Support

Two types of solid supports are generally used in SPOS, namely controlled-pore glass (CPG) and copolymers of polystyrene and divinylbenzene (PS–DVB), as illustrated in Figure 10.2 with the most common linkers and spacers used for the attachment of the first nucleoside.[11]

10.2.1.1 Controlled-pore Glass (CPG)[12]

CPG is commonly used as an insoluble support in oligonucleotide synthesis[7] owing to its rigid structure and non-swelling properties. The deep pores within the glass provide a suitable environment for the synthesis process. Typically, glass supports with 500 Å (50 nm) pores are utilized for the routine synthesis of short oligonucleotides. However, as the length of the oligonucleotide increases, the synthesis yields decrease significantly when using resins of 500 Å pore size, because the growing oligonucleotide strands obstruct the

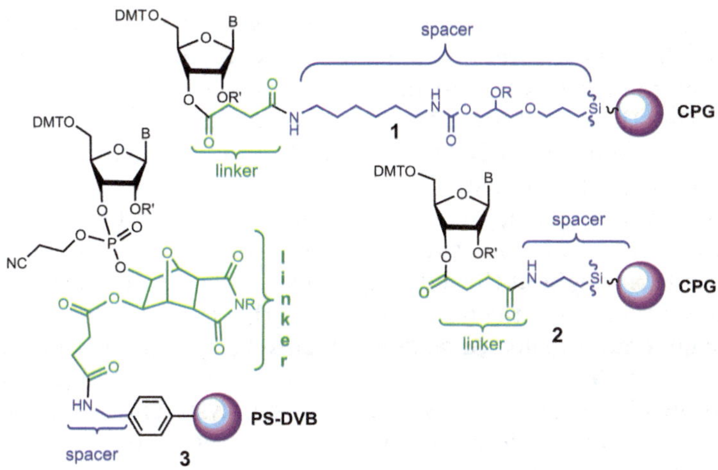

Figure 10.2 Typical spacer and linkers for CPG and PS–DVB.

pores, hindering the diffusion of reagents through the matrix. As a result, longer oligonucleotides encounter difficulties in efficient synthesis when using these matrices. To address this issue, glass beads with larger pore sizes are employed. Although larger pore beads are more fragile, they offer increased diffusion rates for reagents, facilitating the synthesis of longer oligonucleotides. Specifically, 1000 Å CPG has been found suitable for synthesizing oligonucleotides up to 100 mers in length. For even longer oligonucleotides, 2000 Å supports are utilized, providing the necessary space for efficient synthesis without significant obstruction of pores. Moving to large-scale production, CPG technology showed limits in terms of productivity due to the low loading of the support.

10.2.1.2 Copolymer of Polystyrene and Divinylbenzene (PS–DVB)

This class of solid supports is the main alternative to CPG technology and is preferred for large-scale production.[7–10] The porosity of the beads can be modulated by managing the level of crosslinking to afford supports that are physically stable throughout the entire process and exhibit chemical resistance against a wide range of solvents and reagents used in the synthesis of ASOs, including strong acids, bases, and organic solvents. An important advantage of PS–DVB is the good moisture exclusion properties, a key feature taking into consideration the sensitivity of reagents such as phosphoramidites to water. Although the ability of these resins to swell in various solvents can enhance the reaction efficiency, by increasing the accessibility of reactant molecules to the active sites within the beads, it can also lead to technical challenges, especially in applications involving fixed-bed flow synthesis. Inhomogeneity of the column flow and pressure fluctuations often generate inhomogeneity of the concentration distribution of the reactants. However, by moving to a stirred-bed technology, which is very popular in peptide chemistry, all of these issues are overcome. In terms of reaction time, whereas all the SPOS synthetic steps are very rapid when using CPG (1–2 min), on changing to PS–DVB the coupling becomes slower (>30 min). Using PS–DVB resin, Bachem reported batches of >2 mol with a stirred-bed reactor very similar to that used for peptide chemistry.[13]

The target of the spacer, linked to the 3′-position of the first nucleotide, is to keep the growing oligonucleotide chain far from the solid support surface and be stable throughout the entire synthesis.[14] These steps are not part of the iterative process and the resin bound with the first nucleotide can be purchased or generated *ad hoc*. As shown in Figure 10.2, the most common spacers with CPG are long-chain amine (**1**) or aminopropylsilanol (**2**) fragments anchored on silanes of the glass surface. The polystyrene is functionalized with an aminomethyl moiety, such as spacer **3**, linked to the 5′-position of the nucleoside (see Figure 10.2).

The linker between the 3′-OH and the spacer is generally a diacid, with succinic acid being the most common in oligonucleotide synthesis (see **1** or **2**, Figure 10.2), which at the end of the synthesis is cleaved off by treatment with ammonium hydroxide. An alternative spacer that generates the final ASO with no phosphate at the 3′-position is the Unilinker/Unisupport.[15]

The chemistry involved in nucleoside functionalization is based on standard chemistry and green attributes, such as solvents and reagents, that can be easily implemented. After the insertion of the first nucleoside, all amino groups are capped with an acetate moiety.

10.2.1.3 Solid Support Reuse and Recycling

Styrene and divinylbenzene are petrochemical-derived compounds commonly used in the production of various polymers, with wide application also in the consumer segment for packaging. For this reason, recycling technologies and transformation into high-value products are important areas of investigation.[16] The amount of resin used in the manufacture of oligonucleotides is negligible compared with other applications, but it is important to expand the resin portfolio to include more sustainable supports, such as poly(ethylene glycol) (PEG) derivatives.[17]

Another interesting area of investigation to increase the sustainability is the recovery and recycling of CPG solid supports.[18,19] There have been a few reports but this approach is useful in the case of large volumes and still unproven in terms of overall energy (CO_2) consumption.

10.2.2 Phosphoramidites

Remarkably, the SPOS of a 20-mer oligonucleotide offers >70% full-length product with minimal process optimization using a standard protocol. This is primarily due to the high quality of commercial phosphoramidites available from multiple vendors. P(III) chemistry is the cornerstone for phosphoramidite SPOS and LPOS technologies. These substrates are sensitive to both oxygen and humidity, hence manipulations are carried out under nitrogen and super-dry solvents must be used in the final step of the synthesis and for the iterative process. Amidites are the most important starting materials because they contribute a significant fragment to the complete structure of the drug substance and a large part of the overall cost. Comprehensive accounts of the synthesis, analysis, and stability of these starting materials in greater depth can be found elsewhere.[18,19] Here, the process improvements that led to the delivery of sustainable protocols for the synthesis of amidites are summarized. Broadly, there are four components in a phosphoramidite structure: (i) nucleosides; (ii) 5′-*O*-DMT (4,4′-dimethoxytrityl), an acid-labile protecting group; (iii) a base-labile *N*-acyl protecting group; and (iv) a cyanoethyl-protected P(III) functional group (see Figure 10.3).

Today, all 2′-deoxynucleosides are manufactured in ton quantities starting from 2-deoxyribose, which is easily synthesized from natural D-glucose available from sugarcane. The current source of ribonucleosides is the fermentation process routinely applied on a commercial scale. Collectively, both 2′-deoxy- and ribonucleosides are accessible using biocatalysis, shortening the conventional routes *via* chemical synthesis. The use of bio-based raw materials allows a reduction of greenhouse gas emissions, making these protocols more sustainable to support the manufacture of amidites.[7] Phosphoramidites are unstable towards oxygen and water, and the standard synthesis requires a final purification step. In contrast, the use of flow chemistry and an immobilized azole activator allows the building block to be produced on demand, avoiding any purification step.[20] This technology provided a very

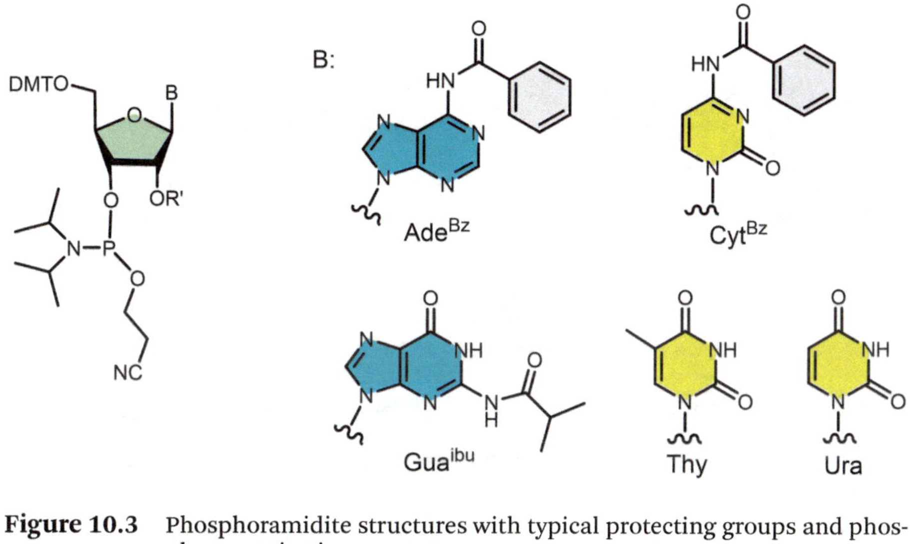

Figure 10.3 Phosphoramidite structures with typical protecting groups and phosphorus activation reagents.

similar crude purity with a significant increase in speed, which is particularly useful when unstable phosphoramidites are to be incorporated. Moreover, since no column is needed to isolate each monomer, a great advance in process sustainability can be realized *via* reduction of the amounts of solvents and gel needed for purification.

10.3 The Four-step SPOS Cycle

The conventional chemical synthesis of an oligonucleotide on a solid phase is accomplished by a computer-controlled automated synthesizer. The desired sequence is assembled by sequential coupling of phosphoramidite monomers following a four-step (detritylation/coupling/capping/oxidation) repetitive elongation cycle of reactions (see Scheme 10.1). An acetonitrile wash step is implemented between each cycle to remove the excess of reagents from the previous reaction. During synthesis, the elongating oligonucleotide remains attached to the solid support. The upstream process is fully automated, and the synthesizer technologies are directly related to the batch size of the oligonucleotide.[7–9]

10.3.1 Detritylation

This is the first step, which removes the 5′-hydroxyl protecting group with an acid. The DMT group is the most frequently used protecting group for oligonucleotide synthesis, for several reasons. First, the DMT group is highly regioselective during the synthesis of nucleoside amidites. Second, it is rapidly cleaved in minutes with an acid treatment such as using dichloroacetic acid. Third, it generates an orange carbocation, allowing

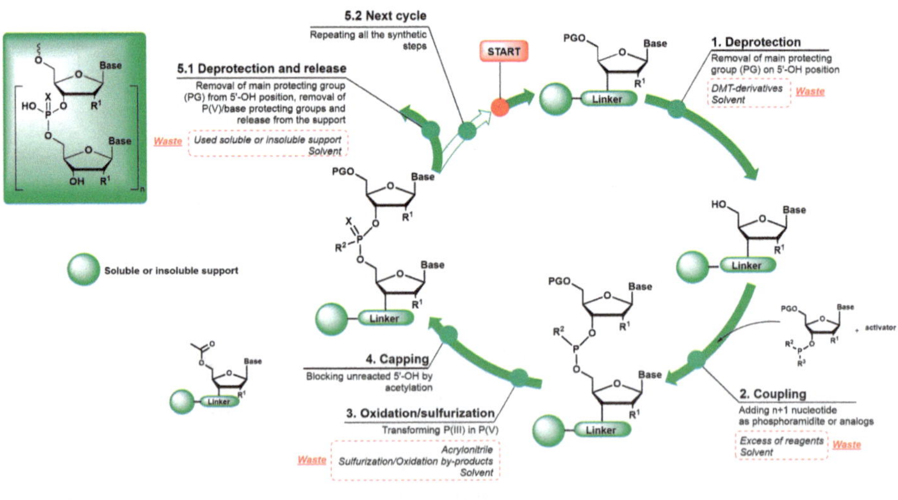

Scheme 10.1 Overview of SPOS and LPOS synthetic cycle.

easy quantification of the coupling efficiency during SPOS. Fourth, it also offers a hydrophobic handle for ease of purification of oligonucleotides using reversed-phase chromatography. Numerous process improvement steps have been implemented to make the detritylation step safer and economical.

Even today, the classical detritylation reagent is prepared as a mixture of a haloacetic acid (typically di- or trichloroacetic acid) in dichloromethane (DCM) and is often used in a small-scale bench-top synthesizer. However, for the large-scale production of therapeutic oligonucleotides, the use of toxic, volatile, and halogenated solvents is very undesirable. Therefore, researchers at Ionis proposed toluene as a green substitute for DCM without compromising the speed of detritylation and the yield of the reaction.[21] In fact, both DCM and toluene were proved to be able to prevent acid binding to the solvent, in comparison with EtOAc and tetrahydrofuran (THF).[22] Nowadays, on a commercial scale the detritylation step is routinely carried out with 10% dichloroacetic acid in toluene. Although this effort mitigated the risk of using a hazardous solvent, the current PMI assessment of the detritylation step alone is about 20% of the total raw materials used for SPOS. The impact of detritylation on PMI is further increased if it is considered that the oxidation step occurring before deprotection is performed with acetonitrile, thus requiring several washings to avoid above-mentioned acid binding to solvent residues, which results in decreased efficiency in this synthetic step.

DMT is a high molecular weight protecting group that is discarded upon detritylation, and for this reason it affects the atom economy of the synthesis cycle. In some protocols, scavenging of the DMT carbocation deriving from acidic deprotection is performed by adding pyrrole, alcohols, silanes, anisole, thioanisole, or thiols.[23–25]

After each deprotection, the solid support must be well rinsed to remove residual reagents and reduce eventual interferences in the subsequent reaction step. It has been reported recently that a 50% reduction in post-detritylation washings with acetonitrile is possible with minimal impact on product quality and results in an overall 70% reduction in solvent consumption for washings. It was reported that for 1 kg of an 18-mer oligonucleotide, the PMI decreased from 887 to 542, reducing the amount of solvent used in operational washing steps.[26]

Another approach could be to prevent the loss of the detritylation by-product in the waste stream by using processes for the capture and regeneration of the DMT group. An example of this possibility was patented by Sanghvi and Guo.[27] Briefly, the DMT cation was treated with base to generate DMT-OH, which upon isolation and chlorination with acetyl chloride furnished DMT-Cl, useful for the preparation of 5′-*O*-DMT-protected phosphoramidites. The recovery and reuse of a protecting group are an important example of improved life-cycle management during SPOS. However, applications of this approach in large-scale oligonucleotide manufacturing have yet to be implemented.

10.3.2 Coupling of P(III) Phosphoramidite Monomer

Selecting the proper reagents and conditions for the coupling step has a great impact on the sustainability of the overall process. Apparently, in the evaluation of the PMI, the synthesis of the oligonucleotide is responsible for 44% of the total value, and only 6% is associated with the coupling reaction.[4] Nevertheless, the efficiency of this step has a great impact on the downstream process and on its contribution to the overall PMI. In iterative protocols, the need to drive coupling to completeness is required to have clean crude oligonucleotide, the purification of which could be simpler and less solvent demanding. For this, the use of an excess of reagents is commonly tolerated in view of striking a balance between the upstream and downstream workflows.

Generally, the coupling cycle on a commercial scale requires use of a slight excess (1.5–2.0 equiv.) of phosphoramidite monomers to drive the reaction to near completion in the presence of an activator agent, utilized to activate an incoming phosphoramidite monomer to react with the 5′-hydroxyl group of a growing oligonucleotide chain to form a phosphite triester intermediate. The activator agent has a dual role, as a proton donor and a nucleophilic catalyst. Under these conditions, the phosphoramidite coupling reaction is extremely fast and facilc, going to completion in a few minutes under strictly anhydrous conditions.

Acetonitrile is considered the optimal solvent, being able to dissolve the reagents for all four steps of oligonucleotide synthesis, thanks to its polar aprotic nature. It is essential that acetonitrile used for this step is anhydrous with a water content of <30 ppm to afford >99% coupling efficiency, and an HPLC-grade solvent, containing 0.001% of water, is suitable for this purpose. However, acetonitrile is listed as a "problematic" solvent in the CHEM21 solvent selection guide, which is the standard reference work for process sustainability assessment.[28] The metabolism of this solvent by cytochrome P450 produces hydrogen cyanide, a high-dose exposure to which can be lethal.[29] Traditionally, acetonitrile was quite cheap and readily available, being obtained by the SOHIO ammoxidation process as a by-product of acrylonitrile synthesis.[30] Since 2008, the reduced demand for acrylonitrile has led to a significant shortage and fluctuations of the price of acetonitrile, highlighting the need for different synthetic routes.[31] A sustainable alternative is preparation from bioethanol and ammonia.[32]

Further, to reduce the volume of acetonitrile in SPOS, acetone has been proposed as an alternative solvent for the washing steps. However, acetone cannot be used in the coupling step owing to its high moisture content (up to 0.3%).[33]

A crucial reagent in the coupling step is the activator. 1*H*-Tetrazole was widely used in the past in phosphoramidite chemistry, but alternative azoles were introduced later to overcome its disadvantages, especially for large-scale applications. Owing to its explosive properties, it is strongly recommended that the use of 1*H*-tetrazole as an activator during SPOS is avoided. The possible crystallization from acetonitrile solutions at low temperatures may lead to occlusion of automated systems and the generation of dry solids from leaking evaporation can generate safety concerns.

The accepted mechanism for 1*H*-tetrazole activation ($pK_a = 4.9$)[34,35] involves a two-step reaction starting with the protonation of the phosphoramidite diisopropylamino group, followed by nucleophilic substitution to form the tetrazolide intermediate active species. Alternative activators can be good proton donors and nucleophiles to guarantee a high rate of reaction. Today, most of the large-scale manufacturing of therapeutic oligonucleotides is performed with 4,5-dicyanoimidazole (DCI) ($pK_a = 5.2$) during the assembly of antisense oligonucleotides. This activator is less acidic than 1*H*-tetrazole and, for this reason, its use decreases the possible formation of double addition products ("longmers"), due to the side reaction of monomer deprotection in solution, thus affording a cleaner crude oligonucleotide for purification. For small interfering RNA (siRNA) production, the preferred activator is 5-ethylthiotetrazole (ETT) ($pK_a = 4.3$), which is much more soluble than 1*H*-tetrazole in acetonitrile.[36]

The search for a safer and cheaper activator led to the discovery of pyridinium trifluoroacetate (PTFA) ($pK_a = 5.3$) reported by researchers at Ionis.[37] Although PTFA is able to activate the coupling by itself, the addition of *N*-methylimidazole affords better yields by increasing the nucleophilicity of the attacking 5-hydroxyl group. Moreover, *N*-methylimidazole decreases the acidity of the solution, thus inhibiting the formation of $n + 1$ longmers. So far, no detailed toxicological data on PTFA have been reported, but the available data on the two components (pyridine and trifluoroacetic acid) suggest caution in large-scale use.

Finally, given the high cost of phosphoramidite building blocks, numerous researchers have developed a process for recovering the excess of unreacted amidites and 1*H*-tetrazole collected from the waste stream. In particular, collection of the acetonitrile solution containing excess reagents in reservoirs containing *N*,*N*-diisopropylamine induces precipitation of *N*,*N*-diisopropylammonium tetrazolide salt, affording a solution of regenerated phosphoramidite. Further treatment of the filtered salt with trifluoroacetic acid also leads to the recovery of 1*H*-tetrazole.[38] In a different protocol, free nucleosides were recovered by treatment of the unreacted monomers with ethylene glycol or glycerol to induce dephosphonylation.[39] Although these recovery protocols have been around for two decades, none are currently used in practice.

However, it is expected that some of the next generation of highly modified amidites may change the situation because long delivery times and high costs make the recovery process worthy of consideration. Clearly, the processes that allow the capture of expensive amidite from waste streams will not only help reduce the environmental burden but also permit cost reductions for the production of these life-saving drugs at scale.

10.3.3 Oxidation or Sulfurization

The phosphite triester formed during the coupling reaction is next oxidized to the more stable P(v) phosphate triester using an oxidizing agent. The most popular oxidizing agent for small-scale synthesis is aqueous iodine under basic conditions. A mixture of 0.05 M iodine in pyridine–water (9 : 1 v/v) is in use for the commercial production of Spiegelmers (L-RNA aptamers). The oxidation cycle is fast and is completed in 5 min under optimized conditions. Interestingly, phosphoramidite chemistry demands the use of anhydrous reagents but for the oxidizing agent an aqueous solution is used, defeating the purpose. For this reason, several non-aqueous approaches have been proposed, using nitrogen dioxide in DCM,[40] *m*-chloroperbenzoic acid in DCM,[41] *tert*-butyl hydroperoxide in toluene,[42] bis(trimethylsilyl) peroxide in DCM or an acetonitrile–DCM mixture,[42] dimethyldioxirane in DCM,[42] (1*S*)-(+)-(10-camphorsulfonyl)oxaziridine (CSO) in acetonitrile,[43,44] and ethyl(methyl)dioxirane in DCM.[45] However, these reagents have several limitations related to toxicity, high cost, and a tendency to induce side reactions. Subsequently, a safer mixture of *N*-bromosuccinimide (NBS) and dimethyl sulfoxide (DMSO) in acetonitrile was proposed.[46] This non-aqueous oxidation cycle minimizes the need for multiple washing steps to remove residual water from the solid support, significantly reducing the impact on the PMI.

In the case of antisense therapeutic oligonucleotides, one of the non-bridging oxygen atoms is replaced with a sulfur atom because of the increased *in vivo* stability. The insertion of sulfur in the phosphate backbone is conveniently accomplished with a sulfur transfer reagent, among which phenylacetyl disulfide (PADS)[47,48] and 3-amino-1,2,4-dithiazole-5-thione (XHS)[49] in the presence of base are in use for commercial production. More recently, DDTT [*N*,*N*-dimethyl-*N*-(3-thioxo-3*H*-1,2,4-dithiazol-5-yl)methanimidamide] was introduced as a more stable and highly soluble sulfur transfer reagent.[50] Additionally, another alternative reagent, dimethylthiuram disulfide (DTD), was reported to have potential for reducing the solvent consumption by *ca.* 20% while reducing the cycle time during SPOS.[51] The use of these reagents is in line with the green chemistry principles, reducing the production cost and cycle time. Currently, DDTT is often the preferred reagent for sulfurization owing to its combination of low cost, reaction kinetics, efficacy, and stability.

10.3.4 Capping

The fourth and last step in each synthesis cycle is the capping reaction. The main purpose of capping is to block the unreacted 5′-hydroxyl groups that failed to couple. An efficient capping step terminates the extension of the oligonucleotide chain, which otherwise results in the formation of an undesired shorter ($n-1$) chain that is difficult to remove post-synthesis. It is important to emphasize that the capping step is not always included in synthetic protocols, thus requiring elaboration in PMI comparisons.

For commercial-scale oligonucleotide production, Cap A [*N*-methylimidazole (NMI)–pyridine] and Cap B (acetic anhydride) are the first choice as a capping cocktail in acetonitrile. The conventional protocol for capping recommends a large excess of reagents to drive it to completion. To make SPOS more sustainable, this is another step in which the total volume of reagents can be reduced without impacting the reaction efficiency. In fact, this task was clearly demonstrated in a study in which only half the volume of capping reagents not only maintained the acylation efficiency but also reduced impurity formation.[52] The efforts made to reduce or eliminate the capping step led to the development of a modified approach in which both sulfurization and capping are accomplished in a single step. Researchers at Biogen described the utility of a three-step cycle for the synthesis of a phosphorothioate oligonucleotide where ~500 L of solvents and reagents were saved for every kilogram of 18-mer drug substance made.[53] Such strides will continue to chip away at the extraordinary resources currently involved during SPOS. In addition, several options for the selection of optimal capping reagents have been proposed over the years. More recently, a lipophilic cholesterol–phosphoramidite has been proposed as a capping reagent.[54] This protocol allowed efficient removal of shortmer impurities due to conjugated hydrophobic moieties by reversed-phase chromatography. These advances may permit the synthesis of long oligonucleotide sequences where purification remains a challenging task.

10.4 Cleavage and Deprotection (C&D)

Upon completion of the full-length oligonucleotide synthesis, a two-step process is applied to cleave the oligonucleotide sequence from the support and remove all protecting groups (see Figure 10.4). Often these steps require hazardous conditions and an elevated reaction temperature. First, amine washing of the column is carried out to remove the 2-cyanoethyl (CE) protecting group from the phosphate backbone. The purpose of this step is to prevent the addition of acrylonitrile to the pyrimidine base. Second, base

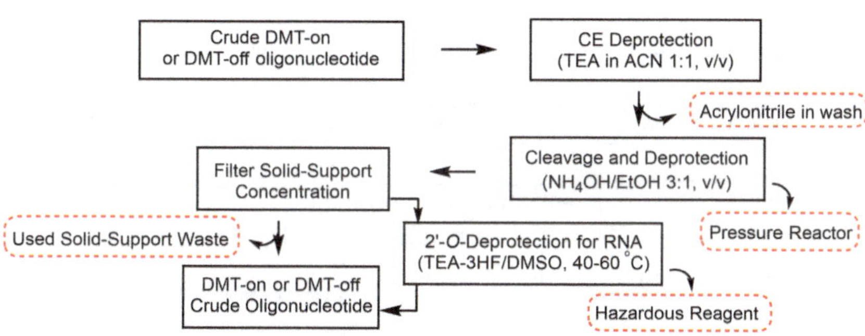

Figure 10.4 An overview of post-synthesis chemical deprotection steps.

deprotection is performed in a pressure vessel for several hours at 50–60 °C with aqueous ammonia. Recently, scientists at Biogen have demonstrated that piperazine is highly efficient in CE deprotection and acrylonitrile scavenging. This protocol is carried out in one pot, eliminating the two-step conventional C&D procedure, thus simplifying the process and making it safer for large-scale production.[55]

For manufacturing siRNA, an additional step is required to deprotect the 2′-*O*-*tert*-butyldimethylsilyl (TBDMS) group used routinely for RNA synthesis during SPOS. The current protocol recommends use of triethylamine trihydrofluoride (TEA·3HF) at elevated temperature. Obviously, this is a hazardous step and known to cause skin burns and hypocalcaemia under extreme exposure. Therefore, NH_4F and KF have been reported as safer alternatives and worthy of consideration for the bulk-scale production of RNA oligonucleotides.[56]

10.5 Purification

Currently, the oligonucleotide purification process accounts for >50% of the materials used in the production of these complex molecules. It is essential that oligonucleotide drugs are purified carefully for their efficacy and patient safety. The balancing act of achieving high yields while keeping a low impurity profile is a formidable task during the purification step (see Figure 10.5).

The waste generated during chromatography is hazardous and currently not recovered during large-scale production. There are several techniques that can be applied, but the most important is to move from single-column to multiple-column purification, in particular the twin column approach developed by Morbidelli and Aumann (see Chapter 15).[57] In analogy with the peptide application, the ETH group developed a new protocol for purification utilizing two in-line chromatographic columns, allowing the continuous separation of the purified product from impurities. Briefly, during a gradient purification, the late- and early-eluting impurities are fed into a second column as a switch and the operation is fully automated.[58] Bachem reported that the multi-column countercurrent solvent gradient purification

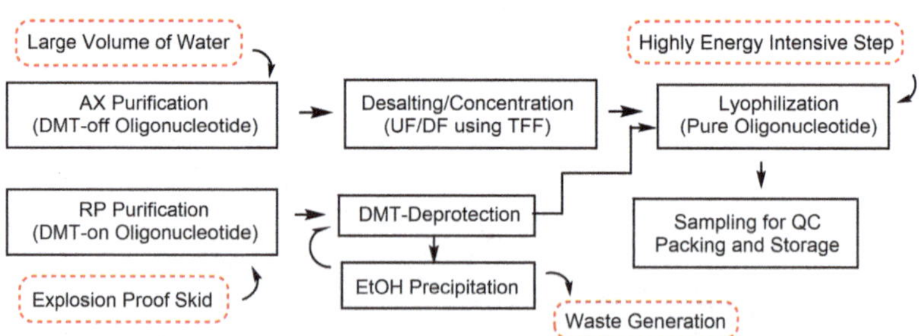

Figure 10.5 An example of the downstream process flow of a drug substance. AX = anion-exchange; RP = reverse-phase.

(MCSGP) protocol permitted the large-scale purification of an oligonucleotide with 93.7% purity and 81.6% yield. This yield is 8% higher than that using the conventional single-column purification protocol while reducing the PMI significantly.[59] Therefore, continuous chromatography is a perfect fit for manufacturing a high-volume product following the principles of green chemistry. Desalting and concentration can be easily achieved by ultrafiltration/diafiltration (UF/DF) by using tangential flow filtration (TFF). The high consumption of water in the diafiltration step can be reduced by using continuous diafiltration technology.[4,60] The aqueous solution can also be concentrated using thin-film evaporation, but water evaporation is an energy-consuming process.

10.6 Lyophilization

Freeze-drying (lyophilization) is the final step in the production of therapeutic oligonucleotides. This is a multi-step process that takes a long time to dry the product under a high vacuum and at low temperature; it is an energy-intensive step lasting several days. Interestingly, most of the oligonucleotide drug products are intended for parenteral administration where a lyophilized powder is dissolved in aqueous buffer for formulation. Since the lyophilization step is a bottleneck in the manufacturing process calling for high energy usage and long cycle times, a group of industrial scientists belonging to the European Pharma Oligonucleotide Consortium (EPOC) proposed skipping it in favour of "solution active pharmaceutical ingredients (APIs)".[61] It was concluded that solution APIs offer not only an economical path but also a sustainable process for oligonucleotide manufacture.

Spray drying in large-volume API manufacturing is considered a competitive alternative. However, the use of standard spray-drying technologies that use high temperatures of >100 °C is hardly compatible with TIDES stability.[4] More recently, a new spray-drying technology has been developed, namely PolarDry®, in which an electrostatic charge is applied at the nozzle of the spray-drying system, allowing good performances to be achieved at consistently lower temperatures of ~70 °C in a short-time using a nitrogen flow. This technology should be explored since it represents an interesting alternative to freeze-drying, being rapid, continuous, and less energy demanding with respect to batch freeze-drying.[62]

10.7 Summary, Challenges, and Future Perspectives

Currently, SPOS using amidite chemistry remains a *tour de force*, advancing hundreds of therapeutic oligonucleotides due to the high speed of delivery and the ability to make thousands of sequences for discovery on a fully automated synthesizer. Even though several process improvements have been implemented during SPOS, the production cost of these drugs is relatively high compared with small-molecule drugs. The main reasons for the high cost are the use of expensive solid supports, amidites, and large volumes of organic solvents. In particular, the current solid support methods for synthesis are not ideal for the high-volume manufacture of drugs for the treatment of common diseases. Some of the challenges associated with SPOS are as follows: (i) limitation of solid support loading for the synthesis of long (*e.g.* 100-mer) sequences; (ii) heterogeneous chemistry demanding high mass transfer resistance through small pores; (iii) inhomogeneity of conventional packed-bed columns; (iv) use of excess amidites to drive the coupling reaction to completion; (v) the need for a large excess of ancillary reagents for deblocking, oxidation, and capping steps: and (vi) large volumes of acetonitrile for the washing step between each cycle to remove unreacted reagents. More recently, Bachem has implemented stirred-bed SPOS for the production of short oligonucleotides, demonstrating the potential to save solvents and reduce the PMI. The stirred-bed approach offers greater flexibility when swelling of the solid support resin is a concern and maximizing the homogeneity during each cycle of synthesis. An improved reactor design for stirred beds may permit large-scale commercial oligonucleotide API manufacture with greater purity.[13]

The high PMI of SPOS is a serious concern, demanding further exploration of alternative oligonucleotide production technologies. It is encouraging that several industrial and academic groups have stepped up their research and development activity to minimize or reduce the environmental impact during oligonucleotide synthesis. However, some of the current limitations of insoluble solid supports could be circumvented by employing a soluble support for production (see Chapter 11).

NATiAS is a Japanese company working on the Blockmer™ approach, where dimeric and trimeric amidites are utilized during the synthesis cycle. The use of Blockmers allows the manufacture of higher quality products compared with the monomeric amidite protocol. These Blockmers are now available as commercial building blocks.[63] The integration of the on-demand synthesis of amidites, already discussed above, can be utilized without purification for automated SPOS, exhibiting a high coupling efficiency.[20] This is in line with green chemistry principles; the flow reactions empower shorter cycle times while eliminating the storage of expensive amidites. In the future, we envision lower production costs due to the smaller footprint and reduced energy consumption of the continuous-flow set-up.

The large amounts of waste produced and poor atom economy in SPOS have encouraged the scientific community to consider enzymatic approaches as a viable alternative. Whereas in peptide chemistry the use of biocatalysis is an established technology for hybrid synthesis (see Chapter 6), application to oligonucleotides is in the early research stage. Lovelock's group in the UK is engaged in the development of engineered enzymes for the assembly of chemically modified nucleoside triphosphates into a therapeutic oligonucleotide on the bench scale.[64] The entire process is performed in aqueous solution to avoid the use of the solid supports and organic reagents currently utilized in SPOS. Similarly, Codexis in the USA is perfecting enzyme-catalysed oligonucleotide synthesis (ECO Synthesis™) utilizing terminal deoxynucleotidyl transferase (TdT) as a linchpin offering both wide applicability and high reactivity with their protocol.[65] Almac is another UK-based enterprise marrying the best of both worlds.[66] Their work allows the utilization of engineered RNA ligase to couple short fragments of oligonucleotides into a full-length siRNA for therapeutics. The short fragments were synthesized by SPOS. These bio-based manufacturing technologies are possible options for a low-carbon, low-waste, and circular economy with societal benefits (see Table 10.2). However, this approach still requires further massive research investments to become industrially relevant. In addition, for this technology, as for any other that is covered by intellectual property protection, the application will be limited by this stumbling block.

Sustainability and environmental protection during oligonucleotide manufacture are not only dominating public discourse, they are also influencing the way in which we think about large-scale pharmaceutical production processes. Although new drugs dramatically increase patients' life expectancy, the costs for ground-breaking oligonucleotide therapies are

Table 10.2 Comparison of SPOS and emerging enzymatic approaches.

Solid-phase oligonucleotide synthesis (SPOS)	Enzymatic oligonucleotide synthesis
Amidite building blocks:	*Nucleoside triphosphate building blocks:*
• High molecular weight	• Moderate molecular weight
• Protecting groups required	• Protecting groups not essential
• Multistep synthesis	• One-pot protocol for synthesis
• Used in excess	• Excess reagents not required
• High purity for P(III) is essential	• Lower purity is tolerated
• Long supply chain	• Shorter supply chain
Synthesis cycle:	*Synthesis cycle:*
• Linear or Blockmer	• Linear or convergent
• Organic solvents required	• Aqueous media for reactions
• Excessive solvent wash required	• Washing between steps not required
• Fast and efficient	• Fast and fewer errors
• Chiral control of PS backbone	• Chiral control of PS backbone
Purification protocol:	*Purification protocol:*
• Several shortmers are formed	• Lower amount of shortmer impurities
• Process optimization required	• Less optimization required
• Isolation of higher purity at the cost of yield	• Complementary base pair allows high purity and good yield
Overall process:	*Overall process:*
• Scale-up limitations	• Scale-up possible
• Complex equipment	• Standard equipment
• High waste disposal cost	• Lower disposal cost
• Safety concern	• Significantly safer

soaring to new heights. Therefore, as chemists, continued rethinking is necessary to deliver processes with a focus on green chemistry. It is hoped that the review in this chapter will provide readers with an in-depth understanding of the challenges that remain and further insightful directions on sustainable SPOS.

List of Abbreviations

API	Active pharmaceutical ingredient
ASO	Antisense oligonucleotide
AX	Anion-exchange
C&D	Cleavage and deprotection
CE	2-Cyanoethyl
CPG	Controlled-pore glass
CSO	10-(Camphorsulfonyl)oxaziridine
DCI	Dicyanoimidazole
DCM	Dichloromethane
DDTT	*N*,*N*-Dimethyl-*N*-(3-thioxo-3*H*-1,2,4-dithiazol-5-yl) methanimidamide
DMSO	Dimethyl sulfoxide
DMT	4,4′-Dimethoxytrityl
DTD	Dimethylthiuram disulfide
EPOC	European Pharma Oligonucleotide Consortium
ETT	5-Ethylthiotetrazole
HPLC	High-performance liquid chromatography
MCSGP	Multi-column countercurrent solvent gradient purification
NBS	*N*-Bromosuccinimide
NMI	*N*-Methylimidazole
PADS	Phenylacetyl disulfide
PMI	Process mass intensity
PS–DVB	Polystyrene–divinylbenzene copolymer
PTFA	Pyridinium trifluoroacetate
RP	Reverse-phase
siRNA	Small interfering RNA
SOHIO	Standard Oil of Ohio
SPOS	Solid-phase oligonucleotide synthesis
SPPS	Solid-phase peptide synthesis
TBDMS	*tert*-Butyldimethylsilyl
TdT	Terminal deoxynucleotidyl transferase
TEA·3HF	Triethylamine trihydrofluoride
TFF	Tangential flow filtration
THF	Tetrahydrofuran
UF/DF	Ultrafiltration/diafiltration
XHS	3-Amino-1,2,4-dithiazole-5-thione

Acknowledgement

The authors thank the PNRR project "National Center for Gene Therapy and Drugs Based on RNA Technology" – Spoke 6 – RNA Chemistry – Task 6.2.1 "Development of New RNA-targeting Compounds RIBOTAC and Stereopure Oligonucleotides and Peptides Conjugates" for financial support.

References

1. S. L. Beaucage and M. H. Caruthers, *Tetrahedron Lett.*, 1981, **22**, 1859.
2. Y. S. Sanghvi, V. T. Ravikumar, A. N. Scozzari and D. L. Cole, *Pure Appl. Chem.*, 2001, **73**, 175.
3. P. T. Anastas and J. C. Warner, *Green Chemistry: Theory and Practice*, Oxford University Press, New York, 1998.
4. B. I. Andrews, F. D. Antia, S. B. Brueggemeier, L. J. Diorazio, S. G. Koenig, M. E. Kopach, H. Lee, M. Olbrich and A. L. Watson, *J. Org. Chem.*, 2021, **86**, 49.
5. M. E. Kopach and B. I. Andrews, *Curr. Opin. Green Sustainable Chem.*, 2022, **37**, 100643.
6. L. Ferrazzano, D. Corbisiero, A. Tolomelli and W. Cabri, *Green Chem.*, 2023, **25**, 1217.
7. B. Roy, A. Depaix, C. Périgaud and S. Peyrottes, *Chem. Rev.*, 2016, **116**, 7854.
8. E. Paredes and T. Konishi, in *Synthesis of Therapeutic Oligonucleotides*, ed. S. Obika, and M. Sekine, Springer, 2018, p. 97.
9. E. Paredes, V. Aduda, K. L. Ackley and H. Cramer, *Compr. Med. Chem. III*, 2017, **6**, 233.
10. Y. S. Sanghvi, in *Advances in Nucleic Acid Therapeutics*, ed. S. Agrawal, & M. J. Gait, Drug Discovery Series, 2019, vol. 68, p. 453.
11. For an extensive account on solid supports seeA. P. Guzaev, *Curr. Protoc. Nucleic Acid Chem.*, 2013, **53**, 3.1.1–3.1.60.
12. J. Brzezinska, S. Trzcinski, J. Strzelec and M. K. Chmielewski, *Bioorg. Chem.*, 2023, **140**, 106806.
13. D. Samson and M. Minuth, *Oligonucleotide Manufacture*, Bachem White Paper, 2023.
14. A. P. Guzaev and R. T. Pon, *Curr. Protoc. Nucleic Acid Chem.*, 2013, **52**, 3.2.1–3.2.23.
15. V. T. Ravikumar, R. K. Kumar, P. Olsen, M. N. Moore, R. L. Carty, M. Andrade, D. Gorman, X. Zhu, I. Cedillo, Z. Wang, L. Mendez, A. N. Scozzari, G. Aguirre, R. Somanathan and S. Bernee, *Org. Process Res. Dev.*, 2008, **12**, 399.
16. Z. Huang, M. Shanmugam, Z. Liu, A. Brookfield, E. L. Bennett, R. Guan, D. E. Vega Herrera, J. A. Lopez-Sanchez, A. G. Slater, E. J. L. McInnes, X. Qi and J. Xiao, *J. Am. Chem. Soc.*, 2022, **144**, 6532.
17. S. Mazzini, F. García-Martin, M. Alvira, A. Aviñó, B. Manning, F. Albericio and R. Eritja, *Chem. Biodiversity*, 2008, **5**, 209.

18. R. T. Pon, S. Yu, Z. Guo and Y. S. Sanghvi, *Nucleic Acids Res.*, 1999, **27**, 1531.
19. R. T. Pon, S. Yu, Z. Guo, R. Deshmukh and Y. S. Sanghvi, *J. Chem. Soc., Perkin Trans. 1*, 2001, 2638.
20. A. F. Sandahl, T. J. D. Nguyen, R. A. Hansen, M. B. Johansen, T. Skrydstrup and K. V. Gothelf, *Nat. Commun.*, 2021, **12**, 2760.
21. A. H. Krotz and V. T. Ravikumar, Methods for synthesis of oligonucleotides, *US pat.*, US7273933B1, 2007.
22. Q. Li, Y. S. Sanghvi and H. Yan, *Synlett*, 2024, **35**, 672.
23. C. B. Reese, H. T. Serafinowska and G. Zappia, *Tetrahedron Lett.*, 1986, **27**, 2291.
24. I. Habus and S. Agrawal, *Nucleic Acids Res.*, 1994, **22**, 4350.
25. V. Ravikumar, M. Andrade, D. Mulvey and D. J. Cole, Carbocation scavenging during oligonucleotide synthesis, *US pat.*, 5510476, 1996.
26. L. Xiao, T. Pickel, Z. Li, D. Luciano, J. Yang, D. Cho, S. Mac, X. Shi, G. Bou-Assaf, F. Antia and Y. Fillon, *Green Chem.*, 2023, **25**, 4292.
27. Y. S. Sanghvi and Z. Guo, Recovery of triarylmethyl halide groups cleaved during oligonucleotide synthesis, *WO pat.*, WO2000004062A1, 1999.
28. D. Prat, A. Wells, J. Hayler, H. Sneddon, C. R. McElroy, S. Abou-Shehada and P. J. Dunn, *Green Chem.*, 2016, **18**, 288.
29. H. Robles, *Encyclopedia of Toxicology*, 2014, vol. 1, p. 40.
30. ACS Chemical Landmark, The Sohio Acrylonitrile Process, INEOS Commemorative Booklet, 2007, https://www.acs.org/education/whatischemistry/landmarks/acrylonitrile.html.
31. A. H. Tullo, *Chem. Eng. News*, 2022, **100**, 13.
32. A. Tripodi, E. Bahadori, D. Cespi, F. Passarini, F. Cavani, T. Tabanelli and I. Rossetti, *ACS Sustainable Chem. Eng.*, 2018, **6**, 5441.
33. S. L. Beaucage and M. H. Caruthers, *Tetrahedron Lett.*, 1981, **22**, 1859.
34. B. H. Dahl, J. Nielsen and O. Dahl, *Nucleic Acids Res.*, 1987, **15**, 1729.
35. S. Berner, K. Mühlegger and H. Seliger, *Nucleic Acids Res.*, 1989, **17**, 853.
36. Glen Report 19.29: Technical Brief – About Activators: Now And Tomorrow, https://www.glenresearch.com/reports/gr19-29.
37. A. Eleuteri, D. C. Capaldi, A. H. Krotz, D. L. Cole and V. T. Ravikumar, *Org. Process Res. Dev.*, 2000, **4**, 182.
38. C. L. Scremin, L. Zhou, K. Srinivasachar and S. L. Beaucage, *J. Org. Chem.*, 1994, **59**, 1963.
39. W. Wang, Q. Song and R. A. Jones, *Tetrahedron Lett.*, 1999, **40**, 8971.
40. G. S. Bajwa and W. G. Bentrude, *Tetrahedron Lett.*, 1978, **19**, 421.
41. K. K. Ogilvie and M. J. Nemer, *Tetrahedron Lett.*, 1981, **22**, 2631.
42. Y. Hayakawa, M. Uchiyama and R. Nayori, *Tetrahedron Lett.*, 1986, **27**, 4195.
43. M. Manoharan, Y. Lu, M. D. Casper and G. Just, *Org. Lett.*, 2000, **2**, 243.
44. T. Wada, A. Mochizuki, Y. Sato and M. Sekine, *Tetrahedron Lett.*, 1998, **39**, 7123.
45. M. Kataoka, A. Hattori, S. Okino, M. Hyodo, M. Asano, R. Kawai and Y. Hayakawa, *Org. Lett.*, 2001, **3**, 815.

46. M. C. Uzagare, K. J. Padiya, M. M. Salunkhe and Y. S. Sanghvi, *Bioorg. Med. Chem. Lett.*, 2003, **13**, 3537.
47. J. L. Scotson, B. I. Andrews, A. P. Laws and M. I. Page, *Org. Biomol. Chem.*, 2016, **14**, 8301.
48. J. L. Scotson, B. I. Andrews, A. P. Laws and M. I. Page, *Org. Biomol. Chem.*, 2016, **14**, 10840.
49. O. Ponomarov, A. O. Laws and J. Hanusek, *Org. Biomol. Chem.*, 2012, **10**, 8868.
50. A. P. Guzaev, *Tetrahedron Lett.*, 2011, **52**, 434.
51. Z. Wang, Q. Song and Y. S. Sanghvi, Dimethylthiuram disulfide, in *Oligonucleotide Synthesis. Methods in Molecular Biology*, ed. P. Herdewijn, Humana Press, 2005, vol. 288.
52. A. A. Rodriguez, I. Cedillo, B. P. Mowery, H. J. Gaus, S. S. Krishnamoorthy and A. K. McPherson, *Bioorg. Med. Chem. Lett.*, 2014, **24**, 3243.
53. J. Yang, J. A. Stolee, H. Jiang, L. Xiao, W. F. Kiesman, F. D. Antia, Y. A. Fillon, A. Ng and X. Shi, *J. Org. Chem.*, 2018, **83**, 11577.
54. Q. Ren, T. Osawa and S. Obika, *Tetrahedron*, 2024, **150**, 133774.
55. X. Zhou, X. Shi, Y. Fillon, F. Antia, T. Pickel, J. Yang, W. Zhang, A. Delavari and J. Zhang, *Nucleic Acid Ther.*, 2024, **34**, 83.
56. D. Zewge, *Curr. Protoc. Nucleic Acid Chem.*, 2012, **49**, 3.21.1–3.21.11.
57. L. Aumann and M. Morbidelli, *Biotechnol. Bioeng.*, 2007, **98**, 1043.
58. F. Steinebach, N. Ulmer, L. Decker, L. Aumann and M. Morbidelli, *J. Chromatogr. A*, 2017, **1492**, 19.
59. Bachem, An oligonucleotide case study, https://www.bachem.com/news/the-era-of-large-scale-peptide-oligo-purification-with-mcsgp-has-begun/, last accessed March 25, 2024.
60. R. S. Gronke, K. Ruanjaikaen, A. Delavari, J. P. Immel-Brown, J. C. Penrod, Y. Lam and F. D. Antia, *Biotechnol. Processes*, 2023, **39**, e3350.
61. J. Muslehiddinoglu, R. Simler, M. L. Hill, C. Mueller, J. H. A. Amery, L. Dixon, A. Watson, K. Storch, C. Gazziola, F. Gielen, S. A. Lange, J. D. Prail and D. P. Nesta, *Nucleic Acid Ther.*, 2020, **30**, 189.
62. (a) P. Jayaprakash, A. Maudhuit, C. Gaiani and S. Desobry, *J. Food Eng.*, 2023, **339**, 111260. (b) https://www.fluidairinc.com/en-us/products/electrostatic-spray-dryers, last accessed March 25, 2024
63. NATiAS.co.jp/blockmer.
64. E. R. Moody, R. Obexer, F. Nickl, R. Spiess and S. L. Lovelock, *Science*, 2023, **380**, 1150.
65. https://www.codexis.com/technology/eco-synthesis-technology, last accessed March 25, 2024.
66. S. Paul, D. Gray, J. Caswell, J. Brooks, W. Ye, T. S. Moody, R. Radinov and L. Nechev, *ACS Chem. Biol.*, 2023, **18**, 2183.

CHAPTER 11

Sustainable Approaches in Liquid-phase Oligonucleotide Synthesis (LPOS)

L. FERRAZZANO* AND L. RAPISARDA

P4i Laboratory, Department of Chemistry "Giacomo Ciamician", Alma Mater Studiorum – University of Bologna, via P. Gobetti 85, 40129 Bologna, Italy
*E-mail: lucia.fcrrazzano4@unibo.it

11.1 Introduction

As oligonucleotides are now recognized as valuable therapeutic agents for the treatment of rare and intractable diseases, research on synthetic methodologies for their production has become a first-line issue to be addressed. In addition, as for peptide synthesis, the manufacturing costs for large-scale production are strictly connected to the synthetic approaches, which need to be optimized to satisfy the market demand. The most reliable and widely applied protocol for oligonucleotide synthesis is, as stated in the previous chapter, solid-phase oligonucleotide synthesis (SPOS), which relies mainly on phosphoramidite chemistry, and allows oligonucleotides to be obtained with high purity on the kilogram scale. The recent expansion of the oligonucleotide therapeutics market emphasizes the necessity for further scale-up, which is difficult to address with classical SPOS as it is based on the use of

Green Chemistry Series No. 85
Sustainability in TIDES Chemistry: Green Approaches to Oligonucleotides and Oligopeptides Synthesis
Edited by Alessandra Tolomelli, Lucia Ferrazzano and Walter Cabri

Published by the Royal Society of Chemistry, www.rsc.org

large excesses of reagents and solvents and is thus characterized by poor environmental and economic sustainability.

In order to overcome these limitations, studies on conducting oligonucleotide synthesis in homogeneous solutions have been carried out over the years, in parallel with the development of liquid-phase peptide synthesis (LPPS) approaches. With solution-phase synthesis methods, scalability, cost-effectiveness, and environmental sustainability are assured by the intrinsic features of the synthetic technique, acknowledged as a viable alternative for the large-scale production of oligonucleotides.

In contrast to SPOS, liquid-phase oligonucleotide synthesis (LPOS) is based on the use of a soluble support, and is also known as solution-phase, classical, or traditional oligonucleotide synthesis. The variability in the nomenclature of the technique is in contrast to peptide synthesis, where a further difference between liquid- and solution-phase synthesis approaches can be made according the chemical structure of the soluble support.[1]

In SPOS, the soluble support must remain unmodified throughout all the synthetic steps and ensure complete solubility of the growing chains. Thanks to the chemical features of the support, purification can be addressed by simply changing the solvent to precipitate or extract the sequences, or by chromatographic approaches.

Before proceeding with optimization of the protocols developed for SPOS to be transferred to LPOS, several years were spent in studying in depth and identifying the best soluble support. Therefore, in the following sections a classification of the various approaches according to the support type is proposed. As the solvent is the main component of the waste produced during the LPOS cycle, it will be stressed in the analysis of each synthetic approach to underline the limits and possibilities in terms of sustainability associated with the techniques.

11.2 Polymeric Soluble Supports

The first studies on LPOS were reported by Hayatsu and Khorana in 1966,[2] introducing the use of a polystyrene support with an average molecular weight of 270 000 Da, functionalized with monomethoxytrityl chloride as anchoring point. Process conditions are reported in Table 11.1. The sequence was

Table 11.1 Summary of synthetic cycle conditions used with polystyrene soluble support.

Synthetic step	Reagents and solvents
Loading	5′-OH-deoxynucleoside in Pyr
Precipitation	MeOH, H_2O
Coupling	Pyridinium 3′-*O*-acetyldeoxyribonucleoside 5′-phosphate in Pyr
Precipitation	$NaCl_{aq}$
Deacetylation	NaOMe
	Pyr–MeOH–DMSO
Cleavage	TFA in $CHCl_3$

assembled in the 5′→3′ direction, hence the soluble support was anchored to the 5′-OH of the first nucleoside. The loading of thymidine was conducted in pyridine (Pyr) and, after addition of a small amount of methanol (MeOH), precipitation of the loaded support was performed from water. For chain elongation, pyridinium 3′-*O*-acetyl-5′-phosphate bases were used together with mesitylene sulfonyl chloride as activating agent, using Pyr as solvent and aqueous sodium chloride ($NaCl_{aq}$) solution for precipitation. After removal of the 3′-*O*-acetyl group with sodium methoxide (NaOMe) in Pyr–MeOH–dimethyl sulfoxide (DMSO) solution and cleavage from the support with trifluoroacetic acid (TFA) in chloroform ($CHCl_3$), the authors were able to synthesize dimers and trimers of ribo- and deoxyribonucleotides, namely T_pT, dT_pG, dT_pC, and T_pT_pT.

Screening other soluble supports, Ishido and co-workers[3,4] proposed the use of a cellulose acetate derivative with 4-(2-hydroxyethylsulfonyl)dihydrocinnamate as linker, which allowed the release of the oligonucleotide sequence by treatment with triethylamine (TEA) in Pyr solution. The sequences were assembled in the 3′→5′ direction (see Table 11.2), with loading of the first 5′-*O*-DMT-guanosine 3′-(3-carboxy)propionate (DMT = dimethoxytrityl) protected on the base, with triisopropylbenzenesulfonyl chloride (TPS) and *N*-methylimidazole (NMI) in Pyr. Capping was performed with acetic anhydride (Ac_2O) in Pyr, 5′-*O*-detritylation was effected with 2% *p*-toluenesulfonic acid ($TsOH{\cdot}H_2O$) in $CHCl_3$–MeOH (7 : 3), and precipitation of the loaded support was achieved from ethanol (EtOH). Couplings were performed using 3′-(*o*-chlorophenyl)phosphate blockmers. Through this approach they were able to synthesize the sequences AAAAAAUUAUG and d[GGAATTCC], proving the applicability to the preparation of both RNA and DNA oligonucleotides. In particular, whereas in 1988 the sequences were assembled as blockmers AAAA+AAUUAU+G and GGAA+TTCC, respectively, in 1990 they proposed a monomer-based assembly approach for the first sequence and GGAA+TTC+C assembly for the second. For both approaches, the solvent of choice was Pyr, which is now classified as advisable for substitution in synthetic processes.

Table 11.2 Summary of synthetic cycle conditions used with cellulose acetate-based support.

Synthetic step	Reagents and solvents
Loading	5′-*O*-DMT-protected nucleoside-3′-(3-carboxypropionate) in Pyr TPS, NMI
Capping	Ac_2O in Pyr
Detritylation	$TsOH{\cdot}H_2O$ in $CHCl_3$–MeOH
Precipitation	EtOH
Coupling	5′-*O*-DMT-polynucleoside-3′-(*o*-chlorophenyl)phosphate
Oxidation	Tetramethylguanidinium (*E*)-2-pyridinealdoximate (TMG-PAO) in Pyr–H_2O
Precipitation	EtOH
Cleavage	NH_4OH_{aq}
Final purification: chromatography	

It was selected as an effective solubilizing agent for polystyrene soluble supports. However, as no further studies on this support have been proposed, any consideration of replacing Pyr with greener alternatives can be suggested.

11.3 PEG Soluble Supports

Among the polymeric soluble supports, poly(ethylene glycol) (PEG) deserves an independent evaluation, being the most explored in the class. The greatest advantage of making use of PEG soluble supports is the easy purification, which can be achieved by rapid crystallization, thus avoiding all of the drawbacks related to the precipitation of gels that is often observed with other soluble supports,[5] and as a consequence shortening the process timing.[6] One of the first examples of PEG soluble supports used for LPOS was proposed by Köster and Heyne,[7,8] who reported reactions in water, pyridine, and dioxane as the best solvents for PEG chains with an average molecular weight of 20 000 Da. As a purification method, they proposed dialysis in a buffer system to remove excess of low molecular weight reagents from the target product. As linkers for first nucleotide loading, monomethoxytrityl and uridine were used. However, this soluble support was abandoned until Bonora *et al.* developed the so-called HELP (high-efficiency liquid-phase) protocol.[9] The synthetic steps were similar to SPOS in series and chemistry (see Table 11.3) but required a smaller excess of reagents. In the first approach, phosphotriester chemistry was adopted. The loading of the first nucleoside on the PEG support occurred by means of a symmetrical anhydride derivative of one of the four possible 5′-*O*-DMT-2′-deoxynucleoside-3′-*O*-succinates, activated

Table 11.3 Summary of synthetic cycle conditions used in HELP protocol.

Synthetic step	Reagents and solvents
Loading	5′-*O*-DMT-2′-deoxynucleoside-3′-*O*-succinate symmetrical anhydride DCC DMAP Pyr–DCM
Precipitation	Et_2O
Capping	Ac_2O in Pyr
Precipitation	Et_2O
Detritylation	TCA in DCM
Recrystallization	DCM–Et_2O
Coupling	5′-*O*-DMT-nucleoside-3′-monophosphate NMI MSNT Pyr–2,6-lutidine
Precipitation	Et_2O
Recrystallization	EtOH
Cleavage	NH_4OH_{aq}
Final purification: ion-exchange chromatography	

by dicyclohexylcarbodiimide (DCC) and dimethylaminopyridine (DMAP) in dichloromethane (DCM) doped with Pyr. Before proceeding further, purification by precipitation with diethyl ether (Et_2O) allowed the isolation of a pure loaded PEG–nucleoside intermediate. The capping of the unreacted PEG chains (and also truncated sequences) was performed with Ac_2O in Pyr, without adding DMAP as for SPOS to avoid unwanted side reactions on the protecting groups of bases, and then repeating precipitation. Detritylation was carried out under standard conditions [3% trichloroacetic acid (TCA) in DCM] and required recrystallization from DCM–Et_2O. With optimized conditions reported for SPOS, in the condensation step NMI and mesitylenesulfonylnitrotriazole (MSNT) were used as activators in Pyr–2,6-lutidine solution. In this case, precipitation with Et_2O and recrystallization with absolute EtOH were necessary to purify the intermediate. Cleavage from the support was performed under classical basic conditions with 30% NH_4OH_{aq} and isolation of the final product was performed by ion-exchange chromatography. In this way, homodinucleotides [d(TpT), d(CpC), d(ApA), d(GpG)] and the longer d(TAGCGCTA) were obtained with coupling yields in the range 90–95% after 60 min. Optimization performed on the 0.6 mmol nucleotide scale was extended to 180 μmol,[10] at the same time replacing the succinate derivative for the coupling of nucleosides with the bifunctional phosphorylating agent 2-chlorophenyl-*O*,*O*-bis(1-benzotriazolyl)phosphate (diOBT). This change avoided the use of MSNT as activator, thus suppressing the side reactions that usually occur especially with the guanidine moiety. This was particularly useful for simplifying the purification of intermediates, but the number of precipitation and recrystallization steps remained the main drawback in terms of sustainability, significantly impacting the generation of huge amounts of solvent-based wastes (namely DCM, Et_2O, and Pyr).

Once the large-scale HELP method had been verified, the authors evaluated for the first time in a liquid-phase approach the application of phosphoramidite chemistry in the so-called HELP Plus method (see Table 11.4).[11] As consequence of solubility and stability issues with phosphoramidite monomers, the solvent for coupling was changed to acetonitrile (ACN) and the activator was replaced with tetrazole. It is important to emphasize that in this case an extra oxidation step was required and always performed after a capping step of eventual truncated sequences. Oxidation was performed using *tert*-butyl hydroperoxide (*t*BuOOH) in ACN, thus solving the purification issues observed when I_2 was used. A complete synthetic cycle required 4 h, with two of them necessary for precipitation, filtration, and final crystallization from EtOH. The use of ACN for couplings and cappings and dichloroethane (DCE) for deprotections required changes to the purification after each step, which involved precipitation with Et_2O and recrystallization from DCE–Et_2O after detritylation and precipitation with Et_2O and recrystallization with ACN–Et_2O after condensation and capping. With this approach, a 20-mer sequence as an example of an antisense-size molecule was successfully synthesized, although no information on the oligonucleotide sequence was reported.

Table 11.4 Summary of synthetic cycle conditions used in HELP-Plus and HELP-S protocols.

Synthetic step	Reagents and solvents
Loading	5′-*O*-DMT-2′-deoxynucleoside-3′-*O*-succinate in DCE DCC DMAP
Precipitation	Et_2O
Recrystallization	DCE–Et_2O or EtOH
Capping	Ac_2O in ACN
Precipitation	Et_2O
Detritylation	TCA in DCE
Precipitation	Et_2O
Recrystallization	DCE–Et_2O
Coupling	5′-*O*-DMT-nucleoside-3′-phosphoramidite in ACN Tetrazole
Sulfurization *(only in HELP-S protocol)*	TETD in ACN
Precipitation	Et_2O
Recrystallization	ACN–Et_2O
Oxidation *(not present in HELP-S protocol)*	TBHP in ACN
Cleavage	NH_4OH_{aq}
Final purification: chromatography	

As phosphorothioate derivatives are important compounds with recognized therapeutic efficacy, the HELP protocol was also defined as HELP-S, replacing the oxidation with sulfurization in the phosphoramidite-based protocol (see Table 11.4).[12] As sulfurization agent, tetraethylthiuram disulfide (TETD) in ACN was used, being added *in situ* after the condensation step to avoid unwanted oxidation during the precipitation step. Therefore, the capping was performed in this case after sulfurization using a standard procedure, with a reaction time of 3 min to reduce partial decyanoethylation of the protected phosphotriester. Also in this case, a 20-mer phosphorothioate (PS)-oligonucleotide was successfully obtained, with an overall yield of 83%.

The HELP-S approach represents a major innovation in oligonucleotide synthesis, and in particular for the large-scale production of antisense oligonucleotides. However, it is still not affordable from a sustainability point of view, especially considering the addition of an extra synthetic step (oxidation or sulfurization). Therefore, to make LPOS a viable alternative to SPOS for this purpose, it is essential to integrate the advantages of both techniques, in particular: (i) the alternative solubility of the support in some solvents and its insolubility in mixtures used for purification addressed by the first technique, and (ii) the possibility of introducing automation reported for the latter technique, making the synthesis a time-saving process even if it has a high energy demand. To make the two approaches really integrated, the final target should be the introduction of automation also in liquid-phase strategies.

In 1998, Bonora and co-workers proposed a new apparatus adapted from peptide and oligonucleotide solid-phase synthesis to fit with the necessities of liquid-phase operation.[13] The core of the apparatus was a batch-type reactor allowing complete solubilization of the PEG–oligonucleotide sequence during the synthesis, with a filter at the bottom to facilitate the purification. In addition, it also included a thermostatic module, for reaction temperature control. However, the test was carried out only on the synthesis of a dimer obtained in 99% yield, and further investigations were not reported.

The processes proposed by Bonora and co-workers, despite being based on lower consumption of reagents and with high efficiency and the possibility of using convergent approaches, involved long and repeated purification steps, which increased the total process time and affected the E-factor and process mass intensity (PMI) due to the large consumption of solvent.[14] This drawback can be overcome by reducing the number of synthetic steps, which is possible by replacing monomers with phosphorothioate dimers during the coupling step. Therefore, in 2000 Bonora *et al.*[15] reported a HELP method for the large-scale synthesis of phosphorothioate oligonucleotides starting from phosphorothioate dimer blocks and using diethyldithiocarbonate disulfide (DDD) as sulfurization agent (see Table 11.5).

After synthesizing a 4-mer sequence, the feasibility of the procedure was evaluated on the first millimolar-scale synthesis of a bench oligonucleotide, namely the sequence d[PS(TTGGGGTT)], and then on the synthesis of a 15-mer sequence. In the latter, only the condensation of nucleotide 13 was performed using a monomer. It is important to emphasize that large-scale synthesis in this case means working on 10 g of soluble support to obtain around 5 g of the target 15-mer oligonucleotide. For a complete assessment of the efficiency and applicability of the protocols, the authors reported an evaluation of the expected amount of solvent and reagents for synthesis on the millimolar scale (see Table 11.6). The data revealed that methyl *tert*-butyl ether (MTBE), used for precipitation and filtration, is the main component that considerably affects the PMI, whereas phosphoramidites and other reagents do not contribute significantly to waste production. It is interesting to underline that MTBE represents the first green solvent proposed in LPOS. Combined with a shorter synthetic process like that described above, it resulted in the first example of attempting sustainability, even if this was outside the scope of the reported study.

11.4 Ion Tag Support

Taking advantage of the promising results obtained in the liquid-phase synthesis of small molecules,[16,17] oligopeptides,[18] and oligosaccharides,[19,20] ionic soluble supports have also been proposed for oligonucleotide synthesis, named ionic liquid-supported synthesis (ILSS) (see Table 11.7).[21–23] This technique is compatible with large-scale synthesis, does not require chromatographic purifications, affords high-purity products, and is compatible with phosphoramidite chemistry. In addition, ionic liquids are generally

Table 11.5 Summary of synthetic cycle conditions used in the large-scale HELP-S protocol.

Synthetic step	Reagents and solvents
Loading	5′-*O*-DMT-dinucleotidephosphorothioate-3′-*O*-succinate
	DCC
	NMI
	DCE–Pyr
Precipitation	MTBE
Recrystallization	EtOH
Capping	Ac_2O in CAN
	2,6-Lutidine
	NMI
Precipitation	MTBE
Detritylation	TCA in DCE
Precipitation	MTBE
Recrystallization	DCE–MTBE
Coupling	5′-*O*-DMT-dinucleosidephosphorothioate-3′-phosphoramidite
	Tetrazole
	ACN
Precipitation	MTBE
Recrystallization	EtOH
Sulfurization	DDD in ACN
Precipitation	MTBE
Cleavage	NH_4OH_{aq}
Precipitation	Acetone
Final purification: chromatography	

Table 11.6 Estimated amounts of solvents and reagents used in the large-scale HELP protocol. The amounts refer to the synthesis of 5.175 g of 15-mer sequence starting from 10 g of soluble PEG support. PMI calculations are reported for the synthesis of 1 kg of target molecule and calculated as kg_{waste} per $kg_{product}$.

Reagents and solvents	Estimated amount (kg)	Percentage of PMI[a] (%)
MTBE	33.3	65.6
ACN	2.0	3.8
DCE	4.4	8.7
EtOH	11.1	21.8
Dimeric amidites	0.015	0.1
Tetrazole	0.004	0.1
DDD	0.012	0.1

[a]PMI: 9816.

included among green solvents, thus representing the first green soluble supports for LPOS.

The ionic liquid was prepared by using NMI and 2-bromoethanol, with ion exchange for the introduction of tetrafluoroborate ion. The first 5′-*O*-DMT-nucleoside was anchored through a succinic linker using DCC and a catalytic

Table 11.7 Summary of synthetic cycle conditions used with an ion tag soluble support.

Synthetic step	Reagents and solvents
Loading	5′-*O*-DMT-2′-deoxyribonucleoside-3′-succinylate in ACN
	DCC
	DMAP
Precipitation	Et_2O–EtOAc
Extraction	$CHCl_3$–water
Detritylation	TFA in DCM or ACN
Precipitation	Et_2O–EtOAc
Coupling	5′-*O*-DMT-2′-deoxyribonucleoside-3′-phosphoramidite in ACN or THF
	DCI
Precipitation	Et_2O–EtOAc
Oxidation	2,4,6-Collidine or Pyr
	I_2 in THF–H_2O
	Sodium bisulfite
Extraction	$CHCl_3$–water

amount of DMAP in ACN. The loaded support was purified by precipitation from Et_2O–EtOAc solution and then extracted with chloroform and water, which allowed the separation of the more lipophilic anchored support from the excess of ionic liquid. The loading of the other nucleosides occurred with only 1.5 equiv. of phosphoroamidite nucleoside with dicyanoimidazole (DCI) in tetrahydrofuran (THF) or ACN. The reaction was quenched with EtOH to facilitate the removal of unreacted phosphoramidite during the precipitation step with EtOAc–Et_2O (1 : 9 v/v), which afforded pure dinucleoside phosphite triester. The oxidation step was carried out with iodine and a small amount of pyridine or collidine in ACN–THF–water, and the product was isolated by organic extraction. Detritylation was performed with 3% TFA in DCM or ACN and the product was precipitated using EtOAc–Et_2O. With this approach, from di- to pentamers were successfully synthesized, but no further explorations of the synthesis of longer sequences or data related to the efficiency of solid support removal were reported.

11.5 Levulinic Acid Support

Exploring H-phosphonate chemistry and the possibility of applying it successfully to LPOS, in 1997 Reese and Song[24,25] proposed a new procedure based on the use of levulinic acid as a soluble support anchored on the 3′-position of deoxyribonucleosides, properly protected on bases (see Table 11.8).

Couplings were conducted using triethylammonium 5′-*O*-DMT-2′-deoxyribonucleoside H-phosphonates and di(2-chlorophenyl)phosphorochloridate as activating agent in pyridine–DCM mixture, leading to the target H-phosphonate dimer quantitatively in 10 min with only a 20% excess of H-phosphonate deoxyribonucleoside. Without isolation of the intermediate, *N*-[(4-chlorophenyl)sulfanyl]phthalimide was added to the reaction mixture to

Table 11.8 Summary of synthetic cycle conditions used with levulinic acid soluble support for H-phosphonate chemistry.

Synthetic step	Reagents and solvents
Coupling	5′-OH-3′-*O*-levulinyl-2′-deoxyribonucleoside 5′-*O*-DMT-2′-deoxyribonucleotide H-phosphonate Di(2-chlorophenyl)phosphorochloridate Pyr–DCM
Sulfur transfer	*N*-[(4-Chlorophenyl)sulfanyl]phthalimide
Detritylation	HCl in DCM–dioxane
Acetylation	Ac_2O in Pyr
Deprotection	(*E*)-2-Benzaldoxime TMG ACN
Cleavage	NH_4OH_{aq} Hydrazine

obtain the corresponding dinucleoside phosphorothioate in quantitative isolated yield as precursor of classical phosphodiester oligonucleotides. If pure phosphorothioate or chimera oligonucleotides had to be synthesized, the latter reagent was replaced with 4-[(2-cyanoethyl)sulfanyl]morpholine-3,5-dione.[26] Differently from SPOS, the reaction occurred at lower temperature (−40 °C) and sulfurization was performed after each coupling step. Detritylation was performed with HCl in dioxane–DCM mixture and, after acetylation of 5′-OH, unblocking of *S*-(4-chlorophenyl)-protected phosphorothioates was performed with (*E*)-2-nitrobenzaldoxime and 1,1,3,3-tetramethylguanidine (TMG) in ACN when partially protected oligonucleotide phosphate was the target. Otherwise, 1,8-diazabicyclo[5.4.0]undec-7-ene (DBU) allowed removal of the 2-cyanoethyl group to give the partially protected phosphorothioate oligonucleotide. The final treatment with NH_4OH_{aq} allowed removal of all the acylating groups, affording pure d[ApC] without further purifications. As proof of the feasibility of the procedure for the synthesis of pure phosphodiesters, pure phosphorothioate, and a combination of both, the syntheses of d[TpG-p(s)ApC], d[Tp(s)Gp(s)Ap(s)C], and d[TpGpApC] were successfully achieved. To remove the 3-*O*-levulinyl group, treatment with hydrazine under mild conditions afforded free 3′-OH functions. Moving from deoxyribonucleotides to ribonucleotides, the same group applied the protocol to the synthesis of linear and cyclic oligoribonucleotides,[27] and also to the synthesis of branched and cyclic branched oligoribonucleotides (called "lariats").[28] The H-phosphonate building block was replaced with the corresponding 2′-O(Fpmp) (Fpmp = 1-(2-fluorophenyl)-4-methoxypiperidin-4-yl) derivative for the synthesis of linear and linear cyclic r(ApCpGpU). Regarding the branched compounds, the "lariat" synthon (see Figure 11.1) was synthesized and included in the synthetic process allowing ramification of sequences.

Following these procedures, in 2002 Reese and Yan reported for the first time the synthesis of a commercially available antisense oligonucleotide, Vitravene (see Figure 11.2).[29] To perform this synthesis in the liquid phase, they

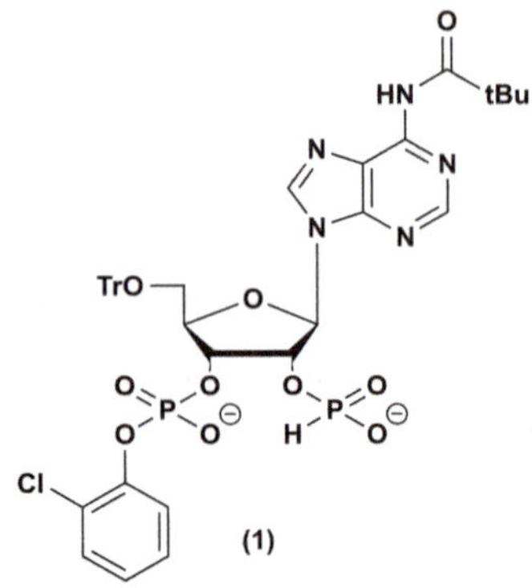

(1)

Figure 11.1 Structure of "lariat" synthon **1** used for the synthesis of branched and branched cyclic oligoribonucleotides *via* H-phosphonate chemistry on a levulinic acid soluble support.

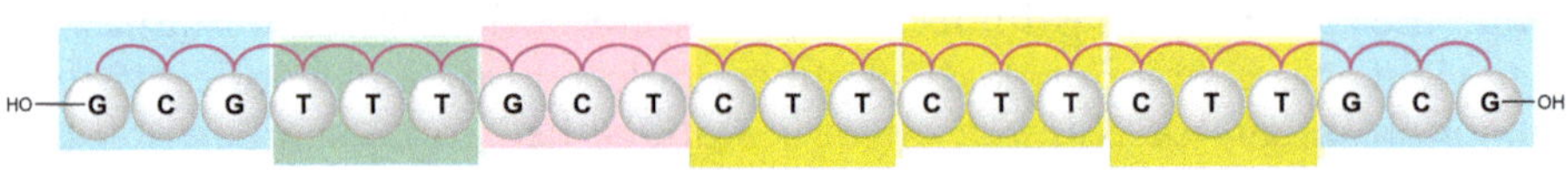

Figure 11.2 Structure of Vitravene, highlighting the trimer units employed for sequence assembly *via* H-phosphonate chemistry on a levulinic acid soluble support.

divided the sequence into four trimer blocks, since two trimers (CTT and GCG) were repeated three times and twice, respectively. In addition to the clear advantages in terms of simplification of synthesis, reduction of costs, and large-scale synthesis, a reduction in the number of impurities was possible, thus simplifying the purification and resulting in an overall greener protocol. The four trimers GCG, TTT, GCT, and CTT were synthesized following the H-phosphonate approach described previously. The coupling reagent bis(2-chlorophenyl)phosphorochloridate was replaced with diphenyl phosphorochloridate as it was commercially available and as effective as the former. In addition, for the synthesis of phosphorothioate linkages, morpholine-3,5-dione reagent was replaced with *N*-[(2-cyanoethyl)sulfanyl]succinimide, being soluble in pyridine and selective for H-phosphonate diesters over H-phosphonate monoesters and coupling reagent. Once the trimers had been synthesized, the Vitravene sequence was assembled in six coupling steps, in which one of the coupling partners was a trimer. The isolated yields of each 5′-OH intermediate and of the final Vitravene sequence were around 91%. The final deprotection was performed in four steps: (i) acetylation of 5′-OH to avoid unwanted reactivity with phosphorothioate triester linkages, (ii) DBU-mediated removal of the *S*-cyanoethyl protecting group from the internucleotide linkages in the presence of chlorotrimethylsilane as scavenger, (iii) deprotection of guanine and thymine residues by (*E*)-2-nitrobenzaldoxime, and (iv) deprotection of all the other acylating protecting groups using aqueous ammonia solution at 55 °C in the presence of 2-mercaptoethanol as scavenger. In this way, Vitravene as the sodium salt was successfully obtained for the first time in solution with a 75% overall isolated yield.

These results proved the efficacy of the application of blockmer-based synthesis in the liquid phase using H-phosphonate chemistry. The process was not fully optimized and no other reports on it were subsequently published. However, the blockmer-based approach verified on the synthesis of a long oligonucleotide confirms the convenience related to the reduction of the number of synthetic steps, consequently improving the greenness of the process, which is still far away from being considered sustainable.

11.6 Tetravalent and Trivalent Clusters as Soluble Supports

Going back to phosphoramidite and phosphotriester chemistry, another interesting approach concerns the use of multi-branched clusters for the simultaneous synthesis of three or four oligonucleotide sequences. This approach has important advantages in terms of greenness related to good atom economy, reduction in the number of precipitations per gram of product (compared with monosubstituted soluble supports), and, consequentially, reduced consumption of solvents, which significantly affected the PMI of the process.

The pioneering use of liquid-phase carriers (LPCs) was introduced in 1999 by Wörl and Köster,[30] who adapted preliminary results reported for the phosphotriester approach to phosphoramidite chemistry, notably being characterized by higher condensation rates. They proposed two symmetric multifunctional cores, pentaerythritol and tris-1,3,5-benzenetricarboxylic acid trimethyl ester (see Scheme 11.1) as examples of tetra- and trivalent soluble supports, respectively.

In both cases, an initial synthetic step was necessary to introduce spacers bearing primary amino terminal groups to be exploited for loading of the first nucleoside by a succinyl linker, as for controlled-pore glass (GPC) resins used in solid-phase approaches. In the case of pentaerythritol, acrylonitrile was used as a linker, since treatment with refluxing HCl in methanol to convert cyano groups into methyl esters and subsequent aminolysis with ethylenediamine afforded four terminal amines. Each arm of the tetravalent soluble platform was acylated with 5′-*O*-DMT-deoxythymidine-3′-*O*-succinyl *p*-nitrophenyl ester in Pyr. After two different and laborious purifications, a highly symmetrical liquid-phase carrier was achieved in 45% isolated yield. Regarding tris-1,3,5-benzenetricarboxylic acid trimethyl ester, functionalization started with aminolysis and proceeded with condensation of the first 5′-*O*-DMT nucleoside in a Pyr–DMF mixture, achieving the target trivalent liquid carrier in 90% yield. However, after exploring different techniques for purifying the products derived from dedimethoxytritylation, no further elongations of the chains were reported. Subsequently, they reported only for the

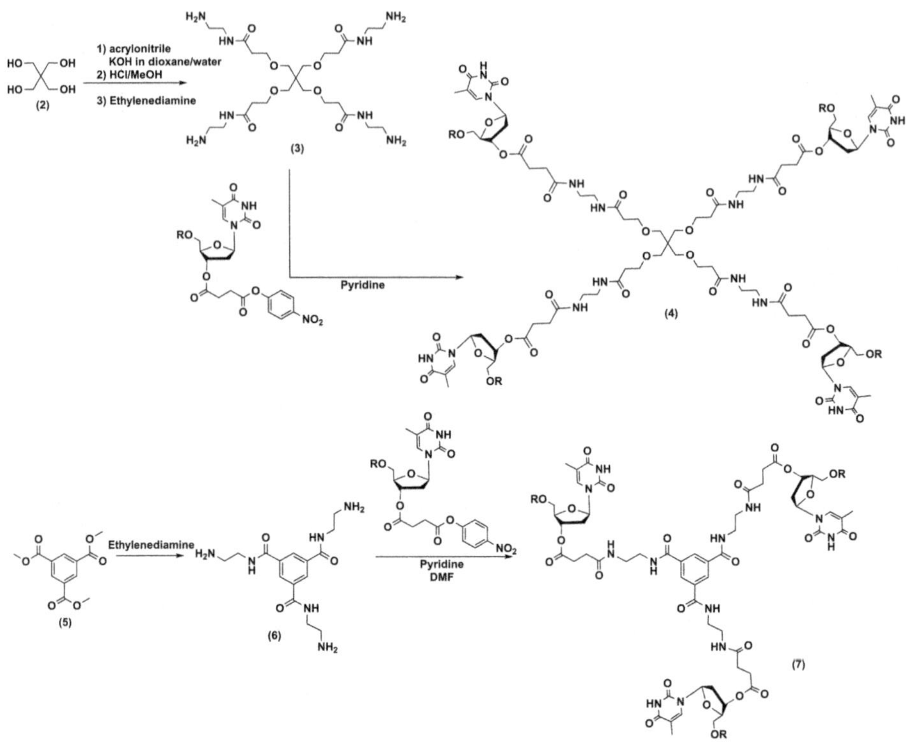

Scheme 11.1 Synthesis of liquid-phase carriers derived from pentaerythritol (**2**) and 1,3,5-benzenetricarboxylic acid (**5**).

second type of LPC the synthesis of a decanucleotide [d(GACGGCCAGT)], using monomeric phosphoramidites, and of a pentanucleotide [d(GCCCT)], using dimeric phosphoramidites.[31] The authors provided further information about the solubility issues related to the LPC used, which is weakly soluble in ACN (see Table 11.9). Thus, for condensations, they replaced ACN with Pyr for solubilizing the LPC and dissolved 1*H*-tetrazole in ACN for use as an activating agent. The purification of the resulting intermediate was achieved by separation on a Sephadex LH20 column with THF–methanol (80 : 20) as eluent. The resulting fractions containing the product were directly oxidized with addition of *t*BuOOH and, after drying, the crude material was directly demethoxytritylated with TFA in DCM–nitromethane–MeOH (80 : 19 : 1) and then quenched with TEA to avoid depurination reactions. By this approach, 5–12 equiv. of phosphoramidites were used to obtain an average yield of 96% in condensation.

Analysing the technique from the point of view of greenness, the extensive use of toxic solvents and too large excesses of reagents needed to bring the process to completion immediately appears clear, hence significant improvements are required for conversion into a greener approach. In addition, purification of each intermediate was performed by chromatography, thus significantly

Table 11.9 Summary of synthetic cycle conditions used with tris-1,3,5-benzenetricarboxylic acid trimethyl ester soluble support.

Synthetic step	Reagents and solvents
Loading	5′-*O*-DMT-2′-deoxyribonucleoside-3′-*O*-succinate *p*-nitrophenyl ester
	Pyr–DMF
Purification	Chromatography
Detritylation	TFA in DCE–nitromethane–MeOH
Purification	Chromatography
Coupling	5′-*O*-DMT-2′-deoxyribonucleotide phosphoramidite
	Tetrazole
	Pyr–ACN
Purification	Chromatography
Oxidation	*t*BuOOH
	Di-*tert*-butyl peroxide

Table 11.10 Summary of synthetic cycle conditions used with pentaerythritol-based soluble support and DMT as 5′-*O*-protecting group.

Synthetic step	Reagents and solvents
Loading	5′-*O*-DMT-3′-*O*-pentynoyl nucleoside
	$CuSO_4$
	Sodium ascorbate
	H_2O–dioxane
Extraction	EtOAc–H_2O
Detritylation	HCl in MeOH–DCM
Coupling	5′-*O*-DMT-2′-nucleotide phosphoramidite
	DCI
	DMF–ACN
Oxidation	I_2 in Pyr–THF–H_2O
Precipitation	MeOH
Detritylation	HCl in MeOH–DCM
Precipitation	MeOH–DCM
Cleavage	NH_4OH_{aq}

affecting the sustainability of the process with the use of large volumes of solvents. However, during the same synthetic cycle, a threefold increase in millimoles of product was obtained, with interesting atom economy.

Studies on the use of pentaerythritol derivatives as soluble supports for the synthesis of clustered oligonucleotides were recently reported by Virta's group, with the aim of exploiting the greater advantage of this type of tetrahedrally branched support with the possibility of using methanol for precipitation during the purification steps.[32] It is important to emphasize that, similarly to the previous example, using these supports 4 equiv. of oligonucleotide can be assembled on 1 equiv. of anchor, thus resulting in very good atom economy. In addition, the greenness of the process is illustrated by the need for only two precipitation steps, after oxidation and dedimethoxytritylation, to purify the intermediates (see Table 11.10).

The soluble support was assembled from pentaerythrityl tetrabromide, introducing benzyl chloride arms further converted into the corresponding azides. The loading of the first nucleosides occurred using 5′-*O*-DMT-thymidine acylated with pentynoic anhydride, exploiting the click reaction with a small excess of alkyne-derived nucleoside. The first dedimethoxytritylation was performed in HCl in MeOH–DCM, while the coupling was carried out with a fourfold excess of phosphoramidites in the presence of DCI in DMF–ACN solution. The resulting P(III) phosphite triesters were oxidized to P(V) phosphotriesters with iodine in Pyr–water–THF solution and then quenched with addition of trimethoxyphosphite. Purification by precipitation with methanol afforded pure phosphotriesters. After repeating the dedimethoxytritylation under the above conditions and neutralizing the reaction mixture by addition of pyridine, the 5′-OH intermediate was purified by precipitation. For testing the feasibility of the procedure, the syntheses of three branched trimers and one pentamer (GCT, AGT, CCT, AGCCT) were conducted by monomer block addition on the 27 and 35 μmol scale, respectively. The yields ranged from 43% for the pentamer to 67–73% for trimers. To adapt the protocol to the synthesis of oligoribonucleotides, specific building blocks bearing a 2-cyanoethyl protecting group at the 2′-OH position and a *tert*-butyldimethylsilyl group at the 5′-OH position were introduced.[33] The chemistry of phosphoramidites was exploited also in this case for the synthesis of a pentamer (UUGCA) with a 54% overall yield.

Recently, this synthetic protocol was re-evaluated to define new protecting groups that can be more compatible with the LPOS scheme. In fact, classic 5′-OH protection with DMT has drawbacks related to the reversible nature of acid-catalysed deprotection, which generates a carbocation that has to be captured by addition of scavengers such as thiols and silanes, and to the lipophilic nature of the by-products generated after deprotection, which require an additional purification step to be removed from the main product. Therefore, Virta and co-workers proposed acetone acetals as a 5′-OH protecting group, able to be removed in an irreversible manner under mild conditions and generating volatile by-products. In particular, they proposed the 2-isopropoxyprop-2-yl (IIP) group, prepared from acetone and 2-propanol on a 100 g scale, even though the yield was moderate (21%).[34] To install this protecting group at the 5′-position, they started from nucleosides 3′-protected with a *tert*-butyldimethylsilyl (TBS) group and proceeded with acylation in the presence of TsOH. Then, removal of 3′-*O*-TBS carried out using tetrabutylammonium fluoride (TBAF) and phosphitylation with 2-cyanoethyl *N*,*N*-diisopropylchlorophosphoramidite with triethylamine (TEA) in DCM afforded the 5′-*O*-IIP-protected phosphoramidites in 86% yield on average. These building blocks were applied to LPOS following the protocol developed by Virta and co-workers on pentaerythritol tetrabranched LPC (see Table 11.11). Couplings were conducted with tetrazole and DCI in DMF–ACN (2 : 1 v/v), while 5′-position deprotection was conducted with 6% formic acid in DCM–MeOH (2 : 1 v/v). Oxidation of phosphite esters to phosphate triesters was performed with 3-chloroperoxybenzoic acid. Two precipitations with 2-propanol were

Table 11.11 Summary of synthetic cycle conditions used with pentaerythritol-based soluble support and IIP as 5′-*O*-protecting group.

Synthetic step	Reagents and solvents
Loading	5′-*O*-DMT-3′-*O*-pentynoyl nucleotide
	$CuSO_4$
	Sodium ascorbate
	H_2O–dioxane
Extraction	EtOAc–H_2O
Detritylation	HCl in MeOH–DCM
Coupling	5′-*O*-IIP-nucleotide phosphoramidite
	Tetrazole
	DMF–ACN
Oxidation	3-Chloroperbenzoic acid
Precipitation	2-Propanol
Deprotection	Formic acid in DCM–MeOH
Precipitation	2-Propanol
Cleavage	NH_4OH_{aq}
Final purification: chromatography	

Table 11.12 Summary of synthetic cycle conditions used with pentaerythritol-based soluble support and MIP as 5′-*O*-protecting group.

Synthetic step	Reagents and solvents
Coupling	5′-*O*-DMT-3′-Ψ-activated-2′-deoxyribonucleotide
	DBU
	Pyr–ACN
Precipitation	2-Propanol
Deprotection	DCA in DCM–MeOH
Precipitation	2-Propanol
Cleavage	NH_4OH_{aq}

performed, after oxidation and deacetalization, respectively. The protocol was applied to the synthesis of dimers and trimers with 86% average yields.

The same group also reported[35] 2-methoxyisopropyl (MIP) protection for the 5′-OH position of 2-deoxynucleosides for the chiral synthesis of PS-oligonucleotides *via* chiral limonene oxathiaphospholane sulfide [(+)- and (−)-Ψ reagent], proposed by Baran and co-workers.[36]

As in the previous study, the MIP protection was installed on a temporary 3′-*O*-TBS-protected 2′-deoxynucleoside, then replaced by the chiral pool. Taking advantage of the use of a tetrapodal soluble support, the couplings were performed in Pyr–ACN (2 : 3 v/v) solution in the presence of DBU, quenched with acetic acid, and the crude material was precipitated with 2-propanol. MIP was removed with 4% DCA in DCM–MeOH (2 : 1 v/v) mixture, neutralized with Pyr, and precipitated again with 2-propanol. Following this procedure (see Table 11.12), di-, tri-, and tetranucleotide phosphorothioates with chiral definition of PS linkages were synthesized, although further optimization is required to increase the overall yield and extend the protocol to the synthesis of longer oligonucleotides.

However, this new protecting group was not further explored and DMT continued to be the 5′-OH protection of choice despite its well-known above-mentioned drawbacks. The above-reported synthetic approaches were characterized by a smaller number of synthetic steps, revealing the intention of moving towards greener protocols.

To overcome the problems related to the premature cleavage of 2-cyanoethyl from classic phosphodiester internucleotide linkages and of benzoyl group/depurination of 2-deoxyadenosine during dedimethoxytritylation, the same group also explored the use of phosphotriester chemistry to avoid an oxidation step, introducing 3′-(benzotriazol-1-yl 2-chlorophenylphosphate) building blocks.[37] The starting materials, in contrast to phosphoramidite-based approaches, need to be synthesized, thus reducing the versatility of the approach. The procedure was verified by synthesizing four trimers and one pentamer (see Table 11.13). In this case, the condensations were performed with NMI in dioxane and the intermediate was purified by precipitation with MeOH. Dedimethoxytritylation was carried out with HCl in MeOH–DCM (1:1 v/v) solution and the mixture was neutralized with Pyr, then precipitated with MeOH. The 2-chlorophenyl protecting groups on the phosphotriester linkages were removed using (*E*)-2-nitrobenzaldoxime and TMG in water–dioxane, and the other protecting groups were removed by classical basic treatment with NH_4OH_{aq} at 55 °C. The crude material was purified by adding ethanol to its solution in ACN and $NaOAc_{aq}$. Trimers (TTT, TdAdA, TdCdC, TdGdG) and the pentamer (TdAdCdGdG) were obtained in 70% and 55% isolated yields, respectively.

The same soluble support was also further modified by derivatization with the so-called Q-linker, namely hydroquinone *O*,*O*′-diacetic acid, as linker (see Figure 11.3).[38] It allowed the release of the oligonucleotide sequence keeping protecting groups on the bases, by means of K_2CO_3 in DCM–dioxane–MeOH

Table 11.13 Summary of synthetic cycle conditions used with pentaerythritol-based soluble support with phosphotriester chemistry.

Synthetic step	Reagents and solvents
Coupling	5′-*O*-DMT-deoxynucleotide bis(benzotriazolyl)-2-chlorophenylphosphate NMI Pyr–dioxane
Precipitation	MeOH
Detritylation	HCl in DCM–MeOH
Precipitation	MeOH
Deprotection	(*E*)-2-Nitrobenzaldoxime TMG Dioxane–H_2O
Cleavage	NH_4OH_{aq}
Precipitation	EtOH

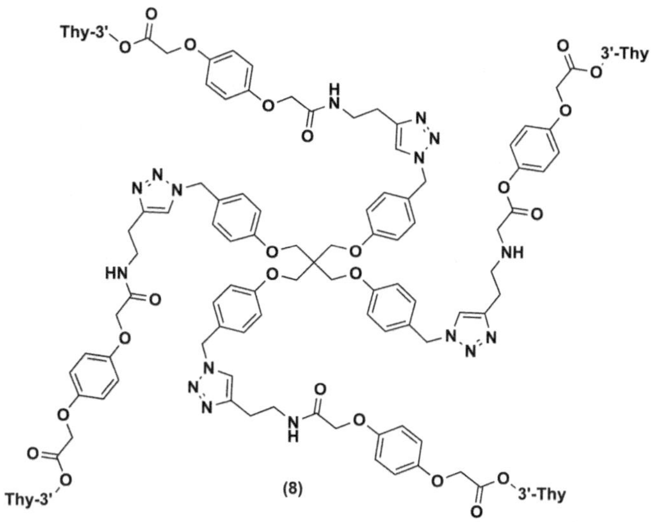

(8)

Figure 11.3 Structure of tetrakis-*O*-(4-azidomethylphenyl)pentaerythritol derivatized with hydroquinone-*O*,*O*′-diacetic acid (Q-linker).

(3 : 10 : 43 v/v/v). When reaction was complete, pyridinium chloride was added to quench the reaction and purification by filtration on silica gel afforded trimers TAT, AGT, TTA, CAT, and GCT in 65–70% overall isolated yield.

In conclusion, these branched soluble supports require further optimization, especially in the reduction of purification steps by precipitation which involve the consumption of huge volumes of solvents, which do not conform to the green chemistry principles. Only a few examples of longer sequences have been reported as synthesized through this approach, so no conclusions on the feasibility and greenness of the large-scale synthesis of long oligonucleotides can be drawn.

11.7 Alkyl Chain Type Support

The advantages of the use of alkyl chain-soluble supports (ASS) are mainly (i) the simple separation of products thanks to hydrophobic interactions, (ii) favourable reaction conditions, (iii) absence of post-synthesis manipulation for analytical purposes, and (iv) low aggregation of growing anchored sequences compared with other types of polymeric supports.[39] From the structural point of view, two types of ASS have been proposed and preliminarily evaluated to ensure complete solubility of oligonucleotides (see Figure 11.4).

Their structures are derived from gallic acid and are functionalized with long alkyl chains on the alcohol positions. The main difference between the two supports is the linker that allows distancing and anchoring of the oligonucleotide sequence. It can be a linear or cyclic diamine functionalized with succinic acid, which allows loading of the first nucleoside through its 3′-OH

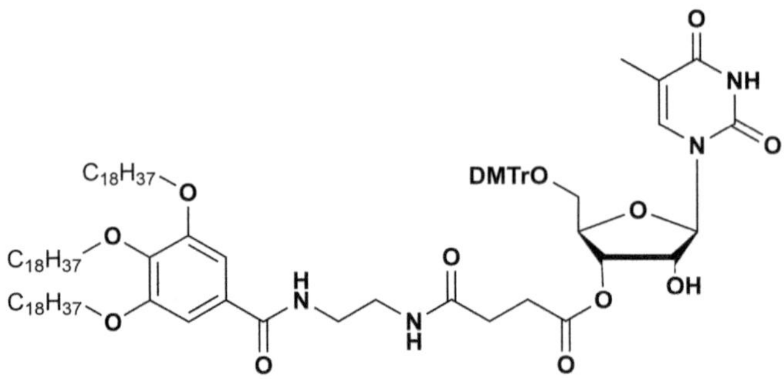

(9)

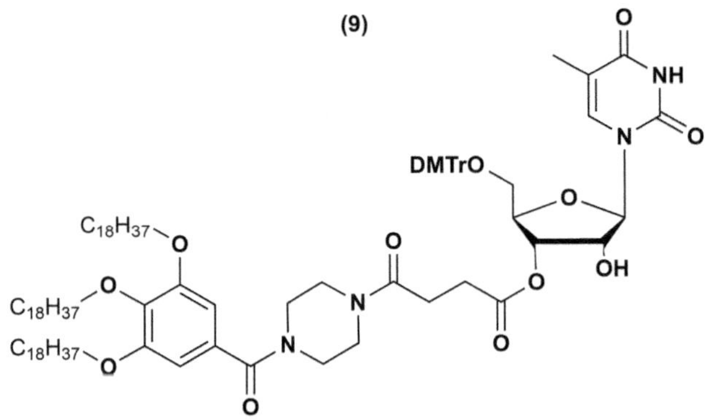

(10)

Figure 11.4 Structures of alkyl chain soluble supports **9** and **10**.

Table 11.14 Summary of synthetic cycle conditions used with alkyl chain-containing soluble support.

Synthetic step	Reagents and solvents
Coupling	5′-*O*-DMT-deoxynucleotide phosphoramidite BMT ACN–DCM
Oxidation	2-Butanone peroxide in DCM
Precipitation	MeOH
Detritylation	DCA in DCM
Precipitation	MeOH
Cleavage	NH_4OH_{aq}
Precipitation	EtOH
Final purification: chromatography	

moiety. Evaluating their solubilities in the most common solvents for LPOS, the ASS bearing a cyclic diamine as linker (specifically piperazine) show high solubility in apolar solvents, namely THF, $CHCl_3$, and DCM, and low or no solubility in polar solvents, namely MeOH, ACN, and EtCN. This behaviour ensures the possibility of purifying intermediates by solvent change, as in liquid–liquid extractions or precipitations (see Table 11.14). Therefore, reactions were performed in DCM–ACN (10 : 1 v/v) solution, the ACN being necessary to solubilize and stabilize the phosphoramidites used in amounts of 1.5–2.0 equiv. and DCM being necessary for solubilization of the supports. For couplings, 5-(benzylmercapto)-1*H*-tetrazole (BMT) was used as activator, followed by the addition of 2-butanone peroxide in DCM for oxidation. The first purification was performed through precipitation by MeOH. Finally, detritylation was carried out, as usual, with 3% DCA in DCM. The reaction was applied to the synthesis of 5′-r(CUUACGCUGACUACUUCGA)-d(TT)-3′, proving the versatility of the procedure to both ribo- and deoxynucleotides.

ASS were later applied also to the convergent synthesis of an 18-mer mixed PO–PS 2′-*O*-MOE (MOE = methoxyethyl) oligonucleotide, with sequence $5'\text{-}{}^{\mathrm{Me}}\mathrm{C}^{\mathrm{Me}}\mathrm{C_OG}^{\mathrm{Me}}\mathrm{U}^{\mathrm{Me}}\mathrm{UTT}^{\mathrm{Me}}\mathrm{CTTA}^{\mathrm{Me}}\mathrm{C}^{\mathrm{Me}}\mathrm{CA}^{\mathrm{Me}}\mathrm{C_O}{}^{\mathrm{Me}}\mathrm{C}^{\mathrm{Me}}\mathrm{C}^{\mathrm{Me}}\mathrm{U}\text{-}3'$.[40] The entire sequence was assembled starting from three tetramers (${}^{\mathrm{Me}}\mathrm{C}^{\mathrm{Me}}\mathrm{C_OG}^{\mathrm{Me}}\mathrm{U}$, ${}^{\mathrm{Me}}\mathrm{UTT}^{\mathrm{Me}}\mathrm{C}$, and $\mathrm{A}^{\mathrm{Me}}\mathrm{C_O}{}^{\mathrm{Me}}\mathrm{C}^{\mathrm{Me}}\mathrm{C}$), one pentamer (${}^{\mathrm{Me}}\mathrm{CTTA}^{\mathrm{Me}}\mathrm{C}^{\mathrm{Me}}\mathrm{C}$), and a monomer building block (${}^{\mathrm{Me}}\mathrm{U}$). The monomer building block was loaded on a soluble support and then detritylated, while the other fragments were temporarily 3′-*O*-protected with the *tert*-butyldiphenylsilyl group (TBDPS), which showed the best stability among various silyl-protecting groups towards TCA used for detritylation. Following the classical reaction scheme, couplings, sulfurization, and detritylation were performed in the same vessel, thus requiring the identification of a common solvent for all these synthetic steps. Since ACN was the solvent of choice for coupling and sulfurization, and DCM was the best performing in detritylation, a mixture of the two was selected to ensure telescoped reactions. Only 1.4 equiv. of phosphoramidites was necessary to complete couplings if the residual water content was strictly monitored and kept below 100 ppm, thus only half of the phosphoramidite equivalents commonly used was necessary. For sulfurizations, xanthane hydride was generally used, but was replaced with (*Z*)-*N*,*N*-dimethyl-*N*′-(3-thioxo-3*H*-1,2,4-dithiazol-5-yl)formimidamide (DDTT) for the last sulfurization to suppress the unwanted detritylation that occurs with xanthane hydride. Detritylation was performed using TFA in DCM with dodecanethiol as scavenger, necessary to form DMT-thioethers that prevent the growth of oligonucleotides from retritylation, which can occur during basic workup (with aqueous $NaHCO_3$) used to quench the reaction.

This reaction scheme optimized for the synthesis of fragments was applied to the entire oligonucleotide assembly, introducing $NC(CH_2)_2OP(NiPr_2)_2$ to convert each fragment into its corresponding phosphoramidite. In this way, it was possible to control the formation of impurities, which resulted in a LC–UV purity of 90% for the final oligonucleotide. The 18-mer sequence was also further elongated to a 34-mer oligonucleotide by repeating the same synthetic scheme and affording the target sequence with an LC–UV purity of 44% and 24% yield. For the first time, following the outlines reported by the American Chemical Society Green Chemistry Institute Pharmaceutical Roundtable (ACS GCI-PR), the authors calculated the PMI of this process with reference to the amounts of phosphoramidites and solvents used for the synthesis of 1 kg of the target 18-mer oligonucleotide. They demonstrated a reduction of 47% and 28%, respectively, from SPOS to convergent LPOS.

11.8 Homostar Soluble Support

All of the previously reported strategies are based on the purification of intermediates by precipitation or crystallization, and of final products by chromatography. Although these approaches are in theory scalable, when applied to the synthesis of multi-kilogram amounts of oligonucleotides they can be impractical. Therefore, Livingston and co-workers developed a novel approach based on the use of organic solvent nanofiltration (OSN) to separate the growing oligonucleotide from other impurities and reagents by size exclusion and geometric selection.[41] The soluble support in this work was a monodisperse tris(octagol) homostar (see Figure 11.5), which guarantees simultaneous growth of three oligonucleotide sequences and retention of macromolecules by the membrane.

The homostar soluble support was made of polybenzimidazole cross-linked with *p*-dibromoxylene, which was compatible with all reagents and by-products formed when phosphoramidite chemistry was used (see Table 11.15). The anchoring of the first nucleoside was performed with 2,6-dichlorobenzoyl chloride (DcbCl) and NMI in DCM, followed by detritylation with dichloroacetic acid (DCA) in DCM with pyrrole as scavenger. The

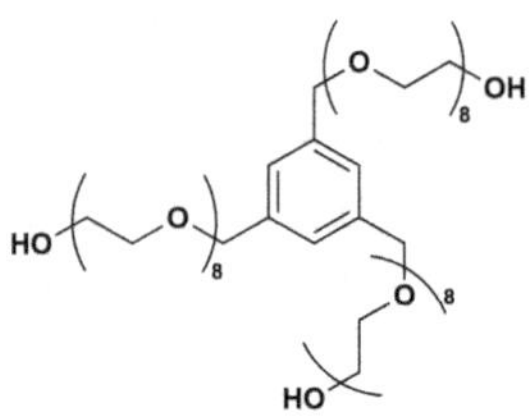

(11)

Figure 11.5 Structure of homostar soluble support **11**.

Table 11.15 Summary of synthetic cycle conditions used with homostar soluble support.

Synthetic step	Reagents and solvents
Loading	5′-*O*-DMT-2′-MOE succinylnucleoside in DCM NMI DcbCl
Extraction	$NaHCO_{3aq}$–DCM
Purification	Chromatography
Detritylation	DCA in DCM
Extraction	$NaHCO_{3aq}$–DCM
Purification	Chromatography
Coupling	5′-*O*-DMT-2′-MOE nucleotide phosphoramidite in ACN ETT
Sulfurization	PADS
Purification	OSN in ACN
Cleavage	NH_4OH_{aq} $MeNH_2$

subsequent couplings were conducted with 1.5 equiv. of phosphoramidite in ACN with ethylthiotetrazole (ETT). Sulfurization was performed with phenylacetyl disulfide (PADS) in Pyr under standard conditions. When sulfurization was complete, the solution diluted with ACN was passed through the OSN apparatus to remove all the small-sized impurities. The crude anchored oligonucleotide sequences were then washed with DCM from the OSN ring, and DCA and pyrrole were added to the solution to perform detritylation. This reaction scheme was repeated for chain elongation, but solubility issues arose from trimers on, which required the use of ACN–DMF (9:1 v/v) mixture for couplings. Similarly, during OSN the solvent was changed from ACN to MeOH–ACN (1:4 or 1:3 v/v), and for detritylation after the second diafiltration 1% pyridinium dichloroacetate was used to promote permeation of thioate salts across the membrane. With this approach, 5–9-mer 2′-methyl RNA homostars were synthesized. After a cumulative yield of 52% for the first three couplings, the combined yield for the next five couplings was about 76%. However, with this approach a large volume of solvent was required for diafiltration, which makes this technique not greener than SPOS. It was suggested that the introduction of a second membrane with a low molecular weight cut-off would allow recovery of the solvent, but no data were reported.

In a subsequent in-depth analysis of the developed protocol, the same group reported an economic analysis of the technique applied to the synthesis of 1 kg of a 23-mer oligoribonucleotide with an overall yield of 50%.[42] The main components of the overall cost were, as expected, the phosphoramidites and the solvents. In fact, phosphoramidites and solvent account for 44% and 79%, respectively, of overall synthetic costs; assuming that 90% recovery of ACN was possible through distillation, the percentages change to 50% and 9%, respectively. This meant that the cost of a 1 kg batch of oligonucleotide decreases from $121 000 to $67 000 if solvent recovery is considered. This costing is in agreement with the principles of green chemistry, which strongly recommend the recovery and reuse of chemicals, which is a priority from the point of view of sustainability and can be translated into an advantage if the cost analysis reveals the benefit of this practice.

11.9 Conclusion

The majority of synthetic protocols developed in LPOS rely on phosphoramidite chemistry, as this scaffold is the most reactive and affords the target sequences with optimal efficacy compared with other chemistries that have been explored. The chemistry of P(III) is the strongest despite the tentative application of P(V) building blocks proposed by Baran and co-workers.[36] Further explorations in this respect would have a great beneficial effect since a reduction in the number of synthetic steps could be achieved, removing oxidation of intermediates. In addition, using P(V) building blocks will significantly improve the impurity profile, allowing the recovery of the more stable monomers that have been used in excess for couplings. From the point of view of process development, approaches based on the use of blockmers instead of monomers appears of great convenience, being time saving thanks to the reduced number of synthetic steps and allowing more efficient purifications, with suppression of all the $n - 1$ truncated sequences. At present, LPOS is still not favourable from the point of view of greenness, being based on the use of large excesses of solvents and reagents. The first attempts to recover reagents and solvents as proposed when using homostar soluble supports need to be further explored to allow the practical large-scale application of LPOS.

List of Abbreviations

Ac_2O	Acetic anhydride
ACS GCI-PR	American Chemical Society Green Chemistry Institute Pharmaceutical Roundtable
ASS	Alkyl chain-soluble supports
BMT	5-(Benzylmercapto)-1*H*-tetrazole
CPG	Controlled-pore glass
DBU	1,8-Diazabicyclo[5.4.0]undec-7-ene
DCA	Dichloroacetic acid
DcbCl	2,6-Dichlorobenzoyl chloride
DCC	Dicyclohexylcarbodiimide
DCE	Dichloroethane
DCI	Dicyanoimidazole
DCM	Dichloromethane
DDD	Diethyldithiocarbonate disulfide
DDTT	(*Z*)-*N*,*N*-Dimethyl-*N*′-(3-thioxo-3*H*-1,2,4-dithiazol-5-yl) formimidamide
diOBT	2-Chlorophenyl-*O*,*O*-bis(1-benzotriazolyl)phosphate
DMAP	Dimethylaminopyridine
DMSO	Dimethyl sulfoxide
DMT	Dimethoxytrityl
ETT	Ethylthiotetrazole
Fpmp	1-(2-Fluorophenyl)-4-methoxypiperidin-4-yl)
HELP	High-efficiency liquid phase
IIP	2-Isopropoxyprop-2-yl
ILSS	Ionic liquid-supported synthesis
LPC	Liquid-phase carrier
LPOS	Liquid-phase oligonucleotide synthesis
LPPS	Liquid-phase peptide synthesis
MIP	2-Methoxyisopropyl
MOE	Methoxyethyl
MSNT	Mesitylenesulfonylnitrotriazole
MTBE	Methyl *tert*-butyl ether
NMI	*N*-Methylimidazole
OSN	Organic solvent nanofiltration
PADS	Phenylacetyl disulfide
PEG	Poly(ethylene glycol)
PMI	Process mass intensity
PO	Phosphodiester

PS	Phosphorothioate
pTSA	*p*-Toluenesulfonic acid
Pyr	Pyridine
SPOS	Solid-phase oligonucleotide synthesis
TBAF	Tetrabutylammonium fluoride
TBDPS	*tert*-Butyldiphenylsilyl
TBS	*tert*-Butyldimethylsilyl
*t*BuOOH	*tert*-Butyl hydroperoxide
TCA	Trichloroacetic acid
TEA	Triethylamine
TETD	Tetraethylthiuram disulfide
TFA	Trifluoroacetic acid
THF	Tetrahydrofuran
TMG	Tetramethylguanidine
TMG-PAO	Tetramethylguanidinium (*E*)-2-pyridinealdoximate
TPS	Triisopropylbenzenesulfonyl chloride
TsOH	*p*-Toluenesulfonic acid

Acknowledgements

The authors thank the PNRR project “National Center for Gene Therapy and Drugs Based on RNA Technology” – Spoke 6 – RNA Chemistry – Task 6.2.1 “Development of New RNA-targeting Compounds RIBOTAC and Stereopure Oligonucleotides and Peptides Conjugates” for financial support.

References

1. L. Ferrazzano, M. Catani, A. Cavazzini, G. Martelli, D. Corbisiero, P. Cantelmi, T. Fantoni, A. Mattellone, C. De Luca, S. Felletti, W. Cabri and A. Tolomelli, *Green Chem.*, 2022, **24**, 975.
2. H. Hayatsu and H. G. Khorana, *J. Am. Chem. Soc.*, 1966, **83**(13), 3182.
3. K. Kamaike, Y. Hasegawa and Y. Ishido, *Tetrahedron Lett.*, 1988, **29**, 647–650.
4. K. Kamaike, Y. Hasegawa, I. Masuda, Y. Ishido, K. Watanabe, I. Hirao and K.-I. Miura, *Tetrahedron Lett.*, 1990, **46**, 163.
5. M. M. Shemayakin, Y. A. Ovchinnikov, A. A. Kirushkin and I. V. Kozhevnikova, *Tetrahedron Lett.*, 1965, **27**, 2323.
6. G. M. Bonora, *Appl. Biochem. Biotechnol.*, 1995, **54**, 3.
7. H. Köster, *Tetrahedron Lett.*, 1972, **16**, 1535.
8. H. Köster and K. Heyne, *Tetrahedron Lett.*, 1972, **16**, 1531.
9. G. M. Bonora, C. L. Scrimin, F. P. Colonna and A. Garbesi, *Nucleic Acids Res.*, 1990, **18**, 3155.
10. F. P. Colonna, C. L. Scrimin and G. M. Bonora, *Tetrahedron Lett.*, 1991, **32**, 3251.
11. G. M. Bonora, G. Biancotto, M. Maffini and C. L. Scrimin, *Nucleic Acids Res.*, 1993, **21**, 1213.

12. C. L. Scrimin and G. M. Bonora, *Tetrahedron Lett.*, 1993, **34**, 4663.
13. J. A. Bagno, S. Bicciato, C. Di Bello and G. M. Bonora, *J. Chem. Technol. Biotechnol.*, 1998, **71**, 77.
14. A. Gimenez Molina and Y. S. Sanghvi, *Curr. Protoc. Nucleic Acid Chem.*, 2019, e82.
15. G. M. Bonora, R. Rossin, S. Zaramella, D. L. Cole, A. Eleuteri and V. T. Ravikumar, *Org. Process Res. Dev.*, 2000, **4**, 225.
16. W. Miao and T.-H. Chan, *Org. Lett.*, 2003, **5**, 5003.
17. J. Fraga-Dubreuil and J. P. Bazureau, *Tetrahedron*, 2003, **59**, 6121.
18. W. Miao and T.-H. Chan, *J. Org. Chem.*, 2005, **70**, 3251.
19. X. He and T.-H. Chan, *Synthesis*, 2006, **10**, 1645.
20. J.-Y. Huang, M. Lei and Y.-G. Wang, *Tetrahedron Lett.*, 2006, **47**, 3047.
21. R. A. Donga, S. M. Khaliq-Uz-Zaman, T.-H. Chang and M. J. Damha, *J. Org. Chem.*, 2006, **71**, 7907.
22. R. A. Donga, M. Hassler, T.-H. Chang and M. J. Damha, *Nucleosides, Nucleotides Nucleic Acids*, 2007, **26**, 1287.
23. R. A. Donga, T.-H. Chang and M. J. Damha, *Can. J. Chem.*, 2007, **85**, 274.
24. C. B. Reese and Q. Song, *Bioorg. Med. Chem.*, 1997, **7**, 2787.
25. C. B. Reese and Q. Song, *J. Chem. Soc., Perkin Trans. 1*, 1999, 1477.
26. J. Klose, C. B. Reese and Q. Song, *Tetrahedron*, 1997, **53**, 14411.
27. C. B. Reese and Q. Song, *Nucleic Acids Res.*, 1999, **27**, 963.
28. C. B. Reese and Q. Song, *Nucleic Acids Res.*, 1999, **27**, 2672.
29. C. B. Reese and H. Yan, *J. Chem. Soc., Perkin Trans. 1*, 2002, 2621.
30. R. Wörl and H. Köster, *Tetrahedron*, 1999, **55**, 2941.
31. R. Wörl and H. Köster, *Tetrahedron*, 1999, **55**, 2957.
32. V. Kungurstev, J. Laakkonen, A. Gimenez Molina and P. Virta, *Eur. J. Org. Chem.*, 2013, **29**, 6687.
33. A. Gimenez Molina, A. M. Jabgunde, P. Virta and H. Lönnberg, *Beilstein J. Org. Chem.*, 2014, **10**, 2279.
34. Z. Liang, P. Rosenqvist, E. Pajuniemi, M. Ora, P. Heinonen, P. Virta and M. Oivanen, *Eur. J. Org. Chem.*, 2023, **26**, e202300614.
35. P. Rosenqvist, V. Saari, E. Pajuniemi, A. Gimenez Molina, M. Ora, A. Hovart and P. Virta, *J. Org. Chem.*, 2023, **88**, 10156.
36. K. W. Knouse, J. N. deGruyter, M. A. Schmidt, B. Zheng, J. C. Vantourout, C. Kingston, S. E. Mercer, I. M. Mcdonald, R. E. Olson, Y. Zhu, C. Hang, J. Zhu, C. Yuan, Q. Wang, P. Park, M. D. Eastgate and P. S. Baran, *Science*, 2018, **361**, 1234.
37. V. Kungurstev, P. Virta and H. Lönnberg, *Eur. J. Org. Chem.*, 2013, **35**, 7886.
38. V. Kungurstev, H. Lönnberg and P. Virta, *RSC Adv.*, 2016, **6**, 105428.
39. S. Kim, M. Matsumoto and K. Chiba, *Chem. – Eur. J.*, 2013, **19**, 8615.
40. X. Zhou, W. F. Kiesman, W. Yan, H. Jiang, F. D. Antia, J. Yang, Y. A. Fillon, L. Xiao and X. Shi, *J. Org. Chem.*, 2022, **87**, 2087.
41. P. R. J. Gaffney, J. F. Kim, I. B. Valtcheva, G. D. Williams, M. S. Anson, A. M. Buswell and A. G. Livingston, *Chem. – Eur. J.*, 2015, **21**, 9535.
42. J. F. Kim, P. R. J. Gaffney, I. B. Valtcheva, G. Williams, A. M. Buswell, M. S. Anson and A. G. Livingstone, *Org. Process Res. Dev.*, 2016, **20**, 1439.

CHAPTER 12

Stereopure Oligonucleotides

L. FERRAZZANO

P4i Laboratory, Department of Chemistry "Giacomo Ciamician", Alma Mater Studiorum – University of Bologna, via P. Gobetti 85, 40129 Bologna, Italy
E-mail: lucia.ferrazzano4@unibo.it

12.1 Introduction

As discussed in Chapter 9, antisense synthetic oligonucleotides (ASOs) are structurally designed and synthesized to afford a high hybridization affinity to selected regions of target oligonucleotide sequences to inhibit the biological processes deriving from this interaction, especially protein synthesis. For this reason, synthetic ASOs need to be resistant to the activity of nucleases while still ensuring adequate cell membrane permeability. Among the possible modifications explored to improve their stability, the replacement of one non-bridging oxygen atom in the internucleotide linkage is particularly useful for enhancing resistance to nuclease degradation, and in particular, the introduction of amine ($-NR^1R^2$), borane ($-BH_3^-$), proton (H), selenium (Se^-), methyl (Me), and sulfur (S^-) opened up new families of modified phosphates, namely phosphoramidites (PNs), boranephosphates, H-phosphonates, phosphoroselenoates, methylphosphonates (MPOs), and phosphorothioates (PSs). The PSs were demonstrated to be particularly effective in promoting the water solubility, nuclease stability, and good specificity for base pairing of oligonucleotides. As proof, several therapeutic oligonucleotides in clinical trials or already on the market for the treatment of rare diseases include these modifications (see Figure 12.1). The challenge

Green Chemistry Series No. 85
Sustainability in TIDES Chemistry: Green Approaches to Oligonucleotides and Oligopeptides Synthesis
Edited by Alessandra Tolomelli, Lucia Ferrazzano and Walter Cabri

Published by the Royal Society of Chemistry, www.rsc.org

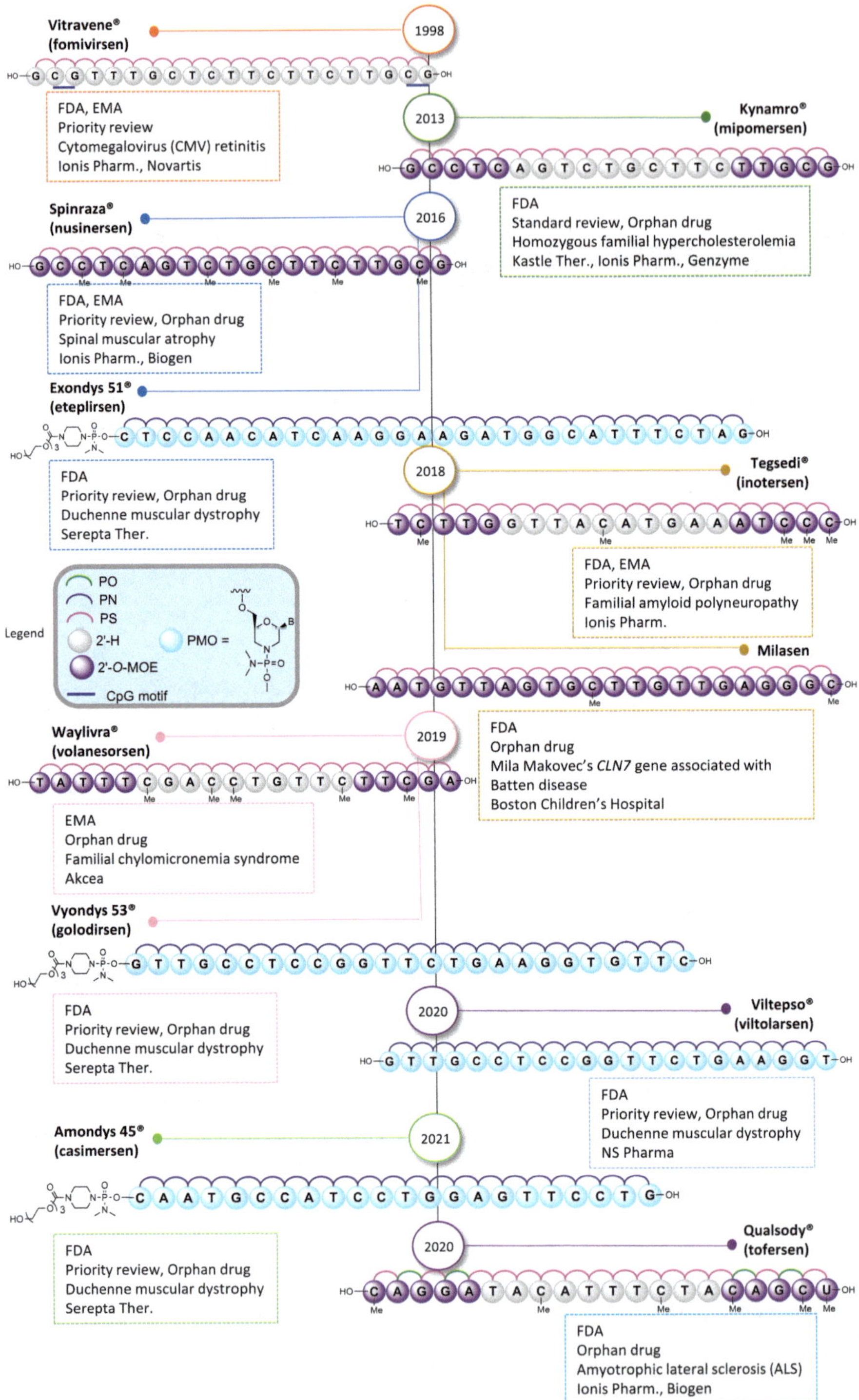

Figure 12.1 Clinically approved commercially available ASOs.

related to the synthetic introduction of these modifications is that the presence of different substituents on phosphorus makes it a chiral centre, thus influencing the way in which oligonucleotides are oriented when paired with their RNA or DNA complementary sequence and, consequently, the properties of the new double construct, in particular thermal stability.

From the very beginning of research studies on *P*-chiral oligonucleotides, the chromatographic approach for obtaining pure diastereoisomers was initially explored, since synthetic methodologies able to afford stereocontrolled products had not yet been developed. To better understand how complicated and unsustainable from the point of view of greenness proceeding by isolating single diastereoisomers is, specific consideration of the case of phosphorothioate oligonucleotides (PSOs) can be useful. PSOs consist of oligonucleotides bearing a phosphorothioate moiety at the 3′-*O*-position of the sugar, which is exploited to build internucleotide linkages. Although the biological relevance and efficacy have been extensively studied and demonstrated by the approval of several phosphorothioate-based ASOs (PS-ASOs) for clinical applications (see the structures in Figure 12.1), their polydiastereoisomerism still remains a challenge. In fact, the replacement of one of the non-bridging oxygen atoms with a sulfur atom results in the generation of a stereocentre at the phosphorus, which is the reason for the existence of 2^n diastereoisomers, where n is the number of PS linkages in a sequence of $n + 1$ bases. Issues regarding the effect of polyisomerism of PS-ASOs on the biological activity of these molecules have been raised[1,2] and studies on stereopure sequences have been conducted,[3,4] considering that reported data demonstrate that a lethal effect on pathogens requires oligonucleotide concentrations in the range 0.25–25 μM, when an average 25-mer sequence is used.[5,6] In the last example, it was found that if only some of the 16384 possible diastereoisomers are actually responsible for the cellular protection, the concentration of a generic ASO in specific formulations could be significantly lower. Therefore, several attempts to obtain stereopure oligonucleotides have been reported and they will be reviewed in the following sections.

All of the reported synthetic methods for achieving stereopure oligonucleotides were not focused on the use of green reagents and solvents. Consequently, the chemical processes described in the literature do not provide sufficient information to be able to assess the sustainability of these synthetic procedures. The environmental viability of stereocontrolled synthetic approaches is intricately connected to the ability to synthesize selectively well-defined oligonucleotide sequences, thereby streamlining the chromatographic purification required for sequence isolation. This synthetic approach has the potential to decrease solvent use, ultimately leading to a more environmentally friendly and sustainable synthetic process.

12.2 Stereopure Oligonucleotides Obtained *via* Enzymatic and Chromatographic Approaches

Considering the recent advances in the application of greener synthetic techniques to the synthesis of peptides and the efforts in transferring these laboratory-scale green procedures to manufacturing practice, it is immediately clear that the same green approach should be applied also to oligonucleotide synthesis, this being the downstream phase of their production process significantly impacting on the large-scale synthesis of these macromolecules and, consequently, on the green metrics.[7,8] However, when green metrics were still not treated as valuable parameters for planning synthetic processes, the approaches explored for the synthesis of stereocontrolled oligonucleotides considered specifically the use of enzymes for sequence assembly and of chromatography for the resolution of diastereomeric mixtures.

Although it is always efficient in providing diastereopure oligonucleotides containing one or more *P*-chiral phosphates and convenient for the synthesis of long stereopure oligonucleotides bearing S, Se, BH_3, and Me, the enzymatic approach suffers several limitations, and polymerase enzymes used for chain elongation have limited specificity for substrates in terms of selectivity for stereoisomers and structural modifications. In addition, the synthetic enzymatic process is based on elongation starting from a primer oligonucleotide to which the growing sequence is covalently bound. One of the early studies in this direction was that carried out in the 1970s by Eckstein's group, who investigated the effect of enzyme-induced polymerization on the stereochemistry of adenosine triphosphate (ATP) analogues.[9] Employing a DNA-dependent RNA polymerase from *Escherichia coli*, they evaluated the stereochemistry of the phosphodiester linkage during the enzymatic polymerization of a single isomer of adenosine 5′-*O*-(1-thiotriphosphate) (ATPαS, namely isomer A) with uridine triphosphate (UTP) to form mixed polymers poly($U_{-S}A$) (see Scheme 12.1). The study was performed exploiting the enzymatic hydrolysis of synthesized oligonucleotides by pancreatic RNase to obtain adenosine 3′-monophosphate (AMP) and uridine 2′,3′-cyclic phosphorothioate, and assign the absolute configuration of the latter. The authors proved that isomer B of ATPαS was not polymerized and evaluated the possible stereochemical pathway. In fact, during polymerization mediated by DNA-dependent RNA polymerase between the 3′-position of U and the 5′-position of A to form a phosphorothioate linkage, only linkages with

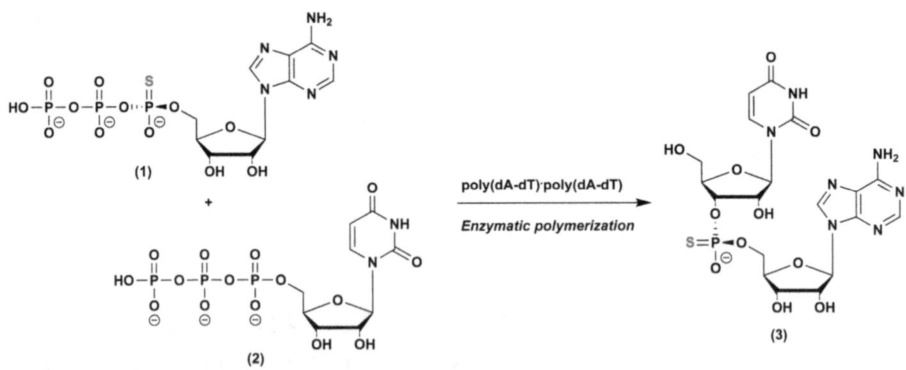

Scheme 12.1 Enzymatic condensation between ATPαS **1** and UTP **2** mediated by poly(dA-dT)-poly(dA-dT), affording selectively (R_P)-poly($U_{-S}A$) **3**.

an R_P configuration were obtained. They subsequently demonstrated also that the same stereocontrol was possible with other polymerases.[10] However, this enzymatic approach retained limitations related to substrate specificity, impractical large-scale synthesis, and unfeasible selective synthesis of (S_P)-PS-oligonucleotides. For these reasons, the approach has been gradually abandoned.

Hence the chemical stereocontrolled synthetic approach for *P*-chiral oligonucleotide synthesis represents a valid alternative to enzymatic synthesis, being free from the limitations mentioned above. However, although several studies have been proposed in the literature and patents, to the best of our knowledge its application to commercial ASOs is still limited to a few examples and further work needs to be done to transform academic synthetic procedures into large-scale industrially applicable protocols.

Considering the stereochemical induction deriving from the presence of chiral centres all around the P-linkage, the chirality of the sugars composing oligonucleotide sequences may themselves be able to induce stereoselectivity in the synthesis of internucleotide linkages. In fact, even without the support of stereoselective synthetic techniques, the two possible *P*-chiral centres (R_P and S_P) of a generic dinucleotide are not obtained in equal proportions, even using the common consolidated synthetic phosphoramidite method. Despite attempts to enhance the stereoselectivity using bulky condensing agents,[11] protections on the 5′-position,[12] and substituents on the 2′-position[13] or on the phosphorus atom,[14] it is clear that the steric hindrance and chirality of sugars are not sufficient to enhance stereoselectivity while creating a highly asymmetric environment.

As mentioned previously, in the first instance the topic of diastereoselectivity was predominantly approached from the point of view of the chromatographic isolation of stereopure monomers for assembling oligonucleotides or directly of the synthesized sequence. In 1984, Stec *et al.*[15] reported that for a successful chromatographic separation of diastereoisomers, it is important to consider the chain length, the position of phosphorothioate bonds in the sequence, and the base composition, and even with careful control of the chromatographic conditions the separation of diastereoisomers is not always completely successful.[16] As an example, they reported that among the 14 possible monophosphorothioate diastereoisomers of d(GGAATTCC) (namely dGGAATTC$_{PS}$C, dGGAATT$_{PS}$CC, dGGAAT$_{PS}$TCC, dGGAA$_{PS}$TTCC, dGGA$_{PS}$ATTCC, dGG$_{PS}$AATTCC, and dG$_{PS}$GAATTCC, in both R_P and S_P configurations), only five pairs were separable *via* high-performance liquid chromatography (HPLC), both R_P and S_P configurations of d(GGA$_{PS}$ATTCC) and d(GGAA$_{PS}$TTCC) being coeluted.[15]

Therefore, as proceeding with the isolation of a single or groups of diastereoisomers from complex mixtures of oligonucleotides is complex, expensive, and environmentally unsustainable, the chromatographic separation of diastereopure monomers to be incorporated in longer oligonucleotide sequences was considered a possibility. For long sequences, *P*-chiral monomers can have R_P and S_P configurations and the diastereoisomers can be selectively isolated through chromatographic purification or obtained synthetically. Obviously, monomers containing a P(V) centre are more stable than those containing P(III), and therefore their purification and synthesis have minor drawbacks related to degradation. However, research on the chromatographic isolation of diastereopure monomers initially focused on P(III) nucleotides. In fact, the separation of these diastereomerically pure nucleotides by chromatography was first reported in 1984 by Stec and Zon, who used isolated chiral phosphoramidite nucleosides for the synthesis of a dinucleotide, which was then sulfurized with elemental sulfur.[17] However, the activator introduced for the condensation, namely 1*H*-tetrazole, induced rapid epimerization at the phosphorus atom, revealing a limitation in the use of this reagent that also occurred more recently in other explorations.

In 2000, Beaucage and co-workers proposed diastereopure 3′-*O*-(3-*N*-acyl-1,3,2-oxaphospholidine) monomers[18] chromatographically separated from diastereomeric mixtures of rotamers to obtain *trans*-R_P and *trans*-S_P scaffolds. The condensation was promoted by 1,1,3,3-tetramethylguanidine (TMG) and the sulfurization was carried out using the Beaucage reagent (3*H*-1,2-benzodithiol-3-one 1,1-dioxide),[19] affording 4–11-mer sequences successfully through the solid-phase oligonucleotide synthesis (SPOS) approach (see Scheme 12.2). The advantage of this method is its compatibility with the phosphoramidite method, thus opening up the possibility of efficient stereopure phosphorothioate/phosphodiester (PS/PO) chimera oligonucleotide synthesis.

A few years later, attention moved to P(V) diastereopure monomers isolated chromatographically. In particular, Cosstick and Williams in 1987 proposed the use of diastereopure 3′-*O*-*S*-alkyl phosphorothioate monomers for the synthesis of PS-dinucleotides mediated by mesitylenesulfonyl-5-(pyridin-2-yl) tetrazole (MSPy) or 1-mesitylenesulfonyl-3-nitro-1,2,4-triazole (MSNT) as activating agents, less prone to induce epimerization (see Scheme 12.3).[20]

Similarly, Lesnikowski and co-workers reported several studies on the stereospecific synthesis of all-R_P and all-S_P 2–5-mers of thymidine and uracil using stereopure P(V) phosphorothioate monomers with 2-nitrobenzyl as a sulfur protecting group (see Scheme 12.4).[21–24] However, all these examples were applied only to the synthesis of short sequences of oligodeoxynucleotides (ODNs) or oligoribonucleotides (ORNs) such as MPO-ODN, PS-ODN, and PS-ORN, and were not transferred and integrated into the usual practices of SPOS.

For further application of diastereomerically pure monomers to the synthesis of longer oligonucleotide sequences, the oligonucleotide community had to wait until 1995 and 1998, when Stec and co-workers first reported the solid-phase stereocontrolled synthesis of 10–28-mer PS-oligonucleotides using diastereomerically pure 5′-*O*-dimethoxytrityl (DMT) nucleoside, 3′-*O*-(2-thio-1,3,2-oxathiaphospholane)s, and 1,5-diazabicycloundec-7-ene (DBU) as activating agent (see Scheme 12.5).[25,26] They demonstrated that when using diastereomerically pure monomers (determined by ^{31}P NMR spectroscopy after chromatographic purification), the target PS-oligonucleotides

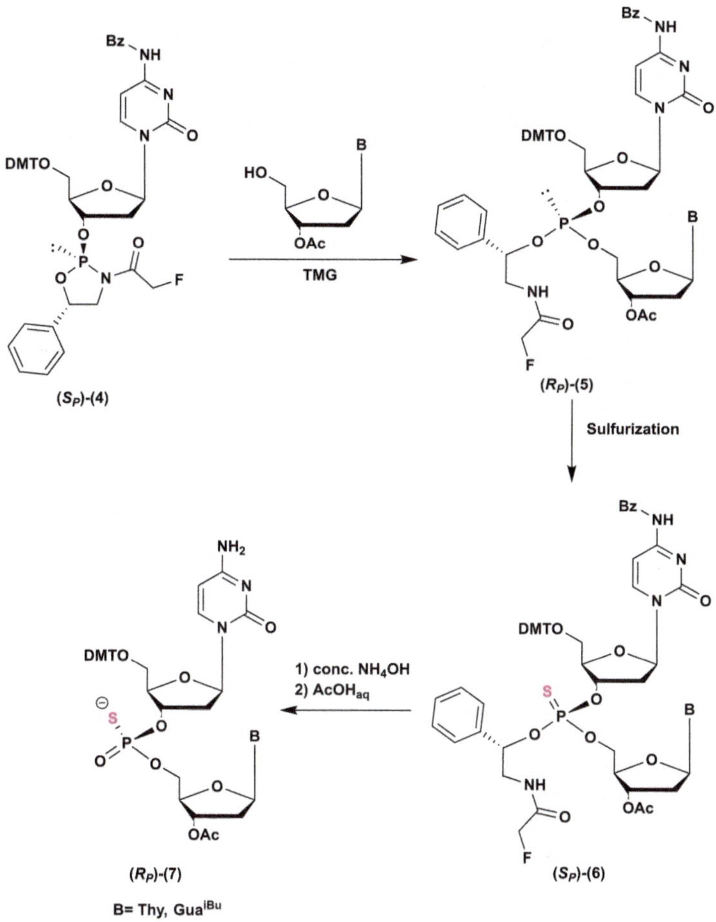

Scheme 12.2 Stereocontrolled synthesis of phosphorothioate dinucleotides **7** starting from 3′-*O*-(3-*N*-acyl-1,3,2-oxaphospholidine) monomers **4**.

were obtained in good purity, proving the stereospecificity of each synthetic step.

As stated by the authors, determining the stereopurity of such a longer sequence by ^{31}P NMR spectroscopy or HPLC is complicated by the poor resolution of single diastereoisomers. Therefore, following the example of Eckstein's group in the 1970s,[9] diastereoselective hydrolysis of PS-oligonucleotides by nucleases or phosphodiesterase allowed the processing of (S_P)- and (R_P)-PS-ODNs, respectively. In fact, processing full S_P or R_P sequences with the respective non-specific enzyme revealed that the amount of unprocessed PS-oligonucleotide was in accord with the diastereomeric purity calculated by HPLC, at the cost of a slight reduction because the incomplete diastereoselectivity of phosphodiesterases and nucleases reduced the selectivity for phosphorothioate linkages when in longer sequences. Although the

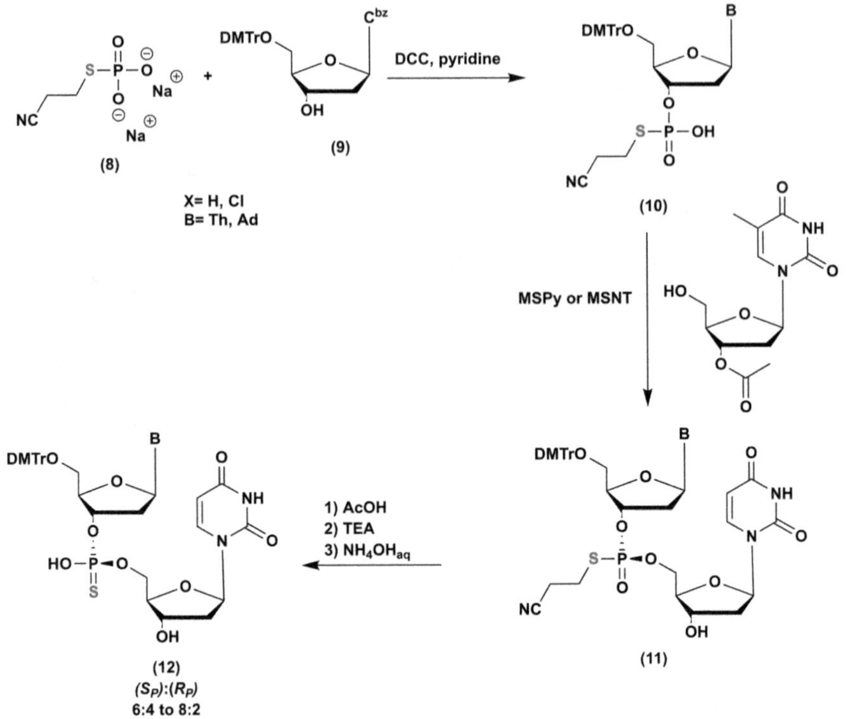

Scheme 12.3 MSPy- and MSNT-mediated synthesis of phosphorothioate dinucleotides **12**.

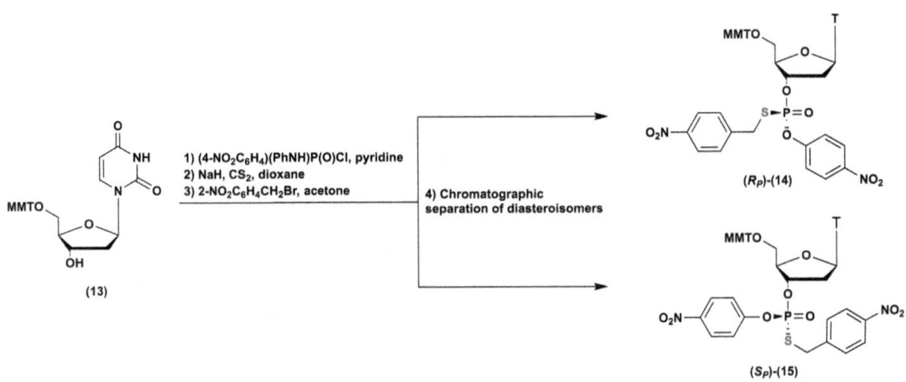

Scheme 12.4 Chromatographic separation of chiral phosphorothioate **15** bearing a 2-nitrobenzyl protecting group on sulfur.

oxathiaphospholane method was applicable to long *P*-chiral oligonucleotide sequences, it has intrinsic disadvantages due to (i) the necessity of separating diastereomeric mixtures of monomers *via* chromatography, (ii) the low efficacy of DBU as a condensing agent when 2′-hindered monomers are used,[27] (iii) the incompatibility of the technique with the classical phosphoramidite

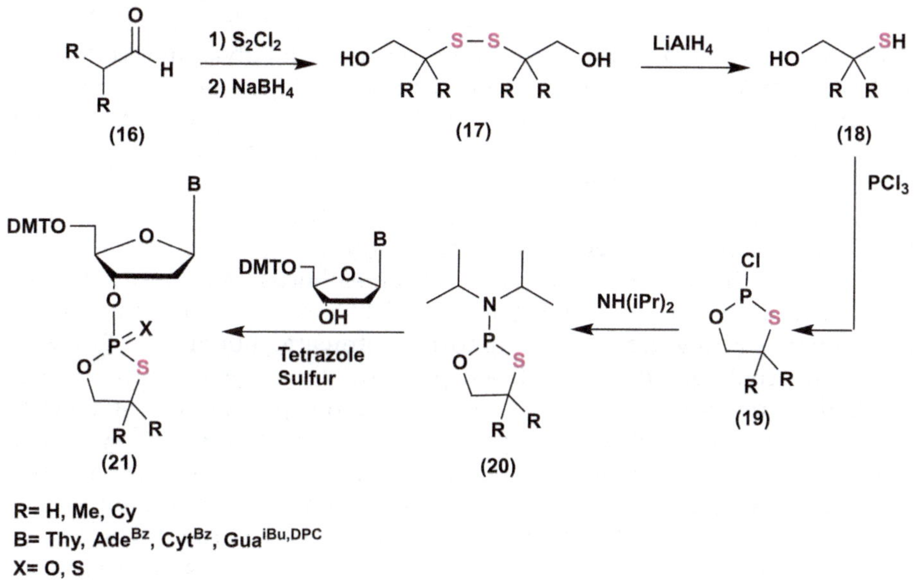

Scheme 12.5 Synthesis of 3′-*O*-(2-thio-1,3,2-oxathiaphospholane)s **21**.

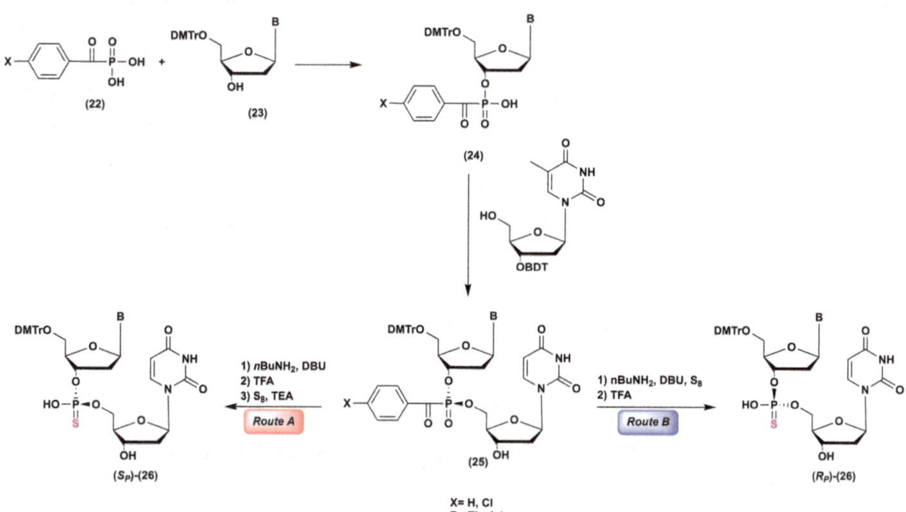

Scheme 12.6 Synthesis of (S_P)- and (R_P)-dinucleotides **26** starting from aroylphosphonates **24**.

method, and (iv) the sensitivity of phosphorothioate diester linkages to oxidation.

Therefore, other approaches also explored the effect of the presence of protecting groups on phosphorus. Fujii and co-workers reported the use of stereopure acylphosphonates and subsequent *in situ* treatment with sulfur for converting the intermediates in phosphorothioates (see Scheme 12.6).[28,29]

They reported that, once the P-aroyl-protected dinucleotides had been obtained, if proceeding with aroyl removal using *n*-butylamine (nBuNH$_2$) and DBU, DMT removal with trifluoroacetic acid (TFA) solution, and then sulfurization with S_8 in triethylamine (TEA), predominantly S_P isomers were obtained. On the other hand, if the removal of aroyl protection on phosphorus was conducted in the presence of S_8 before removing the 5′-*O*-protecting group, the stereocontrol of the process produced mainly the R_P isomers. However, this approach was limited to the synthesis of dimers.

Expanding the scope of protection on phosphorus, Wozniak, Stec and co-workers focused also on the synthesis of 2′-deoxyribose phosphorothioate dimers starting from *O*-aryl phosphorothioate monomers, followed by stereospecific deprotection of phosphorothioate by oximate ion, which resulted in inversion of the configuration at the phosphorus stereocentre as a consequence of oximate attack and β-elimination of an aryl group (see Scheme 12.7).[30] Again, the *O*-aryl group can be considered as a temporary protecting group on the oxygen, inserted to prevent conversion of the unprotected phosphorothioate diesters into phosphate during iodine–water treatment applied during chain elongation. Therefore, the same group proposed an alternative protocol using 3′-*O*-(2-thio-1,3,2-oxathiaphospholane) nucleotides for the synthesis of *S*-2-nitrobenzyl-protected phosphorothioate dinucleotides introduced in 31-mer monophosphorothioate sequences with a predetermined sense of chirality (see Scheme 12.8).[31] Similarly, they also proposed the use of 3′,5′-methylphosphonothioate as a chiral linkage for the synthesis of stereopure chimera MPO-oligonucleotides consisting of both 2′-deoxy- and 2′-*O*-methoxyribonucleotides (see Scheme 12.9).[32,33] The applicability of both the diastereomerically pure *O*-aryl-protected PS-dinucleotides and MPO-2′-OMe-dinucleotides was verified in the successful synthesis of 6–19-mer chimera sequences. Considering the sensitivity to the chromatographic conditions and the impractical isolation of pure monomers from racemic mixtures in the context of large-scale application, attention moved to stereocontrolled synthetic approaches.

12.3 Chemical Stereocontrolled Synthesis of Internucleotide Linkages Based on P(III)-based Chemistry

Studies were focused from the very beginning on exploiting the use of chiral auxiliaries as stereochemical inducers. In this regard, Agrawal and co-workers were pioneers in introducing the concept of chiral auxiliaries as stereochemical inducers of chirality in the internucleotide linkage. In preliminary studies, inspired by the efficiency of β-cyanoethyl phosphoramidite nucleoside widely employed for the non-stereoselective synthesis of oligonucleotides, they proposed oxazaphospholidines including a cyclic β-amino alcohol as a pool. After a first achiral approach based on the use of 3-methyl-1,3,2-oxazaphospholidine,[34] they proposed a variation with the use of 1,2-disubstituted amino alcohols and (1*R*,1*S*)-ephedrine for the

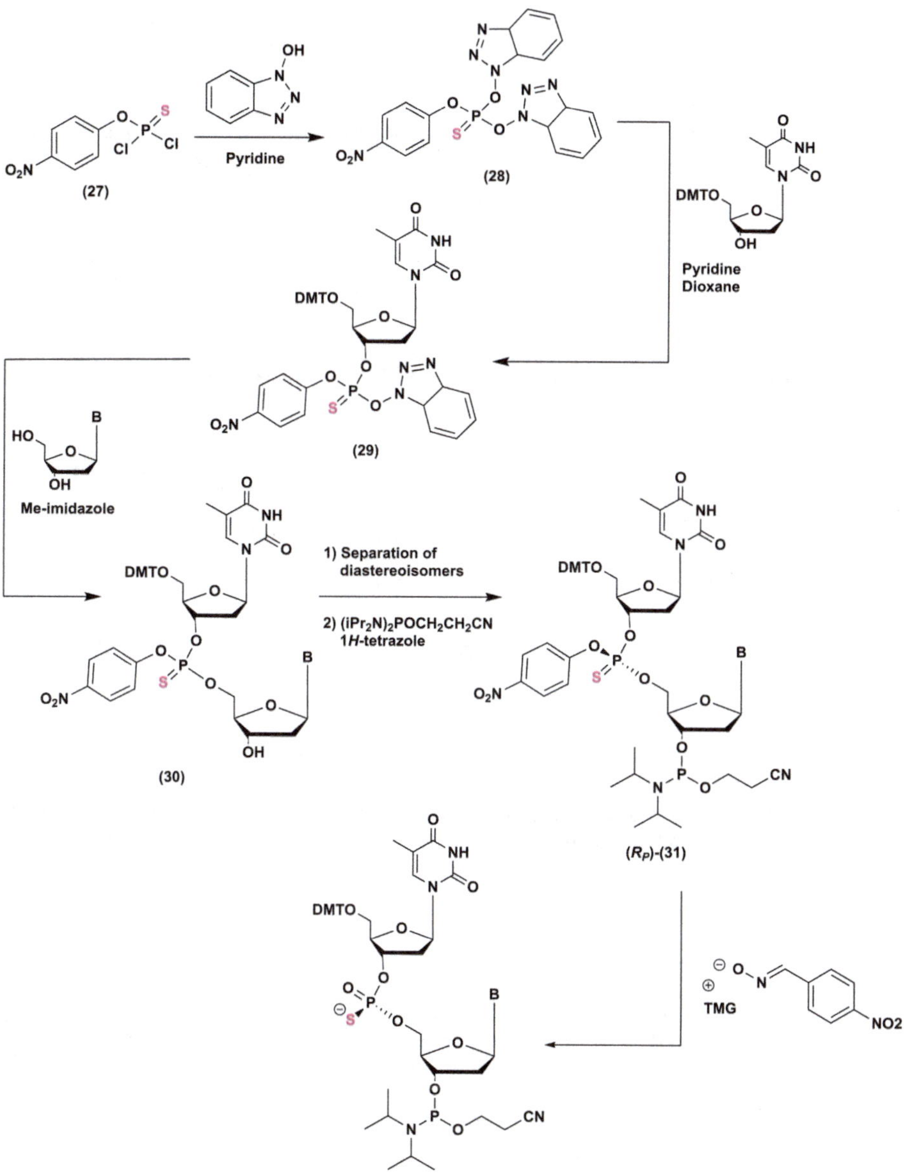

Scheme 12.7 Synthesis of stereopure dinucleotides starting from *O*-arylphosphorothioate monomers **29**.

synthesis of chiral monomers (see Scheme 12.10). Applied to all 5′-protected nucleosides, both of the resulting scaffolds were employed in the SPOS of PO and PS dimers (TT, CT, GT, and AT) and trimers (TTT). However, the results in terms of diastereoselectivity were poor and the effect of 1*H*-tetrazole on it was not further explored.[35]

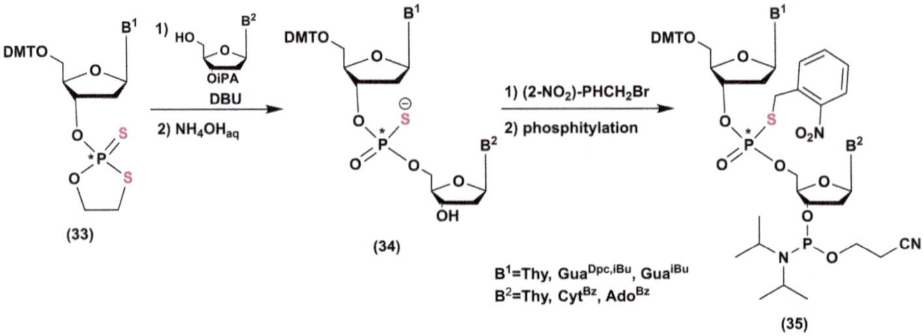

Scheme 12.8 Synthesis of stereopure *S*-2-nitrobenzyl-protected dinucleotides **35**.

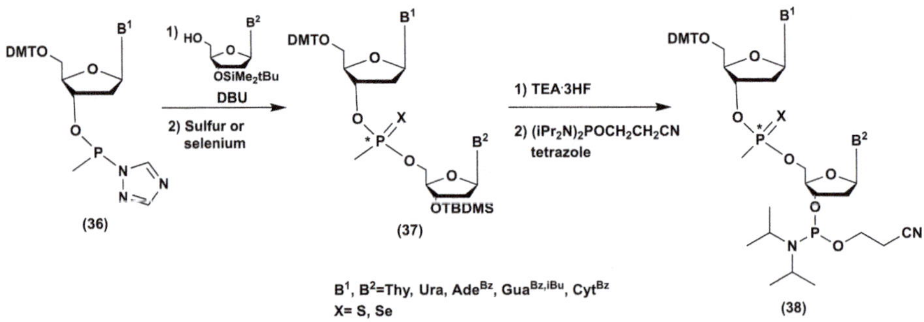

Scheme 12.9 Synthesis of stereopure dinucleotides starting from 3′-methylphosphoramidite **36**.

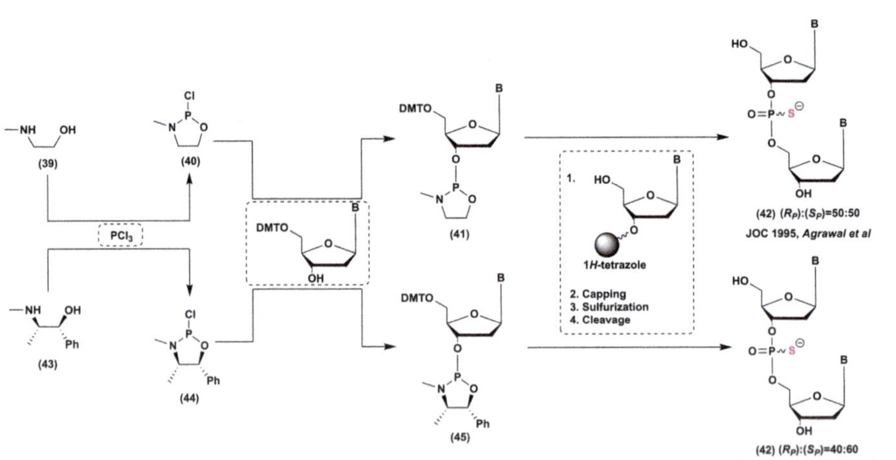

Scheme 12.10 3-Methyl-1,3,2-oxazaphospholidine monomers deriving from 2-(methylamino)ethanol and ephredine for the synthesis of phosphorothioate dinucleotides **42**.

These studies represented the background of the future explorations on the use of chiral auxiliaries for the diastereoselective synthesis of PS-oligonucleotides. In fact, in 1996 Just and co-workers reported a successful approach for the stereoselective synthesis of phosphorothioate dinucleotides by using xylose derivatives as chiral auxiliaries and 2-bromo-4,5-dicyanoimidazole as activator (see Scheme 12.11).[36] They demonstrated that the stereochemistry of the auxiliary deriving from D- or L-xylose had effects on the diastereoselectivity during the coupling reaction, leading to the preferential formation of (R_P)- and (S_P)-dithymidine phosphorothioate, respectively.

Exploring further the use of chiral and cyclic phosphoramidites, they synthesized (*S*)-1-(imidazol-2-yl)propan-2-ol[37] and (*S*)-1-(indol-2-yl)propan-2-ol[38] as single diastereoisomers (see Scheme 12.12A and B). In particular, the results on indole-based chiral auxiliaries opened the way to their evaluation in the stereocontrolled synthesis of phosphorothioate dinucleotides. Further, as the auxiliaries were not removable, they later modified the structure of the indole-based chiral auxiliary by introducing a cyano or an acetamide group on the glycidol backbone, in order to remove it under the same conditions as used for the deprotection of sequences in solid-phase synthesis, namely using ammonium hydroxide solution (see Scheme 12.12C).[39] However, after synthesizing the thymidine monomer or dinucleotide, these approaches were not extended to the context of the synthesis of longer oligonucleotides.

Inspired by these results and giving continuity to their previous pioneering studies, in 1998 Agrawal and co-workers proposed the use of bicyclic

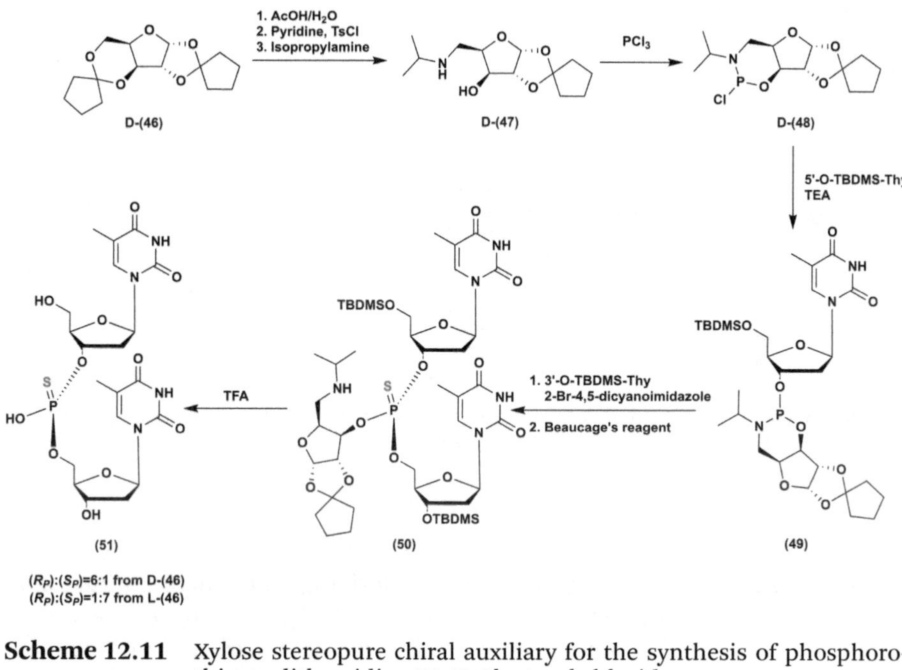

Scheme 12.11 Xylose stereopure chiral auxiliary for the synthesis of phosphorothioate dithymidine **51**. TsCl = tosyl chloride.

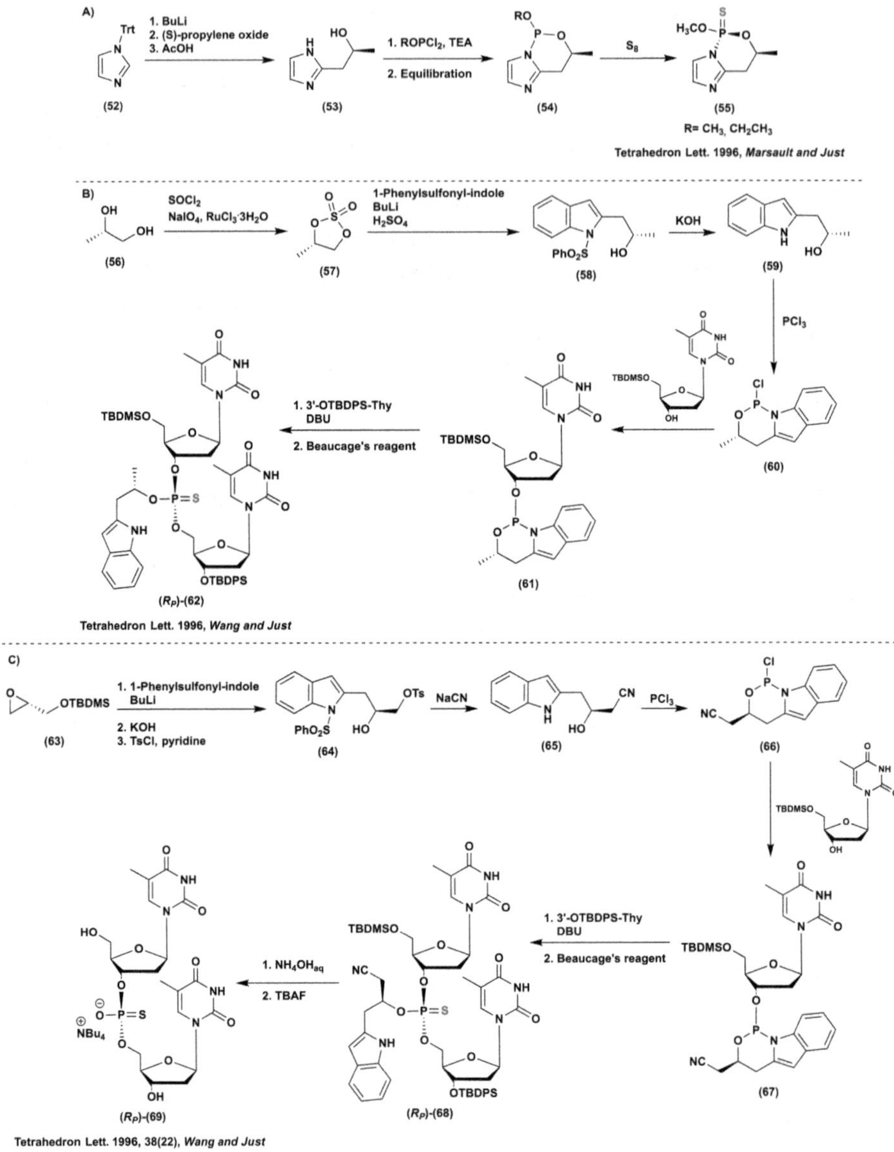

Scheme 12.12 Imidazolylpropanol and indolylpropanol stereopure chiral auxiliaries for the synthesis of phosphorothioate derivatives.

oxazaphospholidine nucleosides as monomers for the stereoselective solid-phase synthesis of PS-oligonucleotides, deriving from the β-amino alcohol (*S*)-(+)-2-pyrrolidinemethanol (see Scheme 12.13).[40] Initially employed to explore its reactivity in solution with oxygen, nitrogen, and carbon nucleophiles which proved retention of configuration after the substitution reaction, they were applied to solid-phase synthesis in the presence of 1*H*-tetrazole

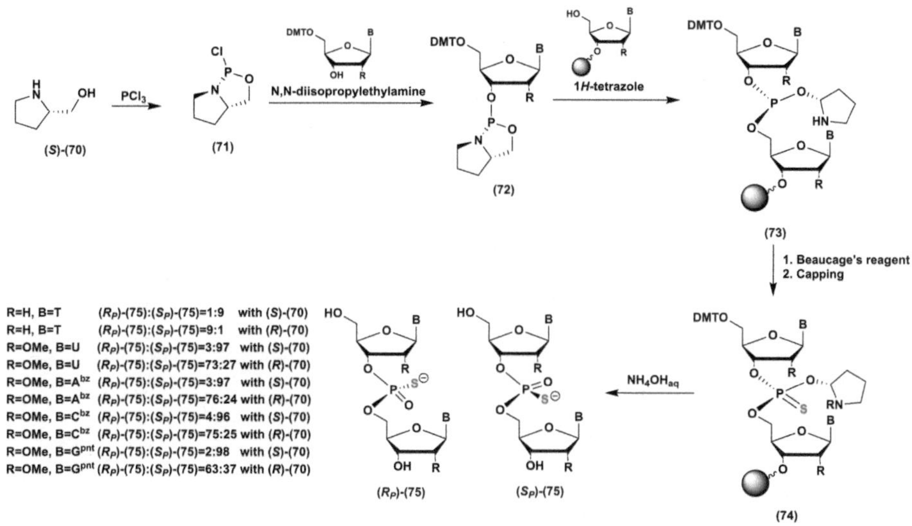

Scheme 12.13 (*S*)-(+)-2-Pyrrolidinemethanol stereopure chiral auxiliaries for the synthesis of phosphorothioate dinucleotides **75**.

as activating agent. When starting from (S_P)-oxazaphospholidine, (S_P)-$T_{PS}T$ was predominantly obtained with 90:10 diastereoselectivity, in agreement with the identification of the corresponding (R_P)-dimer when D-prolinol was used as a chiral pool. Since all these studies were conducted on 2′-deoxyribonucleosides, subsequently they also explored the possibility of transferring the chiral approach to 2′-OMe-ribonucleoside oxazaphospholidine (see Scheme 12.13).[41] This exploration showed great potential if one looks at the structures of marketed oligonucleotides, which often contain 2′-*O*-methoxyethoxy (2′-*O*-MOE) ribonucleosides. The high diastereoselectivity in the synthesis of dimers was confirmed also in this case.

Despite all the reported examples of the synthesis of stereodefined PS-ODNs, the use of 3′-*O*-(2-thio-1,3,2-oxathiaphospholane) nucleoside, proposed by Stec and co-workers, has been considered the most valid available synthetic tool, despite some limitations related to the necessity for chromatographic isolation of monomers from diastereomeric mixtures and the low efficacy of internucleotide linkage formation under basic conditions. Therefore, inspired by these scaffolds, Wada and co-workers proposed the stereoselective synthesis of PS-oligonucleotides, introducing diastereopure presynthesized 3′-*O*-bicyclic-(1,3,2-oxazaphospholidine) derivatives obtained starting from stereopure 1,2-amino alcohols (see Scheme 12.14).[42] The success of this method is similar to that of the classic phosphoramidite method; in fact, after presynthesizing stereopure monomers from the corresponding amino alcohol, using a mild acidic activator for chain elongation, and proceeding with capping, oxidation/sulfurization, and 5′-*O*-detritylation, the stereoregular chimera PO/PS-oligonucleotides can be obtained. The stereopure monomers were synthesized through nucleophilic attack at phosphorus of 2-chloro-1,3,2-oxazaphospholidine to replace chloride,[43] obtaining both R_P

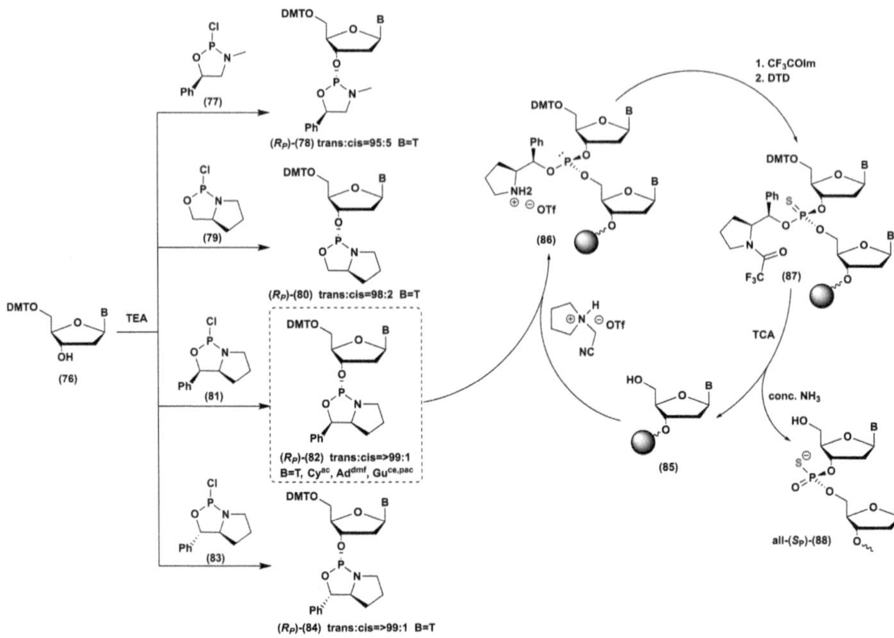

Scheme 12.14 Synthesis of stereopure 3′-*O*-oxazaphospholidine monomers and their application to solid-phase synthesis of stereoregular PS-ODNs **88**.

and S_P monomers according to the stereochemistry of the proline-derived bicyclic oxazaphospholidine rings.

Wada and co-workers later demonstrated that when using a *trans* bicyclic derivative bearing a phenyl group at the 5-position, exclusively the *trans* isomer of the corresponding thymidine was obtained, being stable in the presence of 1-(cyanomethyl)pyrrolidinium trifluoromethanesulfonate (CMPT) as activating agent,[44] and this was confirmed also for the other nucleobases. The thymidine monomer was first applied in solution for the synthesis of dithymidine phosphorothioate, through condensation with 3′-*O*-protected thymidine in the presence of CMPT. The stereopure dithymidine phosphite intermediate, after sulfurization, was converted in the corresponding fully protected dithymidine phosphorothioate triester, which gave, after PS linkage deprotection removing the chiral auxiliary, the 3′,5′-*O*-protected dithymidine phosphorothioate. The stereopurity of the compound was verified also after acidic 3′,5′-*O*-detritylation of the dimer, and the methodology proved equally efficient also in the synthesis of other (R_P)- and (S_P)-dinucleotides, with yields of >96% and diastereoselectivity ≥99 : 1. Applying the procedure to the synthesis of longer PS-oligonucleotides through the solid-phase approach, excellent results were confirmed for 8–12-mer PS-ODNs, obtained with overall yields in the range 12–34% and single coupling yields of >95%. Reversed-phase HPLC (RP-HPLC) allowed the isolation of the diastereoisomers, the diastereopurity of which was confirmed by enzymatic digestion. This approach was more recently confirmed to be compatible also with modifications at the C-5 position of uracil, these substituents being located close to the sulfur atom.[45]

Similarly, Radzikowska and Baraniak reported in 2015 the synthesis of chimera PO/PS-ODNs using commercially available 3′-*O*-(2-cyanoethyl-*N*,*N*-diisopropylamino)phosphoramidites for the insertion of phosphate units and diastereopure nucleoside 2-thio-1,3,2-oxathiaphospholane monomers for phosphorothioate nucleotides.[46] Phosphodiester segments were assembled through classical phosphoramidite chemistry, and phosphorothioate linkages were generated using oxathiaphospholane monomers. They were able to synthesize all-R_P and all-S_P 9-mer sequences, controlling the unwanted oxidation of phosphorothioates by using *tert*-butylperoxytrimethylsilane as oxidizing agent.

The protocol developed by Wada and co-workers[42] still represents the methodology of choice for the synthesis of stereopure PS-ODNs through P(III) chemistry, with only small modifications of the activating agent[36] and oxidation/sulfurization conditions. The strength of this approach was confirmed also by its application to the stereocontrolled synthesis of phosphorothioate oligoribonucleosides (PS-ORNs), maintaining the successful outcome in SPOS. This aspect is particularly important considering the structure of ASOs in clinical trials or commercial introduction, which includes substitutions at the 2′-position of the ribose scaffold. In particular, Wada and co-workers also reported the synthesis of PS-ORNs using diastereopure 2′-*O*-(2-cyanoethoxymethyl) (CEM)-protected nucleoside[47] and 2′-*O*-*tert*-butyldimethylsilyl (TBMDS)-protected nucleoside[48] 3′-*O*-(1,3,2-oxazaphospholidine) monomers (see Scheme 12.15). With the latter monomers, they were able to synthesize all-(R_P)- and all-(S_P)-PS-U_{10} with excellent results in terms of yield and stereoselectivity, but when trying to synthesize PS sequences with mixed bases the coupling efficacy was found to be lower owing to steric hindrance of the TBDMS protecting group. Therefore, by adopting 2′-*O*-CEM-protected nucleosides they were able to overcome this issue.

Similarly, Li *et al.* reported the enantiopure synthesis of 2′-*O*-(2-methoxyethyl)ribonucleotides bearing bicyclic oxazaphospholidines and their application in the SPOS of 12-mer oligonucleotides.[49] The possibility of stereoselectively synthesizing these modified scaffolds bearing a MOE group is particularly interesting, being included in the approved drugs mipomersen, nusinersen, and others in clinical trials. In particular, they synthesized the stereochemically pure PS-MOE lateral segments of mipomersen, but they did not observe differences in mRNA silencing between R_P, S_P, and mixed versions of the drug as a consequence of the poor influence of these parts of the oligonucleotide sequence on the RNase activity, as also demonstrated by Wan *et al.*, who reported complementary synthesizing sequences with stereopure gap regions and stereorandom PS wings.[50]

The most advanced application of these techniques to the stereoselective synthesis of a commercialized ASO was reported by Iwamoto *et al.* in 2017.[51] Applying 3′-oxazaphospholidine for the stereocontrolled synthesis of

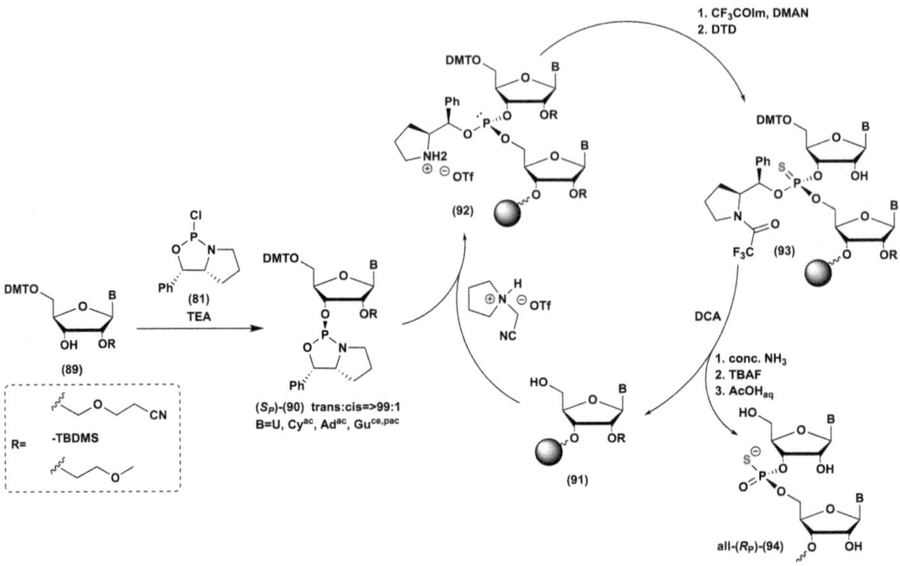

Scheme 12.15 Synthesis of stereopure 2′-*O*-CEM, 2′-*O*-TBDMS, and 2′-MOE oxazaphospholidine monomers and their application to the solid-phase synthesis of stereoregular PS-ORNs **94**.

mipomersen, they explored a different solid-phase protocol based on the use of iterative capping and sulfurization, which in classic SPOS are performed once the sequence elongation is completed. The approach is generally known as SOSICS, standing for stereocontrolled oligonucleotide synthesis with iterative capping and sulfurization. With this protocol, together with sulfurization conducted using *S*-cyanoethyl methylthiosulfonate, spontaneous release of the capped chiral auxiliary occurs. The removal of the *S*-cyanoethyl protecting group on phosphorothioate occurs during cleavage of the sequence from the solid support and removal of base-protecting groups. Following this approach and applying the principles of rational design to define the stereochemically pure sequences to be synthesized, they synthesized seven stereopure sequences of mipomersen. In addition to complete S_P and R_P sequences, they also explored the effect of uniform stereochemistry of the 2′-MOE wings and on the central 10-mer sequence. They demonstrated the possibility of obtaining stereopure sequences with high stereochemical control and chemical purity, and proved that (S_P)-PS linkages are more stable than (R_P)-PS linkages, preventing pharmacological inactivation of mipomersen. In addition, it was also demonstrated that the presence of 3′-S_P,S_P,R_P motifs in the central 10-mer sequence of the drug is able to promote RNase cleavage *in vitro* and the durability of the response in mice *in vivo*.[51]

Starting from these excellent results on Wada and co-workers' auxiliary, Sakamuri *et al.* in 2020 studied the extension of the plethora of possible chiral bicyclic auxiliaries, introducing a phenol or cyclohexane ring on the pyrrolidine side of Agrawal and co-workers' chiral auxiliary, resulting in dihydroindole (DHI) and octahydroindole (OHI) chiral auxiliaries, respectively, obtained through a considerably shorter and cheaper synthetic route.[52] The efficacy of this new protocol was explored with the synthesis of a model 10-mer PO–RNA with a single PS linkage and 2′-substituted nucleosides, and this study revealed the greater versatility of 2′-substitution of chiral DHI amidites, with CMPT as activator. Despite the efficacy even when applied to 2′-deoxynucleosides or in the presence of nucleobase substitutions, the auxiliary deprotection with 30% aqueous ammonia solution was incomplete and needs further exploration, in addition to the possibility of introducing multiple PS linkages in the same sequence.

12.4 Chemical Stereocontrolled Synthesis of Internucleotide Linkages Based on P(v)-based Chemistry

All the chemistry described so far in this chapter was strongly founded on the chemistry of P(III). More recently, greater innovations in the field of stereoselective synthesis of PS-oligonucleotides have come from Baran's group, with collaboration with the research laboratory of Bristol-Myers Squibb. They reported a new P(v)-based series of reagents for the diastereoselective incorporation of PS linkages in both ASOs and cyclic dinucleotides (CDNs),[53] and in 2018 they proposed the Ψ reagent for phosphorus–sulfur incorporation (see Scheme 12.16).[54] It is derived from a limonene backbone anchored to a P(v)-oxathiaphospholane sulfide auxiliary and a perfluorinated thiophenol

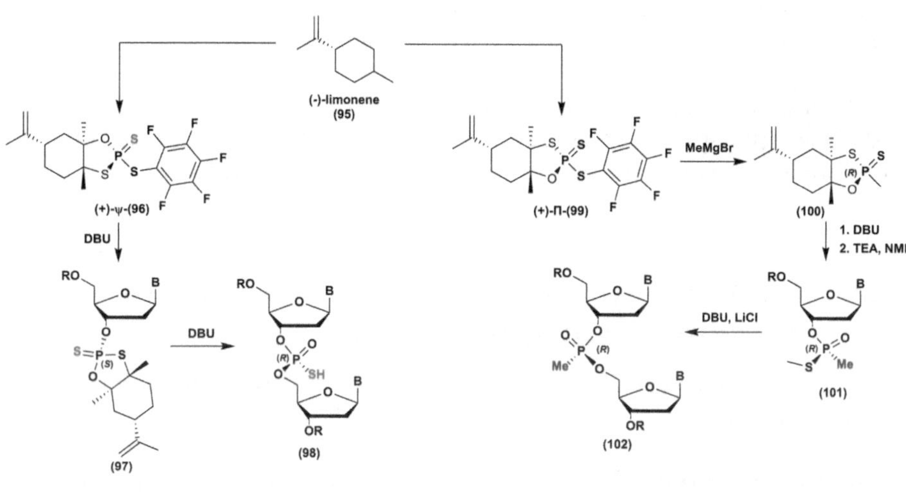

Scheme 12.16 Synthesis of Ψ and Π reagents **96** and **99**, respectively, and application in the synthesis of phosphorothioate and methylphosphonate linkages.

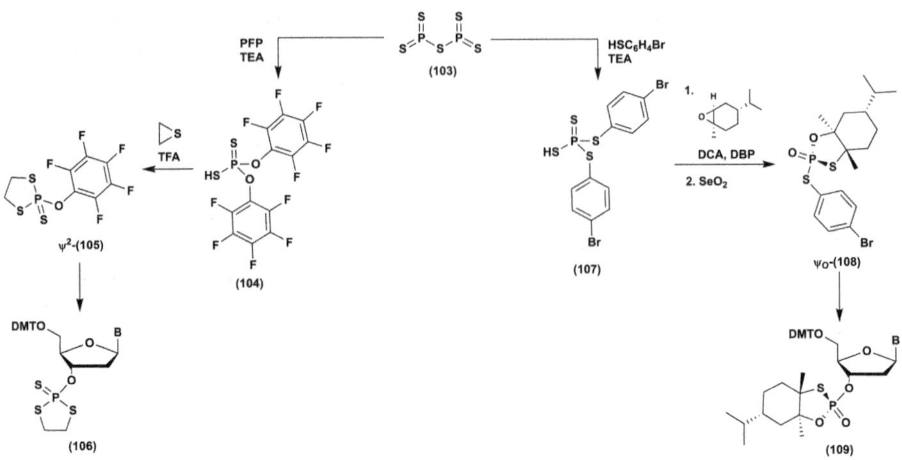

Scheme 12.17 Synthesis of Ψ^2 and Ψ_O reagents **105** and **108**.

as leaving group. After efficient 3′-functionalization of nucleosides with both (+)-Ψ and (−)-Ψ in the presence of DBU as activating agent, under the same conditions they first synthesized dinucleotides as a combination of the four native deoxynucleosides, proving the efficacy of the protocol on the multi-gram scale. The stereochemical outcome of the reactions led to (*S*)-loaded nucleotides when starting from (+)-Ψ, and consequently to *R*-coupled products. In addition to the synthesis of dimeric CDNs, the authors also explored the application of these new chiral monomers to SPOS, synthesizing the thymidine pentamer in 23% yield and with a >99 : 1 diastereomeric ratio (dr). By introducing small structural modifications of Ψ reagent, they were also able to synthesize the new Π reagent for the synthesis of stereopure methylphosphonate linkages (see Scheme 12.16).[55] Applied to the synthesis of dimers, the approach successfully obtained a >20 : 1 dr. Subsequently, they proposed new Ψ^2 and Ψ_O reagents for the incorporation of phosphorodithioates (PS_2) and classic phosphodiesters, respectively, and demonstrated their efficacy along with Ψ reagent for the stereocontrolled incorporation of PS, PS_2, and PO linkages in 16-mer sequences (see Scheme 12.17).[56]

12.5 Catalytic Approaches for the Synthesis of Stereopure Oligonucleotides

In 2021, Featherston *et al.* demonstrated for the first time the possibility of exploiting chiral phosphoric acids (CPAs) as catalysts for the stereocontrolled formation of chiral phosphorus in PS linkages, presenting a new possible approach that prevents the use of temporary chiral auxiliaries, thus enhancing the atom economy of the process (see Scheme 12.18).[57] In this context, the chiral phosphoric acid replaces the classic 1*H*-tetrazole or its derivatives employed for coupling P(III) phosphoramidite nucleosides, but can

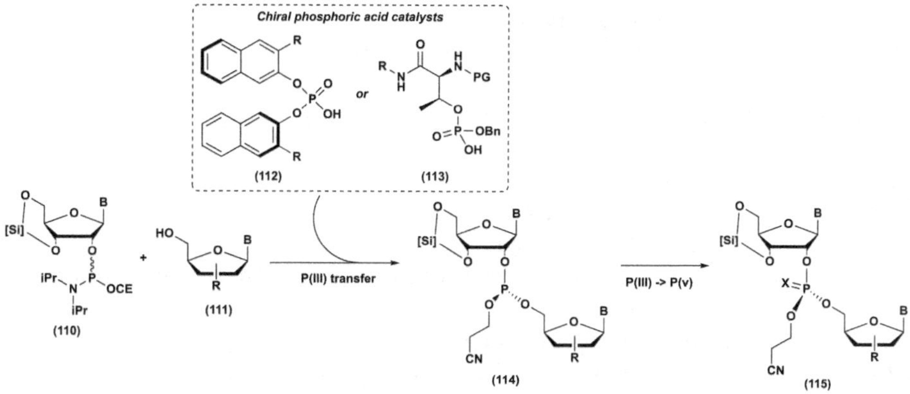

Scheme 12.18 Asymmetric oligonucleotide synthesis mediated by chiral phosphoric acid catalyst **112** or **113**.

be efficiently employed in catalytic amounts. The CPAs explored belonged to two families of catalysts, namely BINOL type and Fmoc-*allo*-pThr type (BINOL, 1,1′-bi-2-naphthol; Fmoc, fluorenylmethoxycarbonyl; pThr, phosphothreonine). In the latter type, the phosphoric acid was anchored to the β-oxygen of threonine, while the carboxylic acid was functionalized with a sequence of ^{D}Pro–Aib–Dap (Aib, 2-aminoisobutyric acid; Dap, diaminopropionic acid). In particular, Dap was introduced to capture further non-covalent interactions occurring between the substrate and the catalyst. The two types of catalysts allowed the selective synthesis of (S_P)- or (R_P)-2′,5′-dimers when choosing the pThr or BINOL type, respectively. However, this approach is still limited to the synthesis of CDNs and to date has not been extended to longer oligonucleotide sequences.

12.6 Conclusion

P-Chiral oligonucleotides with stereoregulation have found applications in biological and therapeutic research, with potential utility in synthesizing oligonucleotide conjugates and commercial drugs. To encourage further exploration of these applications, it is essential to establish user-friendly and environmentally sensitive methods for producing these oligonucleotide analogues, ideally incorporating automated processes. Challenges remain in adopting green synthetic conditions, particularly in SPOS for stereocontrolled approaches, requiring focused efforts to overcome these limitations.

List of Abbreviations

A, Ade	Adenosine
Ac	Acetyl
AcOH	Acetic acid

Aib	2-Aminoisobutyric acid
ASO	Antisense synthetic oligonucleotide
ATP	Adenosine triphosphate
ATPαS	Adenosine 5′-*O*-(1-thiotriphosphate)
BINOL	1,1′-Bi-2-naphthol
C, Cyd	Cytidine
CDN	Cyclic dinucleotide
CEM	Cyanoethoxymethyl
CMPT	1-(Cyanomethyl)pyrrolidinium trifluoromethanesulfonate
CPA	Chiral phosphoric acid
Dap	Diaminopropionic acid
DBU	1,5-Diazabicycloundec-7-ene
DCC	Dicyclohexylcarbodiimide
DHI	Dihydroindole
DMT, DMTr	Dimethoxytrityl
Fmoc	Fluorenylmethoxycarbonyl
G, Gua	Guanidine
HPLC	High-performance liquid chromatography
MMT	Monomethoxytrityl
MOE	Methoxyethoxy
MPO	Methylphosphonate
MSNT	1-Mesitylenesulfonyl-3-nitro-1,2,4-triazole
MSPy	Mesitylenesulfonyl-5-(pyridin-2-yl)tetrazole
$n\mathrm{BuNH_2}$	*n*-Butylamine
NMI	*N*-Methylimidazole
ODN	Oligodeoxynucleotide
OHI	Octahydroindole
ORN	Oligoribonucleotide
PN	Phosphoramidite
PO	Phosphodiester
Pro	Proline
PS	Phosphorothioate
PS_2	Phosphorodithioate
PS-ASO	Phosphorothioate-based antisense synthetic oligonucleotide
PSO	Phosphorothioate oligonucleotide
pThr	Phosphothreonine
RP-HPLC	Reversed-phase high-performance liquid chromatography
SOSICS	Stereocontrolled oligonucleotide synthesis with iterative capping and sulfurization
SPOS	Solid-phase oligonucleotide synthesis
T, Thy	Thymidine
TBDMS	*tert*-Butyldimethylsilyl
TBDPS	*tert*-Butyldiphenylsilyl
TEA	Triethylamine
TFA	Trifluoroacetic acid
TMG	1,1,3,3-Tetramethylguanidine

TsCl	Tosyl chloride
U	Uracil
UTP	Uridine triphosphate

Acknowledgements

The author thanks the PNRR project "National Center for Gene Therapy and Drugs Based on RNA Technology" – Spoke 6 – RNA Chemistry – Task 6.2.1 "Development of New RNA-targeting Compounds RIBOTAC and Stereopure Oligonucleotides and Peptides Conjugates" for financial support.

References

1. W. J. Stec and A. Wilk, *Angew. Chem., Int. Ed. Engl.*, 1994, **33**, 709.
2. A. Wilk and W. Stec, *Nucleic Acids Res.*, 1995, **23**, 530.
3. W. B. Wan, M. T. Migawa, G. Vasquez, H. M. Murray, J. G. Nichols, H. Gaus, A. Berdeja, S. Lee, C. E. Hart, W. F. Lima, E. E. Swayze and P. P. Seth, *Nucleic Acids Res.*, 2014, **42**, 13456.
4. M. A. Islam, A. Fujisaka, J. Kawakami, T. Yamaguchi and S. Obika, *Bioorg. Med. Chem. Lett.*, 2020, **30**, 127166.
5. S. Agrawal and J. J. Lisziewicz, *Biotechnol. Healthcare*, 1994, **7**, 167.
6. F. F. Weichold, J. J. Lisziewicz, R. A. Zeman, L. S. Nerurkar, S. Agrawal, M. S. Reitz Jr. and R. C. Gallo, *AIDS Res. Hum. Retroviruses*, 1995, **77**, 863.
7. L. Ferrazzano, D. Corbisiero, A. Tolomelli and W. Cabri, *Green Chem.*, 2023, **25**, 1217.
8. B. I. Andrews, F. D. Antia, S. B. Brueggemeier, L. J. Diorazio, S. G. Koenig, M. E. Kopach, H. Lee, M. Olbrich and A. L. Watson, *J. Org. Chem.*, 2021, **86**, 49.
9. F. Eckstein, V. W. Armstrong and H. Sternbach, *Proc. Natl. Acad. Sci. U. S. A.*, 1976, **73**, 2987.
10. F. Eckstein, *Annu. Rev. Biochem.*, 1985, **54**, 576.
11. E. Ohtsuka, M. Shiraishi and M. Ikeahara, *Tetrahedron*, 1985, **41**, 5271.
12. T. Wada, N. Kobayashi, T. Mori and M. Sekine, *Nucleosides Nucleotides*, 1998, **17**, 351.
13. (a) M. Sobkowski, *New J. Chem.*, 2010, **34**, 854. (b) H. Almer, J. Stawinski and R. Stromberg, *J. Chem. Soc., Chem. Commun.*, 1994, 1459.
14. K. Seio, K. Kumura, J.-C. Bologna and M. Sekine, *J. Org. Chem.*, 2003, **68**, 3849.
15. W. J. Stec, G. Zon, W. Egan and B. Stec, *J. Am. Chem. Soc.*, 1984, **106**, 6077.
16. J. K. Frederiksen and J. A. Piccirilli, *Methods Enzymol.*, 2009, **468**, 289.
17. W. J. Stec and G. Zon, *Tetrahedron Lett.*, 1984, **25**, 5275.
18. A. Wilk, A. Graikowski, L. R. Phillips and S. L. Beaucage, *J. Am. Chem. Soc.*, 2000, **122**, 2149.
19. R. P. Iyer, W. Egan, J. B. Regan and S. L. Beaucage, *J. Am. Chem. Soc.*, 1990, **112**, 1253.

20. R. Cosstick and D. M. Williams, *Nucleic Acids Res.*, 1987, **15**, 9921.
21. Z. J. Lesnikowski and M. M. Jaworska, *Tetrahedron Lett.*, 1989, **30**, 3821.
22. Z. J. Lesnikowski, *Nucleosides Nucleotides*, 1992, **11**, 1621.
23. Z. J. Lesnikowski, M. Jaworska and W. J. Stec, *Nucleic Acids Res.*, 1988, **16**, 11675.
24. M. M. Jaworska-Maslanka, W. Kacperczyk, D. Kroczynski and Z. J. Lesnikowski, *Antisense Nucleic Acid Drug Dev.*, 1997, **7**, 23.
25. W. J. Stec, A. Grajkowski, A. Kobylańksa, B. Karwowski, M. Koziolkiewicz, K. Misiura, A. Okruszek, A. Wilk, P. Guga and M. Boczkowska, *J. Am. Chem. Soc.*, 1995, **117**, 12019.
26. W. J. Stec, B. Karwowski, M. Boczkowska, P. Guga, M. Koziolkiewicz, M. Sochachi, M. W. Wieczorek and J. Blaszczyk, *J. Am. Chem. Soc.*, 1998, **120**, 7156.
27. A. Sierzchala, A. Okruszeck and W. J. Stec, *J. Org. Chem.*, 1996, **61**, 6713.
28. M. Fujii, K. Ozaki, A. Kume, M. Sekine and T. Hata, *Tetrahedron Lett.*, 1986, **26**, 935.
29. M. Fujii, K. Ozaki, M. Sekine and T. Hata, *Tetrahedron Lett.*, 1987, **43**, 3395.
30. L. A. Wozniak, M. Góra, M. Bukowiecka-Matusiak, S. Mourgues, G. Pratviel, B. Meunier and W. J. Stec, *Eur. J. Org. Chem.*, 2005, 2924.
31. B. Nawrot, B. Rebowska, K. Cieślińkska and W. J. Stec, *Tetrahedron Lett.*, 2005, **46**, 6641.
32. L. A. Wozniak, M. Janika and M. Bukowiecka-Matusiak, *Eur. J. Org. Chem.*, 2005, 5189.
33. L. A. Wozniak, M. Janika and M. Bukowiecka-Matusiak, *J. Org. Chem.*, 2005, **690**, 2658.
34. R. P. Iyer, D. Yu, T. Devlin, N.-H. Ho and S. Agrawal, *J. Org. Chem.*, 1995, **60**, 5388.
35. R. P. Iyer, D. Yu, N.-H. Ho, W. Tan and S. Agrawal, *Tetrahedron: Asymmetry*, 1995, **6**, 1051.
36. Y. Jin, G. Biancotto and G. Just, *Tetrahedron Lett.*, 1996, **37**, 973.
37. E. Marsault and G. Just, *Tetrahedron Lett.*, 1996, **37**, 977.
38. J.-C. Wang and G. Just, *Tetrahedron Lett.*, 1997, **38**, 705.
39. J.-C. Wang and G. Just, *Tetrahedron Lett.*, 1997, **38**, 3797.
40. R. P. Iyer, M.-J. Guo, D. Yu and S. Agrawal, *Tetrahedron Lett.*, 1998, **39**, 2491.
41. M. Guo, D. Yu, R. P. Iyer and S. Agrawal, *Bioorg. Med. Chem. Lett.*, 1998, **8**, 2539.
42. N. Oka, M. Yamamoto, T. Sato and T. Wada, *J. Am. Chem. Soc.*, 2008, **130**, 16031.
43. N. Oka, T. Wada and K. Saigo, *J. Am. Chem. Soc.*, 2003, **125**, 8307.
44. Y. Nukaga, N. Oka and T. Wada, *J. Org. Chem.*, 2016, **81**, 2753.
45. R. I. Hara, R. Yoshino, Y. Nukaga, Y. Maeda, K. Sato and T. Wada, *RSC Adv.*, 2020, **10**, 34006.
46. E. Radzikowska and J. Baraniak, *Org. Biomol. Chem.*, 2015, **13**, 269.
47. Y. Nukaga, K. Yamada, T. Ogata, N. Oka and T. Wada, *J. Org. Chem.*, 2012, **77**, 7913.

48. N. Oka, T. Kondo, S. Fujiwara, Y. Maizuru and T. Wada, *Org. Lett.*, 2009, **11**, 967.
49. M. Li, H. L. Lightfoot, F. Halloy, A. L. Malinowska, C. Berk, A. Behera, D. Schumperli and J. Hall, *Chem. Commun.*, 2017, **53**, 541.
50. W. B. Wan, M. T. Migawa, G. Vasquez, H. M. Murray, J. G. Nichols, H. Gaus, A. Berdeja, S. Lee, C. E. Hart, W. F. Lima, E. E. Swayze and P. P. Seth, *Nucleic Acids Res.*, 2014, **42**, 13456.
51. N. Iwamoto, D. C. D. Butler, N. Svrzikapa, S. Mohapatra, I. Zlatev, D. W. Y. Sah, Meena, S. M. Standley, G. Lu, L. H. Apponi, M. Frank-Kamenetsky, J. J. Zhang, C. Vargeese and G. L. Verdine, *Nat. Biotechnol.*, 2017, **35**, 845.
52. S. Sakamuri, D. Liu, L. Eltepu, B. Liu, L. J. Reboton, R. Preston and C. W. Bradshaw, *ChemBioChem*, 2020, **21**, 1298.
53. K. W. Knouse, D. T. Flood, J. C. Vantourout, M. A. Schmidt, I. M. Mcdonald, M. D. Eastgate and P. S. Baran, *ACS Cent. Sci.*, 2021, **7**, 1473.
54. K. W. Knouse, J. N. deGruyter, M. A. Schmidt, B. Zeng, J. C. Vantourout, C. Kingston, S. E. Mercer, I. M. Mcdonald, R. E. Olson, Y. Zhu, C. Hang, J. Zhu, C. Yuan, Q. Wang, P. Park, M. D. Eastgate and P. S. Baran, *Science*, 2018, **361**, 1234.
55. D. Xu, N. Rivas-Bascon, N. M. Padial, K. W. Knouse, B. Zheng, J. C. Vantourout, M. A. Schmidt, M. D. Eastgate and P. S. Baran, *J. Am. Chem. Soc.*, 2020, **142**, 5785.
56. Y. Huang, K. W. Knouse, S. Qiu, W. Hao, N. M. Padial, J. C. Vantourout, B. Zheng, S. E. Mercer, J. Lopez-Ogalla, R. Narayan, R. E. Olson, D. G. Blackmond, M. D. Eastgate, M. A. Schmidt, I. M. Mcdonald and P. S. Baran, *Science*, 2021, **373**, 1265.
57. A. L. Featherston, Y. Kwon, M. M. Pompeo, O. D. Engl, D. K. Leahy and S. J. Miller, *Science*, 2021, **371**, 702.

CHAPTER 13

Impact of Impurities on the Quality and Safety of Oligonucleotides as Drug Substances

J. KAUFMANN AND T. M. RUPP*

Axolabs Berlin GmbH, Wolfener Strasse 23, 12681 Berlin, Germany
*E-mail: thomas.rupp@lgcgroup.com

13.1 Introduction

This chapter covers chemistry, manufacturing, and controls (CMC) aspects based on the quality vision of synthetically manufactured oligonucleotides as active pharmaceutical ingredients (APIs) in RNA therapeutics and related control strategies with respect to process- and product-related impurities. Biopharmaceuticals, *i.e.* biologically manufactured RNA therapeutics, such as mRNA or DNA vaccines, are not discussed in this chapter.

To date, 26 single- or double-stranded synthetic RNA therapeutics have gained approval by regulatory authorities as medications for the treatment of previously incurable clinical indications. The manufacturing process of oligonucleotides by the sequential addition of nucleoside phosphoramidites to a solid support derivatized either with the 3′-terminal nucleoside of the desired sequence or with a universal linker molecule has remained largely the same since the last major innovative changes in the early 1980s.[1] The

Green Chemistry Series No. 85
Sustainability in TIDES Chemistry: Green Approaches to Oligonucleotides and Oligopeptides Synthesis
Edited by Alessandra Tolomelli, Lucia Ferrazzano and Walter Cabri

Published by the Royal Society of Chemistry, www.rsc.org

historic development of the manufacturing process resulting in today's state-of-the-art automated solid-phase synthesis process dates back to 1955, when Michelson and Todd reported the first chemical dinucleotide synthesis using the phosphotriester approach,[2] followed by Khorana and co-workers in 1956 using a phosphodiester method.[3,4] In the following years, Khorana and co-workers further refined the phosphodiester method by the introduction of transient protecting groups allowing for directed sequential synthesis of both DNA and RNA polynucleotide sequences.[5–7] Following the introduction of solid supports in peptide synthesis by Merrifield,[8,9] Letsinger and Mahadevan developed support-bound phosphotriester synthesis on a non-polar polystyrene-based support resin in 1965.[10,11] The introduction of a novel nucleoside-derivatized polymeric solid support suitable for use in polar reaction media in 1977 by Gait and Sheppard brought the advantages of reproducibility, shorter reaction times, and easy reagent washing.[12] Shortly thereafter, Beaucage and Caruthers published the initial phosphoramidite method that afforded syntheses in high yields by applying highly reactive P(III) phosphoramidite building blocks with good chemical stability, allowing for convenient storage and handling.[13] Further refinement of the phosphate protecting groups led to the β-cyanoethyl phosphoramidite chemistry established today as the standard method for both laboratory-scale and large-scale oligonucleotide synthesis.[1]

The general synthesis cycle builds the oligonucleotide sequentially from the 3′-end towards the 5′-end and consists of four consecutive chemical reaction steps. Starting with detritylation of the solid support, and consecutively of the 5′-end of the growing oligonucleotide chain, a free 5′-hydroxy group is left behind. The subsequent incoming phosphoramidite monomer is protonated at the 3′-diisopropylamino group by an appropriate activator. The highly reactive intermediate salt that is formed is subsequently coupled to the 5′-hydroxy group by nucleophilic substitution.[14–20] Typically, the stepwise coupling efficiency achieves yields between 98.5 and 99.8%, depending on the monomer being added. The remaining uncoupled 5′-hydroxy groups are acetylated ("capped") in the following capping step. Finally, the coupled monomer needs to be oxidized from the acid-unstable P(III) phosphite triester to the stable P(V) phosphate triester using aqueous iodine solution. Alternatively, the phosphite triester can also be oxidized by a sulfur transfer reagent. This results in one of the most prominent modifications in RNA therapeutics, the so-called phosphorothioate (PS) modification. It is introduced by oxidation of the phosphite triester using a sulfur transfer reagent by insertion of one non-bridging sulfur atom into the bridging phosphate backbone. This modification increases the systemic half-life of oligonucleotides, and is one of the earliest introduced into oligonucleotides.[21–24] Using this process, the synthesis of oligonucleotides up to ~200 nt in length is possible.[25]

After synthesis, the oligonucleotide chain is liberated from the solid support (cleavage) and the persistent protecting groups on the nucleobases and the phosphate backbone must be removed (deprotection). Both cleavage and deprotection are carried out under basic conditions. Prior to cleavage from the solid support and nucleobase deprotection, a separate deprotection step with a less nucleophilic base such as diethylamine (DEA) is suggested to cleave and scavenge the cyanoethyl group from the phosphate group and therefore avoid the formation of acrylonitrile side-products (CNETs).[26,27] For RNA, a second deprotection step using an F^- source, typically triethylamine trihydrofluoride (TEA·3HF),[28] must be performed to remove the silyl-protected 2′-OH groups.

Downstream, the crude product is purified and isolated, mostly by high-performance liquid chromatography (HPLC), followed by an ultrafiltration step for desalting. After freeze-drying (lyophilization) of the desalted oligonucleotide, the final product is typically harvested as a solid bulk material. The bulk API is then stored under controlled conditions until it is sent to a "fill and finish" facility for reconstitution in an aqueous buffer and final formulation as the drug product (see Figure 13.1). Recent approaches also omit the final freeze-drying step, yielding an API in solution, as freeze-drying is seen as a bottleneck in the manufacturing process and consumes large amounts of energy.[29]

Depending on their mechanism of action, RNA therapeutics can make use of a variety of different modifications. Despite the presence of numerous modifications and differences in the related molecular structures, all classes of synthetic RNA therapeutics share an identical manufacturing platform, resulting in the formation of similar process- and product-related impurities. This general manufacturing process was laid out in 1984 and improved and standardized in subsequent decades.[1] Given the fact that oligonucleotide

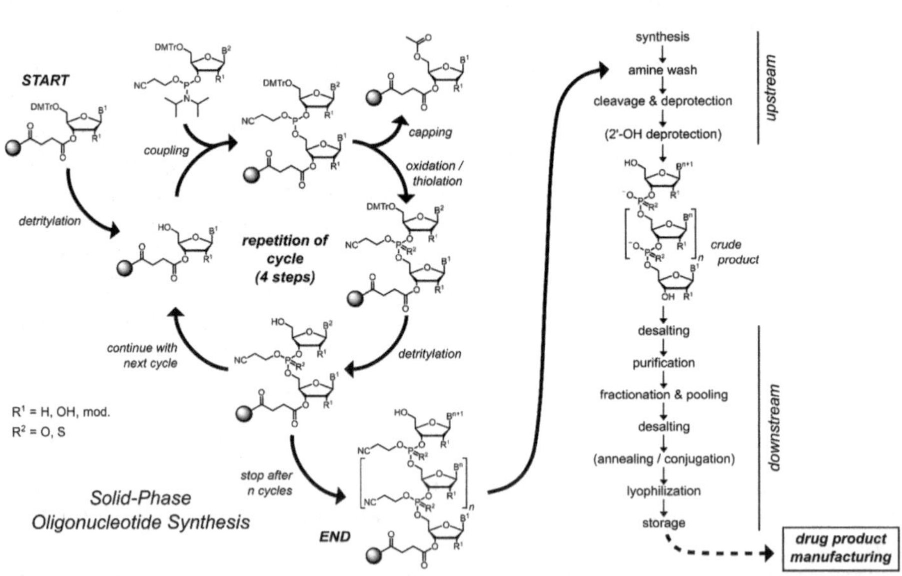

Figure 13.1 General solid-phase oligonucleotide synthesis (SPOS) cycle and further processing of the synthetic oligonucleotide.

drugs have the potential for long-lasting effects using only minimal dosages,[30] their production processes must follow strict current good manufacturing practice (cGMP) regulations, as is the case for any other small-molecule API.

The by-products and impurities that are formed therefore require good understanding and characterization. Most of the impurities are formed during the upstream synthesis and deprotection process steps and are either raw material or process related. These organic impurities can be detected and identified on a crude level by high-performance liquid chromatography-mass spectrometry (HPLC-MS) and successively traced back to the root cause. In return, the gathered knowledge is a powerful tool for controlling and minimizing impurity formation, which can then be addressed in specific process development.

In this chapter, advances in the field of oligonucleotide synthesis and the control and detection of impurities are presented together with progress that has been made in terms of regulatory guidance and related safety aspects. The interplay of all these topics has a direct impact on oligonucleotide control strategies that are outlined in the last section.

13.2 RNA Therapeutics

13.2.1 Perspectives and Challenges of Synthetic RNA Therapeutics

The journey of synthetic oligonucleotides as novel drugs against genetically originated diseases started about 50 years ago with the discovery of antisense technology in the late 1970s,[31,32] and this groundbreaking work of Stephenson and Zamecnik laid the foundation for their later introduction as therapeutic modalities acting on a molecular level. Nowadays, synthetic RNA therapeutics are a well-established class of drugs that have gained wide attention since the approval of the first RNA drug (Vitravene®) in 1998. Early developed and approved RNA therapeutics were based on the so-called antisense mechanism. Here, an antisense oligonucleotide (ASO) is designed complementary to an accessible region of a disease-related mRNA. After binding the target mRNA, the ASO either arrests translation or degrades the target mRNA through RNase H recruitment. Further research in the field led to the discovery of a number of different mechanisms of action related to other classes of therapeutic oligonucleotide moieties such as aptamers (1990),[33,34] cytosine–phosphate–guanine (CpG) motifs as vaccine adjuvants (1995),[35] transcription factor decoys (TFDs) (1995),[36,37] splice-switching oligonucleotides (SSOs) (1995),[38] small interfering RNA (siRNA) (1998),[39] anti-miRNA oligonucleotides (AMOs) (2004),[40–42] and lately single guide RNA (sgRNA) molecules as guide and scaffold moiety for clustered regularly interspaced short palindromic repeats/CRISPR-associated protein (CRISPR/Cas) applications (2012).[43] In total and to date, 26 synthetic RNA therapeutics have been approved by regulatory authorities for a variety of diseases. Of the 26 approved molecules, three were designed as so-called (*n*-of-1) therapeutics for single-digit patient populations, and one is a single guide RNA, Casgevy™, which has been granted conditional market authorization for the treatment of two autologous *ex vivo* CRISPR/Cas gene therapies: sickle cell disease (SCD) and transfusion-dependent β-thalassemia (TDT).[44] A summary of the mechanisms of action of all currently approved RNA therapeutics is given in Table 13.1. Their drug names are summarized in Table 13.2 together with the respective years of approval.

The rapid progress in identifying new mechanisms of action and their immediate application in new therapeutic approaches was promoted by the perspectives that arose with this new class of drugs. In contrast to small-molecule drugs, synthetic RNA therapeutics allow for the independent control of the dianophore and the pharmacophore, meaning that their distribution and delivery can be adjusted independently of their efficiency in target knockdown.[45] In conventional small-molecule drugs, the structure of the entire small molecule determines both parameters. In oligonucleotide drugs, the backbone is mainly responsible for the distribution and delivery whereas the base sequence itself in combination with high-affinity sugar modifications mostly influences target recognition and binding. Only about 15% of the proteins encoded in the human genome carry clefts and pockets for small-molecule binding and are expected to be druggable with conventional drugs.[46] RNA therapeutics have been proven to enhance this therapeutic window since they intervene directly at the genome level. By this nature, RNA therapeutics also have the opportunity to become extremely efficient designer drugs against hereditary diseases. The steadily increasing in-depth knowledge of the human genome and the associated understanding of potential genetic origins of given diseases enable the rapid and highly specific development of oligonucleotide drugs. Among other potential uses, this makes oligonucleotides ideal candidates for orphan disease treatment, giving new hope to patients suffering from diseases that would otherwise remain untreated. Pioneering approaches were the development of individualized antisense therapies such as *n*-of-1 or *n*-of-few drugs, *e.g.* atipeksen, milasen, and valeriasen.[47–50]

In addition to the positive perspectives listed above, there are open challenges to RNA therapeutics in general, which still remain not fully addressed. Except for a few modifications such as phosphorodiamidate morpholino oligomers (PMOs) and peptide nucleic acids (PNAs), RNA therapeutics are negatively charged molecules and therefore highly hydrophilic. Without appropriate modification, they exhibit poor protein binding capacities, which results in unfavorable pharmacokinetics and fast urinary excretion.[51–55] RNA therapeutics are based on naturally occurring molecules, so they are recognized by endogenous nucleases. Degradation in the body occurs primarily by enzymatic hydrolysis of the phosphodiester (PO) backbone starting from either the 5′- or 3′-end (exonucleases) or at internal positions (endonucleases). Therefore, RNA therapeutics require chemical stabilization to prevent fast recognition and resulting degradation by these enzymes.[56] This is mainly achieved by thiolation of the phosphate backbone or by sugar modification.[57,58]

Table 13.1 Mechanisms of action and chemical properties of RNA therapeutics.

Mechanism of action	No. of approved molecules	Typical length (nucleotides)	Typical sugar chemistry	Backbone chemistry	Structure	Conjugation
Anti-miRNA	0	17–27	LNA, 2′-*O*-MOE, 2′-F, 2′-OMe	PS	Single strand	–
Antisense (translational arrest, RNase H)	6	18–20	2′-OMe/DNA, 2′-*O*-MOE/DNA	PS/PO mixed	Single strand	–/+
Antisense (splice switching)	8	18–24	PMO, 2′-OMe, 2′-*O*-MOE	PS	Single strand	–
Aptamer	2	20–90	DNA, RNA, 2′-F, 2′-OMe	PO	Structured single strand	–/+
CpG motif	2	22–30	DNA	PS	Single strand	–
DNAzyme	0	33–35	DNA	PO	Structured single strand	–
sgRNA	1	100–120	RNA, 2-OMe	PO, PS	Single strand	–
siRNA	6	19–23	RNA, 2′-OMe, 2′-F	PO, PO with individual PS	Duplex	–/+
siRNA	0	50–60	RNA, 2′-OMe, 2′-F	PO, PO with individual PS	Folded single strand	–/+
Transcription factor decoy	0	22–50	DNA, LNA	PO, PS	Structured single strand, duplex DNA	–

Table 13.2 Overview of the currently approved RNA therapeutics.

Trade name	Other name	Year approved	Type
Vitravene®	Fomivirsen	1997	Antisense oligonucleotides (translational arrest, RNase H)
Kynamro®	Mipomersen	2013	
Tegsedi®	Inotersen	2018	
Waylivra®	Volanesorsen	2019	
Wainua™	Elontersen	2023	
Qalsody®	Tofersen	2023	
Exondys 51®	Eteplirsen	2016	Antisense oligonucleotides (splice switching)
Spinraza®	Nusinersen	2016	
(*n*-of-1)	Milasen	2017	
Vyondys 53®	Golodirsen	2019	
Viltepso™	Viltolarsen	2020	
(*n*-of-1)	Atipeksen	2020	
(*n*-of-1)	Valeriasen	2020	
Amondys 45®	Casimersen	2021	
Macugen®	Pegaptanib	2004	Aptamers
Izervay™	Avacincaptad pegol	2023	
Heplisav-B®	ISS-1018	2017	CpG motif (immune stimulatory)
Cyfendus™	AV-7909	2023	
Casgevy™	Exagamglogene autotemcel (exa-cel)	2023	sgRNA (CRISPR/Cas9)
Onpattro®	Patisiran	2018	siRNA
Givlaari®	Givosiran	2019	
Oxlumo®	Lumasiran	2020	
Leqvio®	Inclisiran	2021	
Amvuttra®	Vutisiran	2022	
Rivfloza™	Nedosiran	2023	
Rytelo™	Imetelstat	2024	Antisense oligonucleotide (telomerase inhibitor)

However, RNA therapeutics can utilize these endogenous systems as, for example, in the case of RNase H-recruiting antisense gapmers to downregulate genes by enzymatic degradation of the target mRNA.[30,45,55,59,60] Another challenge is the immunogenic potential of some specific sequence motifs such as CpG motifs.[35] On the other hand, this immunostimulatory mechanism can purposefully be activated to enhance the body's reaction to the applied drug as in the case of Heplisav-B® and Cyfendus™.

Finally, and in addition to the challenges with application, the challenges in production should also be pointed out. Two-thirds of the currently approved RNA therapeutics entered the market within the last 6 years. With this tremendous increase in approved therapeutics and hundreds more in clinical trials, the global need for large-scale good manufacturing practice (GMP) production sites has dramatically gained momentum. Current solid-phase flow-through synthesizers have a maximum capacity of up to 2 mol per batch (around 8–10 kg of oligonucleotide). However, the upscaling of micromole-scale or even low millimole-scale high-throughput manufacturing processes is not trivial. Small-scale processes are mostly designed as universal manufacturing processes for rapid access to all different classes of oligonucleotides, independent of reagent consumption, product yield, or product quality. They typically use large volumes of wash solvents and a high number of equivalents of critical process components, such as phosphoramidite monomers. Yield and purity are not critical for small-scale approaches, whereas large-scale manufacturing processes are designed as low-throughput procedures with quality of product and cost of goods in mind. Such processes need to deliver products reproducibly in high yield and purity at reasonable cost and are optimized for quality and reagent consumption using scale-down modeling approaches during process development. Ideally, the process development should start in-house with a focus on critical process parameters such as linear flow rates, column bed heights, and contact times, keeping the large-scale equipment and its capacities in mind. During scale-up, the overall solvent consumption becomes one of the greatest concerns for the manufacture of RNA therapeutics. Compared with small molecules, the process mass intensity (PMI) to produce oligonucleotides is about 20 times higher.[61] As an example, about 30 000 L of acetonitrile and 11 000 L of deblock solution [dichloroacetic acid (DCA) in toluene] are needed to produce 10 kg of oligonucleotide. Currently, the search for alternative, more sustainable synthesis methods and manufacturing processes to reduce solvent consumption is ongoing and of top priority. Several methods, such as liquid-phase synthesis and enzymatic approaches, have been reported recently, but so far have not been applied to production-scale processes.[62–66]

Despite all the challenges, RNA therapeutics are more popular than ever. Their unique properties make them promising candidates to target not only rare diseases but also diseases that have been considered undruggable before. With hundreds more in clinical trials, global production capacities will have to expand further and the innovations in the field will continue.

13.2.2 Recent Trends in Modification and Conjugation

In modern RNA therapeutics, the number of modifications continues to increase, which directly impacts the type of impurities formed during synthesis or deprotection. The most common modification is the PS backbone, where one non-bridging oxygen is replaced with a sulfur atom. Owing to the larger size of the sulfur atom and its higher polarizability, the PS modification leads to an increase in lipophilicity, which results in better plasma protein binding and better pharmacokinetics.[51–55] Additionally, the nuclease stability is drastically increased by the PS modification.[57,58] Early generations of ASOs were based on DNA stabilized with PS linkages. Vitravene® was the first ASO drug approved by the US Food and Drug Administration (FDA), back in 1998, having a backbone completely made of PS–DNA.[30,59] The following generations of RNA therapeutics combined the PS backbone with sugar modifications, most prominently bridged analogs such as a bridged nucleic acid (BNA) or locked nucleic acid (LNA),[67,68] in addition to 2′-OMe,[69] 2′-*O*-MOE (methoxyethyl),[70] and 2′-F modifications.[30,45,71] Depending on the molecule and its mechanism, the PS content can vary from just a few linkages to a fully thiolated backbone. Whereas the latest generation of RNase H-recruiting ASOs are designed as gapmers with 2′-*O*-MOE wings and a DNA center, splice-modulating ASOs are designed as either full PMOs or full 2′-*O*-MOE phosphorothioates.[30,72] In PMOs, the sugar is replaced with a morpholine ring. A phosphorodiamidate group then connects two morpholino units.[73] Owing to this phosphorodiamidate group, PMOs are neutrally charged molecules and therefore exhibit properties that are different from those of PO or PS oligonucleotides. They show a higher binding affinity for target RNA and also good resistance to nucleases.[73,74] The named modifications (see Figure 13.2) and their use in different oligonucleotide designs have been reviewed by various authors[30,45,59,75,76] and therefore will not be discussed further here. Despite the vast number of chemical modifications to either sugar, base, or backbone, all of them – with the exception of PMOs and PNAs – share the same synthesis platform, hence similar impurity profiles can be expected.

In addition to the sugar and backbone modifications, RNA therapeutics can be conjugated to a variety of small molecules to fine-tune their properties, usually targeting, uptake, delivery, or biodistribution.[77] This is done by the attachment of either lipophilic molecules such as cholesterol or tocopherol or of receptor-targeting molecules such as *N*-acetylgalactosamine (GalNAc). The trivalent GalNAc moiety is one of the most prominent conjugates mainly used in siRNAs. Initially introduced by Alnylam Pharmaceuticals, their presence strongly enhances uptake and targeted delivery.[78] Conjugation can be carried out in a post-synthetic step at a

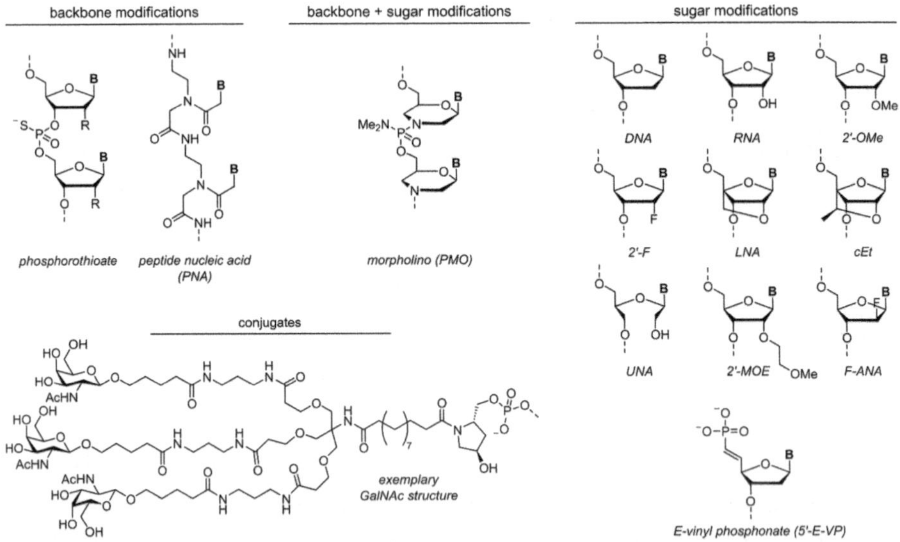

Figure 13.2 Overview of exemplary oligonucleotide modifications. Modifications can be introduced in the backbone, at the sugar moiety, or as conjugates. B = nucleobase.

reactive site previously introduced during solid-phase synthesis. Alternatively, conjugates can be directly incorporated *via* the corresponding phosphoramidites or solid supports. Encapsulation in lipid nanoparticles (LNPs) is another uptake-enhancing strategy to permit more efficient delivery and has been applied to the first approved siRNA therapeutic, Onpattro®.[79,80]

13.3 Impurities – Definitions, Origin, and Categorization

The International Council on Harmonisation of Technical Requirements for Registration of Pharmaceuticals for Human Use (ICH) has issued two complementary quality guidelines, ICH Q3A (R2) *Impurities in New Drug Substances*[81] and ICH Q3B (R2) *Impurities in New Drug Products*,[82] to provide guidance on the classification of impurities in new drug substances produced by chemical syntheses and new drug products produced from chemically synthesized new drug substances, respectively. Even though synthetic oligonucleotides are explicitly excluded from both guidelines, the basic concepts of both are expected to be applied when entering lead candidates into the clinical pipeline. In this chapter, organic impurities are further subdivided into process-related and product-related impurities. In analogy with ICH Q6B *Specifications: Test Procedures and Acceptance Criteria for Biotechnological/Biological Products*,[83] impurities arising during the manufacturing process are referred to as "process-related". "Product-related" impurities, on the other hand, are those formed upon degradation during storage. In some cases, the same structural impurity can either be process-related or product-related.

As ICH Q3B addresses only impurities in drug products classified as degradation products of the drug substance or reaction products of the drug substance with an excipient and/or immediate container closure system,[82] the focus of this section will be ICH Q3A, which suggests the classification of non-genotoxic impurities in drug substances into the following three main categories:[81]

a. Organic impurities, both process-, and product-related

"*Organic impurities [typically] arise during the manufacturing process and/ or during storage of the new drug substance. They can be [known and] identified or [unknown and] unidentified, volatile, or non-volatile [impurities], and include:*

- *Starting materials*
- *By-products*
- *Intermediates*
- *Degradation products*
- *Reagents, ligands, and catalysts.*"[81]

b. Inorganic impurities

"Inorganic impurities can result from the manufacturing process. They are normally known and identified and include:

- *[Inorganic] reagents, ligands, and catalysts*
- *Heavy metals or other residual metals*
- *Inorganic salts*
- *Other materials (e.g. filter aids, charcoal).*"[81]

c. Residual solvents

"Solvents are inorganic or organic liquids used as vehicles for the preparation of solutions or suspensions in the synthesis of a new drug substance. Since these are generally of known toxicity, the selection of appropriate controls is easily accomplished [following ICH Guideline Q3C]."[81]

Each of these categories is described in more detail in the following sections: Section 13.3.1 details sources of inorganic, elemental impurities, Section 13.3.2 describes residual solvent origins, and Section 13.3.3 gives more insight into the actual formation and classification of organic impurities of RNA therapeutics.

13.3.1 Sources of Inorganic Impurities

According to the definition provided in quality guideline ICH Q3D *Guideline for Elemental Impurities*,[84] inorganic elemental impurities are traces of metals that can be found in finished drug products, with the main potential sources being the production process of the drug product.

The risk assessment performed during drug substance manufacture should address the potential for inclusion of elemental impurities in the drug product and therefore elemental impurities should already be controlled at a drug substance level. Considering the manufacturing process for drug products, there are broad categories of potential sources of elemental impurities:

- *"Residual impurities resulting from elements intentionally added (e.g. catalysts) in the formation of the drug substance, excipients, or other drug product components. The risk assessment of the drug substance should address the potential for inclusion of elemental impurities in the drug product."*[84]

This is typically not a high risk in oligonucleotide manufacture, as no catalysts are being used. Nevertheless, elemental impurities could be introduced if molecular sieves are being used to dry solvents.

- *"Elemental impurities that are not intentionally added and are potentially present in the drug substance, water or excipients used in the preparation of the drug product."*[84]

To mitigate this risk, oligonucleotide drug substances should be tested for elemental impurities. Water and excipients used in the manufacture of the drug product should be of the highest quality possible.

- *"Elemental impurities that are potentially introduced into the drug substance and/or drug product from manufacturing equipment."*[84]

A list of wetted part material of construction of the equipment shall be part of the risk assessment.

- *"Elemental impurities that have the potential to be leached into the drug substance and drug product from container closure systems."*[84]

Container closure systems should come with the respective certificates to mitigate the risk.

Each of these sources may contribute elemental impurities to the drug product, through any individual or any combination of the potential sources listed above. During the risk assessment, the potential contributions from each of these sources should be considered to determine the overall contribution of elemental impurities to the drug product. Table 13.3 gives an overview of elements to be considered during the risk assessment.

Since oligonucleotide APIs typically are described as sodium salts, sodium must not be seen as an elemental impurity. However, the amount should be controlled and therefore is typically determined by inductively coupled plasma mass spectrometry (ICP-MS),[85] or other appropriate quantitative analytical methods and presented in the release Certificate of Analysis (CoA). The theoretical sodium content can be calculated and should be the base for the specification, as the amount of sodium counter ion has an impact on the pH of the molecule. Additionally, the assay of the oligonucleotide API must be corrected for the measured sodium content.

13.3.2 Sources of Residual Solvent Impurities

Focusing USP 467 *Residual Solvents* and ICH Q3C *Guideline for Residual Solvents* on synthetic oligonucleotides only,[86,87] residual solvents can be defined as organic volatile chemicals that are used or produced as by-products in the manufacture of drug substances, or excipients, or in the preparation of drug products. Solvents used during the synthesis process of the drug substance may not be completely removed by the downstream purification and isolation processes and might therefore reside with the final product. As residual solvents generally do not provide any therapeutic benefit, they should be removed to the greatest extent possible, and controlled against the safety-based limits or other quality-based requirements.

Strictly following ICH Q3C, the level of residual solvents needs to be specified, controlled, and limited on a drug product level.[87] The control and

Table 13.3 ICH Q3D – Elements to be considered during risk assessment. Reproduced from ref. 84.

Element	Class	If intentionally added (all routes)	If not intentionally added		
			Oral	Parenteral	Inhalation
Cd	1	Yes	Yes	Yes	Yes
Pb	1	Yes	Yes	Yes	Yes
As	1	Yes	Yes	Yes	Yes
Hg	1	Yes	Yes	Yes	Yes
Co	2A	Yes	Yes	Yes	Yes
V	2A	Yes	Yes	Yes	Yes
Ni	2A	Yes	Yes	Yes	Yes
Tl	2B	Yes	No	No	No
Au	2B	Yes	No	No	No
Pd	2B	Yes	No	No	No
Ir	2B	Yes	No	No	No
Os	2B	Yes	No	No	No
Rh	2B	Yes	No	No	No
Ru	2B	Yes	No	No	No
Se	2B	Yes	No	No	No
Ag	2B	Yes	No	No	No
Pt	2B	Yes	No	No	No
Li	3	Yes	No	Yes	Yes
Sb	3	Yes	No	Yes	Yes
Ba	3	Yes	No	No	Yes
Mo	3	Yes	No	No	Yes
Cu	3	Yes	No	Yes	Yes
Sn	3	Yes	No	No	Yes
Cr	3	Yes	No	No	Yes

limitation of residual solvents nevertheless should already happen on a drug substance level according to a risk-based approach, especially because solvents of class 2 (for classification see Table 13.4) are essential during the oligonucleotide drug substance manufacturing process. According to the guideline, class 1 solvents should not be used for the manufacture of pharmaceuticals unless their use is unavoidable in order to produce an official product with a significant therapeutic advance. For the analysis of residual solvents, typically headspace gas chromatography (HS-GC) is used. Solvents used in the manufacturing process of oligonucleotides and their classification are additionally given in Table 13.5. Some reagents used during oligonucleotide synthesis are not classified in Q3C, *e.g.* acetic anhydride, 2,6-lutidine, and *N*-methylimidazole. Some suggest the classification of 2,6-lutidine similar to that of pyridine due to their structural similarity (see also Chapter 14 by M. Lemaitre). Since purge factors during the entire process are typically high, residual reagents and reactive or mutagenic by-products such as acrylonitrile are efficiently cleared by the process itself. Usually, no additional controls of the drug substance are required.[88] This is in agreement with multidisciplinary ICH guideline M7 *Guideline on Assessment and Control of DNA Reactive (Mutagenic) Impurities in Pharmaceuticals to Limit Potential Carcinogenic Risk*.[89]

Table 13.4 ICH Q3C – Classification of residual solvents. Reproduced from ref. 87.

Residual solvent class	Comment	Assessment
Class 1	Solvents to be avoided	Known or strongly suspected human carcinogens Solvents known to have ozone-depleting properties
Class 2	Solvents to be limited	Non-genotoxic animal carcinogens or possible causative agents of other irreversible toxicity, such as neurotoxicity or teratogenicity Solvents suspected of other but reversible toxicities
Class 3	Solvents with low toxic potential; no health-based exposure limit is needed	Solvents with low toxic potential to humans

Table 13.5 USP 467 – Exemplary solvents used in the oligonucleotide manufacturing process and their classification and limits. Data from ref. 86.

Class	Solvent	Process step of origin	PDE/mg per day	Concentration limit/ppm
2	Acetonitrile	Main solvent during synthesis, purification buffers, conjugations	4.1	410
2	*N*,*N*-Dimethylformamide (DMF)	Co-solvent for amidites	8.8	880
2	Pyridine	Capping, oxidizer/thiolation	2.0	200
2	Tetrahydrofuran (THF)	Co-solvent for amidites, conjugations	7.2	720
2	Toluene	Solvent during detritylation	8.9	890
3	Acetic acid	Purification buffers, detritylation	3200	5000[a]
3	2-Butanol	Precipitation	300	5000[a]
3	Dimethyl sulfoxide (DMSO)	Cleavage and deprotection	50	5000[a]
3	Ethanol	Cleavage and deprotection, precipitation	166.7	5000[a]
3	2-Propanol (IPA)	Instrument cleaning	138	5000[a]

[a]Amounts of Class 3 solvents of up to 50 mg per day (corresponding to 5000 ppm or 0.5% assuming a daily administration of the drug product of 10 g) are generally acceptable without justification.

As oligonucleotides are hygroscopic molecules by nature, water is not seen as an impurity, but rather as the reflection of its current environment. The water content may not be seen as an indicator of quality or stability of the oligonucleotide drug substance, but it should be analyzed by Karl Fischer (KF) titration and the result presented in the release CoA. Additionally, the assay of the oligonucleotide API must be corrected for the water content.

13.3.3 Sources of Organic Impurities – Formation During Manufacturing and Classification

According to the definition provided in quality guideline ICH Q3A, organic impurities can arise during the manufacturing process (process-related) and/or during storage (product-related) of the new drug substance.[81]

Process-related impurities are generated mainly during the synthesis and deprotection of oligonucleotides. As the average stepwise coupling efficiency achieves yields of 98.5–99.8% per synthesis cycle, overall impurity levels increase with increase in oligonucleotide length. One of the most critical steps during oligonucleotide manufacture is the deprotection step since it is typically conducted under harsh conditions, *i.e.* basic conditions with strong nucleophiles for an extended period and at elevated temperature. As described previously, a similar manufacturing process is used for all classes of oligonucleotide drug substances. Nevertheless, minor differences in reagent compositions, equivalents, length of reactions, flow rates, and other critical process parameters are subject to individual choices and will have a direct impact on the observed impurity profile. These impurities include shortmers, longmers, incomplete removal of protecting groups, base or sugar modifications, and impurities generated as a result of acid or base treatment during synthesis. Since the SPOS protocol and its individual reaction steps are well characterized, impurities can be traced back to their specific origin (raw material, synthesis steps, or cleavage and deprotection) by HPLC-MS analysis of the crude product. Degradation of the oligonucleotides during the downstream processing or during storage will generate new product-related impurities, which are also mainly major metabolites or degradants as well as new impurities.

All identified organic impurities should be classified according to the Oligonucleotide Safety Working Group (OSWG) White Paper (see below)[90] and need to be controlled through a well-defined control strategy that supports scalability of the developed process. Even though quality guideline ICH Q11 *Development and Manufacture of Drug Substances*[91] is applicable only to drug substances as defined in ICH Q6A and therefore excludes oligonucleotides, the spirit of the guideline nevertheless shall be applied. The process development program performed during process upscaling should identify which material attributes (*e.g.* of raw materials, starting materials, reagents, solvents, process aids, intermediates) and process parameters should be controlled to reduce impurities. It provides further clarification of the principles and concepts described in quality guidelines ICH Q8 *Pharmaceutical Development*, ICH Q9 *Quality Risk Management*, and ICH Q10 *Pharmaceutical Quality Systems* as they pertain to the development and manufacture of drug substances.[92–94]

From a manufacturing perspective, it is useful to further specify impurities formed during SPOS as either raw material-related or process-related. Raw material-related impurities can arise from either the phosphoramidites, conjugates, linkers used in conjugation, base-loaded solid supports, or ancillary synthesis reagents. Among the raw materials used, special attention needs to be paid to starting materials. Starting materials according to ICH Q11 are considered those incorporated as a significant structural fragment into the drug substance.[91] This mainly affects phosphoramidites and base-loaded solid supports. Small-molecule conjugate precursors and linkers are also considered significant structural fragments of the drug substance and therefore need to be controlled in a similar manner. Nucleoside-loaded solid supports represent the starting point in SPOS, and the structure of the loaded support will be incorporated as the 3′-end of the final oligonucleotide. The same applies for phosphoramidite monomers. The sugar and base will be fully incorporated into the oligonucleotide, in addition to parts of the amidite group. Universal supports, on the other hand, are used as the starting point for synthesis, but they are entirely cleaved during the cleavage and deprotection step so that no structural element is introduced into the oligonucleotide.

Detritylation and capping do not add any structural element to the growing oligonucleotide chain. Also, capping groups are removed during the later deprotection step. It is debatable if oxidation or thiolation does add a significant structural element to the overall structure of the drug substance as either step adds only a single atom per bridging phosphate to the oligomer. Furthermore, the oxygen atom added during oxidation does not originate from the oxidizer itself. During thiolation, the sulfur is transferred from the thiolation reagent to the oligonucleotide phosphate backbone and introduces a new stereocenter. Even though the two diastereomers show different properties regarding nuclease resistance, binding affinity, and RNase H recruitment,[95–97] there is so far no clear opinion on the impact of the R_P/S_P distribution of the sulfur atom on safety and efficacy.[98] The authors suggest that neither the oxidizer nor thiolation reagents should be considered as starting materials.

Focusing on phosphoramidite starting materials, their impurities can be subdivided into three classes: (1) non-reactive and non-critical, (2) reactive but non-critical, or (3) reactive and critical.[99] The first class includes impurities that do not interfere with oligonucleotide synthesis and cannot be incorporated, *e.g.* non-phosphorylated nucleosides or oxidized species. Class 2 includes, for example, amidites with modified protecting groups that can still be cleaved during deprotection and will not be present in the final product. This section further discusses class 3 phosphoramidite impurities. Among all starting material impurities, these are especially critical since they are incorporated repeatedly into the oligonucleotide. By repeated incorporation, small percentages of impurity in the small molecule easily sum up to reach high impurity levels in the crude product. Rupp and Cramer gave the example of a 0.05% critical and reactive impurity in a phosphoramidite that would reach a level of 0.25% in the crude product when incorporated five times into a 20-mer oligonucleotide. Similarly, an only slightly higher percentage of impurity in the starting material of 0.25% would lead to a 1.25% impurity level in the crude product.[59] Exemplary class 3 impurities in raw materials are anomeric forms of the phosphoramidites, 5′-phosphoramidites with a 3′-(4,4′-dimethoxytrityl) group (3′-DMTr group), and phosphoramidite building blocks with either a 2′,5′-linkage (RNA monomers), missing or modified cyanoethyl groups, or missing protecting groups at the base. A more detailed explanation of the different phosphoramidite impurity classes was given by Cramer *et al.*[99] Also noteworthy, a White Paper addressing the quality attributes of starting materials is currently in preparation by Lemaitre and others.

Critical raw material impurities have also been reported for reagents other than phosphoramidites. Examples are chloral impurities in dichloroacetic acid and 1,3,5-trimethylhexahydro-1,3,5-triazine levels in 1-methyl-1*H*-imidazole in capping solutions, both of which can lead to the formation of critical impurities in the final oligonucleotide product.[99–101]

Although adjustment of the process can help reduce the levels of process-related impurities, it cannot prevent the formation of raw material-related impurities since their presence is strictly dependent on their levels in the starting materials.[88] It is therefore especially important to control the level of potential impurities in starting materials and to work exclusively with the highest quality commercially available to keep impurity levels as low as possible.

The following text gives a brief overview of process-related impurities that potentially can be formed by either over-reaction, under-reaction, or side reactions. Each impurity arises during a specific process step and therefore the process parameters have an overall impact on the impurity profile of the crude product. This means that a profound knowledge of the manufacturing process and adequate control of individual process steps directly impact and improve product quality and yield.

Looking from a CMC perspective, process-related impurities can form in every step of the synthesis cycle if the latter is not controlled properly. For oligonucleotide manufacture, it has been shown that process-related impurities are mainly formed during the four reactions of each synthesis cycle and during deprotection. For more in-depth information, the interested reader is referred to recently published reviews dedicated to this topic.[59,99]

The initial step of the synthesis cycle, detritylation, removes the DMTr group from the 5′-end of the solid support, and consecutively the 5′-end of the growing oligonucleotide chain. It is carried out under harsh acidic conditions, typically using between 3 and 15% dichloroacetic acid or 3% trichloroacetic acid in a dry organic solvent, typically toluene.[102] An over-reaction resulting from excessive detritylation equivalents or prolonged reaction times can, for example, result in depurination. In combination with insufficient oxidation, DMTr-C-phosphonate esters can form where during the subsequent detritylation step the DMTr group is attacked by the phosphorus atom.[103] An under-reaction resulting from insufficient detritylation equivalents fails to deprotect all 5′-ends of the growing oligonucleotide chains. Coupling of the next monomer is therefore not possible, resulting in $n - x$ impurities in the crude product. Following the completion of the detritylation reaction, a free

5′-hydroxy group is left behind to allow for the next reaction in the cycle, the coupling reaction.

During the coupling step, the incoming phosphoramidite monomer requires protonation by an appropriate acidic and nucleophilic organic activator. The reaction takes place in a dry organic solvent, typically acetonitrile. Tetrazole derivatives such as 5-ethylthio-1*H*-tetrazole (ETT) or 5-benzylthio-1*H*-tetrazole (BTT) are typically used as activators.[16–18] Approaches using 4,5-dicyanoimidazole (DCI) as an alternative activator that is less acidic than tetrazole but much more nucleophilic have also been successfully described.[19] A mixture of saccharine and *N*-methylimidazole has been proposed as an activator in more sustainable synthesis approaches on an industrial scale.[20] In all cases, a reactive intermediate salt is formed that is subsequently attacked by the free 5′-hydroxy group of the preceding nucleotide.[14,15] The main impurities formed during the coupling step are $n - 1$ or $n + 1$ impurities. Since the reaction mechanism of this step involves the nucleophilic attack at the phosphorus center, any nucleophile other than the free 5′-hydroxy group of the growing oligonucleotide chain can potentially interfere. Attack of other nucleophiles, such as residual traces of water in solvents, quench the reactive phosphoramidite intermediate and render it unreactive. This would lead to the formation of $n - 1$ or $n - x$ impurities, so-called shortmers. These same impurities could also be provoked by insufficient activation of the phosphoramidite monomer, leading to an under-reaction. Since activators are not only nucleophiles but also at the same time weak acids, over-reaction might remove the 5′-DMTr group of either the incoming or the recently coupled monomer and promote double coupling, resulting in $n + 1$ or $n + x$ impurities, so-called longmers. This can especially be critical when long coupling times are applied. The $n - 1$ and $n + 1$ impurities elute close to the main peak and need to be considered during analytical development.

Incomplete coupling leads to shortmers, which are then capped at the unreacted 5′-hydroxy group in the next step during the synthesis cycle. The capping mixture typically consists of acetic anhydride and *N*-methylimidazole with pyridine or 2,6-lutidine in acetonitrile. Acetic anhydride acetylates the remaining 5′-hydroxy groups, rendering them inactive for the following cycles. The acetyl group is removed during the later deprotection step, leaving a shortmer behind. Even though the capping step is seen as an insurance policy to prevent the formation of products with internal base deletions, more impurities can form under excess capping.[104] One of them is the modification of guanosine. The formation of different O^6-modified analogs has been reported, and also solutions to prevent their formation.[105,106] Furthermore, protecting groups on guanosine can be exchanged, which results in impurity formation during the later deprotection step.[107]

Finally, the phosphotriester formed needs to be oxidized from the acid-labile P(III) to the acid-stable P(V) form by addition of an oxygen (PO backbone) or a sulfur atom (PS backbone) to the phosphorus center. Commonly, 0.05 M iodine in pyridine with small amounts of water (<8%) is used as an oxidizing agent. 3*H*-1,2-Benzodithiol-3-one 1,1-dioxide, also known as Beaucage reagent, was used as a sulfur transfer reagent in the early days.[23] Later developments in this field led to the introduction of more efficient and stable sulfur transfer reagents such as phenylacetyl disulfide (PADS), 3-ethoxy-1,2,4-dithiazoline-5-one (EDITH), 3-[(dimethylaminomethylidene)amino]-3*H*-1,2,4-dithiazole-3-thione (DDTT), and 3-amino-1,2,4-dithiazole-5-thione (xanthane hydride, XHD) that are used in solutions of acetonitrile (ACN), pyridine, or mixtures thereof.[108–112] Common impurities during thiolation include incomplete sulfur transfer resulting in a divergent number of PS linkages. Incomplete oxidation/thiolation can result in impurity formation during the detritylation step of the next cycle where the released DMTr can be attacked by the unoxidized P(III) atom and form DMTr-C-phosphonate.[103] Prolonged oxidation phases, on the other hand, can lead to the formation of overly oxidized species. Especially deoxyguanosine is prone to oxidative damage.[113,114] Also, PS–PO conversion is a commonly observed impurity in oligonucleotides with mixed PS and PO backbones. In such mixmers, sulfur replacement by oxygen most likely occurs at the 3′-end linkages since they will have repeated contact with the oxidizer solution.

Another critical, potentially impurity-generating step during the manufacture of RNA therapeutics is the deprotection step. Harsh conditions are applied using highly concentrated aqueous or ethanolic solutions of ammonium hydroxide or ammonium hydroxide–methylamine (1 : 1 v/v) (AMA). Partial deprotection of only the cyanoethyl protecting groups from the phosphate backbones with secondary amines is usually performed before cleavage from the solid support to prevent CNET adduct formation.[26,27] Since the kinetics of solid support cleavage and nucleobase deprotection are different,[115,116] the subsequent cleavage and deprotection steps can be carried out using either a combined one-step procedure or separate procedures. For large-scale manufacturing processes, the oligonucleotide is typically first cleaved from the support at slightly elevated ambient temperature, followed by transfer of the oligonucleotide-containing deprotection solution into a temperature-controlled vessel for full deprotection. This separation of cleavage from the support and deprotection avoids the accumulation of support-related impurities in the oligonucleotide, *i.e.* dissolution of elemental impurities out of the glass beads in controlled-pore glass or leaching of small-molecule impurities out of polystyrene-based solid supports. The temperature and time to accomplish full deprotection need to be chosen individually for each oligonucleotide, depending on base composition, chemistry, length, and the modifications present. Under-reaction leads to incomplete deprotection of the nucleobase, leaving protecting groups behind.[116,117] The exchange of sulfur with oxygen (PS loss) has also been reported as one of the abundant impurities that arise during deprotection over-reactions.[118,119] Additionally, transamination reactions can occur at the cytidine base, leading to base modifications.[120] Depurinated sites generated during detritylation promote strand breaks under basic conditions that finally lead to fragmentation of the oligonucleotide chain.[102,121] Another known impurity created during the deprotection step is deamination of cytosine and adenosine residues.[122,123]

Finally, the use of universal supports can be mentioned. In addition to their obvious advantages, namely minimization of the risk of human error by loading of the wrong support and cost-effectiveness regarding warehousing, universal supports have the disadvantages of less efficient coupling of the first base and steric hindrance during the first detritylation. Thus, the widely used UnyLinker™ support requires higher concentrations of DCA during detritylation and double coupling to add the first base efficiently. Additionally, owing to the slower cleavage reaction kinetics, longer reaction times or harsher conditions are required, making the UnyLinker™ support unsuitable for more base-labile chemistries. Critical UnyLinker™-related impurities originating from incomplete cleavage need to be considered when carrying out syntheses on such a solid support.[59]

Exemplary raw material-, process-, and product-related impurities are listed in Table 13.6 and their potential origins are given together with measures to prevent their formation.

Degradation is an additional pathway for creating product-related impurities in oligonucleotides. Exposure to acids, bases, light, temperature, or oxidative conditions can catalyze degradation during processing or storage.

As described in the previous sections, ICH Q3A classifies impurities into new drug substances produced by chemical synthesis [new chemical entities (NCEs)] and regulates their reporting and qualification limits (see Table 13.7).[81] Suitable analytical methods need to be chosen for establishing the quantitation limit to ensure proper detection of the respective impurities (see also Section 13.4).

Oligonucleotides are explicitly excluded from ICH Q3A, but the spirit of the guideline shall be applied when controlling oligonucleotide impurity levels.[81] They fall in a regulatory gray area since they are manufactured through chemical reactions similar to those for small-molecule drugs but are polymeric biologically active products. Because of the sequential synthetic manufacturing process, most of the impurities are structurally closely related to the product such as the $n - 1$ impurity being the most abundant impurity and the major metabolite. Owing to their close relation in structure, a similar pharmacokinetic profile and similar safety profile to those of the parent molecule can be expected. As specified in ICH Q2 *Validation of Analytical Procedures: Text and Methodology*, the analytical procedure must be validated to demonstrate its suitability for the detection and quantification of impurities.[124] Separation of oligonucleotides by chromatographic methods is more challenging compared with small molecules owing to the similar structural and physicochemical properties of the impurities and product. In addition, diastereomeric PS linkages result in peak broadening, leading to lower signal-to-noise ratios and more difficult baseline separation. In their 2017 White Paper, the OSWG therefore argued that the proposed thresholds from ICH Q3A are not transferable to oligonucleotides.[81,90] Owing to the limitations of HPLC-MS, they suggested a reporting threshold of 0.2%; 1.0% was suggested to be the minimum level to which the control of unspecified impurities is possible and therefore the identification threshold should be set accordingly.

Table 13.6 Exemplary process- and product-related organic impurities and their origins and classification.

Impurity	Description	Origin	Classification	Control strategy
− Nucleotide (terminal)	Terminal $n-x$ impurity at 5′- or 3′-end	Process (detritylation, coupling, capping, deprotection), raw material (nucleoside-loaded solid support), or storage (degradation)	I	Adjustment of process parameters; control of raw materials; stability studies
PS–PO conversion	PS backbone conversion to PO	Process (capping, deprotection, oxidation) or storage (degradation)	II	Adjustment of process parameters
2′/3′-Isomer	2′,5′-Linked sugar in RNA	Raw material	II	Impurity control in phosphoramidites
2′-*O*-Alkyl	2′-*O*-Alkyl incorporation due to impurity in 2′-*O*-MOE phosphoramidites	Raw material	II	Impurity control in phosphoramidites
Deamination	Deamination at A, C, or G base site	Process (deprotection)	III	Adjustment of deprotection parameters
+ Nucleotide	$n+x$ impurity	Process (coupling, capping)	III	Adjustment of process parameters
− Nucleotide (internal)	Internal $n-x$ impurity	Process (detritylation, coupling, capping) or raw material	III	Adjustment of process parameters or impurity control in phosphoramidites
+ DMTr	Incomplete deprotection of final 5′-OH	Process (detritylation)	IV	Adjustment of process parameters
+ CNET	Addition of acrylonitrile at nucleobase site	Process (deprotection)	IV	Introduction of amine wash before deprotection
+ Chloral	Adduct formation due to chloral impurity in DCA	Raw material	IV	Impurity control in DCA
+ Protecting group	Insufficient deprotection of 2′-OH or protecting group	Process (deprotection)	IV	Adjustment of deprotection parameters
Depurination	Depurination at purine base site	Process (detritylation)	IV	Adjustment of process parameters
+ UnyLinker™	Incomplete cleavage from UnyLinker™	Process (deprotection)	IV	Adjustment of deprotection parameters

Table 13.7 Comparison of suggested impurity thresholds for RNA therapeutics in comparison with the existing guideline for small molecules, ICH Q3A. Reproduced from ref. 81.

Guideline	Maximum daily dose/g	Reporting threshold/%	Identification threshold	Qualification threshold
ICH Q3A (NCE) (oligonucleotides excluded)[81]	≤2	0.05	0.10% or 1.0 mg per day intake (whichever is lower)	0.15% or 1.0 mg per day intake (whichever is lower)
	>2	0.03	0.05%	0.05%
FDA Peptide Guideline 2003 (withdrawn)	—	>0.2	>0.5%	>1.0%
Ph. Eur. Monograph 2034 (synthetic peptides)[125]	—	>0.1	>0.5%	>1.0%
Recommended by authors and others[90]	—	>0.2	>1.0%	>1.5%

The qualification threshold was then suggested to be 1.5% since this agrees with small-molecule impurities where the qualification threshold is set 50% higher than the identification threshold. The thresholds for organic impurities in peptide drugs have already been adjusted in Ph. Eur. Monograph 2034.[125] They are significantly higher than those given in ICH Q3A for small molecules and therefore follow the same trend as the suggestions made for oligonucleotides (see Table 13.7).

Furthermore, the OSWG proposed the categorization of oligonucleotide impurities into four distinct classes depending on their type and occurrence in natural nucleic acids (see Table 13.8).[90] Class I–III impurities are considered non-critical. Impurities of Class I comprise metabolites of the parent molecule. They share the same structure and sequence and are typically of product-related origin. Examples are truncated forms of the parent oligonucleotides with nucleotides missing at either end or loss of conjugate or linker.

Class II impurities include naturally occurring structural elements but still represent a modified parent compound. They can be either raw material- or process-related. Phosphodiester impurities formed by PS/PO conversion are typical Class II impurities. Another example is 2′,5′-linkages in RNA or substitution of 2′-*O*-MOE with 2′-OMe due to impurities in the raw material.

Sequence variants that do not show any other structural modification are considered Class III impurities. They are all process related and include internal $(n-1)$ and $(n+1)$ variants in addition to substitutions of bases such as deaminated species.

Class IV impurities are considered critical and include all structural elements that neither are found in the parent compound nor are naturally occurring. Unidentified impurities are automatically assigned to Class IV. They can either be raw material-related, for example reactive and critical amidite

Table 13.8 Classification of impurities as suggested by the OSWG. Data from ref. 90.

Impurity class	Examples	Safety assessment suggested?	Data type
Class I Impurities that are also major metabolites (structure and sequence are the same as for the parent)	• Impurities that lack single or multiple nucleotides from the 3′- or 5′-end of the parent oligonucleotide • Impurities formed by incomplete conjugation of the (parent) conjugated oligonucleotides • Parent single-stranded impurity of double-stranded oligonucleotides	No	N/A
Class II Impurities that contain only structural elements found in naturally occurring nucleic acids	• Phosphate diester impurity of phosphorothioate oligonucleotides • (2′,5′)-Linked sugar in RNA • 5-Methylcytosine degradation to cytosine	No	N/A
Class III Impurities that are sequence variants of the parent molecule	• Nucleotide removal, $(n-1)$, missing a single internal base • Nucleotide addition, $(n+1)$, one additional base • Nucleotide substitution, deaminated impurities, *i.e.* (T → C)	No	N/A
Class IV Impurities that contain structural elements not found in the parent oligonucleotide or in naturally occurring nucleic acids	• Base modifications • Backbone modifications • Depurinated species • Unidentified impurities	Yes	Non-clinical safety studies

impurities, reagent-related, or process-related. Examples of Class IV impurities are chloral and CNET adducts, residual protecting groups, branchmers, depurinated species, and UnyLinker™-related impurities, to mention only a few.

The suggested classification of exemplary impurities is provided together with their origins and possible control strategies in Table 13.6. Qualification of the impurities formed should be considered according to their class. Since Classes I–III are considered non-critical, safety assessment is not considered necessary. Class I impurities include major metabolites of the parent compound. These impurities are therefore covered in routine toxicological studies and considered qualified.[90] This is also in agreement with ICH Q3A.[81] Class II impurities are expected to have low toxicity since they contain only naturally occurring structural elements and they do not differ in sequence. Class III impurities are sequence variants of the parent compound. They might theoretically be able to induce off-target effects or show signs of immunostimulation. However, their concentration is considered too low to induce these effects. Therefore, additional safety assessment is not suggested. Class IV impurities, on the other hand, include critical impurities that contain unnatural structural elements and unidentified impurities. If their specification limit is set above the qualification threshold, their safety should be evaluated in safety studies.[90]

ICH M7 also lists *N*-nitroso compounds in their "cohort of concern". However, the risk of nitrosamine impurities in RNA therapeutics is overall considered low unless they are specifically introduced as small-molecule impurities during the process. If a high risk, *i.e.* through a specific modification or conjugate, is identified, additional risk assessment may be required.[126]

13.4 Analytics of RNA Therapeutics

As described in the previous sections, the SPOS process has mostly remained unchanged since 1984.[1] However, the number of modifications used in today's RNA therapeutics has increased drastically.[30,45,59,75,76] Together with this increase in modifications came an increase in complexity of the molecules that had a direct impact on oligonucleotide analysis.

ICH Q11 suggests approaches to support the development and manufacture of drug substances.[91] As part of the development, steps designed to decrease impurities are mentioned. The guideline focuses on the understanding and knowledge of the process and the resulting impurities carried over to the final substance as critical quality attributes (CQAs). Control elements such as in-process controls (IPCs) should be implemented to ensure that CQAs are kept within their defined limits. In addition to critical process parameters (CPPs), the control strategy should cover the entire manufacturing process including raw materials, starting materials, reagents, solvents, process aids, and intermediates. Establishing appropriate controls for impurities is one critical aspect following this strategy. Even though ICH Q11 only applies to the drug substance, its principles should be implemented early in development and put into practice throughout the entire manufacturing process, ensuring that the CQA and the overall quality target product profile (QTPP) of the drug product as defined according to ICH Q8 are maintained.[92] In the case of an siRNA this means the implementation of at least three control points, starting with analysis of the starting material, the purified single strands, and finally the double-stranded drug substance.

The quality of the starting material is crucial, especially for long oligonucleotides with multiple couplings per amidite. Impurities can otherwise easily accumulate during synthesis and reach critical levels. To prevent the formation of raw material-related impurities, incoming goods should be tested for purity and the criticality of observed impurities should be evaluated. All raw materials except for solid supports are small molecules and can easily be analyzed by gas chromatography (GC), nuclear magnetic resonance (NMR) spectroscopy, and HPLC, either alone or directly coupled to MS. For amidites, ^{31}P NMR spectroscopy is a powerful complementary analytical method to detect P(III) impurities and show non-UV active but reactive impurities in phosphoramidites, such as those resulting from phosphitylating agents. The use of HPLC-MS is the easiest way to confirm phosphoramidite identity. Also, low moisture levels are essential for good coupling. KF titration is routinely used to check the moisture content in incoming phosphoramidites. An exemplary overview of parameters tested for starting materials used during oligonucleotide synthesis is given in Table 13.9.

HPLC has been the main separation method for oligonucleotides for decades and remains the best choice to date to test and assure drug substance quality. While the separation principle of anion-exchange (AEX) chromatography makes use of charges in the analyte, reversed-phase (RP) chromatography relies on the hydrophobic interactions between the molecule and the stationary phase. By adding ion-pairing (IP) reagents to the RP buffers, interaction between the IP reagent and the oligonucleotide backbone leads to a charge-based separation character.[127–129] Both techniques have their own advantages and should be used in an orthogonal fashion to allow a thorough characterization of the drug substance-formed impurities. Whereas AEX chromatography offers advantages in the resolution of the phosphodiester impurity in fully phosphorothiolated oligonucleotides, (IP-)RP-HPLC,

Table 13.9 Starting materials and suggestions for their analysis.

Raw material	Parameter	Analytical procedure	Comment
Phosphoramidites (might also be applicable to conjugates)	Water content	KF titration	Water decreases the coupling efficiency and must be strictly excluded
	Purity	HPLC, GC, ^{31}P NMR spectroscopy	Class 3 (reactive and critical) phosphoramidite impurities need to be kept to the minimum
	Identity	UV/vis spectroscopy, LC-MS, GC-MS, ^{1}H NMR spectroscopy	Identity of the monomer is critical
Base-loaded solid support	Loading	UV/vis spectroscopy after detritylation	Parameter is crucial for the synthesis planning
	Purity	HPLC, GC after cleavage	See above
	Identity	UV/vis spectroscopy, LC-MS, GC-MS, ^{1}H NMR spectroscopy after cleavage	See above
	Residual OH groups	Coupling experiments without preceding detritylation	Residual OH groups could act as synthesis starting points and potentially lead to 3′-($n-1$) impurities

due to the added hydrophobic interactions, is able to better resolve small structural changes to the nucleobases. These impurities include depurinated species, incompletely removed protecting groups or adducts such as CNET, or species formed due to excessive capping.[26,27,107,130] Both methods should ideally be used in an orthogonal manner to maximize the information on impurities and therefore also process knowledge. Recent developments have led to the introduction of hydrophobic interaction chromatography (HIC) or hydrophilic interaction liquid chromatography (HILIC) for the analysis of oligonucleotides.[131–136] However, these methods have not yet been widely implemented.

The combination of optimized (IP-)RP methods coupled with high-resolution MS analysis reveals insights into the synthesis process and helps identify critical process parameters that impact the purity of the final product. The implementation of powerful analytics in early stages of product development is therefore an important step to gain knowledge on the influence of specific synthesis parameters on drug substance quality and further optimize the process for large-scale production (see Figure 13.3).

In addition to the purity analysis of the individual single strands, siRNA requires additional analysis of the duplex formed, as according to the definition in the OSWG publication an excess of one of the strands must be considered an impurity.[90] Conventional HPLC methods need to be adjusted

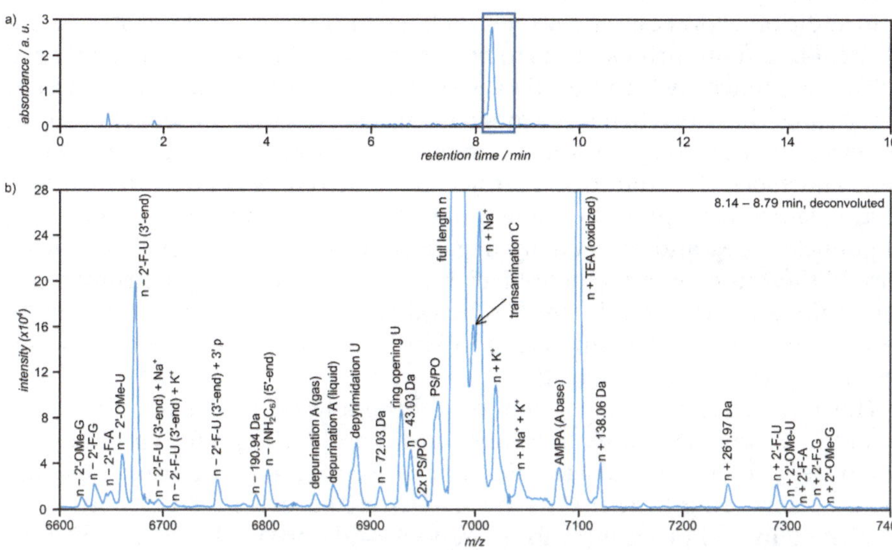

Figure 13.3 Exemplary impurity characterization by HPLC-MS. The method should be optimized to allow fully resolved impurity characterization and give insights into the process for further optimization. (a) The ultra high performance liquid chromatography (UHPLC) trace is shown with the integrated window used for MS analysis. (b) MS analysis for the depicted integration window. Identified impurities are annotated.

since they often run under denaturing conditions using high temperatures or chaotropic reagents that tend to disrupt duplexes. Milder conditions at low temperatures and neutral pH are typically used in non-denaturing methods. Size-exclusion chromatography (SEC) is an excellent tool for this specific analytical problem since it is typically run under non-denaturing conditions and both strands differ sufficiently in size to give good separation.

All analytical HPLC methods should be developed to allow for a complete picture of the occurring impurities. The insights gained into the process by impurity analysis can in return be used during process development to control impurity formation. Good analytical development is therefore essential.

HPLC and MS methods are routinely used for purity and identity analysis, but they do not provide information on the oligonucleotide sequence. MS sequencing is one possibility to confirm the specific sequence, but even with today's modern analytical equipment it is still time consuming and expensive and requires special training. Thermodynamic profiling, *i.e.* melting temperature determination, is increasing in popularity and is especially useful for the characterization of siRNA duplexes or secondary structure-forming RNA therapeutics as aptamers or transcription factor decoys. In addition to environmental factors such as the buffer type and its concentration,[137] the melting temperature of the duplex is sensitive to changes in the sequence such as permutations or base deletions.[138,139] Melting temperature determinations can therefore be considered as a complementary method to HPLC and MS analyses with which changes to the sequence are hardly identifiable.

In addition to process- and product-related impurities, the level of microbiological-based impurities needs to be kept to the minimum in the final drug substance. Since they can easily accumulate during aqueous downstream processing, they need to be controlled from early stages onwards. This is best achieved by a thorough risk assessment of the downstream process. Pyrogens and especially endotoxins cannot easily be removed after product harvesting since they are in the same molecular weight range as 20–30 nt long oligonucleotides and are equally negatively charged.[140–142] Endotoxin tests should therefore be implemented after sanitization of every instrument, or the sanitization step would need to be validated. Convenient testing can be carried out using endotoxin detection kits with photometric readout [limulus amebocyte lysate (LAL) test].

The final drug substance should additionally be tested for water and sodium content. The latter can be determined by the preferred method,[85] and water content can be determined by conventional KF titration. Residual solvents according to ICH Q3C are mainly tested using HS-GC.[87] An overview of analytical procedures for oligonucleotide drug substance testing by experts in the field is given elsewhere.[85] The authors also recommend two books specifically focused on oligonucleotide analytics.[143,144]

Special focus during analytical method development needs to be aimed at the development of stability-indicating methods (SIMs) as part of the overall CMC control strategy. SIMs describe quantitative analytical methods that are used to unambiguously identify the degree of degradation in the drug substance. HPLC methods are highly suitable for this purpose owing to their high resolution, sensitivity, and specificity.[145] Stability testing is covered by ICH guidelines Q1A *Stability Testing of New Drug Substances and Products* and Q1B *Stability Testing: Photostability Testing of New Drug Substances and Products*.[146,147] Stability tests "*provide evidence on how the quality of a drug substance or drug product varies with time under the influence of a variety of environmental factors such as temperature, humidity, and light, and to establish a re-test period for the drug substance or a shelf life for the drug product and recommended storage conditions.*"[146] This includes stability-challenging experiments under acidic, basic, oxidative, photolytic, and thermal conditions over time. For analysis of the outcome of these experiments, validated SIMs should be applied according to ICH Q1A.[146] A SIM should be able to measure the drug substance accurately and precisely free from interferences with impurities.[145] Therefore, it needs to be validated with regard to the following attributes, as specified in ICH Q2: specificity, linearity, range, precision, accuracy, robustness, detection limit, and quantitation limit.[124] Ideally, baseline resolution of all degradation products formed and the drug substance should be achieved. However, in oligonucleotides, this is not always feasible, especially when high numbers of PS linkages are present.[145] During the development of SIMs, the choice of solvent, column, pH, and temperature might heavily influence the resulting data. The mass balance should be checked where appropriate to validate the developed SIM.[148] For more detailed information on SIMs, the reader is referred to the relevant literature.[145,148]

13.5 Conclusion

To date, 26 single- or double-stranded synthetic RNA therapeutics have gained approval by regulatory authorities, with increasing speed. Apart from the four PMO splice-modulating antisense oligonucleotides, all drug substances were manufactured using standardized SPOS and phosphoramidite chemistry. Even though this process was established in the early 1980s,[1] it is well characterized, but the challenges of developing a suitable and unified CMC strategy remain, since there still are no harmonized regulatory guidelines for the development of oligonucleotides as RNA therapeutics. CMC strategies are crucial activities during new pharmaceutical product development. The main objective of CMC control strategies is the quality assurance of the final product during all phases of development by defining manufacturing processes and product specifications that must be met to ensure product safety and consistency between manufacturing batches.

The large-scale manufacture of RNA therapeutics typically starts with process development work using a small-scale model. With CQAs being the central element of ICH Q11,[91] the process development should already be focused on CMC objectives. Additionally, it should be driven by the general goal of obtaining a good understanding of the process to gain control over the impurity profile and the characteristics of the final drug substance with respect to the QTPP. After creation of a robust, reliable process at the developmental scale, the process then can be transferred and scaled up. For a successful technological transfer, it is important to implement already a scale-down model of the full-scale process during the initial process development. CQAs need to be kept constant throughout the entire process from small-scale to pilot-scale to large-scale syntheses.

Although the harmonized ICH quality guidelines on impurity thresholds and specifications exclude oligonucleotides in their preamble, their spirit should still be applied. Process- and product-related impurities generated by the established manufacturing process are well understood and state-of-the-art analytical techniques support their identification and characterization.

In a CMC-related review article, Carmody suggested acceleration of oligonucleotide drug development by early assurance that a drug is safe and meets appropriate quality standards. Using a front-loaded CMC strategy allows informed decisions to be made early in the overall process of drug substance manufacture, rather than the traditional back-ended approach during CMC development.[149] This supports accelerated drug development while at the same time ensuring the production of drug substances that meet appropriate safety and quality standards. This approach can already be implemented during candidate design and selection, toxicological studies, and all steps of quality control development including early characterizations, adequate IPCs, and control of starting materials. Impurities should be identified early and appropriate controls should be implemented supported by analytical data. Identification of degradation products in stability studies gives additional knowledge about the drug substance and can help further acceleration of development, guide drug design, formulation, manufacture, or clinical study design if conducted in the early stages.[149] A good control strategy should be established throughout the entire process, from development stages through to large-scale manufacturing to assure constant quality throughout. Development of a CMC strategy in an early stage with focus on safety, quality, and efficacy will support and accelerate the clinical development of oligonucleotide drug substances.

List of Abbreviations

ACN	Acetonitrile
AEX	Anion-exchange
AAS	Atomic absorption spectrometry
AMA	Ammonium hydroxide–methylamine (1 : 1 v/v)
AMO	Anti-miRNA oligonucleotide
API	Active pharmaceutical ingredient
ASO	Antisense oligonucleotide
BNA	Bridged nucleic acid
BTT	5-Benzylthio-1*H*-tetrazole
Cas	CRISPR-associated protein
CMC	Chemistry, manufacturing, and controls
cGMP	Current good manufacturing practice
CNET	Acrylonitrile side product

CoA	Certificate of Analysis
CpG	Cytosine–phosphate–guanine
CPP	Critical process parameter
CQA	Critical quality attribute
CRISPR	Clustered regularly interspaced short palindromic repeats
DCA	Dichloroacetic acid
DCI	4,5-Dicyanoimidazole
DDTT	3-[(Dimethylaminomethylidene)amino]-3*H*-1,2,4-dithiazole-3-thione
DEA	Diethylamine
DMTr	4,4′-Dimethoxytrityl
EDITH	3-Ethoxy-1,2,4-dithiazoline-5-one
ETT	5-Ethylthio-1*H*-tetrazole
FDA	US Food and Drug Administration
GalNAc	*N*-Acetylgalactosamine
GC	Gas chromatography
GMP	Good manufacturing practice
HIC	Hydrophobic interaction chromatography
HILIC	Hydrophilic interaction liquid chromatography
HPLC	High-performance liquid chromatography
HPLC-MS	High-performance liquid chromatography-mass spectrometry
HS-GC	Headspace gas chromatography
IC	Ion chromatography
ICH	International Council for Harmonisation of Technical Requirements for Pharmaceuticals for Human Use
ICP-MS	Inductively coupled plasma mass spectrometry
IP	Ion-pairing
IPC	In-process control
KF	Karl Fischer
LAL	Limulus amebocyte lysate
LNA	Locked nucleic acid
LNP	Lipid nanoparticle
miRNA	MicroRNA
MS	Mass spectrometry
MOE	Methoxyethyl
NCE	New chemical entity
NMR	Nuclear magnetic resonance
OSWG	Oligonucleotide Safety Working Group
PADS	Phenylacetyl disulfide
PMI	Process mass intensity
PMO	Phosphorodiamidate morpholino oligomer
PNA	Peptide nucleic acid
PO	Phosphodiester
PS	Phosphorothioate
QTPP	Quality target product profile
RP	Reversed-phase

SCD	Sickle cell disease
SEC	Size-exclusion chromatography
sgRNA	Single guide RNA
siRNA	Small interfering RNA
SIM	Stability-indicating method
SPOS	Solid-phase oligonucleotide synthesis
SSO	Splice-switching oligonucleotide
TDT	Transfusion-dependent β-thalassemia
TFD	Transcription factor decoy
UHPLC	Ultra high performance liquid chromatography
UV	Ultraviolet
XHD	Xanthane hydride, 3-amino-1,2,4-dithiazole-5-thione

Acknowledgements

The authors would like to acknowledge Axolabs Kulmbach GmbH, especially Ingo Röhl, Julia Paulus, and Timo Schierling for their contributions to the analytics section, and Hagen Cramer of QurAlis Corporation for his critical peer review.

References

1. H. Köster and N. Sinha, Process for the Preparation of Oligonucleotides, *US pat.*, 4725677, 1984.
2. A. M. Michelson and A. R. Todd, *J. Chem. Soc.*, 1955, **22**, 2632.
3. H. G. Khorana, G. M. Tener, J. G. Moffatt and E. H. Pol, *Chem. Ind.*, 1956, 1523.
4. G. M. Tener, H. G. Khorana, R. Markham and E. H. Pol, *J. Am. Chem. Soc.*, 1958, **80**, 6223.
5. M. Smith and H. G. Khorana, *J. Am. Chem. Soc.*, 1958, **80**, 1141.
6. H. Schaller, G. Weimann, B. Lerch and H. G. Khorana, *J. Am. Chem. Soc.*, 1963, **85**, 3821.
7. H. Schaller and H. G. Khorana, *J. Am. Chem. Soc.*, 1963, **85**, 3828.
8. R. B. Merrifield, *Fed. Proc., Fed. Am. Soc. Exp. Biol.*, 1962, 412.
9. R. B. Merrifield, *J. Am. Chem. Soc.*, 1963, **85**, 2149.
10. R. L. Letsinger and V. Mahadevan, *J. Am. Chem. Soc.*, 1965, **87**, 3526.
11. R. L. Letsinger and V. Mahadevan, *J. Am. Chem. Soc.*, 1966, **88**, 5319.
12. M. J. Gait and R. C. Sheppard, *Nucleic Acids Res.*, 1977, **4**, 1135.
13. S. L. Beaucage and M. H. Caruthers, *Tetrahedron Lett.*, 1981, **22**, 1859.
14. E. J. Nurminen, J. K. Mattinen and H. Lönnberg, *J. Chem. Soc., Perkin Trans. 2*, 1998, 1621.
15. E. J. Nurminen, J. K. Mattinen and H. Lönnberg, *J. Chem. Soc., Perkin Trans. 2*, 2000, 2238.
16. B. Sproat, F. Colonna, B. Mullah, D. Tsou, A. Andrus, A. Hampel and R. Vinayak, *Nucleosides Nucleotides*, 1995, **14**, 255.

17. F. Wincott, A. DiRenzo, C. Shaffer, S. Grimm, D. Tracz, C. Workman, D. Sweedler, C. Gonzalez, S. Scaringe and N. Usman, *Nucleic Acids Res.*, 1995, **23**, 2677.
18. R. Welz and S. Müller, *Tetrahedron Lett.*, 2002, **43**, 795.
19. C. Vargeese, *Nucleic Acids Res.*, 1998, **26**, 1046.
20. M. A. Russell, A. P. Laws, J. H. Atherton and M. I. Page, *Org. Biomol. Chem.*, 2008, **6**, 3270.
21. F. Eckstein, *J. Am. Chem. Soc.*, 1966, **88**, 4292.
22. F. Eckstein, *Annu. Rev. Biochem.*, 1985, **54**, 367.
23. R. P. Iyer, L. R. Phillips, W. Egan, J. B. Regan and S. L. Beaucage, *J. Org. Chem.*, 1990, **55**, 4693.
24. R. McKay, L. L. Cummins, M. J. Graham, E. A. Lesnik, S. R. Owens, M. Winniman and N. M. Dean, *Nucleic Acids Res.*, 1996, **24**, 411.
25. S. Palluk, D. H. Arlow, T. de Rond, S. Barthel, J. S. Kang, R. Bector, H. M. Baghdassarian, A. N. Truong, P. W. Kim, A. K. Singh, N. J. Hillson and J. D. Keasling, *Nat. Biotechnol.*, 2018, **36**, 645.
26. D. C. Capaldi, H. Gaus, A. H. Krotz, J. Arnold, R. L. Carty, M. N. Moore, A. N. Scozzari, K. Lowery, D. L. Cole and V. T. Ravikumar, *Org. Process Res. Dev.*, 2003, **7**, 832.
27. C.-A. Chang and T. Horn, *Nucleosides Nucleotides*, 1999, **18**, 1205.
28. E. Westman and R. Strömberg, *Nucleic Acids Res.*, 1994, **22**, 2430.
29. J. Muslehiddinoglu, R. Simler, M. L. Hill, C. Mueller, J. H. A. Amery, L. Dixon, A. Watson, K. Storch, C. Gazziola, F. Gielen, S. A. Lange, J. D. Prail and D. P. Nesta, *Nucleic Acid Ther.*, 2020, **30**, 189.
30. M. Egli and M. Manoharan, *Nucleic Acids Res.*, 2023, **51**, 2529.
31. M. L. Stephenson and P. C. Zamecnik, *Proc. Natl. Acad. Sci. U. S. A.*, 1978, **75**, 285.
32. P. C. Zamecnik and M. L. Stephenson, *Proc. Natl. Acad. Sci. U. S. A.*, 1978, **75**, 280.
33. C. Tuerk and L. Gold, *Science*, 1990, **249**, 505.
34. A. D. Ellington and J. W. Szostak, *Nature*, 1990, **346**, 818.
35. A. M. Krieg, A.-K. Yi, S. Matson, T. J. Waldschmidt, G. A. Bishop, R. Teasdale, G. A. Koretzky and D. M. Klinman, *Nature*, 1995, **374**, 546.
36. R. Morishita, G. H. Gibbons, M. Horiuchi, K. E. Ellison, M. Nakama, L. Zhang, Y. Kaneda, T. Ogihara and V. J. Dzau, *Proc. Natl. Acad. Sci. U. S. A.*, 1995, **92**, 5855.
37. M. Hecker and A. H. Wagner, *Biochem. Pharmacol.*, 2017, **144**, 29.
38. Y. Takeshima, H. Nishio, H. Sakamoto, H. Nakamura and M. Matsuo, *J. Clin. Invest.*, 1995, **95**, 515.
39. A. Fire, S. Xu, M. K. Montgomery, S. A. Kostas, S. E. Driver and C. C. Mello, *Nature*, 1998, **391**, 806.
40. G. Hutvágner, M. J. Simard, C. C. Mello and P. D. Zamore, *PLoS Biol.*, 2004, **2**, e98.
41. G. Meister, M. Landthaler, Y. Dorsett and T. Tuschl, *RNA*, 2004, **10**, 544.
42. J. Weiler, J. Hunziker and J. Hall, *Gene Ther.*, 2006, **13**, 496.

43. M. Jinek, K. Chylinski, I. Fonfara, M. Hauer, J. A. Doudna and E. Charpentier, *Science*, 2012, **337**, 816.
44. H. Frangoul, D. Altshuler, M. D. Cappellini, Y.-S. Chen, J. Domm, B. K. Eustace, J. Foell, J. de la Fuente, S. Grupp, R. Handgretinger, T. W. Ho, A. Kattamis, A. Kernytsky, J. Lekstrom-Himes, A. M. Li, F. Locatelli, M. Y. Mapara, M. de Montalembert, D. Rondelli, A. Sharma, S. Sheth, S. Soni, M. H. Steinberg, D. Wall, A. Yen and S. Corbacioglu, *N. Engl. J. Med.*, 2021, **384**, 252.
45. A. Khvorova and J. K. Watts, *Nat. Biotechnol.*, 2017, **35**, 238.
46. S. J. Dixon and B. R. Stockwell, *Curr. Opin. Chem. Biol.*, 2009, **13**, 549.
47. J. Kim, C. Hu, C. Moufawad El Achkar, L. E. Black, J. Douville, A. Larson, M. K. Pendergast, S. F. Goldkind, E. A. Lee, A. Kuniholm, A. Soucy, J. Vaze, N. R. Belur, K. Fredriksen, I. Stojkovska, A. Tsytsykova, M. Armant, R. L. DiDonato, J. Choi, L. Cornelissen, L. M. Pereira, E. F. Augustine, C. A. Genetti, K. Dies, B. Barton, L. Williams, B. D. Goodlett, B. L. Riley, A. Pasternak, E. R. Berry, K. A. Pflock, S. Chu, C. Reed, K. Tyndall, P. B. Agrawal, A. H. Beggs, P. E. Grant, D. K. Urion, R. O. Snyder, S. E. Waisbren, A. Poduri, P. J. Park, A. Patterson, A. Biffi, J. R. Mazzulli, O. Bodamer, C. B. Berde and T. W. Yu, *N. Engl. J. Med.*, 2019, **381**, 1644.
48. L. E. Burbano, M. Li, N. Jancovski, P. Jafar-Nejad, K. Richards, A. Sedo, A. Soriano, B. Rollo, L. Jia, E. V. Gazina, S. Piltz, F. Adikusuma, P. Q. Thomas, H. Kopsidas, F. Rigo, C. A. Reid, S. Maljevic and S. Petrou, *JCI Insight*, 2022, **7**, e146090.
49. M. Synofzik, W. M. van Roon-Mom, G. Marckmann, H. A. van Duyvenvoorde, H. Graessner, R. Schüle and A. Aartsma-Rus, *Nucleic Acid Ther.*, 2022, **32**, 83.
50. J. Kim, S. Woo, C. M. de Gusmao, B. Zhao, D. H. Chin, R. L. DiDonato, M. A. Nguyen, T. Nakayama, C. A. Hu, A. Soucy, A. Kuniholm, J. K. Thornton, O. Riccardi, D. A. Friedman, C. M. El Achkar, Z. Dash, L. Cornelissen, C. Donado, K. N. W. Faour, L. W. Bush, V. Suslovitch, C. Lentucci, P. J. Park, E. A. Lee, A. Patterson, A. A. Philippakis, B. Margus, C. B. Berde and T. W. Yu, *Nature*, 2023, **619**, 828.
51. E. E. Swayze and B. Bhat, in *Antisense Drug Technology - Principles, Strategies, and Applications*, ed. S. T. Crooke, CRC Press, Boca Raton, FL, USA, 2nd edn, 2007, vol. 6, p. 143.
52. R. S. Geary, R. Z. Yu and A. A. Levin, *Curr. Opin. Invest. Drugs*, 2001, 562.
53. A. A. Levin, *Biochim. Biophys. Acta, Gene Struct. Expression*, 1999, **1489**, 69.
54. S. T. Crooke, M. J. Graham, J. E. Zuckerman, D. Brooks, B. S. Conklin, L. L. Cummins, M. J. Greig, C. J. Guinosso, D. Kornbrust, M. Manoharan, H. M. Sasmor, T. Schleich, K. L. Tivel and R. H. Griffey, *J. Pharmacol. Exp. Ther.*, 1996, **277**, 923.
55. T. C. Roberts, R. Langer and M. J. A. Wood, *Nat. Rev. Drug Discovery*, 2020, **19**, 673.
56. J. Kurreck, *Eur. J. Biochem.*, 2003, **270**, 1628.

57. C. A. Stein, C. Subasinghe, K. Shinozuka and J. S. Cohen, *Nucleic Acids Res.*, 1988, **16**, 3209.
58. B. P. Monia, J. F. Johnston, H. Sasmor and L. L. Cummins, *J. Biol. Chem.*, 1996, **271**, 14533.
59. T. M. Rupp and H. Cramer, in *RNA Therapeutics*, ed. P. H. Giangrande, V. de Franciscis, and J. J. Rossi, Elsevier, 2022, vol. 9, p. 263.
60. P. Hair, F. Cameron and K. McKeage, *Drugs*, 2013, **73**, 487.
61. B. I. Andrews, F. D. Antia, S. B. Brueggemeier, L. J. Diorazio, S. G. Koenig, M. E. Kopach, H. Lee, M. Olbrich and A. L. Watson, *J. Org. Chem.*, 2021, **86**, 49.
62. E. R. Moody, R. Obexer, F. Nickl, R. Spiess and S. L. Lovelock, *Science*, 2023, **380**, 1150.
63. S. Paul, D. Gray, J. Caswell, J. Brooks, W. Ye, T. S. Moody, R. Radinov and L. Nechev, *ACS Chem. Biol.*, 2023, **18**, 2183.
64. J. F. Kim, P. R. J. Gaffney, I. B. Valtcheva, G. Williams, A. M. Buswell, M. S. Anson and A. G. Livingston, *Org. Process Res. Dev.*, 2016, **20**, 1439.
65. X. Zhou, W. F. Kiesman, W. Yan, H. Jiang, F. D. Antia, J. Yang, Y. A. Fillon, L. Xiao and X. Shi, *J. Org. Chem.*, 2022, **87**, 2087.
66. P. Rosenqvist, V. Saari, E. Pajuniemi, A. Gimenez Molina, M. Ora, A. Horvath and P. Virta, *J. Org. Chem.*, 2023, **88**, 10156.
67. A. A. Koshkin, S. K. Singh, P. Nielsen, V. K. Rajwanshi, R. Kumar, M. Meldgaard, C. E. Olsen and J. Wengel, *Tetrahedron*, 1998, **54**, 3607.
68. S. Obika, D. Nanbu, Y. Hari, K. Morio, Y. In, T. Ishida and T. Imanishi, *Tetrahedron Lett.*, 1997, **38**, 8735.
69. H. Inoue, Y. Hayase, A. Imura, S. Iwai, K. Miura and E. Ohtsuka, *Nucleic Acids Res.*, 1987, **15**, 6131.
70. P. Martin, *Helv. Chim. Acta*, 1995, **78**, 486.
71. A. M. Kawasaki, M. D. Casper, S. M. Freier, E. A. Lesnik, M. C. Zounes, L. L. Cummins, C. Gonzalez and P. D. Cook, *J. Med. Chem.*, 1993, **36**, 831.
72. O. V. Sergeeva, E. Y. Shcherbinina, N. Shomron and T. S. Zatsepin, *Nucleic Acid Ther.*, 2022, **32**, 123.
73. J. Summerton and D. Weller, *Antisense Nucleic Acid Drug Dev.*, 1997, **7**, 187.
74. R. M. Hudziak, E. Barofsky, D. F. Barofsky, D. L. Weller, S.-B. Huang and D. D. Weller, *Antisense Nucleic Acid Drug Dev.*, 1996, **6**, 267.
75. T. Yamamoto, M. Nakatani, K. Narukawa and S. Obika, *Future Med. Chem.*, 2011, **3**, 339.
76. G. F. Deleavey and M. J. Damha, *Chem. Biol.*, 2012, **19**, 937.
77. J. Winkler, *Ther. Delivery*, 2013, **4**, 791.
78. J. K. Nair, J. L. S. Willoughby, A. Chan, K. Charisse, M. R. Alam, Q. Wang, M. Hoekstra, P. Kandasamy, A. V. Kel'in, S. Milstein, N. Taneja, J. O'Shea, S. Shaikh, L. Zhang, R. J. van der Sluis, M. E. Jung, A. Akinc, R. Hutabarat, S. Kuchimanchi, K. Fitzgerald, T. Zimmermann, T. J. C. van Berkel, M. A. Maier, K. G. Rajeev and M. Manoharan, *J. Am. Chem. Soc.*, 2014, **136**, 16958.
79. S. M. Hoy, *Drugs*, 2018, **78**, 1625.

80. T. Kalita, S. A. Dezfouli, L. M. Pandey and H. Uludag, *Pharmaceutics*, 2022, **14**, 2520.
81. International Council for Harmonisation of Technical Requirements for Pharmaceuticals for Human Use (ICH), *ICH Q3A (R2): Impurities in New Drug Substances*, 2006.
82. International Council for Harmonisation of Technical Requirements for Pharmaceuticals for Human Use (ICH), *ICH Q3B (R2): Impurities in New Drug Products*, 2006.
83. International Council for Harmonisation of Technical Requirements for Pharmaceuticals for Human Use (ICH), *Q6B: Specifications: Test Procedures and Acceptance Criteria for Biotechnological/Biological Products*, 1999.
84. International Council for Harmonisation of Technical Requirements for Pharmaceuticals for Human Use (ICH), *ICH Q3D (R2): Guideline for Elemental Impurities*, 2022.
85. D. Capaldi, K. Ackley, D. Brooks, J. Carmody, K. Draper, R. Kambhampati, M. Kretschmer, D. Levin, J. McArdle, B. Noll, R. Raghavachari, I. Roymoulik, B. P. (Bob) Sharma, R. Thürmer and F. Wincott, *Drug Inf. J.*, 2012, **46**, 611.
86. United States Pharmacopeia, *USP 467: Residual Solvents*, 2007.
87. International Council for Harmonisation of Technical Requirements for Pharmaceuticals for Human Use (ICH), *ICH Q3C (R8): Impurities: Guideline for Residual Solvents*, 2021.
88. Y. A. Fillon, N. Akhtar, B. I. Andrews, D. Benstead, S. Breitler, R. S. Gronke, M. Olbrich, J. A. Stolee and T. Vandermeersch, *Org. Process Res. Dev.*, 2022, **26**, 1130.
89. International Council for Harmonisation of Technical Requirements for Pharmaceuticals for Human Use (ICH), *ICH M7 (R2): Assessment and Control of DNA Reactive (Mutagenic) Impurities in Pharmaceuticals to Limit Potential Carcinogenic Risk*, 2023.
90. D. Capaldi, A. Teasdale, S. Henry, N. Akhtar, C. den Besten, S. Gao-Sheridan, M. Kretschmer, N. Sharpe, B. Andrews, B. Burm and J. Foy, *Nucleic Acid Ther.*, 2017, **27**, 309.
91. International Council for Harmonisation of Technical Requirements for Pharmaceuticals for Human Use (ICH), *ICH Q11: Development and Manufacture of Drug Substances (Chemical Entities and Biotechnological/Biological Entities)*, 2012.
92. International Council for Harmonisation of Technical Requirements for Pharmaceuticals for Human Use (ICH), *ICH Q8(R2): Pharmaceutical Development*, 2009.
93. International Council for Harmonisation of Technical Requirements for Pharmaceuticals for Human Use (ICH), *ICH Q9 (R1): Quality Risk Management*, 2023.
94. International Council for Harmonisation of Technical Requirements for Pharmaceuticals for Human Use (ICH), *ICH Q10: Pharmaceutical Quality Systems*, 2008.

95. M. Koziolkiewicz, A. Krakowiak, M. Kwinkowski, M. Boczkowska and W. J. Stec, *Nucleic Acids Res.*, 1995, **23**, 5000.
96. M. Koziolkiewicz, M. Wójcik, A. Kobylanska, B. Karwowski, B. Rebowska, P. Guga and W. J. Stec, *Antisense Nucleic Acid Drug Dev.*, 1997, **7**, 43.
97. M. Boczkowska, P. Guga and W. J. Stec, *Biochemistry*, 2002, **41**, 12483.
98. M. E. Østergaard, C. L. De Hoyos, W. B. Wan, W. Shen, A. Low, A. Berdeja, G. Vasquez, S. Murray, M. T. Migawa, X. Liang, E. E. Swayze, S. T. Crooke and P. P. Seth, *Nucleic Acids Res.*, 2020, **48**, 1691.
99. H. Cramer, K. J. Finn and N. D. Sinha, in *Analysis of Oligonucleotides and their Related Substances*, ed. G. Okafo, D. Elder, and M. Webb, ILM Publications, St. Albans, Hertfordshire, UK, 2013, vol. 2, p. 21.
100. H. Gaus, P. Olsen, K. Van Sooy, C. Rentel, B. Turney, K. L. Walker, J. V. McArdle and D. C. Capaldi, *Bioorg. Med. Chem. Lett.*, 2005, **15**, 4118.
101. S. Lorenz, J. Przybytek and W. Olive, Synthesis of Oligonucleotides or Phosphorothioate Oligonucleotide with a Capping Agent of N-Methylimidazole Free of 1,3,5-Trimethylhexahydro-1,3,5-triazine, *US pat.*, 8039612 B2, 2008.
102. M. Septak, *Nucleic Acids Res.*, 1996, **24**, 3053.
103. D. C. Capaldi, H. J. Gaus, R. L. Carty, M. N. Moore, B. J. Turney, S. D. Decottignies, J. V. McArdle, A. N. Scozzari, V. T. Ravikumar and A. H. Krotz, *Bioorg. Med. Chem. Lett.*, 2004, **14**, 4683.
104. A. Rodriguez, *Problems Associated with Capping during Oligonucleotide Synthesis*, Isis Pharmaceuticals Inc., TIDES USA, Rhode Island, 2014.
105. R. T. Pon, N. Usman, M. J. Damha and K. K. Ogilvie, *Nucleic Acids Res.*, 1986, **14**, 6453.
106. J. S. Eadie and D. S. Davidson, *Nucleic Acids Res.*, 1987, **15**, 8333.
107. A. A. Rodriguez, I. Cedillo and A. K. McPherson, *Bioorg. Med. Chem. Lett.*, 2016, **26**, 3468.
108. P. C. J. Kamer, H. C. P. F. Roelen, H. van den Elst, G. A. van der Marel and J. H. van Boom, *Tetrahedron Lett.*, 1989, **30**, 6757.
109. Q. Xu, K. Musier-Forsyth, R. P. Hammer and G. Barany, *Nucleic Acids Res.*, 1996, **24**, 1602.
110. Q. Xu, *Nucleic Acids Res.*, 1996, **24**, 3643.
111. A. P. Guzaev, *Tetrahedron Lett.*, 2011, **52**, 434.
112. J.-Y. Tang, Y. Han, J. X. Tang and Z. Zhang, *Org. Process Res. Dev.*, 2000, **4**, 194.
113. C. J. Calvitt, D. S. Levin, B. T. Shepperd and C. J. Gruenloh, *Oligonucleotides*, 2010, **20**, 239.
114. S. K. Singh, M. W. Szulik, M. Ganguly, I. Khutsishvili, M. P. Stone, L. A. Marky and B. Gold, *Nucleic Acids Res.*, 2011, **39**, 6789.
115. J. Boal, A. Wilk, N. Harindranath, E. E. Max, T. Kempe and S. L. Beaucage, *Nucleic Acids Res.*, 1996, **24**, 3115.
116. R. A. Johnsson, J. J. Bogojeski and M. J. Damha, *Bioorg. Med. Chem. Lett.*, 2014, **24**, 2146.
117. J. C. Schulhof, D. Molko and R. Teoule, *Nucleic Acids Res.*, 1987, **15**, 397.
118. J. T. Kodra, J. Kehler and O. Dahl, *Nucleic Acids Res.*, 1995, **23**, 3349.

119. C. B. Reese and Q. Song, *Nucleic Acids Res.*, 1997, **25**, 2943.
120. M. P. Reddy, F. F. Farooqui and N. B. Hanna, *Tetrahedron Lett.*, 1996, **37**, 8691.
121. J. W. Efcavitch and C. Heiner, *Nucleosides, Nucleotides Nucleic Acids*, 1985, **4**, 267.
122. M. Ehrlich, K. F. Norris, R. Y. Wang, K. C. Kuo and C. W. Gehrke, *Biosci. Rep.*, 1986, **6**, 387.
123. L. Frick, J. P. Mac Neela and R. Wolfenden, *Bioorg. Chem.*, 1987, **15**, 100.
124. International Council for Harmonisation of Technical Requirements for Pharmaceuticals for Human Use (ICH), *ICH Q2(R1): Validation of Analytical Procedures: Text and Methodology*, 2005.
125. European Directorate for the Quality of Medicines and Health Care, *Ph. Eur.*, 2022, vol. 11.2, p. 949.
126. C. J. Borths, T. Burr, A. Figuccia, J. G. Ford, B. Guan, M. T. Jones, D. Klingeleers, S. Lochner, A. A. Rodriguez and C. Wetter, *Org. Process Res. Dev.*, 2023, **27**, 1693.
127. T. Fornstedt and M. Enmark, *J. Chromatogr. Open*, 2023, **3**, 100079.
128. E. D. Close, A. O. Nwokeoji, D. Milton, K. Cook, D. M. Hindocha, E. C. Hook, H. Wood and M. J. Dickman, *J. Chromatogr. A*, 2016, **1440**, 135.
129. A. Goyon and K. Zhang, *Anal. Chem.*, 2020, **92**, 5944.
130. A. A. Rodriguez, I. Cedillo, B. P. Mowery, H. J. Gaus, S. S. Krishnamoorthy and A. K. McPherson, *Bioorg. Med. Chem. Lett.*, 2014, **24**, 3243.
131. J. P. Vernille and J. W. Schneider, *Biotechnol. Prog.*, 2004, **20**, 1776.
132. J. M. Savard and J. W. Schneider, *Biotechnol. Bioeng.*, 2007, **97**, 367.
133. A. J. Alpert, *J. Chromatogr. A*, 1990, **499**, 177.
134. P. Holdšvendová, J. Suchánková, M. Bunček, V. Bačkovská and P. Coufal, *J. Biochem. Biophys. Methods*, 2007, **70**, 23.
135. R. N. Easter, K. K. Kröning, J. A. Caruso and P. A. Limbach, *Analyst*, 2010, **135**, 2560.
136. M. Huang, X. Xu, H. Qiu and N. Li, *J. Chromatogr. A*, 2021, **1648**, 462184.
137. C. Schildkraut and S. Lifson, *Biopolymers*, 1965, **3**, 195.
138. L. M. Oliveira, A. S. Long, T. Brown, K. R. Fox and G. Weber, *Chem. Sci.*, 2020, **11**, 8273.
139. N. C. Harris and C.-H. Kiang, *J. Phys. Chem. B*, 2006, **110**, 16393.
140. K. J. Sweadner, M. Forte and L. L. Nelsen, *Appl. Environ. Microbiol.*, 1977, **34**, 382.
141. E. M. White, *Appl. Occup. Environ. Hyg.*, 2002, **17**, 606.
142. P. O. Magalhães, A. M. Lopes, P. G. Mazzola, C. Rangel-Yagui, T. C. V. Penna and A. Pessoa, *J. Pharm. Pharm. Sci.*, 2007, **10**, 388.
143. *Handbook of Analysis of Oligonucleotides and Related Products*, ed. J. V. Bonilla, and G. S. Srivatsa, CRC Press, Boca Raton, FL, USA, 2011.
144. G. Okafo, D. Elder and M. Webb, *Analysis of Oligonucleotides and Their Related Substances*, ILM Publications, St. Albans, Hertfordshire, UK, 2013.

145. V. Murugaiah, in *Handbook of Analysis of Oligonucleotides and Related Products*, ed. J. V. Bonilla, and G. S. Srivatsa, CRC Press, Boca Raton, FL, USA, 2011, vol. 15, p. 403.
146. International Council for Harmonisation of Technical Requirements for Pharmaceuticals for Human Use (ICH), *ICH Q1A(R2): Stability Testing of New Drug Substances and Products*, 2003.
147. International Council for Harmonisation of Technical Requirements forPharmaceuticalsforHumanUse(ICH),*ICHQ1B:StabilityTesting:Photostability Testing of New Drug Substances and Products*, 1996.
148. M. Bakshi and S. Singh, *J. Pharm. Biomed. Anal.*, 2002, **28**, 1011.
149. J. Carmody, *Assuring Quality of Oligonucleotide APIs and DPs*, 2002, https://www.americanpharmaceuticalreview.com/Featured-Articles/587162-Assuring-Quality-of-Oligonucleotide-APIs-and-DPs/, last accessed 22 December 2023.

CHAPTER 14

Oligonucleotide Regulatory Affairs and Challenges

M. LEMAITRE

ML Consult LLC, 881 Clifton Crest Terrace, Cincinnati, OH 45220, USA
E-mail: marc.m.lemaitre@gmail.com

14.1 Introduction

It must be clear from the beginning that this chapter will only address the chemistry, manufacturing, and controls (CMC) for synthetic nucleic acid oligomers, also known as oligonucleotides. It will apply to synthetic small interfering ribonucleic acids (siRNAs), antisense oligonucleotides (ASOs), or guide RNAs used for clustered regularly interspaced short palindromic repeat (CRISPR) gene editing, but recombinant or enzymatically produced oligonucleotides (such as mRNAs) are not included. More specifically, this chapter applies to oligonucleotides manufactured by solid-phase synthesis followed by some chromatographic purification. It also tries to elaborate or imagine some future way of making oligonucleotides using some different manufacturing method that could be desirable to make very large quantities of any given oligonucleotide in ton quantities at a cost per gram in the low double-digit dollars range or to make synthesis more sustainable. Although this is just hypothetical now, the chapter points to the impact it may have on CMC. Further, this chapter summarizes the author's own views, based on his experience, and these are not necessarily the views of the US Food and Drug Administration (FDA), or any other agencies, or any of his present or

Green Chemistry Series No. 85
Sustainability in TIDES Chemistry: Green Approaches to Oligonucleotides and Oligopeptides Synthesis
Edited by Alessandra Tolomelli, Lucia Ferrazzano and Walter Cabri

Published by the Royal Society of Chemistry, www.rsc.org

past clients or former employers. The first investigational new drug (IND) application for an oligonucleotide type of new chemical entity (NCE) drug was filed in 1992 and, since the first IND filing for a 20-mer DNA phosphorothioate antisense oligonucleotide, ISIS 2105, administered locally to genital warts for the treatment of human papilloma virus (HPV), a lot more has been done and there are now 21 approved synthetic oligonucleotides, as of mid-2023, and there are many more in various phases of clinical development. The approved oligonucleotides are briefly described in Table 14.1.

Table 14.1 List of approved synthetic oligonucleotide therapeutics as of 1st January 2024.

Therapeutic[a]	INN[b]	Class of oligonucleotide	Mechanism	Year approved
Vitravene	Fomivirsen	Antisense	ASO	1998 (now withdrawn)
Macugen	Pegatanib sodium	RNA aptamer	Aptamer	2004
Kynamro	Mimopersen	Antisense	ASO/RNAse H	2013
Spinraza	Nusinersen	MOE antisense	Splice modulation	2016
Exondys-51	Eteplirsen	Morpholino	Ex51 splice modulation	2016
Heplisav (hepatitis B vaccine)	CpG1018	CpG oligonucleotide	CpG oligonucle-otide	2017
Onpattro	Patisiran	siRNA	RNAi	2018
Tegsedi	Inotersen	Gapmer antisense	ASO/RNAse H	2018
Waylivra	Volanesorsen	Gapmer antisense	ASO/RNAse H	2019
Vyondys-53	Golodirsen	Morpholino	Ex53 splice modulation	2019
Givlaari	Givosiran	siRNA–GalNAc	RNAi	2019
Oxlumo	Lumasiran	siRNA–GalNAc	RNAi	2020
Viltepso	Viltolarsen	Morpholino	Ex53 splice modulation	2020
Leqvio	Inclisiran	siRNA–GalNAc	RNAi	2020–2021
Amondys-45	Casimersen	Morpholino	Ex45 splice modulation	2021
Amvuttra	Vutrisiran	siRNA–GalNAc	RNAi	2022
Qalsody	Tofersen	Gapmer antisense	ASO/RNAse H	2023
Cyfendus (anthrax vaccine)	CpG7909	CpG oligonucleotide	CpG oligonucle-otide	2023
Izervay	Avacincaptad pegol	RNA aptamer	Complement C5 inhibition	2023
Wainua	Eplontersen	Gapmer antisense	GalNAc/ASO/ RNase H	2023
Rivfloza	Nedosiran	siRNA–GalNAc loop RNA	RNAi	2023

[a]Trade name.
[b]International non-proprietary name.

In 1998, Vitravene (fomivirsen sodium, ISIS 2922, also from the company then named ISIS, now known as IONIS) became the first oligonucleotide drug to receive market approval in the USA and Europe for the treatment of cytomegalovirus (CMV) retinitis. Vitravene is an antisense molecule that blocks translation of viral mRNA by binding to a coding segment of a key CMV gene. So far, six ASOs, five splice-modulating oligonucleotides, six siRNAs, two CpG oligonucleotide adjuvants, and two aptamers have been approved.

Table 14.1 shows that in this growing number of approved products there are various classes of molecules, based on either their structures, chemistry, or mechanism of action. For a better understanding of all the possibilities and how the chemistry evolved over time, readers are directed to a review by Khvorova and Watts[1] for more information regarding the chemical modifications that have been used so far.

14.2 High-level View of CMC Process

Synthesis of an oligonucleotide is typically performed through the sequential addition of nucleoside phosphoramidites[2] to a solid support derivatized with either the 3′-terminal nucleoside of the desired sequence or with a universal linker molecule,[3] which can be used to manufacture any sequence regardless of the identity of the 3′-terminal nucleoside (see Scheme 14.1).

The synthetic steps of solid-phase synthesis are conducted using an automated oligonucleotide synthesizer, which delivers starting material (nucleoside phosphoramidites) and reagent solutions and solvents to a column reactor containing the solid support. The synthesizer is controlled by a computer that executes a program specifying the quantities, flow rates, and order of addition of starting materials, reagents, and solvents to the column.[4]

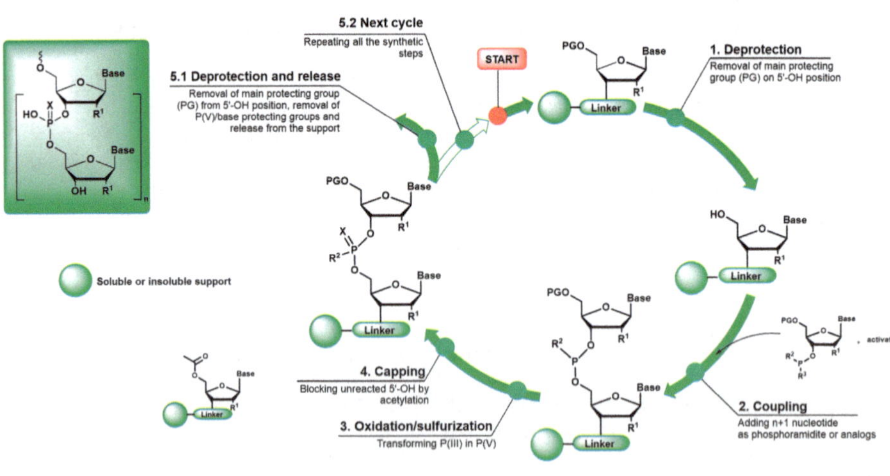

Scheme 14.1 Oligonucleotide synthetic cycle.

Once the synthesis is completed, the product is cleaved from the support and deprotected using a basic solution that is usually concentrated ammonium hydroxide, then purified, desalted, and usually lyophilized. This gives the drug substance (DS), or active pharmaceutical ingredient (API). If the oligonucleotide therapeutic is a single-strand antisense or splice-modulating oligonucleotide, that could be it. That is the simplest case. If the sponsor is developing an siRNA or a duplex oligonucleotide, each strand must be produced using the same process, and they must be annealed before obtaining the API. Should the DS have a ligand (*e.g.* GalNAc) attached to the API, it can be added during the synthesis of the oligonucleotide, or after, then adding one step to the synthesis process, or two steps if the sponsor decides to purify the oligonucleotide strand before a post-synthesis conjugation. Further, the process must be controlled throughout all of the steps.

Once the API is ready, the process can move on to the manufacture of the drug product (DP), that is, the presentation of the product that will be used in the clinic by the doctors and nurses handling the clinical trials (usually a vial or a syringe).

In general, the oligonucleotide API will be dissolved in some liquid compatible with the kind of administration selected, and the solution will be filtered through 0.2 μm filter to sterilize it. Then an appropriate volume of liquid will be introduced in an appropriate vial or syringe. The vial will be closed with a stopper and a cap when the syringe will be competed with an appropriate plunger. Finally, an appropriate label is affixed to the container, and the DP is ready to go to the clinic, provided that the product passes all the release tests.

As for the DS, the DP may include many variations. A delivery system can be used. If the oligonucleotide is encapsulated in some lipid nanoparticles (LNPs) or in other sorts of nanoparticles, or in any type of new excipient, that adds some more steps and controls. This will be discussed later.

Now that we have a certain high-level view of our manufacturing process, we have to address the compliance aspect of it. Making an oligonucleotide in a laboratory for some research project is one thing, but making it for a clinical trial and eventually for commercial use is somewhat different, so we must describe the regulatory aspects of the CMC universe.

14.3 Regulatory Aspects of CMC

Owing to their size and complexity of manufacture, therapeutic oligonucleotides, like synthetic peptides, are sometimes defined as "large small molecules", falling in between small molecules and biologics. That is why oligonucleotides are specifically excluded from the scope of some guidelines for drug manufacture (see Table 14.2).

Since the first IND filing in 1992 there is still no regulatory guideline that specifically addresses the quality expectations for oligonucleotide drug substances. Recognizing the issues sometimes generated by such an absence, the Oligonucleotide Safety Working Group (OSWG) and others

Table 14.2 ICH (International Council for Harmonisation of Technical Requirements for Pharmaceuticals for Human Use) guidelines.

Guideline	Topic	Application to oligonucleotides
Q1	Stability	Yes
Q2	Method validation	Yes
Q3a Q3b	Impurities in drug substance and drug product	Principles of the guidelines apply Limits for reporting, ID, and qualification are reviewed on a case-by-case basis
Q3c Q3d	Residual solvents and elemental impurities in drug substance and drug product	Yes
Q4	Pharmacopeia harmonization	Yes
Q6a (b)	API specification	Principles of the guideline applies but should be adapted for an oligonucleotide case-by-case
Q7	GMPs for API	Yes
Q8	Pharmaceutical development	Yes
Q9	GMP risk management	Yes
Q10	Pharmaceutical quality system	Yes
Q11	Development and manufacturing of drug substances	Yes
Q14	Analytical procedure development (in preparation)	Will apply

have published a series of position papers providing scientific advice where current regulatory guidance is lacking, going well beyond the CMC domain. Eight position papers focusing on non-clinical issues[5–12] and two chemistry-related articles[13,14] have been published. Recently, scientists from a group of pharmaceutical companies interested in developing therapeutic oligonucleotides established the European Pharma Oligonucleotide Consortium (EPOC), the mission of which is to share knowledge and harmonize CMC strategies for oligonucleotide development in the hope of expediting access to life-changing medicines.[15] The Consortium's initial objectives were to identify a shortlist of CMC topics and to prepare and publish scientific advice in the form of technical White Papers written by teams of subject matter experts to discuss current and best practices, and several papers have been published reporting their recommendations on the use of oligonucleotide solution API,[16] on oligonucleotide drug substance (DS) and drug product (DP) identity testing based on scientific rationale and their cumulative experience,[17] on nitrosamine risk assessments in oligonucleotides,[18] considerations for the terminal sterilization of oligonucleotide DPs,[19] sustainability in oligonucleotide manufacture,[20,21] designation of starting materials for oligonucleotide synthesis,[22] and purge factors for use in oligonucleotide control strategies.[23] Anyone involved in oligonucleotide control strategy and CMC should read those publications.

14.4 IND Module 3 – Quality: Content

Most of the CMC activities are focused on the topics that must be described to the authorities in the document named IND (Investigational New Drug) Module 3 for the FDA or IMP (Investigational Medicinal Product) dossier for the European Medicines Agency (EMA) that describes all the manufacturing and quality aspects of a DS and DP. Readers can also look at the ICH-M4Q document ("Common Technical Document for the Registration of Pharmaceuticals for Human Use: Quality Overall Summary of Module 2/Module 3: Quality.")

For an example of the definition of these concepts, Article 2(d) of European Directive 2001/20/EC defines an IMP as follows:

> *A pharmaceutical form of an active substance or placebo being tested or used as a reference in a clinical trial, including products already with a marketing authorisation but used or assembled (formulated or packaged) in a way different from the authorised form, or when used for an unauthorised indication, or when used to gain further information about the authorised form. The IMP dossier (IMPD) gives information related to the quality of any IMP (i.e. including reference product and placebo), manufacture and control of the IMP, and data from non-clinical studies and from its clinical use.*

A simplified view of the content of Module 3 is as follows:

- *3.1 ToC OF MODULE 3*
- *3.2 BODY OF DATA*
- *3.2.S DRUG SUBSTANCE (DS)*
  - *3.2.S.1 General Information*
  - *3.2.S.2 Manufacture*
    - *3.2.S.2.1 Manufacturer(s)*
    - *3.2.S.2.2 Description of Manufacturing Process and Process Controls*
    - *3.2.S.2.3 Control of Materials*
    - *3.2.S.2.4 Control of Critical Steps and Intermediates*
    - *3.2.S.2.5 Process Validation and/or Evaluation*
    - *3.2.S.2.6 Manufacturing Process Development*
  - *3.2.S.3 Characterization*
    - *3.2.S.3.1 Elucidation of Structure and Other Characteristics*
    - *3.2.S.3.2 Impurities*
  - *3.2.S.4 Control of Drug Substance*
    - *3.2.S.4.1 Specification*
    - *3.2.S.4.2 Analytical Procedures*
    - *3.2.S.4.3 Validation of Analytical Procedures*
    - *3.2.S.4.4 Batch Analyses*
    - *3.2.S.4.5 Justification of Specification*
  - *3.2.S.5 Reference Standards or Materials*

3.2.S.6 Container Closure System
3.2.S.7 Stability
3.2.P DRUG PRODUCT
3.2.P.1 Description and Composition of the Drug Product
3.2.P.2 Pharmaceutical Development
3.2.P.2.1 Components of the Drug Product
2.1.1 Drug Substance
2.1.2 Excipients
3.2.P.2.2 Drug Product
2.2.1 Formulation Development
2.2.2 Overages
2.2.3 Physicochemical Properties
3.2.P.2.3 Manufacturing Process Development
3.2.P.2.4 Container Closure System
3.2.P.2.5 Microbiological Attributes
3.2.P.2.6 Compatibility
3.2.P.3 Manufacture
3.2.P.3.1 Manufacturer(s)
3.2.P.3.2 Batch Formula
3.2.P.3.3 Description of Manufacturing Process and Process Controls
3.2.P.3.4 Control of Critical Steps and Intermediates
3.2.P.3.5 Process Validation and/or Evaluation
3.2.P.4 Control of Excipients
3.2.P.4.1 Specifications
3.2.P.4.2 Analytical Procedures
3.2.P.4.3 Validation of Analytical Procedures
3.2.P.4.4 Justification of Specifications
3.2.P.4.5 Excipients of Human or Animal Origin
3.2.P.4.6 Novel Excipients
3.2.P.5 Control of Drug Product
3.2.P.5.1 Specification(s)
3.2.P.5.2 Analytical Procedures
3.2.P.5.3 Validation of Analytical Procedures
3.2.P.5.4 Batch Analyses
3.2.P.5.5 Characterization of Impurities
3.2.P.5.6 Justification of Specification(s)
3.2.P.6 Reference Standards or Materials
3.2.P.7 Container Closure System
3.2.P.8 Stability
3.2.P.8.1 Stability Summary and Conclusion
3.2.P.8.2 Postapproval Stability Protocol and Stability Commitment
3.2.P.8.3 Stability Data

It is not possible to describe in detail all that should be in those sections, nor could we just copy/paste a real IND, because first it is confidential and second it would be much too long. However, this section will try to explain in general terms what should be presented in some of the chapters – in the author's opinion and based on the author's experience. The chapters or sub-chapters not mentioned in the rest of this section have been omitted because of the obviousness of the topic.

3.2.S DRUG SUBSTANCE (DS)

3.2.S.2 Manufacture

In this part of the CMC Module 3 part of the IND, authorities expect a description of the synthesis process. As phosphoramidite chemistry is the standard chemistry, it is repetitive from one oligonucleotide to another. There is not much to say about that. However, it is important to know that not all agencies are in agreement about the level of detail that is required in that part. It is recommended to be quite precise and specific in this part, providing enough information to convince reviewers that the process is well understood and, in particular, that the critical nature of each step is well understood and that the sponsor has defined and applied a control strategy that allows them to limit the risk of mistake or problem. Usually, the critical parts of the synthesis process are cleavage and deprotection, chromatography, final ultrafiltration, and annealing in the case of a duplex oligonucleotide. For a comprehensive study of the manufacturing process and the relationship between the chemistry applied in it and the impurities readers can consult ref. 2 and 29.

3.2.S.2.3 Control of Materials

A phase-appropriate approach applies to this section. Critical starting materials (*i.e.* phosphoramidites and supports) will receive special attention from the agencies. The reason is that part of those starting materials will end up in the final drug product, so impurities in amidites or on what goes from the support to the oligonucleotide will be delivered to the patient, if not purified out. At an early stage, it is necessary to list the amidites, the supports, the solvents, and the reagents and the main specifications for them, and also the names of the vendors. Note that addition of an alternative vendor to the starting materials list will need to be approved as a variation. However, as development progresses, this part of the CMC Module 3 section will provide more information on those starting materials and the impurities in those listed, and they may be asked to provide information on the batches that have been used during the development of the product, so certificates of analysis and QC data should be kept. This is why it is critical to work with suppliers of amidites and supports that have a deep understanding of their product and excellent analytical capabilities to provide necessary information in a timely manner. Table 14.3, adapted from Cramer *et al.*,[27] summarizes the major oligonucleotide impurities that may come from phosphoramidites and supports.

Table 14.3 List of potential impurities originating from phosphoramidites and solid supports. Data from ref. 27. Reproduced from ref. 33 with permission from Elsevier, Copyright 2022.

Origin	Impurity	Type of impurity	Typical level in final API	Control strategy
Amidite starting material	α-Anomeric phosphoramidite	Incorrect orientation of nucleobase (no mass difference)	Trace	Tight amidite specification (HPLC, ^{1}H NMR)
Amidite starting material	3′-*O*-DMT, 5′-*O*-phosphoramidite	Incorrect linkage (5′,5′) (no mass difference)	Trace	Tight amidite specification (HPLC, ^{31}P NMR)
Amidite starting material	*N*-Acylated-3′-*O*-methoxydiisopropylaminophosphoramidite-5′-*O*-DMT-nucleosides	Formation of 2′,5′ internucleotidic linkage (no mass difference)	Trace	Tight amidite specification (HPLC, ^{31}P NMR)
Amidite starting material	Cyanoethyl amiditophosphite	Internucleotidic diphosphate (+80 Da) or dithioate linkage (+96 Da)	Trace	Tight amidite specification (HPLC, ^{31}P NMR)
Amidite starting material	3′-*O*-*t*-BDMSi/OMe-2′-*O*-phosphoramidite	Incorrect linkage (2′,5′) (no mass difference)	Trace	Tight amidite specification (HPLC, ^{31}P NMR)
Amidite starting material	5′-*O*,*N*-Di-DMT-adenosine	Branchmers	Small	Tight amidite specification (RP-HPLC)
Solid support	$n-1$ (3′) (typically with terminal phosphate/phosphorothioate)	3′-Nucleoside missing	Variable	Proper capping of support
Solid support	$n-1$ (3′)	3′-Nucleotide missing	Variable	Detritylation time of Unylinker support
Solid support	$n+[n-1$ (3′)]	Branchmer	Variable, depending on age of support	Use fresh support to minimize loss of base-protecting group during storage of nucleoside-loaded support

In addition to nucleobase phosphoramidites and solid supports, every special molecule that the sponsor would include in their oligonucleotide, such as, but not limited to, GalNAc or cholesterol, will need to be included in the starting material in order to list the potential impurities coming from them, to assess the criticality of the impurity profile, and to set and justify the limits. This has to be provided for each amidite and other types of starting material. As stated in René Thürmer's (BfArM – German Federal Institute for Drugs and Medical Devices) presentation at TIDES in 2018 (Boston, 8th May 2018): "many development programs are now including linkers and conjugates and justification for those products should be in line with ICH Q11 Q&A" (February 2018). Regarding the impurity limits in amidite, it is important to remember that they should be set keeping in mind the number of incorporations of each amidite in the oligonucleotide. Remember that those amidite impurities will add up in the oligonucleotide. Hence the more of one amidite is present in the DS, the lower the level of impurity for that amidite should be. Here, note that Capaldi *et al.*[14] wrote: "It is the authors' experience that by using sensitive and selective liquid chromatography-mass spectrometry (LC-MS) methods, it is possible to control reactive unspecified starting material impurities to a level of ~0.1%. However, because of the iterative nature of SPS, it will be apparent that even this level of control is insufficient to control unspecified starting material-derived drug substance impurities to much less than 1%. For example, the presence of 0.1% of a new reactive impurity in half of the starting materials used to make a 20-mer oligonucleotide may lead to 10 × 0.1% = 1.0% of an unspecified drug substance impurity." That is why it is critically important to choose the starting materials with great care.

3.2.S.2.4 Control of Critical Steps and Intermediates

This is a critical part of the dossier. It concerns process control, those controls used during manufacturing to monitor and, when necessary, adjust the process. This is where the sponsor will include necessary details explaining the control strategy, how they will demonstrate the robustness and reproducibility of the process, and how these controls will notice all drift. It is not unusual that *Section S.2.4* is longer than *Section S.2.2*. Also, this part will evolve considerably from a pre-IND document to a New Drug Application (NDA) document. This is where the batch history will be important to demonstrate the robustness, the stability, and, when possible, the improvements in the process. It is also in this part that the sponsor will explain to the reviewers what the impurities are and how they are controlled.

At an early stage, impurities can be "reported", with just a few "identified" on the basis of analysis by LC-MS and/or using some impurity markers. During the development of the process, the sponsor is expected to gain a better understanding of those impurities, and this part will be more and more detailed as development progresses. The most important aspect is to explain and demonstrate that the limits for the impurities as defined at an early stage are based on (1) manufacturing experience and (2) non-clinical control of those impurities and that those limits will be progressively tightened as the process improves.

This chapter will detail the control strategy for the manufacture of the DS and the associated single-strand intermediates in the case of the duplex. The sponsor will describe the quality attributes (QAs) of the active molecule, the tests used to measure those QAs, their criticality, and their justification.

The sponsor should also describe the process steps and outline the criticality of those steps (see Scheme 14.1). This is obviously part of the package that will develop a lot during the development, from the IND, when experience is limited to the manufacture of the IND-enabling and phase 1 batch or batches, and the NDA, when validation of the process has been carried out. These steps will be submitted to a risk analysis later during the development that will lead to the establishment of the proven acceptable range for the critical steps in the process.

Chromatographic analysis is critical for the above and, although chromatography cannot perform miracles, it did improve considerably from 1992 to now. Readers who want to know more about what is possible can consult papers from the Ionis group,[24,25] or Matthias Kretschmer's presentation at the DIA meeting in 2017 describing the use of an internal standard in denaturing chromatography of duplexes.[26] Further knowledge about the topic of impurities, their limits, and how to group them in some cases, can be found in the 2017 paper by Capaldi *et al.*,[14] and Cramer *et al.* provided comprehensive information on the origins of oligonucleotide impurities.[27] The last two references may be helpful during the preparation of a clinical trial application (CTA)/IND. It is important to highlight here the growing importance of using LC-MS for the analysis of the impurities. With the growing complexity of oligonucleotides used in clinical development, it is more frequent to have difficulties in separating completely all the impurities from the full-length product (FLP). In such cases, the purity of the API cannot be determined just by using the UV chromatogram, and a good quantitative LC-MS analysis is necessary.[25]

3.2.S.2.5 Manufacturing Process Validation

This part of Module 3 comes late during the development cycle and needs to be ready for the marketing application (NDA). For someone who has no experience in this topic, it is recommended to read the FDA guidance for industry entitled *Process Validation: General Principles and Practices.*[34] It is also recommended to start discussing this validation with the contract manufacturing organization (CMO) soon enough because it takes "some time". Before validation itself [the making of the process performance qualification (PPQ) batches], the prevalidation or pre-PPQ work needs to be planned and carried out, aiming at understanding all the critical steps of the process – once they have been defined – as precisely as possible, then making a confirmation batch and finally the three PPQ batches. Those four batches must be performed at the scale intended to be the commercial scale.

Considering a risk-analysis approach, at least four parts in the pre-PPQ activities can be considered:

- *In-process and specification – method optimization and validation.* When this process starts, the sponsor already has qualified analytical purity methods for the DS [usually high-performance liquid chromatography/ultra-performance liquid chromatography (HPLC/UPLC); anion-exchange (AEX) chromatography as an alternative orthogonal HPLC method], as these methods have been used for the preclinical and early-phase clinical batches. It could be assumed, however, that some level of method optimization may be required to put these methods in a good place for validation. It is important to make sure that important impurities are already well separated and identified. If it is not possible to separate all impurities from the active FLP, then when some impurities co-elute with the main peak(s) MS must be used for identification and quantification of those impurities. Also, when the methods are optimized, validation protocols can be developed for each of the methods and they will be validated according to ICH Guidelines.
- *Non-purity specification – methods validation or verification.* At some point during the later stages of clinical development for the DS, regulatory guidance will require all of the methods listed in the Product Specification to be validated or verified, specifically for the methods specific to the product (*i.e.* non-compendial). Verification is necessary for a United States Pharmacopeia (USP) method or other pharmacopeial method as that will demonstrate that the method is suitable for the DS. The specific methods, *i.e.* developed exclusively for the DS, will need to be qualified. These methods include identity methods, MS, ^{31}P NMR spectroscopy (if relevant for the specific DS), assay method, moisture content (USP ⟨921⟩), sodium content, elemental analysis (ICP-MS), pH, residual solvents, endotoxin, and microbial limits (USP ⟨85⟩ and USP ⟨61⟩). Each method validation will be carried out according to an approved protocol.
- *Raw material release – methods optimization and validation.* The CMOs and companies having already made current good manufacturing practice (cGMP) oligonucleotides have already validated methods for the release of the common phosphoramidites [A, C, G, T, U, 5-Me-C bases and 2′-deoxy-, 2′-OH, 2′-OMe-, 2′-F-, and 2′-methoxyethoxy (MOE)-sugars]. For all more uncommon nucleotides and oligonucleotide modifiers [say, for example, but not limited to, GalNAc reagent, special linker, locked nucleic acids (LNAs), *etc.*] a specific validation may be needed. This applies, in particular, to customer-supplied starting raw materials. The CMO will need to optimize and validate the release method for these uncommon raw materials.
- *Process optimization.* Before entering into the process characterization (pre-PPQ) activities, it may be necessary to perform some process optimization work, especially if PPQ batches are to be manufactured at a higher, commercial, scale than the development batches.

Validation of the commercial manufacturing process for the drug substance will then be based on the approved validation protocol. The validation will follow the process described in *Section 3.2.S.2.2* and the controls outlined in *Section 3.2.S.2.4* of the NDA. The validation will be conducted at the expected commercial manufacturing site, at the scale expected for commercial production. If two manufacturing sites are to be used, it is recommended to perform the validation at both sites, and the DS so obtained can be used for launch and commercial use. The validation process is ideally split into several modules, such as (i) synthesis and C&D (cleavage and deprotection), (ii) purification and subsequent ultrafiltration, (iii) annealing (for siRNAs), (iv) concentration (if needed), (v) lyophilization (if needed), and (vi) packaging. However, the sponsor can split the validation as they prefer, in agreement with the applicable regulations. The use of smaller "slices" allows more flexibility to modify and revalidate the process during the commercial life of the product if need be.

3.2.S.2.6 Manufacturing Process Development

This section contains the description of the process development if there has been such a development; and there is usually some. Nobody can start with a perfect process from the pre-IND stage, except, maybe, a sponsor that has already developed many oligonucleotides using the very same chemistry and scales from start to commercial stage. The sponsor should provide a description and discussion of the developmental history of the manufacturing process as described in the *Description of Manufacturing Process and Process Controls* (*Section 3.2.S.2.2*) section of the common technical document (CTD).

Note that it is important to retain samples of the DS and manufacturing intermediates, when possible, in the event that comparability studies may be requested during future product development or after licensure of the investigational DP. This section will include a summary of the DS batch history, describe and justify the various changes based on the results of the experiments that were performed to optimize, for instance, cleavage and deprotection reagents, time, and/or temperature, or a change in purification resin, or a change in annealing conditions for the duplex DS, or a change in the concentration method. It is important to provide data that demonstrate the scientific basis for such changes and how they maintain, and ideally improve, the quality attributes of the DS.

3.2.S.3 Characterization

The purpose of this part of the IND and IMP dossier is to describe the methods used to demonstrate by analytical testing that the batches to be used for clinical studies are representative of batches use in non-clinical toxicological studies. To achieve this, relevant analytical quality test results should be sufficient to allow comparisons of different batches of the product. Tests to accomplish this include the following:

- Identity [sequencing the single strands, molecular weight (MW) determination using MS (high-resolution MS is preferred when available; retention time in chromatography compared with a reference standard)].
- Purity [chromatography – ion-pair reversed-phase (IPRP) and/or AEX chromatography – agencies are now expecting orthogonal methods to be available when development progresses, but orthogonal methods are not necessarily two HPLC methods].
- Counter-ion content (usually sodium by ICP-MS or ICP-OES).
- Structure and physical characteristics [*e.g.* optical rotation (for chiral compounds), melting temperature (T_m), differential scanning calorimetry (DSC) (for duplex), and spectroscopic methods such as NMR, UV, Fourier transform infrared (FTIR), and circular dichroism – as appropriate].
- Assay for purity (UV spectrophotometry or chromatography).
- Impurity profile (*e.g.* product- and process-related impurities, moisture, residual solvents, heavy metals).
- Assay for potency (for oligonucleotides whose structure is difficult to assess by physicochemical methods, *e.g.* aptamers).
- Microbiological characteristics (as appropriate).

Regulatory advice on specifications for chemical substances is provided in ICH Guideline Q6A, which includes a list of universal tests applicable to all types of drug substance and drug product. Among the list is identity testing, which aims to confirm the identity of material claimed to be the active ingredient. For oligonucleotides, agencies expect to see some confirmation of the sequence in the dossier. Also, using direct sequencing of each strand by tandem mass spectrometry (MS/MS) during release testing of the DS is usually required.

This topic is currently under discussion. In a recent paper,[17] a group of scientists from EPOC discussed that requirement, and claimed that a survey of EPOC member companies indicated that although a standard approach to routine identity testing of oligonucleotide drug substances and products was not yet established, some general trends were evident. The paper indicated the following: (i) the consensus opinion (among EPOC members) is that the major concern in identity testing of an oligonucleotide DS is confirmation of sequence; (ii) there appears to be an expectation that the methods used routinely to confirm the sequence of an oligonucleotide DS are selective, and perhaps even specific, for the intended oligonucleotide among all other oligonucleotides that might possibly be synthesized from the same starting materials; and (iii) consistent with recommendations given in Q6A, there is an expectation that oligonucleotide DS identity testing includes a test to confirm counter-ion identity and quantity. In the case of DPs, expectations are less clear, but in general there does not seem to exist the same method specificity requirement around sequence confirmation, and the use of two identification techniques, at least one of which can be defended as sensitive to oligonucleotide sequence, is normally accepted without comment. In the same paper,[17] the authors examined the basis for the current practices regarding oligonucleotide DS and DP identity testing. After enumerating likely failure modes that lead to synthesis of the incorrect sequence and describing the controls employed to prevent failure, they affirmed that *de novo* confirmation of a DS sequence as part of routine batch release is overly conservative, and that more conventional approaches that rely on identity confirmation by comparison with a fully characterized reference standard provide equally effective control. They argued that the practice of *de novo* sequencing of oligonucleotide DSs is also beyond standard practices employed for routine identity testing of peptide DSs, which are also manufactured by solid-phase synthesis. They suggested that their proposed risk-based framework described in the paper[17] should be widely applicable to synthetic oligonucleotide DSs and DPs and help establish a harmonized approach to identity testing in support of most marketing applications. It will be stimulating to see the developments around this interesting paper.

In the meantime, it is prudent for sponsors to include a sequencing by MS/MS at some point, and early, in their plan and to discuss this question with agencies before a definitive decision. We can state that some IND applications have been accepted without MS/MS sequencing in the early dossier but that it was requested by the agency concerned to include that method later during the development.

In addition to MS/MS analysis, identity testing may also include comparison with a well-characterized reference of the DS (or its single-strand constituents in the case of a duplex) of (1) the retention time of the test item in some HPLC or UPLC analysis (ID by retention time of single strand or duplex), (2) the MW of the single strand (or strands for a duplex) using MS (ID by single-strand MW), or (3) the melting temperature of the duplex or of the single strand annealing to a fully complementary sequence using thermal UV spectrophotometry (ID by T_m). Considering that some specification lists for oligonucleotides contain up to four ID tests, and given all the controls inherent to solid-phase synthesis, combined with the risk assessment described in ref. 17, it is conceivable that, one day, it could be possible to organize the ID around comparison of retention time, MW and T_m values with those of a well-characterized reference standard.

3.2.S.3.2 Impurities

We can separate impurities in an oligonucleotide DS into five categories, listed in Table 14.4.

We will not speak much about the residual solvent and the elemental impurities, as they are clearly defined in ICH Q3C and Q3D. For a complete list of solvents, one can consult the Q3C tables and lists on the FDA website.

Regarding the residual solvents, there is no class 1 solvent (benzene, carbon tetrachloride, 1,2-dichloroethane, 1,1-dichloroethene, 1,1,1-trichloroethane) used in oligonucleotide synthesis.

Table 14.4 Impurity types in oligonucleotide drug substances.

Type of impurity	Description	Control
Product-related	Oligonucleotide impurities generated during the manufacturing process or storage (includes degradants)	Optimization of the process, optimal long-term storage, measured by chromatographic methods
Process-related	Low molecular weight organics such as reagents or by-products (*e.g.* protecting groups)	Usually purged during the process – historical data to demonstrate
Genotoxic	Potentially mutagenic organic impurities	ICH M7 assessment
Residual solvents	Volatile organic impurities – mostly solvents	Usually purged during the process – historical data to demonstrate ICH Q3C (R6)
Inorganic impurities	Reagents during the process and/or coming from environmental contamination	ICH Q3D assessment

Usually, the solvents used in oligonucleotide manufacture are used in synthesis, cleavage, and deprotection, and sometimes in purification. It is also possible to consider 2,6-lutidine to be similar in classification and limits to pyridine (which is a class 2 solvent), based on the similarity of their structures.

The ultrafiltration steps are effective in removing inorganic materials from the DS and levels are usually very low, mostly below the detection level. Historical data are useful to demonstrate the purging of them during the development, allowing the tests to be limited to the minimum. Fillon *et al.*[23] studied how to apply the purge factor for the risk analysis.

Regarding the elemental impurities, there is no heavy metal salt or metal catalyst intentionally used in the synthesis or purification of a DS. Sometimes, Rh, Pd, or others could be used in the synthesis of some conjugate or linker. It is prudent to make sure that those metals are checked at the starting material level. Authorities may ask about those catalysts. Here again, purging happens but it is helpful to have historical data to justify the final list of elemental impurities to be tested at the NDA stage.

It is also important to consider impurities with structural alerts, as stated in ICH M7. In the case of oligonucleotides, given that product-related impurities are structurally similar to the DS or its component's single strands, it is reasonable to consider those as being of no concern for genotoxicity. However, some products are formed during manufacture that are classified as possibly carcinogenic agents to humans (Group 2B), *e.g.* dichloroacetic acid, acrylonitrile, and acetamide. Risk assessment should be conducted using the approach of Teasdale *et al.*[28] cited in M7 and to evaluate the purging in any specific manufacturing process.

Concerning product-related impurities, we would first direct readers again to ref. 14 and 27 for extensive and comprehensive reviews of that topic. In summary, those product-related impurities are what we usually name $n - 1$,

$n - 2$, $n + x$, deletion (shortmers), addition (longmers), species with a sugar that differs from the intended structure [such as, but not limited to, minus an F (fluoride) if the DS contains 2′-*O*-F nucleotides], incompletely deprotected oligonucleotides, phosphodiester impurities (those oligonucleotides in which one, or more, non-bridging S atom in the phosphorothioate link is replaced with an O atom), and if applicable, DS with no or incomplete "linker" or conjugate. These impurities are efficiently removed or reduced during the chromatographic purification by using an appropriate fraction pooling strategy that ensures product specifications. The pooling strategy is based on and confirmed by in-process controls of the chromatographic fractions. Impurities in the drug substance are analyzed and controlled as part of the drug substance release specifications. According to ICH Q3A (R2), limits on the content and qualification of impurities in new DSs produced by chemical synthesis should be progressively set while the development of the drug progresses towards registration applications. Even though ICH Q3A (R2) does not specifically cover oligonucleotides, the author is of the opinion that it is suitable to identify and characterize impurities.[14]

The identification of the chemical nature of peaks corresponding to major impurities will be realized during the next phases of development of the product by means of a qualified analytical method specific for the DS. The MWs of the most abundant impurities will be measured by UPLC (with an ion-pair reversed-phase method) coupled with MS to determine their most likely structure. The most abundant impurities can also be synthesized and used for validation of the UPLC methods [*e.g.* for impurities such as $n - 1$, $n + 1$, CNEt, FLP-1PS (PO impurities), FLP without linker if appropriate]. The specifications of impurities will thus be set at this point, and will always be set at a lower level than the values of the impurities observed in the batches used for the toxicity studies.

In early stages, and for "qualification" of impurities by preclinical data, it is important to be covered by the findings from the toxicity studies already performed and the impurity profile of the toxicology batches. The peaks corresponding to the major impurities can be identified by their relative retention times (RRTs) in the chromatographic analysis. It is usual to present in the first IND a table that compares the quantities of impurities above 0.2% of the total area in the chromatographic analysis of the API batch. Typical chromatograms and overlays for representative batches can be provided. Also, it is important to show that there are no new impurities in the clinical batches compared with the toxicology batch(es).

Finally, we have the process-related organic impurities, those small-molecule impurities possibly coming from the process such as phosphoramidites, activators, sulfurizing agents, reagents or components of the capping reagent, and deprotection residues. As they are washed during the synthesis process, subsequent purification, and ultrafiltration step(s), the purge factor is important. It will be important to work on that purging to better understand and document how to apply it to oligonucleotide synthesis. The EPOC group reported some work on that purging.[22]

3.2.S.4 Control of the Drug Substance

3.2.S.4.1 Specifications

The sponsor should provide a list of DS specifications in the original IND submission. Specifications are defined as a list of tests, references to analytical procedures, and appropriate acceptance criteria used to assess quality. Acceptance criteria should be established and justified, based on data obtained from lots used in preclinical and/or clinical studies, data from lots used for demonstration of manufacturing consistency, data from stability studies, and relevant development data.

For products in the early stages of clinical development, very few specifications are finalized, and some tests may still be under development. However, the testing plan submitted in the original IND should be adequate to describe the physical, chemical, or biological characteristics of the DS necessary to ensure that the DS meets acceptable limits for identity, strength (potency), quality, and purity. A typical extensive specification list would include the following:

- appearance (visual inspection)
- identity by T_m (UV spectrophotometry – thermal)
- identity by mass (MW – USP ⟨736⟩)
- identity by retention time against reference (HPLC-UV) for FLP
- identity by relative retention time (HPLC-UV) for major impurities
- identity by sequencing (MS/MS)
- purity and total impurities
- impurities (individual when possible or by grouping)
- unspecified impurities/unidentified impurities (total)
- assay/content (UV spectrophotometry or HPLC)
- sodium counter-ion content (ICP-MS or OES – USP ⟨852⟩)
- pH (USP ⟨791⟩)
- water [Karl Fischer (KF) titration – USP ⟨921⟩]
- residual solvents (headspace GC – USP ⟨467⟩)
- elemental impurities (ICP-MS) – USP ⟨232, 233⟩
- bacterial endotoxins (USP ⟨85⟩)
- bioburden (USP ⟨61⟩).

Limits will progressively be introduced and adjusted as the process develops and is better understood and controlled. NDAs will contain limits for all the tests and all analytical procedures will be well defined. Today, authorities expect some limits for each specification, so "report results" instead of limit might be challenged by reviewers.

3.2.S.4.2 Analytical Procedures

The sponsor should provide a description of the analytical procedures used during manufacturing to assess the manufacturing process and product quality. This part of the IND is also phase appropriate. In the original IND submission, descriptions should have sufficient detail, including a description of system suitability control, so that agencies can understand and evaluate the adequacy of the procedures. The FDA recommends that sponsors develop detailed standard operating procedures (SOPs) for how analytical procedures are conducted at early stages of product development as a part of the control strategy. It is generally acknowledged by agencies that, during product development, analytical methods may be modified to improve control and suitability. However, assay control is critical during all phases of clinical development to ensure product quality and safety and to allow for comparability studies, following manufacturing changes.

Documentation submitted in support of the analytical procedures should describe in detail how a procedure is performed and should specify any reference standards, equipment, and controls to be used. Submission of information, such as individual SOPs or batch records, will generally not be necessary, provided that descriptions of the analytical procedures are sufficiently detailed in the IND. Contractor test reports are acceptable, provided that there is adequate description of the analytical procedure, test sensitivity, specificity, and controls.

Agencies are asking more and more for orthogonal methods for the analysis of DSs. It is clear that the complexity of oligonucleotides makes it difficult to achieve complete resolution of all components in such a complex mixture and perfect elimination of most of the impurities would reduce the yield to such an extent that making oligonucleotides would be uneconomic. It is also practically impossible to make reference standard for all possible impurities. Finally, there will be impurities that we cannot even predict until they have been observed once. If an anion-exchange method is used to analyze the oligonucleotide that has been purified on anion-exchange column, the risk is the impossibility of seeing, and of quantifying, impurities that co-elute with the main peak. That is why orthogonal (or complementary) methods are necessary. Orthogonal means things that cross at a right-angle, *i.e.* at 90°, or, in statistical terms, that are dependent on totally different parameters so that the results will be different. In terms of purity, it is likely to be the case when using an ion-exchange and a reversed-phase method, where charges direct the separation in the former method and hydrophobicity drives the separation in the latter.

For example, an siRNA strand was purified using a TSKgel SuperQ-5PW column, and in-process control was performed using a DNAPac PA200 ion-exchange column. Everything looked good until the sponsor decided to add an additional IPRP UPLC test method in parallel. When testing a certain batch, some "new impurities" appeared, after the FLP. These three impurities were close to each other, and they were also found in various small quantities in previous batches. All three had a similar MW, +98 Da compared with the parent FLP, which corresponded exactly to the dA + 98 described by Capaldi in his presentation at the DIA oligonucleotide therapeutics conference in October 2017. Without an orthogonal RP method, such an impurity may remain invisible.

3.2.S.4.3 Validation of Analytical Procedures

Complete validation of analytical procedures is not required for original IND submissions for phase 1 studies; however, the sponsor should demonstrate that specific test methods are suitable and appropriately controlled. This means that tests should be specific, sensitive, and reproducible and include appropriate controls or standards. The FDA recommends that sponsors use compendial methods when appropriate and qualify safety-related tests prior to initiation of clinical trials. This topic is described in more detail in ICH Q2 and related documents.

The sponsor and CMO or CRO will agree on a protocol and execute it. It is recommended to provide qualification or validation reports as addenda to the filing. Tables showing the characteristics of the methods that have been qualified/validated, the limits, and the results are expected in the IND application.

At an early stage, a minimum requirement is to provide data showing that the (U)HPLC method(s) used for the measurement of the purity of the oligonucleotides is/are stability indicating using a forced degradation protocol (acid, base, oxidation, high temperature). Then, providing a qualification for suitability, accuracy, specificity, linearity, range, and detection limit/quantification limit should be considered. The same qualified method should be used to compare the clinical batches with the impurity profile of the batch used for good laboratory practice (GLP)/toxicological studies.

3.2.S.4.4 Batch Analysis

This is a very important part of the control strategy. The sponsor should include a table with test results for batches (or lots) of the DS that have been manufactured at the time of submission (first IND and updates). For early-stage INDs, this may include at least toxicology lot(s) and the clinical-grade material. The FDA recommends that the sponsor gains adequate experience with a new clinical manufacturing processes prior to making clinical material. This is especially critical following technology transfer to a new manufacturing facility, when manufacturing changes occur during development, and when multiple manufacturing facilities will be utilized. Note that batches manufactured in different ways should be clearly identified in the submission. Information regarding process development of these materials should be outlined in the *Manufacturing Process Development* (*Section 3.2.S.2.6*) section of the CTD. The FDA also recommends that the sponsor updates this section of the IND as new batches are produced. The sponsor should indicate any batches that fail to meet release specifications and any action taken to investigate such failures.

3.2.S.4.5 Justification of Specification

The sponsor should provide justification for the DS specifications in the IND. It is recognized that acceptance criteria may be adjusted throughout the product development stages, based on both manufacturing and clinical experience. For early-stage clinical studies, assays used to characterize production lots may be more variable than those used in later-stage investigations. It is noteworthy that authorities seem less agreeable to the absence of limits for the quality attributes and, if asked, the sponsor would set broad limits (*e.g.* 6–9 for pH) because of the lack of experience and limited number of batches. At least the pH limits can be supported and justified based on the acid–base results of the forced degradation experiments.

For compendial tests, the justification can refer to USP guidelines, *e.g.* the bacterial endotoxin limit has been defined in accordance with the principles of USP ⟨85⟩ and the intended clinical doses to be used.

This is also a section that will change considerably from the first IND submission to the filing of the NDA.

3.2.S.5 Reference Standard Materials

The sponsor should provide information on the reference standards or reference materials used for testing the DS in the original IND submission. The FDA recommends that the sponsor provides the source and lot number, expiry date, Certificate of Analysis, when available, and/or internally or externally generated evidence of identity and purity for each reference standard.

It is recommended that the sponsor reserves and maintains enough material (*e.g.* part of a production lot) to serve as a reference material. The most appropriate batch to use as a reference standard in early phases is the batch used for GLP safety/toxicological preclinical investigation. That would provide a benchmark for the level of impurities in any further batch and provide an indication that those impurities have been tested in terms of potential toxic effects or safety issues.

The sponsor should have a protocol to evaluate the stability of the reference standard as agencies may ask, at some point, for evidence that it is still "in the specs", *e.g.*, when an update is submitted. Plan well in advance for the production and characterization of a new reference material. It is also critical to retain sufficient material to bridge to a new reference material as needed during product development.

3.2.S.6 Container Closure System

In the author's experience, oligonucleotide DSs are packaged in high-density polyethylene (HDPE) bottles and closed with an HDPE screw-cap. The entire bottle is also usually sealed in a laminated foil pouch. The pouch overwrap is also labeled with the appropriate product information, such as the bottle. All filled containers are quarantined for final release and stored at −20 ± 5 °C.

3.2.S.7 Stability

This is an important part of the IND, in which the sponsor should describe the type of container and closure used for the DS in the original IND submission, including the identity of materials used in the construction of the container closure system. It is recommended that the sponsor determines whether the containers and closures are compatible with the DS. For an original IND submission, compatibility with the DS may be evaluated during stability studies or may be based on historical data and experience, using similar products. The best type of primary container and second container, if applicable, should be similar to the type of container and closure described in *Section 3.2.S.6* (*Container Closure System*). For oligonucleotide DSs it is a good way to control how well the container and closure protect the DS from moisture. The most common oligonucleotide container type is HDPE bottles with caps made of either HDPE or polypropylene.

There are several ICH Q1 documents (A–F). In the author's opinion, the most important are the Q1A (R2) revision in 2003, which contains the main recommendations on protocols, and Q1E, which explains situations where extrapolation of retest periods or shelf lives beyond real-time data may be appropriate and provides examples of statistical data analysis. The author recommends that CMC persons should be familiar with the latter.

The first ICH-compliant stability study for an oligonucleotide DS will usually be carried out at three temperatures:

- −20 °C – long-term storage temperature for most oligonucleotides;
- 5 ± 3 °C – intermediate temperature – and possible long-term storage when enough data supporting that decision are available;
- 25 ± 2 °C [60 ± 5% relative humidity (RH) – accelerated temperature].

Another interesting condition could be

- 40 ± 2 °C (60 ± 5% RH – exaggerated temperature). As mentioned by DeCollibus *et al.*, 8 weeks at 40 °C simulate 2 years at 5 °C, and 20 weeks at 40 °C simulate 5 years at 5 °C.[19]

This approach gives important information, at the beginning of clinical trials, on the long-term stability of oligonucleotides, and good stability for a few months at such high temperatures will give confidence if the sponsor needs to extend the shelf life beyond real-time data. This will be discussed further in the *Drug Product* part of the chapter (*Section 3.2.P*).

A question that is often asked by sponsors working on their first oligonucleotides about stability studies is: what do we need at submission? The author's experience is that not much is necessary if the sponsor has the ability to justify the claims, so it is advisable to start to collect stability data as soon as possible, for instance for the first non-GMP engineering batch; 3 months' real-time data at −20, 5, and 40 °C help a lot. If the sponsor is worried that 40 °C would be too high, then 6 months at 25 °C could be an alternative. It is also possible to justify, say, a 12-month stability/shelf-life assessment based on historical data and experience, using a similar product, if available. Also, having stability data at accelerated temperatures for a considerable time is important for dealing with temperature excursions above the recommended storage or shipping temperature during transportation. Despite all efforts and precautions, it may happen that a deviation occurs during shipment of a DS (or DP). Stability data at 25 or 40 °C showing good resistance of the material will be helpful to justify that the deviation is not a serious event.

3.2.P Drug Product

Again, I will summarize the DP part, omitting the self-explanatory parts.

3.2.P.2 Pharmaceutical Development

At NDA filing, this section can be one of the largest. This *Pharmaceutical Development* section should contain information on the development studies conducted to establish that the product formulation, manufacturing process, container closure system, microbiological attributes, and instructions for use are appropriate for the stage of clinical development. The studies described in this part of the package are distinguished from routine control tests conducted according to specifications. Additionally, this section should identify and describe the formulation and process attributes (critical parameters) that can influence batch reproducibility, product performance, and DP quality. Agencies acknowledge that this information may be limited in the early phases of development and recommend that sponsors provide additional information and updates as product development proceeds. Supportive data and results from specific studies or the published literature can be included within or attached to the *Pharmaceutical Development* section. Additional supportive data can be referenced to the relevant non-clinical or clinical sections of the application.

3.2.P.2.2.1 Formulation Development

In this part, the sponsor should briefly describe the development of the DP formulation, taking into consideration the proposed route of administration and use in their IND.

At an early stage, especially if the formulation is very simple, this part is quite small. Most of the approved oligonucleotides and also those in development are simply dissolved in either water for injection (at a concentration of *ca.* 200 μg mL^{-1}) or in a saline buffer or in artificial cerebrospinal fluid. That is why not much development is necessary.

It is recommended to describe any other formulations that have been used in clinical or preclinical studies and provide a reference to such studies, if applicable (change in concentration, change of excipient, vials to syringe, solid to liquid, whatever is applicable). If formulation changes were needed for stability, device compatibility, or safety concerns, this information might be reported here. This part usually contains a list of the different batches of the DP that have been made, together with explanations about the choice of solvent, studies on viscosity if the DP is liquid, osmolality, pH, and container evolution during the development. One point to consider in the development is the question of overfill. The USP *General Chapter ⟨1⟩ Injections* provides that each container of an injectable product is filled with a volume that slightly exceeds the content indicated in the labeling. This allowable excess volume may also be referred to as "overfill", not to be confused with "overage", which is addressed in a separate guidance. Taking this overfill into account is important when planning for execution as when small volumes are filled in the vial (1 mL or less) that overfill (0.1 to 0.15 mL, according to ⟨USP⟩ 1151) can represent a significant proportion of it and, as such, significantly increase the quantity of drug substance needed for the filling.

3.2.P.2.2.3 Physicochemical Properties

For oligonucleotides presented as a liquid solution, the properties are generally pH, osmolality, density, and viscosity at 25 °C.

3.2.P.2.3 Manufacturing Process Development

The sponsor should describe the selection and optimization of the DP manufacturing process [described in the *Description of Manufacturing Process and Process Controls* (*Section 3.2.P.3.3*) section of the CTD] if development studies have been performed. Agencies are conscious that this information may be limited in the early phases of development and recommend that the sponsor provides additional information and updates as product development proceeds.

This section may include paragraphs dealing with parameters such as mixing speed and time studies, compatibility with manufacturing parts, bulk hold time evaluation, filter retention studies, fill volume studies, fill weight monitoring, and impact of order of addition of excipients if there are excipients other than salts or buffer in particular.

3.2.P.2.6 Compatibility

In theory, the sponsor should discuss the compatibility of the DP with the diluent used for reconstitution or the delivery device, as appropriate. For oligonucleotides, the DP is usually a solution in water for injection or a saline aqueous buffer, which limits the work considerably, compared with small molecules. The author recommends that compatibility studies include measures of both product quantity and product integrity, or even activity in the case of biologics or products whose structure is important in the activity of the DP. These in-use and in-device stability data should support recommended hold times and conditions outlined in the clinical protocol for patient administration. The absence of an understanding of in-use and in-device stability, and the potential impacts on product performance, may not justify risks associated with clinical study treatment(s). Therefore, it is recommended that the sponsor carefully controls and assesses DP compatibility and the final steps of product preparation and administration.

Attributes and parameters usually tested and validated in this part of the process are extractables, analytical evaluation threshold calculation, for a single-use system in manufacturing, for tubing, filters, syringes, *etc.* It is also in that section that the sponsor should describe their evaluation of the DP stability under terminal sterilization conditions.

Since 1st October 2019, the new European *Guideline on the Sterilization of the Medicinal Product, Active Substance, Excipient and Primary Container*[30] has been in effect. In general, terminal sterilization is preferred to sterilization by filtration and/or aseptic processing because it is lethal to microorganisms and it is possible to calculate, validate, and control a reliable sterility assurance level, and it thus incorporates a safety margin. For sterile filtration followed by aseptic processing, this is not applicable as accidental contamination caused by inadequate technique cannot be reliably eliminated by monitoring and control. Therefore, terminal sterilization provides the highest assurance of sterility and should be used whenever possible.

Those familiar with meetings discussing oligonucleotide CMC and regulatory topics will recognize in the list below many messages delivered several times by experts, including some working for regulatory agencies:

- Aseptic processing cannot be accepted based solely on an increase in impurity levels upon terminal sterilization without further justification.
- An increased level of impurities above the ICH Q3B identification or qualification limit does not necessarily preclude terminal sterilization of the medicinal product.
- Risks induced by the degradation should be balanced with the risk induced with an aseptic manufacturing method.
- Attempts to find terminal sterilization conditions adjusted to give acceptable impurity levels based on degradation mechanisms of the active substance and the actual bioburden should be described in the quality dossier.
- Applicants intending to use sterile filtration should have comprehensive data in place on the impurity profile after treatment at 121 °C for 15 min or other conditions described in the guideline.
- When a small change in impurity profile is observed, a change in the manufacturing process should be considered and sound justification for sterile filtration should be provided.
- Scientific advice/presubmission meetings with agencies are recommended for applicants planning to submit the Marketing Authorization Application (MAA) in the near future.

Hence the sponsor needs to present some data to support the aseptic filling, at least in Europe for the moment. At the DIA "Oligonucleotide-based" Therapeutics Conference (28–30th October 2019, North Bethesda, MD, USA), two scientists working for two different pharmaceutical companies presented data showing that it seems possible for some oligonucleotides to undergo some final sterilization cycle, meaning that degradation during the sterilization could be limited to 1–2%, and in those two examples, P=S to P=O. This requires efforts to develop an adequate formulation and container capable of undergoing such a cycle and they should be presented at submission of the MAA. It seems obvious that syringes, cartridges, and plastic containers present additional challenges and it is recognized that glass delamination is a potential risk for glass containers. Container closure should also be evaluated. However, it remains that all this will not be possible for every oligonucleotide therapeutic. The author is aware of other situations where the terminal sterilization cycle increased the level of impurity by more than 10% and up to 30% for some strands. Hence it is clearly not a method applicable to all nucleic acid (or peptide) therapeutics, and it is a fact that sterilization experience on small batches among some vendors is limited.

Now, the "gray area" is somewhat concerning: what is "limited degradation" in the eyes of Agency X? Will Agency Z share the same view? And should toxicological assessment now be carried out using "autoclaved" material to cover for impurities, now possibly considered as degradants?

Finally, as the authorities ask for consideration of "risk induced by the degradation should be balanced with the risk induced with an aseptic manufacturing method", the author would like to share his experience that, at the time of writing this chapter, since the first oligonucleotide was approved, any adverse event due to a possible issue related to aseptic filling has never been experienced. Hence we have to compare some new problematic process with what looks like a "non-existent" problem. At a time when citizens and organizations are voicing their negative opinion about the cost of drugs, the author wonders why one should address a non-existent problem in a way that will inevitably increase the cost of development?

3.2.P.3 Manufacture

3.2.P.3.3 Manufacturing Process and Control

The sponsor should provide here a detailed description of the DP manufacturing process and identify process controls and intermediate tests. The description should include both flow diagram(s) and narrative description(s), in addition to packaging, product contact materials, and equipment used.

This part of the IND is obviously one that is likely to see changes from initial submission to the end of the process, hopefully at NDA. Updated flow diagram(s) will be provided as needed and a brief step-by-step description of the manufacturing process should be given. The description can focus on the unit operation (*e.g.* blending) rather than the individual manufacturing steps of the unit operation. Information need not be provided on either the flow diagram or description of the equipment used (*e.g.* mixing system) and packaging and labeling process.

For sterile products, updates on the manufacturing process from that provided for phase 1 studies should be submitted. The phase 2 or 3 information should include changes in the drug product sterilization process (*e.g.* aseptic processing to terminal sterilization, or vials to syringes).

A special section near the end of this chapter will specifically address the question of "change control" (Section 14.6).

3.2.P.3.5 Process Validation and Evaluation

Process validation is not required for early-stage manufacturing, hence most original IND submissions will not include this information. Similarly, information related to the validation of the sterilization process need not be submitted until the end of phase 3, based on the FDA guideline on phase 2/3 studies of CMC information. On the other hand, as those validations require some time to perform, the sponsor should disclose its timelines to the NDA with the CMO in order to decide when the process should start to get everything completed on time.

It is recommended in guidelines that early-stage INDs have information on methods used to prevent contamination, cross-contamination, and product mix-ups. This description is usually provided by the CMO in charge of fill and finish, describing how they handle the validation of equipment, providing information about water for injection systems, monitoring of air pressure, particulate monitoring, and sterilization of materials like containers and stoppers. And again, the more the process develops and moves towards the NDA, the more details will be provided. The sponsor should discuss this with the CMO, ideally during the initial quality audit. Also, if there is a change of CMO, or a change inside the same CMO from one "line" to another "line" during development, it is critical to make sure that there is no gap at that level.

At an early stage, a typical description of the media fill validation may look like this: "Media fills that simulate filling operations are validated and are conducted at routine intervals – for instance every 6 months – to provide additional assurance of aseptic processing capability of staff, facility, and equipment. Associated programs that support aseptic processing, the proficiency of the aseptic processing personnel, and the compliance with cGMP requirements all link together with aseptic process simulation to assure and control the sterility of materials produced."

"The rooms are maintained under positive pressure with HEPA (high-efficiency particulate-absorbing) filtered air to reduce the possibility of airborne contamination. The performance of the filters is monitored periodically. Access to filling rooms is restricted to personnel associated with the filling operation who have completed the necessary training in aseptic operations. During aseptic operations such as filling and stoppering, the environment is monitored for microbial content and particulate content. An air sampler monitors the microbial quality of air. Quality assurance at the CMO in charge of the filling must establish appropriate action and alert levels for both the total particulate levels and microbial quality of the filling environment air. Operators, equipment, and surfaces are monitored routinely. All results are compared with established action limits for the area and, if the limits are exceeded, an investigation is conducted. Appropriate actions are taken based on the investigation findings."

All of this should be discussed with the CMO before the preparation of the IND, and even reviewed during the quality audit.

3.2.P.4 Control of Excipients

This part can be limited (as short as one page), or quite large if the DP contains new excipient(s), *e.g.* lipids for LNP formulation.

The sponsor should provide specifications for all excipients listed in *Section 3.2.P.2.1.2 Excipients* of the IND. In guidance documents, an excipient is any component, other than the active ingredient, that is intended to be part of the final product (*e.g.* water for injection, human serum albumin, dimethyl sulfoxide). If the excipient is compendial, information in the "*Specifications (3.2.P.4.1)*" section of the IND may not be required.

3.2.P.4.6 Novel Excipients

If the sponsor uses a new excipient [*e.g.* new lipid(s), new peptide], this part will look slightly like a part of *Section 3.2.S*. This part, describing excipients used for the first time in a DP or used for the first time in a route of administration, should provide full details of manufacture, characterization, and controls, with cross-references to supporting safety data (non-clinical and/or clinical) in a regulatory file submitted to the agencies, if available. Also, all new excipients will be manufactured and tested under cGMP guidelines, meaning validation of the manufacturing process and analytical methods at the appropriate time. The IND should contain one section for every new excipient present in the final DP. Also, everything written earlier in *Section 3.2.S.2* applies here too. Similarly, additional testing or qualification of compendial excipients should be described.

3.2.P.5 Control of Drug Product

This part will contain the specification of the DP. For most oligonucleotides, the quality attributes will contain tests also used for the DS [appearance, identity test for the DS, purity of the DS and impurities, assay, pH, elemental impurities (or not, depending on the formulation and manufacturing), bacterial endotoxins, and others specific to the DP]. The tests specific for the DP are also phase related.

A typical extensive specification list could include the following:

- appearance (visual inspection)
- identity by HPLC-MS (MW – USP ⟨736⟩ – single strands)
- identity by HPLC – duplex (HPLC-UV)

- purity and total impurities
- impurities by grouping
- unspecified impurities/unidentified impurities
- assay/content (UV or HPLC)
- pH (USP ⟨791⟩) (when solution)
- osmolality (USP ⟨785⟩) (when solution)
- water (KF titration – USP ⟨921⟩) (when dry powder)
- elemental impurities (when applicable)
- bacterial endotoxins (USP ⟨61⟩)
- sterility (USP ⟨71⟩)
- uniformity of dosage units (USP ⟨905⟩)
- container content for injection (USP ⟨697⟩)
- container closure integrity test (USP ⟨85⟩)
- syringe tests according to ISO.

Ideally, the sponsor will use the same UPLC method to analyze the DS in the DP. Depending on the complexity of the formulation, a suitability study or even a specific validation might be required. If some of the analytical procedures are the same as those for the DS, the sponsor does not need to repeat this information unless there is a matrix effect from the DP on assay performance.

Obviously, the greatest difference with the specification for the DS is that it is mandatory to test the DP for sterility. Also, validation of the sterility method will require a substantial number of vials or syringes. The first time the author had to face that was also the first time for the sponsor, and unexpected withdrawal from the batch was a problem. Hence a sponsor moving to fill and finish for the first time should start to discuss their needs with the CMO as soon as possible.

If the sponsor has different DP strengths, as may happen in early clinical phases, for dose escalation for example, a table of specifications per formula is expected. Otherwise, the structure of this *Section P5* is similar to that of *Section S5*, with descriptions of the validation of the analytical procedures, batch analysis, and characterization of impurities.

For the characterization of the impurities, when the formulation is very simple, such as a simple solution in water for injection or in saline, for example, it is possible to use the information in *Section S.3.2 Impurities* on the basis that no impurities beyond those described in *Section S.3.2* are expected as a result of the fill process. That, of course, is supposed to be supported by data coming from the drug development work.

3.2.P.6 Reference Standards

For simple formulations, such as a solution of DS in water for injection, or saline, the same reference standard as for the DS can be used. For more complex formulations, that practice could first be discussed with agencies. If the formulation contains new excipients, it will be necessary also to describe the reference standards for those components.

3.2.P.8 Stability

Everything here is similar to what was described in *Section S.7 Stability* for the DS part. The only specific item is that the stability of the DP will include analyses of sterility, particulate matter, container closure integrity, and deliverable dose/container content/volume in container.

During the development, *Section 3.2.P.8.1 (Stability Summary and Conclusion)* will contain more detailed analysis of the degradation products.

14.4.1 Placebo

If a placebo treatment is used in the clinical trial, a separate DP section should be provided for the placebo. In addition, the sponsor should provide a description of any accompanying reconstitution diluents and a description of the container and closure used for the dosage form and accompanying reconstitution diluent in a separate DP section, if applicable.

14.5 Assessment of the Oligonucleotide Manufacturing Process Against the Principles of Green Chemistry

A group of researchers from various pharmaceutical companies made an evaluation of this in a study published by the EPOC group.[20,21]

Clearly, oligonucleotide manufacture generates large amounts of organic and aqueous wastes from synthesis and purification, and the atom economy is quite poor. Some hazardous reagents are used [*e.g.* dichloroacetic acid, iodine (if oxidation is needed), ammonium hydroxide]. Many starting materials and reagents are from non-renewable sources. However chromatographic resin can be reused. The chemistry involved in the manufacturing processes makes extensive use of protecting groups. No catalytic process is used in oligonucleotide synthesis. Further, there is no real-time analysis of pollution prevention. In terms of safer solvents, there are some possible issues with acetonitrile and toluene. In terms of design for energy efficiency, most reactions proceed at room temperature. Energy use can be reduced if deprotection is made with ammonium hydroxide–methylamine (1 : 1 v/v) at room temperature, and if the purification is conducted at room temperature, better than at 40 or 60 °C. Working at room temperature also makes the process safer. Finally, oligonucleotides in general have a low toxicity and oligonucleotides generally break down into naturally occurring degradation products.

New technologies (sugar base and phosphate modifications, conjugation of oligonucleotides with functional or targeting ligands, and techniques useful in assembling, purifying, and isolating the desired species) have advanced the field to the point of commercial approval.

With this scientific maturation come opportunities to drive further efficiency into the process of oligonucleotide discovery, development, and manufacture. These include improving the sustainability of all operations for greener and more economical processes. Practices and tools that can be used to measure and improve the sustainability footprint and allow further advances in delivering high-quality oligonucleotide APIs to the market for the benefit of patients can be found in the literature.[20,21]

Some technologies, such as liquid-phase oligonucleotide synthesis or enzyme-catalyzed template ligation, could open ways for improvement, but they are still in the development stage. It is not easy to replace a chemical process that allows a complex molecule to be made after 80 consecutive reactions or more with an overall yield of 80% (or more) of FLP in the crude.

14.6 Change Control

During the development of a new therapeutic, several changes in the manufacturing process can be introduced. Needless to say, such changes must be described, documented, and justified. Change control aims to minimize risks and ensure that any modifications made within the pharmaceutical manufacturing process do not negatively impact product quality, safety, or efficacy. Change management is part of the quality system (ICH Q10 and related guidance in various agencies such as the FDA, EMA, or similar elsewhere). Change management is defined in ICH Q10 as a systematic approach to proposing, evaluating, implementing, and reviewing changes.

Steps to consider in change control are planning/scope, analysis and assessment of risks related to change, testing, results evaluation, implementation, and closure. It is necessary to write down why some change is needed, what the goals are and what will make a change successful, how the change will be tested, what the results are, and justification of the decision. It follows the Deming cycle in the quality improvement model: "Plan, Do, Study, and Act". Also, change control must be performed according to a SOP.

Every change control must have an owner, someone who will follow all the steps to make sure that the plan proceeds as planned. If the oligonucleotide DS or DP is manufactured by a contractor, both the contractor and the sponsor must work closely together, according to the signed quality agreement. When the change happens in a commercial stage, that may require a new validation before implementation in commercial production.

Change control may affect several steps in the oligonucleotide manufacturing process such as replacing one solid support for another one, scaling up one or more steps, replacing a type of equipment with a different one, changing the primary container, changing the supplier of the starting material, changes coming after corrective and preventive actions (CAPA) or a request from authorities, changing the manufacturing site, or changing the manufacturing route. In all cases, one of the most important expectations of authorities is to see a demonstration of the comparability of DS or DP between both processes to confirm acceptable toxicology coverage for impurities by both routes/processes.

14.7 Interactions with Authorities

In the author's experience, authorities do not ask many questions related to CMC modules in the early stages. This is usually due to the lack of precision of certain statements, to limited justification for some specifications, and to the absence of limits for some release tests. Most regulatory agency questions and requests fall into a handful of categories with a limited number of possible response types:

- Request for information; then provide the requested information.
- Request for clarification; then provide the clarification.
- Request for change; then there are a few options: (1) comply with the request, or (2) commit to complying with the request, or (3) push back or defend current practice (for instance, provide/generate additional supporting data), or (4) negotiate. Only the type of request for change will define the sponsor's position.

Questions from some agencies can arise because some have much less experience with oligonucleotides than others. Explaining and clarifying are then the key.

Based on the experience shared by some companies with, now, broad experience with IND and NDA submissions, we learned that the key is to make a realistic assessment of all the questions, and achieving a full understanding of the impact of the response commitment is critical, including the ability to obtain the information, the impact it may have on the program (sometimes on the platform if the sponsor has multiple similar products in their pipeline), the regulatory burden, *etc.* The key point is to make the decisions that will optimize the chance of success.

Based on public data, the main issues (>10% occurrence) in a review of oligonucleotides are the following, from highest to lowest: control of materials (starting materials and reagents), control of the DS and/or DP (mostly impurities), manufacturing process of the DP, pharmaceutical development of the DP, and control of excipients in the DP.

14.8 Some Take-home Messages by Way of Conclusion

Based on my experience, I would like to share some suggestions, outlined below.

- During development and early clinical phases, the CMC activities should find the best balance between respecting timelines, providing enough material for pre-clinical and clinical needs and reaching the best possible scientific excellence. I would recommend to executives of companies developing oligonucleotides with limited experience in drug development to make sure that all their "promises" to their boards in terms of timeline to first get a sense of what is possible with CMOs or consultants before committing to timelines that are almost impossible to match.

- The question of the reference material or authentic substance coming from CROs is sometimes a great issue when not included in the plan soon enough.
- Improving the oligonucleotide manufacturing in an early stage saves time and money later.
- Using cleaner raw materials from the beginning means less impurities, better crude, fewer purification issues, and eventually better yields and lower costs.
- Good development of synthesis and C&D also induces less impurities, better crude, better yields, and fewer issues.
- Process and resulting product quality can only be as good as the analytical methods used for testing the product. It is important to understand analytical method limits and their resolution power during process development and even when preparing early tox batches. Overall purity of the very early batches can be lower than what would be needed for late clinical phase or commercial supply as long as impurity profile is well understood.
- Try to harmonize analytical methods for DS and DP at any time possible and manage the transfer of the analytical methods seriously and carefully. It is always embarrassing to have to explain why the purity of the DS is a couple of percent higher in the DP than at the release of the DS. For that it is important to have a common management of CMC at the DS and DP level.
- Have good collaboration between toxicological, CMC, and medical teams to align all the needs and timelines.
- As a CMC consultant, I need to understand what non-clinical and clinical aspects need, and when, to build for success. One-third of my work consists in solving issues due to a lack of forward thinking and to emergencies such as "out-of-spec" due to poor technological transfer from the DS level and DP level.
- Each product and situation is different and early input from the FDA/agencies is always advisable (*e.g.* pre-IND meeting).
- One general EMA guidance that helps is the *Guideline on the Requirements for the Chemical and Pharmaceutical Quality Documentation Concerning Investigational Medicinal Products in Clinical Trials.*[31] In the absence of FDA guidance for oligonucleotide CMC, an online presentation, *CMC Regulatory Considerations for Oligonucleotide Drug Products: FDA Perspective*, can be suggested.[32]

List of Abbreviations

AEX	Anion-exchange (chromatography)
API	Active pharmaceutical ingredient = DS
ASO	Antisense oligonucleotide

C&D	Cleavage and deprotection
CAPA	Corrective and preventive actions
cGMP	Current good manufacturing practice
CMC	Chemistry, manufacturing, and controls
CMO	Contract manufacturing organization
CMV	Cytomegalovirus
CpG	Cytosine–phosphate–guanine
CRISPR	Clustered regularly interspaced short palindromic repeat
CRO	Contract research organization
CTA	Clinical trial application
CTD	Common technical document
DP	Drug product
DS	Drug substance = API
DSC	Differential scanning calorimetry
EMA	European Medicines Agency
EPOC	European Pharma Oligonucleotide Consortium
FDA	US Food and Drug Administration
FLP	Full-length product (API, DS)
FTIR	Fourier transform infrared
GalNAc	*N*-Acetylgalactosamine
GC	Gas chromatography
GLP	Good laboratory practice
GMP	Good manufacturing practice
HDPE	High-density polyethylene
HEPA	High-efficiency particulate-absorbing
HPLC	High-performance liquid chromatography
HPV	Human papilloma virus
ICH	International Council for Harmonisation of Technical Requirements for Pharmaceuticals for Human Use
ICP-MS	Inductively coupled plasma mass spectrometry
ICP-OES	Inductively coupled plasma optical emission spectrometry
ID	Identification
IMP	Investigational Medicinal Product
IND	Investigational New Drug
IPRP	Ion-pair reversed-phase (chromatography)
ISO	International Organization for Standardization
KF	Karl Fischer
LC-MS	Liquid chromatography-mass spectrometry
LNA	Locked nucleic acid
LNP	Lipid nanoparticle
MAA	Marketing Authorization Application – EU equivalent to US NDA
MOE	Methoxyethoxy
MS	Mass spectrometry
MS/MS	Tandem mass spectrometry
MW	Molecular weight
NCE	New chemical entity
NDA	New Drug Application – US equivalent to EU MAA

NMR	Nuclear magnetic resonance
OSWG	Oligonucleotide Safety Working Group
PPQ	Process performance qualification
QA	Quality attribute
RH	Relative humidity
RNAi	RNA interference
RRT	Relative retention time
siRNA	Small interfering ribonucleic acid
SOP	Standard operating procedure
SPS	Solid-phase synthesis
T_m	Melting temperature
UPLC	Ultra-performance liquid chromatography
USP	United States Pharmacopeia
UV	Ultraviolet

References

1. A. Khvorova and J. Watts, *Nat. Biotechnol.*, 2017, **35**, 238.
2. E. Paredes, V. Aduda, K. Ackley and H. Cramer, in *Comprehensive Medicinal Chemistry III*, Elsevier, 2017, ch. 6.1, p. 233.
3. For example, see:V. T. Ravikumar, R. K. Kumar, P. Olsen, M. N. Moore, R. L. Carty, M. Andrade, D. Gorman, X. Zhu, I. Cedillo, Z. Wang, L. Mendez, A. N. Scozzari, G. Aguirre, R. Somanathan and S. Bernees, *Org. Process Res. Dev.*, 2008, **12**, 399.
4. D. C. Capaldi and A. N. Scozzari, in *Antisense Drug Technology: Principles, Strategies and Applications*, ed. S. T. Crooke, CRC Press, Boca Raton, 2nd edn, 2008, p. 401.
5. E. W Alton, H. A. Boushey, H. Garn, F. H. Green, M. Hodges, R. J. Martin, R. D. Murdoch, H. Renz, S. B. Shrewsbury, R. Seguin, G. Johnson, J. D. Parry, J. Tepper, P. Renzi, J. Cavagnaro and N. Ferrari, *Nucleic Acid Ther.*, 2012, **22**, 246.
6. D. Kornbrust, J. Cavagnaro, A. Levin, J. Foy, P. Pavco, C. Gamba-Vitalo and A. Guimond, *Nucleic Acid Ther.*, 2013, **23**, 21.
7. M. Lindow, H.-P. Vornlocher, D. Riley, D. J. Kornbrust, J. Burchard, L. O. Whiteley, J. Kamens, J. D. Thompson, S. Nochur, H. Younis, S. Bartz, J. Parry, N. Ferrari, S. P. Henry and A. A. Levin, *Nat. Biotechnol.*, 2012, **10**, 920.
8. C. L. Berman, K. Cannon, Y. Cui, D. J. Kornbrust, A. Lagrutta, S. Z. Sun, J. Tepper, G. Waldron and H. S. Younis, *Nucleic Acid Ther.*, 2014, **24**, 291.
9. J. Cavagnaro, C. Berman, D. Kornbrust, T. White, S. Campion and S. Henry, *Nucleic Acid Ther.*, 2014, **24**, 313.
10. S. P. Henry, R. Seguin, J. Cavagnaro, C. Berman, J. Tepper and D. Kornbrust, *Nucleic Acid Ther.*, 2016, **26**, 210.
11. C. L. Berman, S. A. Barros, S. M. Galloway, P. Kasper, F. B. Oleson, C. C. Priestley, K. S. Sweder, M. J. Schlosser and Z. Sobol, *Nucleic Acid Ther.*, 2016, **26**, 73.

12. J. L. Marlowe, V. Akopian, P. Karmali, D. Kornbrust, J. Lockridge and S. Semple, *Nucleic Acid Ther.*, 2017, **27**, 183.
13. D. C. Capaldi, K. Ackley, B. Brooks, J. Carmody, K. Draper, R. Kambhampati, M. Kretschmer, D. Levin, J. McArdle, B. Noll, R. Raghavachari, I. Roymoulik, B. P. Sharma, R. Thurmer and F. Wincott, *Drug Inf. J.*, 2012, **46**, 611.
14. D. C. Capaldi, A. Teasdale, S. Henry, N. Akhtar, C. den Besten, S. Gao-Sheridan, M. Kretschmer, N. Sharpe, B. Andrews, B. Burm and J. Foy, *Nucleic Acid Ther.*, 2017, **6**, 309.
15. A. Tivesten, N. Akhtar, J. Stolee, Y. Fillon, R. Orr, M. L. Hill, L. Van den Bergh, T. Huybrechts, J. Muslehiddinoglu, P. McCormac, J. Amery, A. Brecht, A. Geiser, T. Metzenthin and T. Eichinger, *Ther. Innovation Regul. Sci.*, 2019, **52**, 687.
16. J. Muslehiddinoglu, R. Simler, M. L. Hill, C. Mueller, J. H. A. Amery, L. Dixon, A. Watson, K. Storch, C. Gazziola, F. Gielen, S. A. Lange, J. D. Prail and D. P. Nesta, *Nucleic Acid Ther.*, 2020, **30**, 189.
17. D. C. Capaldi, N. Akhtar, T. Atherton, D. Benstead, A. Charaf, T. De Vijlder, C. Heatherington, J. Hoernschemeyer, H. Jiang, U. Riede, F. Ring, R. Peter, J. Stolee and R. Wechselberger, *Nucleic Acid Ther.*, 2020, **30**, 249.
18. C. J. Borths, T. Burr, A. Figuccia, J. Gair Ford, G. Bing, M. T. Joes, D. Klingeleers, S. Lochner, A. A. Rodriguez and C. Wetter, *Org. Process Res. Dev.*, 2023, **27**, 1693.
19. D. P. DeCollibus, J. Searcy, A. Tivesten, N. Akhtar, C. Lindenberg, N. Abarrou, S. Pradhan, M. Fiandaca, J. Franklin, G. Govindan, H. Y. Liu, D. Royle, P. Lim Soo and K. Storch, *Nucleic Acid Ther.*, 2023, **33**, 159.
20. B. I. Andrews, F. D. Antia, S. B. Brueggemeier, L. J. Diorazio, S. G. Koenig, M. E. Kopach, H. Lee, M. Olbrich and A. L. Watson, *J. Org. Chem.*, 2021, **86**, 49.
21. L. Ferrazzano, D. Corbisiero, A. Tolomelli and W. Cabri, *Green Chem.*, 2023, **25**, 1217.
22. W. F. Kiesman, A. K. McPherson, L. J. Diorazio, L. Van den Bergh, P. D. Smith, J. M. Northall, A. Fettes, T. Wang, M. Mehlmann, S. Raza and G. Held, *Nucleic Acid Ther.*, 2021, **31**, 93.
23. Y. A. Fillon, N. Akhtar, B. I. Andrews, D. Benstead, S. Breitler, R. S. Gronke, M. Olbrich, J. A. Stolee and T. Vandermeersch, *Org. Process Res. Dev.*, 2022, **26**, 1130.
24. S. G. Roussis, C. Koch, D. Capaldi and C. Rentel, *Rapid Commun. Mass Spectrom.*, 2018, **32**, 1099.
25. C. Rentel, H. Gaus, K. Bradley, N. Luu, K. Kolkey, B. Mai, M. Madsen, M. Pearce, B. Bock and D. Capaldi, *Nucleic Acid Ther.*, 2022, **32**, 206.
26. M. Kretschmer, *Control of Double Stranded siRNA Therapeutics*, conference – session 2, track A, at DIA/FDA Oligonucleotide-Based Therapeutics Conference, October 25–27, 2017.
27. H. Cramer, F. J. Finn and N. D. Sinha, in *Analysis of Oligonucleotides and Their Related Substances*, ed. G. Okafo, D. Elder, and M. Webb, ILM Publications, St. Albans, 2013, p. 21.

28. A. Teasdale, D. Elder, S.-J. Chang, S. Wang, R. Thompson, N. Benz and I. Sanchez Flores, *Org. Process Res. Dev.*, 2013, **17**, 221.
29. T. Rupp and H. Cramer, in *RNA Therapeutics: The Evolving Landscape of RNA Therapeutics*, Elsevier, 2022, p. 263.
30. *EMA Guideline EMA/CHMP/CVMP/QWP/850374/2015*, 6th March 2019.
31. *EMA Guideline EMA/CHMP/QWP/545525/2017*, Rev. 2, 31st January 2022.
32. M. Sapru, CMC Regulatory Considerations for Oligonucleotide Drug Products: FDA Perspective, 2017, http://pqri.org/wp-content/uploads/2017/02/3-SapruPQRI-FDA-Conference-Oligo-2017-Presentation.pdf.
33. M. Lemaitre, CMC: Regulatory landscape, in *RNA Therapeutics*, ed. P. H. Giangrande, V. de Franciscis, and J. J. Rossi, Academic Press, 2022, pp. 321–352.
34. https://www.fda.gov/files/drugs/published/Process-Validation--General-Principles-and-Practices.pdf

CHAPTER 15

Greening Chromatographic Purification Processes

C. DE LUCA, M. CATANI*, A. CAVAZZINI AND S. FELLETTI

Department of Chemical, Pharmaceutical and Agricultural Sciences, University of Ferrara, via Luigi Borsari 46, 44121 Ferrara, Italy
*E-mail: martina.catani@unife.it

15.1 Introduction

Over the past few years, there has been growing awareness in modern society regarding global threats concerning environmental issues, such as climate change and environmental pollution. In response to these challenges, governments and authorities have implemented specific programs for sustainable development, such as the *European Green Deal*, with the ambitious goal of stimulating economic growth by investing in green technologies. From the perspective of decreasing the footprint of human activities on the environment, green chemistry, applied in different areas, plays a pivotal role. For instance, a Pharmaceutical Roundtable established in 2005 by the ACS Green Chemistry Institute was created with the purpose of incorporating green chemistry principles into the pharmaceutical industry, starting from small molecules to biotherapeutics.[1]

The term *biotherapeutics* refers to active substances deriving from a biological source, but it is currently employed, by extension, to embrace proteins, hormones, monoclonal antibodies, antibodies, gene cell therapy products,

Green Chemistry Series No. 85
Sustainability in TIDES Chemistry: Green Approaches to Oligonucleotides and Oligopeptides Synthesis
Edited by Alessandra Tolomelli, Lucia Ferrazzano and Walter Cabri

Published by the Royal Society of Chemistry, www.rsc.org

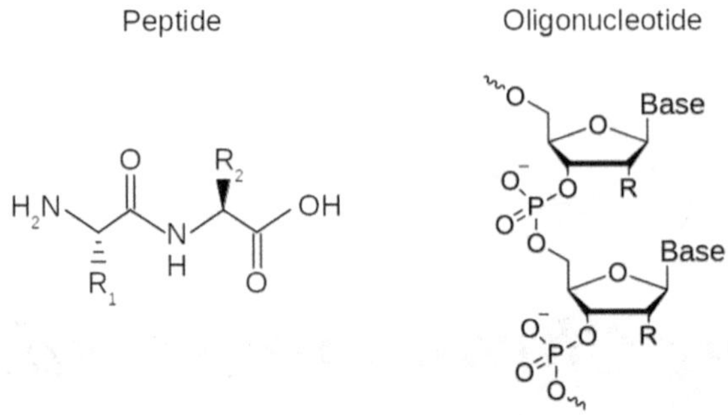

Figure 15.1 Basic structures of peptides and oligonucleotides.

vaccines, and others, including peptides and oligonucleotides (ONs).[2] Peptides and ONs, especially, are commonly referred to as "TIDES".

Peptides are short chains of amino acids, usually with molecular weight between 500 and 5000 Da.[3] On the other hand, ONs are short nucleic acid sequences, such as small single- or double-stranded segments of DNA or RNA made of 10–50 nucleotides, that can modulate gene expression (see Figure 15.1).[4,5] This class includes compounds with very different chemistries: small interfering RNA (siRNA), microRNA (miRNA), antisense oligonucleotides (ASOs), aptamers, and others.

Both peptides and ONs play important roles in biological processes and have attracted significant attention as potential therapeutics for several diseases; therefore, their use is being explored for therapeutic applications. The main driving areas for the pharmaceutical employment of peptides, for instance, are cancer and metabolic diseases (mainly diabetes), but also gastrointestinal, neurological, and heart issues. During the Covid-19 pandemic, 21 peptide drugs were used for disease treatment. Some peptides were even included in the formulation of some Covid vaccines.[6–8] The potential of ONs, on the other hand, lies particularly in areas such as gene therapy, oncology, rare genetic disorders, and neurodegenerative diseases (such as amyotrophic lateral sclerosis, Huntington's disease, and spinal muscular atrophy).

In recent decades, the interest in both of these classes has continued to grow, after overcoming some issues related to poor pharmacokinetics or bioactivity.[9] The value of the global market for therapeutic peptides reached US$ 43 billion in 2022, and it is expected to increase to more than US$ 75 billion by 2032.[10] As of 2018, more than 60 peptides were already available on the market, and almost 400 were in clinical or preclinical phases.[11–13] In a report by Bachem in October 2023, it was stated that more than 40 peptide-based drugs are currently in phase 3 of clinical development, which represents the step that definitively assesses if the drug is safe and effective or not. On the other hand, at the end of 2022, 15 ONs for the treatment of neurological, cardiovascular, hepatic, and ophthalmic disorders had been approved by the US Food and Drug Administration (FDA),[4] and 180 ONs were under clinical trials.[9]

The reasons for the popularity of these biotherapeutics lie in the main advantages that they bring about, such as the following:

- they can modulate a number of disease targets;
- synthetic strategies are robust and well established;
- safety aspects, since biomolecules do not tend to accumulate in the human body.

To satisfy the growing market demand, environmentally and economically sustainable methods for the production of ONs and peptides are needed. The production of biomolecules includes upstream processing, where the target product is produced by means of chemical synthesis or a recombinant approach, and downstream processing, which includes one or more purification steps. Nowadays, both peptides and ONs are generally produced through chemical synthesis, in either the liquid or solid phase, by exploiting a cyclic approach in which one monomer (amino acid or nucleoside) is added, in the desired order, at the end of the growing chain.[14] Functional groups that must not react are usually protected to minimize side reactions: a 4,4′-dimethoxytrityl (DMT) protecting group is used for ONs, while solid-phase peptide synthesis, developed by Merrifield in 1963, involves the use of *tert*-butoxycarbonyl (Boc) and 9-fluorenylmethoxycarbonyl (Fmoc) for peptides.[15] The protecting groups are removed at the end of the synthetic process to release the product. An alternative approach to obtain the product is to employ the recombinant strategy, which involves the use of organisms such as bacteria (*e.g. Escherichia coli*) for the production of the target biomolecule.[16,17]

Concerning downstream processing, the aim of the purification procedures is to remove impurities contained in the synthetic mixture to meet the strict purity requirements imposed by regulatory agencies for every active pharmaceutical ingredient (API). Regardless of the technique chosen, it is inevitable that the crude reaction mixture produced during the upstream step will be characterized by the presence of several impurities. Solid-phase synthesis, for instance, produces diverse impurities with chemical structures similar to that of the target product, resulting from incomplete coupling reactions or from side reactions. Some examples of ON impurities are the so-called "failure sequences" (shortmers, longmers) and the by-products resulting from deamination, depurination, and oxidation, including diastereomers deriving from the chirality of sulfur atoms contained in phosphorothioate ONs.[18,19] Despite the adoption of strategies to make the synthetic route more selective, all production methods lead to a crude mixture containing a series of unwanted impurities, which must subsequently be removed during the purification steps. Obviously, the longer the target peptide or ON, the greater the number of synthetic steps necessary, and the higher the number of impurities that will be produced together with the main product. Sometimes, these biomolecules also need to be modified or functionalized to improve their pharmacokinetic properties, such as in the case of liraglutide, which contains a palmitoyl chain on one amino acid.[20] This requires even more synthetic steps.

Particularly, impurities can be denoted "process-related impurities", deriving from the production process itself (*e.g.* pieces of host cells, DNA, viruses, *etc.*), or "product-related impurities", which are modifications of the main product naturally occurring during the upstream processing (*e.g.* isoforms, fragments, aggregates, shorter peptides, or ONs where the monomers have not completely reacted or have reacted in a wrong position, *etc.*). Whereas the former group of impurities can be relatively easily removed, the latter present difficulties due to the chemical similarity between the target and the unwanted species. Therefore, developing an analytical method that could selectively discriminate between a biomolecule of interest and all its related impurities is of the utmost importance. The challenge is to separate species that differ in only a very small portion of their structure, even by just one amino acid or nucleoside.[21,22]

At an industrial level, preparative liquid chromatography (LC) performed on a single column (batch) is the technique of choice for downstream processing, owing to its practical advantages such as its flexibility, versatility, and selectivity. A suitable modification of the experimental conditions can make it possible to distinguish between molecules that are slightly different in size, charge, or hydrophobicity. Several methods based on LC are now well established and robust, and for this reason they have been approved by regulatory agencies. However, despite being widely used for purifying (bio) products, preparative LC stands out as one of the most resource-intensive techniques in terms of energy consumption and, especially, waste generation. This is attributable, on the one hand, to the energy expended by the instrumentation, the significant dilution that occurs during elution, and the subsequent energy needed to evaporate the solvent from the pool containing the purified product. On the other hand, in preparative applications, large volumes of crude feed are injected into columns that are larger than those generally used for standard analytical LC and, therefore, they are operated at higher flow rates, ranging from 10 to 50 mL min^{-1} under preparative conditions to several liters per minute in production-scale chromatography.[23,24] Consequently, huge volumes of aqueous and organic mixtures are generated daily as waste, posing challenges with their disposal from the point of view of associated costs and environmental impact. Indeed, organic solvents used in chromatographic applications are often toxic towards humans and/or the environment. All these aspects make the downstream processing stage a critical bottleneck in terms of sustainability within the overall production of a target (bio)pharmaceutical.[25]

Also, preparative LC suffers from an intrinsic drawback concerning a trade-off between purity and yield, as will be explained later. Basically, this implies that it is not possible to achieve at the same time very good purity and a very high recovery, owing to the overlap between the main peak and those of impurities. Coelution regions need to be reprocessed, which leads to a decrease in productivity and an increase in solvent consumption.

The necessity to perform more productive methods goes hand-in-hand with the need to find green alternatives to the traditional organic solvents employed in LC.[26] The relationship between these two research lines is decreases in both the amount of waste generated and its toxicity in order to achieve sustainable pharmaceutical process development. Modern trends focusing on these two aspects are reviewed in this chapter, after an overview on the current state-of-the-art, and include the scale-up to continuous multicolumn chromatography and the introduction to innovative green solvents as chromatographic mobile phases. Also, the possibility of using innovative columns and exploiting theoretical modeling during process development is considered.

15.2 State-of-the-Art

In this section, current purification techniques used for the isolation of peptides and ONs are discussed.

15.2.1 Current Purification Methods

As already stated, the upstream process often leads to a starting crude feed composed of a very complex mixture, due to the presence of a series of impurities with physicochemical properties similar to those of the target product. These undesired compounds need to be carefully removed through chromatographic step(s).

The purification method should be chosen based on its ability to discriminate and provide adequate selectivity between molecules with similar characteristics in terms of size, hydrophobicity, or charge. In this context, the main technique used for the purification and isolation of biomolecules, such as peptides or ONs, is preparative LC under batch conditions, thanks to its versatility and flexibility. Owing to the complexity of the feed mixture, purification methods generally involve different and subsequent steps consisting of (see Figure 15.2) (i) equilibration, where the stationary phase is prepared to adsorb the analyte of interest, under conditions that favor adsorption, (ii) loading, where the feed mixture is injected into and adsorbed on the column, (iii) washing, where non-retained or weakly retained impurities are eluted from the column, (iv) elution, where the target peak is eluted by increasing the elution strength of the mobile phase, such as through a gradient of salt or organic modifier, and (v) stripping, where the mobile phase used shows such a high elution strength that even strongly retained impurities or contaminants are eluted, so the column is cleaned and ready for the beginning of a new run. Along the gradient, fractions are collected and successively analyzed offline [by high-performance liquid chromatography (HPLC)] to decide which portion of the peak satisfies the purity requirements imposed. The outcome and performance of the separation can then be quantified through four parameters: purity, recovery, productivity, and solvent consumption.

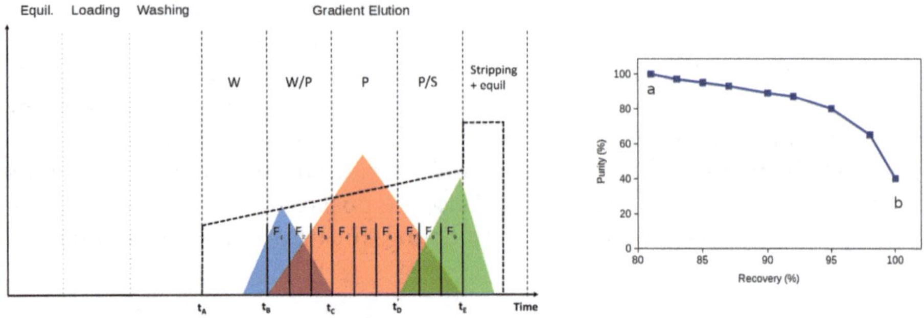

Figure 15.2 Left: General purification scheme consisting of equilibration, loading, washing, (gradient) elution, and stripping steps. W, weak impurities; P, product; S, strong impurities; W/P and P/S indicate overlapping regions between impurities and product; F_1–F_9, fractions. Bold dashed lines indicate the variation of the mobile phase composition during the elution and stripping steps. Right: Pareto curve obtained with fractions F_1–F_9. Point a indicates the purest fraction; point b was obtained by pooling all the fractions.

Chromatographic purity is calculated as the ratio of the area of the target peak (A_{target}) to the area of all the peaks in the analytical chromatogram (A_{tot}), which also includes impurities:

$$\text{Purity} / \% = \frac{A_{\text{target}}}{A_{\text{tot}}} \times 100 \tag{15.1}$$

Recovery is obtained as the ratio of the mass of the target contained in one or more fractions (m_{target}) to the total mass injected (m_{inj}):

$$\text{Recovery} / \% = \frac{m_{\text{target}}}{m_{\text{inj}}} \times 100 \tag{15.2}$$

Productivity is defined as the mass of target in one or more fractions per unit time and volume of the stationary phase (estimated as the geometrical volume of the column, CV):

$$\text{Productivity} / \text{g L}^{-1}\ \text{h}^{-1} = \frac{m_{\text{target}}}{\text{time} \times \text{CV}} \tag{15.3}$$

Finally, solvent consumption is expressed as the ratio of the total volume of mobile phase used during the whole run to the mass of the target collected in one or more fractions:

$$\text{Solvent consumption} / \text{mL mg}^{-1} = \frac{\text{volume}}{m_{\text{target}}} \tag{15.4}$$

Based on the off-line fraction analysis results, the so-called Pareto curve can be obtained. The graph, shown in Figure 15.2, is constructed by plotting purity *versus* recovery values starting from the purest fraction (point a) and then adding adjacent fractions of decreasing purity, until reaching the complete recovery of the target peak (point b). These points correspond to hypothetical fraction pools, indicating the optimal compromise between purity and recovery. Indeed, recovery can only be increased by lowering the purity and *vice versa*. This is also called the *purity–yield trade-off*, related to preparative batch chromatography. The main peak (orange) elutes as intermediate between more weakly adsorbed (blue) and more strongly adsorbed (green) groups of impurities, with which it overlaps on its front and rear side, respectively. This case is commonly referred to as center-cut or ternary separation.

Traditional reversed-phase (RP) chromatography, *i.e.* without the use of ion-pairing reagents, is not the optimal choice when separating peptides or ONs, owing to their net charge. Indeed, as an example, in aqueous solutions with pH >4, ONs present negative charges, since the pK_a of the phosphodiester linkage is 2.[27] Hence the main techniques used to purify peptides and ONs are RP chromatography with the addition of ion-pairing reagents [ion-pair chromatography (IPC)], hydrophobic interaction chromatography (HIC), and ion-exchange chromatography (IEX). A schematic representation of these elution modes is illustrated in Figure 15.3. Both IPC and HIC separate molecules based on their hydrophobicity, whereas IEX separates based on the net charge.

Since the retention of biomolecules is highly influenced by the percentage of organic modifier or the salt concentration in the mobile phase, gradient elution is usually preferred to improve peak shapes and resolution with respect to the isocratic mode, meaning at constant mobile phase composition. In gradient elution, the elution strength of the mobile phase is increased by increasing (or decreasing for HIC) the percentage of the organic modifier or the salt concentration.

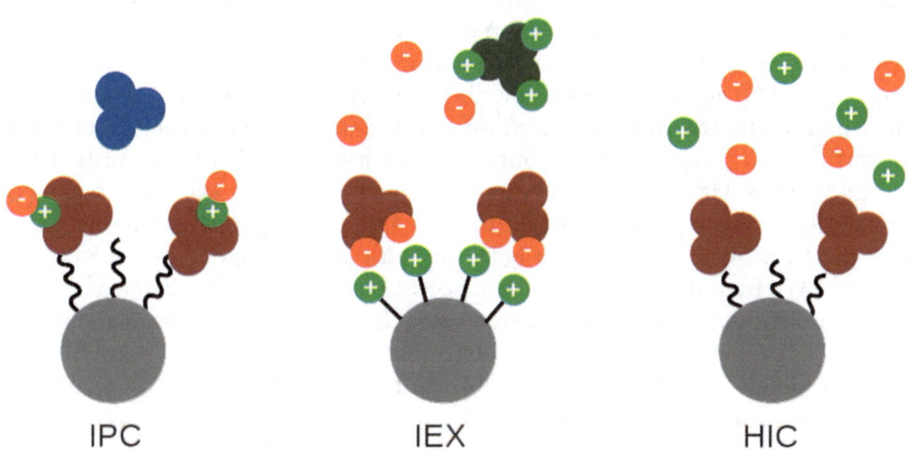

Figure 15.3 Schematic representation of the main elution modes used for peptide and oligonucleotide purification. IPC, ion-pair chromatography; IEX, ion-exchange chromatography; HIC, hydrophobic interaction chromatography.

Most of the time, however, a single chromatographic step, *i.e.* based on one elution mode, is unsuitable to meet the strict purity requirements needed to guarantee product safety and quality, owing to the complexity of the initial crude mixture. Therefore, multiple chromatographic steps, combining different elution modes or different ion-pairing reagents, are routinely required to isolate pure amounts of target products from other impurities. This approach is referred to as orthogonality.

15.2.2 Ion-pair Chromatography

In IPC, an ion pair is added to the mobile phase, consisting of a mixture of water and an organic solvent, to minimize mixed-mode and ionic interactions, and interacts with charged analytes through the formation of electrostatic bonds, leading to neutral complexes that can be separated on neutral stationary phases based on hydrophobicity. Hydrophilic analytes will be less retained and will then elute first from the column.

IPC represents the most appropriate technique for the purification of ONs and peptides in terms of buffers, sample concentration, and instrument requirements.[27] The typical purification conditions in IPC consist of gradient elution with acetonitrile (ACN)–water mixtures with additives (buffers or acids alone) using silica-based achiral stationary phases (C_4–C_{18}). The main concern with these stationary phases is the limited working pH range, usually 2.0–8.0, due to the dissolution of silica at higher pH.

For ONs, the most common ion-pairing reagent is triethylammonium acetate (TEAA). Other possible ion-pairing reagents include tetrabutylammonium hydrogensulfate, tetrabutylammonium phosphate, tetrabutylammonium bromide (TBAB), tributylammonium acetate (TBAA), tetrabutylammonium acetate, triethylammonium acetate (TEAB), hexylammonium acetate (HAA), and trimethylamine in combination with hexafluoroisopropanol (HFIP). The use of HFIP and TEAB could also be advantageous in the case of mass spectrometric (MS) detection coupled with HPLC. With IPC the DMT group is left attached to the ON to favor retention and the separation of the target from impurities without the DMT group. The two main disadvantages of IPC are that only ONs with 8–40 bases can be resolved and that a salt exchange step is usually necessary to use the purified ON in living organisms. Indeed, after the purification step, the target is in the ion-pairing reagent salt form, usually TEAA, which needs to be removed owing to its moderate toxicity.[27]

Perfluorinated homologous series of acids can be used as ion-pairing reagents for peptide purification [trifluoroacetic acid (TFA), pentafluoropropionic acid (PFPA), and heptafluorobutyric acid (HFBA)]. Among them, TFA is the most commonly used ion-pairing reagent for the cleavage step after solid-phase synthesis and separation of peptides. This leads to the recovery of cationic peptides as trifluoroacetate salts. However, perfluorinated compounds are chemically very stable and resistant to biological degradation.[28] TFA may have potential negative effects on the physicochemical characteristics of peptides, both in biological experiments and also in the environment.[25] Hence, when used as a counterion for peptides it needs to be carefully removed and replaced with other less toxic alternatives, such as chlorides and acetates, prior to lyophilization. Other acids, such as acetic, formic, phosphoric, and hydrochloric acid, can be used as alternatives to TFA.

15.2.3 Ion-exchange Chromatography

IEX allows the separation of charged analytes based on their net charge, with the help of stationary phases functionalized with negatively or positively charged groups, the counterions of which are exchanged with analytes bearing the same charge. The retention mechanism is then regulated by electrostatic interactions.

IEX can be divided into cation-exchange (CEX) chromatography, if the stationary phase has a negative charge, and anion-exchange (AEX) chromatography, if the stationary phase has a positive charge. Commonly used CEX packings consist of carboxyl (weak) or sulfonate (strong) groups, whereas AEX packings are made of primary, secondary, and tertiary (weak) or quaternary amine (strong) groups. IEX is often used to purify both basic (net positive charge) and acidic (net negative charge) peptides and charged ONs.

ONs with up to 100 bases can be separated based on the number of charged phosphate groups (see Figure 15.1) using the salt-gradient elution mode and a quaternary ammonium-functionalized stationary phase in the AEX elution mode. The protecting DMT group can be removed either on-column with an acid wash or prior to injection. Several anions can be used in the mobile phase, such as sulfate, phosphate, hydroxide, and chloride. IEX is advantageous, with respect to IPC, in the case of mixed-base ONs and in application to living cells, since the purified ON can be used directly, being in the Na^+ salt form.

Denaturing conditions, consisting of the use of organic solvents (such as ethanol or ACN), high pH, and high temperatures, are necessary when dimers and aggregates are present, even if the use of high pH is not applicable to all ONs owing to stability issues. Owing to the complexity of the crude mixture, yield is often sacrificed to obtain higher product purity.[27]

The orthogonality of IEX and IPC elution modes could be exploited in subsequent purifications when dealing with challenging peptide mixtures. A generic IEX method for peptide purification that can be combined with IPC consists of an acidic mobile phase, usually containing TFA, with the addition of a small proportion of ACN (5–30%) and a strong cation-exchange stationary phase. ACN will enhance the orthogonality, avoiding mixed-mode interactions, thus preventing non-specific hydrophobic interactions.[29] Elution can then be performed with a salt gradient (*e.g.* with NaCl). An IEX prepurification step prior to IPC can also be advantageous since acidic reagents used in peptide cleavage, which can damage RP silica columns, can be removed prior to IPC.

15.2.4 Hydrophobic Interaction Chromatography

An alternative technique to IPC is HIC. As an RP chromatographic mode, HIC permits the separation of DMT-protected ONs based on their hydrophobicity using salts in aqueous buffers. Typical stationary phases in HIC consist of hydrophobic ligands [*e.g.* ethyl, butyl, hexyl, phenyl, or poly(propylene glycol) groups].

HIC makes use of high salt concentrations in an aqueous buffer for the adsorption step, since non-polar regions are exposed and can bind to the stationary phase, and low salt concentrations for the elution of the target. The decreasing ionic strength leads to elution, owing to a larger number of hydrophilic regions on the target. This technique can be considered as a greener alternative than IPC, since it avoids the use of organic solvents, leading to less hazardous waste and lower costs for solvent disposal, reducing the environmental impact of the chromatographic step.

With this technique, a second purification step can be performed directly by loading the product eluting from the HIC column into an AEX column without the need for buffer adjustment.

15.2.5 Hydrophilic Interaction/Cation-exchange Chromatography

Hydrophilic interaction liquid chromatography (HILIC), in contrast to IPC and HIC, separates molecules based on their hydrophilicity. This elution mode makes use of polar stationary phases and a high concentration of organic modifier in the mobile phase (generally 50–80% ACN), to promote hydrophilic interactions between the analytes and the stationary phase through the formation of a water-rich layer on the surface of the porous particles. In this context, especially strong cation-exchange resins have intrinsic hydrophilic character and can then be used in mixed HILIC–CEX mode, by simply adding a high organic modifier content in the mobile phase.[29]

Analogously to IEX, elution occurs with a salt gradient, where peptides are first separated by their net charge. All peptides having the same charge are then separated in order of increasing hydrophilicity.

15.3 Innovative Green Approaches

Owing to the growing interest of the pharmaceutical industries in peptides and ONs, kilogram-scale syntheses have become fairly common, resulting in huge amounts of crude mixtures to be purified through LC on a preparative scale. Given the large volumes of feed to be treated, highly productive methods are needed.[11]

Baseline separation of species with very similar chemical structures is still a difficult task, even though progress in chromatographic hardware, including both column technology and instrumentation, has been achieved in recent years. This is mainly due to the difficulty of finding highly selective chromatographic methods that are able to distinguish between compounds that differ only in a monomer or functional group. This situation becomes even worse when large amounts of feed are loaded into the column, owing to the non-linearity of the adsorption isotherm. The coelution of the main peak with a closely related impurity peak, especially in its front and tail, raises the question of whether to discard those portions (and collect the target product with the highest degree of purity) or collect them together with the main peak (to increase the overall recovery of the process). This shortcoming related to purity–yield trade-off is intrinsic to single-column chromatography. Usually, the requirement to obtain ultrapure compounds is a compelling and inescapable necessity for pharmaceutical ingredients. This means that overlapping regions containing the target product contaminated with impurities must be collected and reprocessed (recycled) in a following run. Clearly, the more product is contained in the overlapping areas, the lower is the productivity of the process and also the higher is the consumption of solvent necessary for the purification.

To overcome the purity–yield trade-off, multicolumn chromatographic processes have been developed, where the recycling of overlapping windows is automatically performed within the system. Thanks to the presence of a multicolumn setup, the recycling regions eluting from one column can be directly adsorbed and taken up into another column, without intermediate treatments. As a result, the duration of the process decreases, with a consequent increase in productivity. Another beneficial aspect from the viewpoint of the process greenness is that also the volume of solvent waste generated is lower compared with the corresponding batch method.

To enhance the sustainability of purification processes, another route involves replacing toxic materials with greener and safer alternatives. In the case of polar biomolecules, such as some peptides and ONs, purification steps commonly employ ACN–water mixtures as mobile phases with the addition of an ion-pairing reagent.[25] ACN shows excellent chromatographic properties, but it can cause toxic effects on the human body, attributable primarily to the metabolic release of cyanide. Consequently, the need to replace ACN with more environmentally sustainable alternatives has become urgent.

The demand for more efficient and productive methods aligns with the necessity to discover environmentally friendly alternatives to the conventional organic solvents used in LC. The synergy between these two research directions aims to reduce both the volume of waste generated and its toxicity, to boost the overall greenness of the process. In the following sections, recent innovations in the field of continuous chromatography and in the solvents employed for the chromatographic purification of biopharmaceuticals (especially peptides and ONs) are reviewed.

15.3.1 Multicolumn Continuous Chromatography

The outcome of the separation, *i.e.* the resolution between the main peak and closely eluting species, significantly influences the performance of the purification process. Consequently, improving the resolution is necessary in order to enhance the process performance, even though this represents a challenging task.

As it can be seen from Figure 15.2, if the entire target peak is collected (t_B–t_E), the maximum yield is achieved at the expense of purity, owing to the presence of undesired compounds. Conversely, if overlapping regions between the target and its impurities are completely discarded (t_B–t_C and t_D–t_E), considerable loss of the expensive product is observed but high purity is achieved. The usual fate of overlapping windows is to be collected, stored, and eventually reprocessed manually by the operator, which implies long times and risks of errors. Because of this, batch chromatography sometimes becomes impractical and not very productive.

To resolve this issue, the amount of feed loaded onto the column, the gradient slope, or the stationary phase particle size could be decreased. However, this will have a detrimental effect on the run time, productivity, and solvent consumption, hence none of these options can effectively be a solution to the problem.[21]

An alternative, appealing approach is to replace single-column techniques with multicolumn countercurrent chromatography. The term "countercurrent" describes a group of techniques in which the stationary phase moves in the opposite direction to the mobile phase. The term virtually indicates that the movement of the columns is not physical but is simulated thanks to the switching of inlet and outlet column valves.[30,31] To do that, the system must be equipped with at least two columns, identical in terms of column and particle geometry and functionalization. This approach is based on the internal recycling, into another column, of the impure portions of the chromatogram that need to be reprocessed, and on the continuous or semicontinuous feeding of the sample into the system. Theoretically, it is as if the peak experienced an increase in the number of theoretical plates within the system, thus leading to an increase also in the chromatographic resolution. This could represent a solution for the mentioned purity–yield trade-off obtained in batch chromatography.

The simplest version of countercurrent chromatography is represented by the simulated moving bed (SMB), which allows the separation of binary mixtures but only in the isocratic mode, exploiting the different affinities of the two compounds for the stationary phase. For more complex, ternary mixtures, gradient separations are required to modulate the retention of all the analytes, so SMB does not represent a feasible solution.

In 2006, a process called multicolumn countercurrent solvent gradient purification (MCSGP) was developed by Morbidelli and Aumann to fill the gap between continuous chromatography and challenging ternary separations.[74] It is a continuous process conducted under linear gradient conditions, specifically designed for center-cut separations where numerous product-related impurities are present. Its key advantages lie in the automation of the process and the internal recycling of impure fractions, which are reintroduced into the unit for reprocessing through a system of valves. These are fundamental aspects to consider when it comes to the purification of valuable biotherapeutics.

In other recycling techniques based on single-column approaches, such as steady-state recycling (SSR) and closed-loop recycling (CLR), the processes can only be operated under isocratic conditions and the chromatographic profile passes through a pump at every cycle, with risks for the product stability. An exception to the isocratic processes is gradient steady-state recycling (GSSR), which comprises a multicolumn open-loop system with cyclic steady-state operation that simulates a solvent gradient moving countercurrent with respect to the solid phase. Here, the feed is always injected into the same column and the product is always collected from the same column, similarly to what happens in batch.[32] Differently, in MCSGP, impure side fractions that are recycled from one column to the other are completely adsorbed and retained in the second column until the gradient restarts the separation; also, the columns exchange positions during the process.[33]

In its first set-up, the MCSGP process was conducted with the use of six or three columns. Nowadays, in its simplest version, MCSGP is equipped with two identical columns (twin-column MCSGP). When a batch method is scaled up to the continuous mode, the same steps present in the batch method are now performed in both columns but at different moments. Basically, they perform the same tasks but shifted by half a cycle; also, the columns work either in an interconnected way or in a disconnected way, depending on the task.

In Figure 15.4, the scheme of the batch steps translated to MCSGP is illustrated. Let us consider the situation of column-2 at the beginning of the gradient (zone 4), after being loaded. The first to elute are the weak impurities (represented in blue), which are to be discarded. At some point, together with weak impurities, also some target product (orange) will start to elute (zone 5): this is the first overlapping region, which does not leave the system but is redirected to column-1 for the uptake of the product. Zone 6, on the other hand, represents the window where the target product elutes with the highest degree of purity, and because of that it can be collected and stored without further purification steps. Symmetrically to what happens on the front of the main peak with weak impurities, zones 7 and 8 correspond, respectively, to the windows where the product overlaps with strong impurities (recycled into column-1) and where strong impurities (green) are desorbed from the column, for instance by using a very high percentage of organic modifier. After stripping, also the equilibration is performed in zone 8.[34,35]

To sum up:[36]

- In zones 5 and 7 (overlapping windows), the eluent coming from that column is recycled into the other column (zones 1 and 3). Therefore, they work in an interconnected way.
- In zone 6, pure product elutes and is collected. At exactly the same time, a portion of fresh feed is loaded into the other column (zone 2). Hence at this point they are disconnected.
- During every washing step (zones 4 and 8) the eluent goes to waste, so also in this case the columns are disconnected.

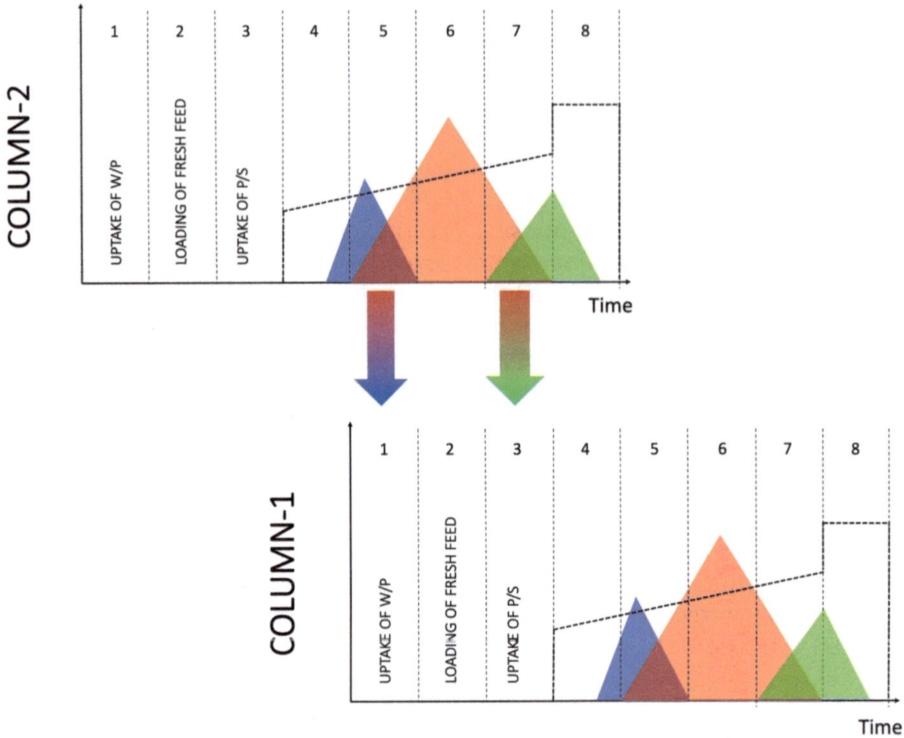

Figure 15.4 Schematic representation of the twin-column MCSGP process and the tasks accomplished by each column. Reproduced from ref. 36 with permission from Elsevier, Copyright 2022.

Once column-2 has been stripped and equilibrated, a *switch* has been accomplished, meaning that it is ready to receive both the two overlapping regions eluting from column-1 and some fresh feed. When the columns come back to the initial position, one *cycle* has been completed (two switches = one cycle).

To set up an MCSGP process, the starting point is to design a batch chromatogram where at least a portion of the main peak has an acceptable purity (*e.g.* with a partial separation between the three regions reported in Figure 15.4). The outcome of the batch separation can be evaluated by offline analysis of the fractions collected during the gradient elution, based on which the corresponding Pareto curve can be built. This curve is necessary to choose the recycling and collection windows (expressed in terms of the times where the inlet and outlet column valves switch position, namely t_B–t_E, as in Figure 15.2), depending on the purity and recovery requirements. It is important to note that the choice of the recycling and collection windows strictly affects the performance of the MCSGP process: by slightly changing the switching times, the performance could become totally different.

From the offline analysis of the batch fractions, it is also possible to estimate the mass balance of the product, namely the percentage of target compound mass contained in the two overlapping regions and in the pure collection window, with respect to the total amount of product injected for the batch run. Indeed, whereas in the batch technique the sample is loaded in a single step, in this countercurrent chromatographic process the feed is loaded in three phases into the column. Two portions of the product loaded derive from the recycling of the overlapping region W/P and P/S eluting from the other column (zones 1 and 3 for both columns), and the fresh feed injected represents only a given percentage of the total mass loaded (zone 2).[37–39]

In particular, the percentage of target product contained in the collection window corresponds to the mass of fresh feed that is injected into each column in an MCSGP cycle (zone 2). The total amount of target compound loaded per unit volume of stationary phase is therefore kept constant in the two processes. For these reasons, from the schematic representation of MCSGP shown in Figure 15.4, it can be deduced that sample feeding into the system is not continuous but rather semicontinuous.[34,35]

A strong point of the MCSGP process is that, after a couple of cycles, the system reaches a steady-state regime, which means that, cycle by cycle, the purity and recovery of the pools collected from zone 6 do not change, so in principle the system could continue to operate for a very long time, processing large volumes of feed without affecting the performance.

The boost in the recovery, at a high degree of purity, certainly represents the most immediate impact introduced with the MCSGP process, which goes hand-in-hand with an increase in productivity, since more product is purified per unit time and column volume. For the same reason, the volume of solvent necessary to purify a given mass of product is significantly reduced, with great savings from the viewpoint of both resources and economics. The extent of the performance gain, as already indicated, is strongly influenced by the times chosen to define the windows.

When advances in technology allow a substantially smaller, cleaner, safer, and more energy-efficient process to be obtained, through a reconfiguration of existing operational schemes, the resulting enhancement is called *process intensification*. Hence it can be stated that MCSGP produces an overall intensification in those purification processes where the batch approach fails or gives only mediocre results.

In addition to these major benefits, other so-called "soft" advantages arise, which are not directly related to the process performance parameters such as yield, productivity, and solvent consumption. Because of the steady-state condition reached by the system, the performance of the process is preserved cycle by cycle: two immediate consequences are that the quality control (QC) burden is reduced, since the variability of the chromatographic outcome decreases, and that the operation can continue 24/7 without interruption, thanks to the continuous loading of the feed. Also, when working in batch mode, a large number of tasks performed manually by operators in the long term can cause risks of errors and longer processing times. With MCSGP, the automation of all the steps, and especially the internal recycling of impure windows, contribute to shortening times and avoid hold steps, since no manual handling of overlapping side fractions is necessary. Avoiding hold steps is particularly convenient for unstable or labile components, such as for some peptides and proteins that tend to aggregate, to form fibrils, or become a gel quite rapidly.

All these aspects contribute to matching the establishment of MCSGP with the principles of Industry 4.0.[75]

15.3.1.1 Applications of MCSGP to Peptides and Oligonucleotides

MCSGP finds a wide range of applications in the purification of high-value biomolecules where difficult ternary separations are involved, especially protein-based biopharmaceuticals, such as mixtures of PEGylated proteins, monoclonal antibodies, peptides, ONs, and cannabinoids. Since the main topics of this chapter are peptides and ONs, only the applications related to these classes of biotherapeutics will be reviewed here, and for different pharmaceutical ingredients the interested reader is directed to the previous literature (*e.g.* mAbs,[30,40–45] PEGylated proteins,[46–49] cannabinoids[50]).

Starting with the peptides, the first one to undergo MCSGP purification was calcitonin, consisting of 32 amino acids. In 2007, Aumann and Morbidelli, the two inventors of the MCSGP technology, tested this peptide under RP conditions and demonstrated that, by collecting the whole peak (almost 100% recovery), the purity increased by more than 40% when using three- or six-column MCSGP with respect to the batch mode, reaching a value of 93% with the continuous technique.[37,51,52] Three-column MCSGP has been successfully applied for peptide purification also in other cases. In 2013, Müller-Späth *et al.* were able to purify an unknown peptide (probably covered by a patent) to 99% purity with increases in yield and productivity of four- and tenfold, respectively, whereas the solvent consumption decreased from 3.5 to 1 L g^{-1} (−70%).[40] In another study, the productivity increase was even higher at 25 times the value obtained in batch mode, with a considerable decrease in solvent consumption of −60%.[39]

Using three columns actually is not essential to obtain outstanding results: decreasing the number of columns to two allows all the advantages of continuous chromatography to be preserved but with simplified equipment.

In recent years, the present authors and colleagues have demonstrated the applicability of twin-column MCSGP to purify two peptide mixtures of industrial relevance, synthesized by means of solid-phase peptide synthesis. In both cases, the technique used was RP-LC, with a 250 × 4.6 mm i.d. column for the batch reference and two 150 × 4.6 mm i.d. columns for twin-column MCSGP. For the larger peptide (glucagon, 29 amino acids) a stationary phase with lower retention was chosen (C_8) whereas for the smaller peptide (icatibant, 10 amino acids) a C_{18} stationary phase was employed. Clearly, the larger the peptide, the more synthetic steps are required, and the higher is the number of impurities produced. Therefore, it is no surprise that the initial feed purity was around 30% for glucagon and 70% for icatibant. At the end of the purification with MCSGP, for glucagon a purity of 89% was obtained with an increase in recovery of 23%, even though in this case the productivity did not benefit from the continuous operation since very low flow rates were used during the interconnected steps, which lengthened the duration of the process.[34] However, in the case of very expensive products, the principal aim is to save as much of the target product as possible, and an increase in recovery can compensate for the loss in productivity.

For icatibant, on the other hand, the performance obtained with the continuous process significantly exceeded that with batch operation, with a purity of >99%. The recovery increased almost sevenfold (up to 95%), and the productivity improved sixfold (up to 7.65 g L^{-1} h^{-1}). Thanks to this performance enhancement, the solvent consumption decreased by 80%, a remarkable result from the point of view of environmental impact and sustainability.[35]

Also for ONs, MCSGP has already been tested, although to a smaller extent than for peptides. In general, the literature on the preparative purification of ONs is still very sparse and limited mainly to single-column (batch) chromatography. YMC ChromaCon in particular demonstrated, as a proof-of-concept, that MCSGP allowed the yield to be increased from 55% to more than 90% with respect to single-column operation, thus alleviating its characteristic purity–yield trade-off.[53] Recently, Weldon *et al.*[33] tested twin-column MCSGP under RP conditions for the purification of an *N*-acetylgalactosamine-cluster-conjugated DNA–LNA (locked nucleic acid) gapmer ON. In that case, despite an increase in yield of 74% from batch to continuous mode, the other parameters (purity, productivity, and buffer consumption) were found to be comparable for the two processes. However, the purity target of 90%, acceptable for clinical-grade products, was exceeded. It must be noted that, in this particular case, the column used for the batch operation was not longer that those used for MCSGP, so the comparison of the two processes is not completely fair because the total bed heights were different.

15.3.2 Innovative Reagents

When considering a route towards sustainable procedures to be applied to the manufacture of pharmaceuticals, it stands out that the downstream processing stage represents the main bottleneck towards the green transition of API production, owing to the huge amount of liquid waste generated daily, containing organic solvents, salts, and reagents.[23] Apart from the development of innovative techniques (described in Section 15.3.1), another solution could be the use of alternative solvents with a higher elution strength and/or lower toxicity in comparison with traditional solvents. The use of water-based techniques (*e.g.* HIC) may be a green and environmentally friendly alternative to traditional elution modes (IPC, IEX), but their applicability is limited and depends on the analyte chemistry.

Solvents can be classified into several classes indicating their greenness ranking, depending on various characteristics, including environmental impact, health, stability, and flammability.[54]

Acetonitrile is classified as a "problematic" solvent, despite being the most commonly employed organic solvent in RPLC, IPC, and IEX thanks to its advantageous properties, such as good miscibility with water, low viscosity, and high elution strength.[23,55,56] Its metabolism leads to toxic effects in humans due to the release of cyanide. Hence, for health reasons, its replacement is a priority. Other problematic solvents include toluene, chlorobenzene, and *n*-heptane.

The use of alcohols may be advantageous in terms of greenness. Methanol is classified between a "problematic" and "recommended" solvent and is one of the most often used alternative solvents to ACN in RP-LC applications, having a similar elution strength and lower boiling point. Nevertheless, exposure to it can be extremely dangerous owing to its toxicity.[57] Fully "recommended" solvents include water, ethanol, isopropyl alcohol (IPA), propanol, *n*-butanol, and ethyl acetate, to name just a few. Ethanol represents an optimal green choice in terms of toxicity and environmental impact. Moreover, it can be obtained from bio-based sources (*e.g.* agricultural wastes and wood processing), representing a further step towards the green transition of preparative chromatography.

The use of alcohols as organic modifiers may have some drawbacks. Their higher viscosity with respect to ACN leads to instrumental limitations in terms of the back-pressures generated. This necessitates the use of lower flow rates, which overall have a negative influence on the total run time and productivity,[58] or larger particle sizes, with a negative effect on peak resolution. Alcohols may also induce conformational changes in peptides. Indeed, it has been observed that, independently of the alcohol type, α-helix or β-sheet structures are formed.[59] Nevertheless, alcohols are commonly employed as storage solutions for ONs to lower the exposure to microbial contamination prior to the purification step, but also as precipitation agents. Indeed, after chromatographic step(s), ONs are dissolved in dilute aqueous solutions containing salts and possibly organic solvents used during the elution step. Alcohols can then be added to promote aggregation and precipitation to perform desalting and concentration of ONs prior to further processing or isolation of the final product.[60,61]

For these reasons, the use of alcohols for preparative purposes is limited to a few cases. Ethanol and IPA have been used recently to purify a therapeutic peptide under IPC conditions using triethylamine phosphate (TEAP) buffer, resulting in slightly higher productivity (alcohols 14 mg L^{-1} h^{-1} *versus* ACN 12 mg L^{-1} h^{-1}) at maximum recovery and slightly lower purity (alcohols 96% *versus* ACN 98%) in comparison with ACN. The Pareto curves of the alcohols lay below that of ACN, as shown in Figure 15.5. Moreover, the volume of alcohols contained in the liquid waste was significantly smaller than that of ACN.[23] MeOH has been used as an organic modifier for peptide purification using flash chromatography, achieving similar final product purity.[62]

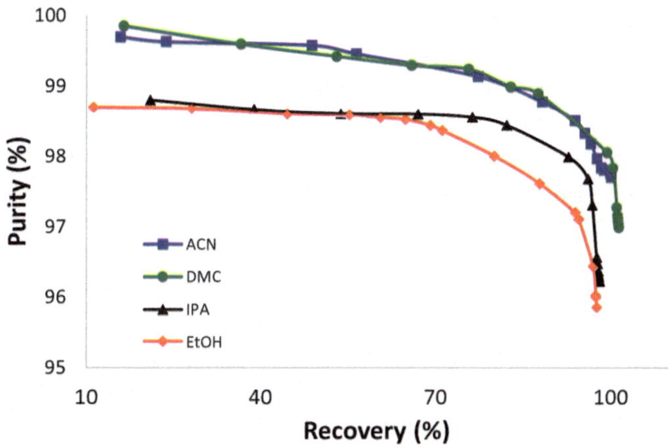

Figure 15.5 Pareto curves obtained with different organic modifiers for the purification of a therapeutic peptide (icatibant). Reproduced from ref. 23 with permission from Elsevier, Copyright 2024.

Owing to some intrinsic drawbacks of alcohols as organic modifiers, the interest of scientists has moved to organic carbonates, a less popular but yet interesting class of solvents with low toxicity and high biodegradability.[63] Among them, recently dimethyl carbonate (DMC) was tested under both analytical and preparative conditions for the separation of small molecules[56,64,65] and a therapeutic peptide[23] as a green substitute for organic solvents in RPLC, IPC, and also normal-phase LC and HILIC.

The results were very promising. DMC showed a higher elution strength than ACN or MeOH, indicating that a lower percentage of DMC is necessary for the elution of analytes, with clear advantages from the environmental and disposal viewpoints. Furthermore, it has been demonstrated that the chromatographic efficiencies and purification performances of DMC and ACN were similar, as clearly demonstrated by comparison of the Pareto curves in Figure 15.5. During peptide purification, the use of four times less DMC than ACN as solvent was required. At the maximum achievable peptide recovery, slightly higher productivity with less than a 1% difference in purity was obtained with DMC.

In conclusion, the application of DMC in chromatography is still very limited, but in the near future increasing interest in this green solvent is expected.

15.3.3 Other Experimental and Theoretical Approaches

In addition to very efficient continuous chromatography and the use of green organic solvents, other strategies could be envisaged for implementation in the production of biopharmaceutical products in the near future. These approaches, which are not very common nowadays, are both experimental and theoretical and could help boost the performance and reduce the times required for process development.

Sometimes, a single chromatographic mode is not sufficient to fulfill the purity requirements imposed for APIs, since different kinds of impurities are present, so usually two chromatographic modes with different selectivities (*orthogonal*) are applied consecutively. Otherwise, orthogonal separations can be accomplished within a single column by functionalizing the stationary phase particles with two different ligands. This allows the integration of two separation mechanisms into a single chromatographic step. The technique in which the stationary phase bears two different ligands with different reactivities is called mixed-mode chromatography.[22,66]

The two most commonly employed chromatographic modes for ONs and peptides, as already mentioned, are RP (or IPC) and IEX. By combining them, some RP stationary phases have been modified with small amounts of ion-exchange ligands, forming a so-called doped reversed phase (DRP).The charge on biomolecules such as peptides depends on their isoelectric point (*i.e.* the pH value at which the molecule carries no net electrical charge) and on the pH of the solution/mobile phase, hence the mobile phase used in mixed-mode chromatography must be carefully chosen to obtain the desired peptide charge. The charged ligands exhibit repulsion towards the analytes with the same charge sign and attraction towards those with an opposite charge sign, while the RP ligands act in an attractive manner towards hydrophobic molecules, and, therefore, the whole stationary phase can work in "attractive–attractive" or "attractive–repulsive" modes.[67,68] Usually, the attractive–repulsive mode is more appealing for challenging non-selective separations, since the repulsion experienced between the dopants and the analytes with the same charge sign pushes them to elute earlier and narrower. On the other hand, the retention of the hydrophobic species is not much influenced by the presence of doping ligands. The overall result is an increase in resolution between hydrophilic charged species (less retained) and more retained hydrophobic compounds. To work in the attractive–repulsive mode, for mobile phases where $pH < pI$, AEX materials must be chosen; otherwise, for phases with $pH > pI$, CEX resins must be used. Since peptides are both charged and hydrophobic, their behavior is affected by both attraction and repulsion with the two different types of ligands.

The use of these doped stationary phases can be potentially beneficial for preparative applications in the field of biomolecules. The increase in the resolution between the main target and other groups of impurities could represent a turning point for purifications at the preparative scale, to minimize the overlap between similar species. Further, from an industrial perspective, it is advantageous for the target peak to exhibit minimal tailing to reduce the side portions of the peak that need to be recycled in further purification steps, in order not to waste large amounts of the valuable target product. This approach boosts productivity and overall process performance. Additionally, it contributes to the environmental sustainability of the process by reducing solvent consumption.

A proof-of-concept study for the specific purification of liraglutide was presented in 2022 with the aim of comparing, for preparative applications, a traditional C_8 column with a column with C_8 doped with AEX groups.[20] It was demonstrated that, when collecting the entire peak or when isolating the purest fractions, the mixed-mode column exhibited slightly superior performance to the C_8 column. Notably, when only the target peak was collected and the tailed portion was discarded, the mixed-mode column led to an increase in recovery of +15%, and a slightly higher purity was achieved. A direct consequence was that a smaller percentage of the target product contained in the tail needed to be reprocessed in a following purification step when using the mixed-mode column (−50% in this particular case). Alternatively, if the whole peak satisfies purity constraints, with this mixed-mode column the concentration of the pool would be higher due to the narrower peak. Moreover, owing to the lower retention of charged species, the duration of the method could be reduced, with a considerable saving of time and solvents.

All these aspects contribute to decreasing the amount of buffer consumed per gram of product purified, since smaller amounts of the target compound would require subsequent solvent-intensive purification steps. Even though this research was specifically intended for liraglutide, similar outcomes can be envisaged also for other mixtures of biomolecules when employing these resins.

A further way to improve the sustainability, especially of the process development phase, is to employ theoretical approaches.[26] The experimental conditions chosen for the purification process are often defined based on a trial-and-error approach, which may cause a significant waste of time and resources. Also, it is not trivial either to deduce the chromatographic behavior under preparative conditions (meaning when the column is overloaded with the sample) starting from the analytical scale, or to understand the mechanisms and fundamentals of the separation. One of the reasons is that, when moving from the analytical to the preparative scale, the peak shape changes from gaussian to asymmetric; the origin of this tailing or fronting must be ascribed to the adsorption isotherm that, in contrast to the analytical scale where it is linear, under overloading conditions becomes non-linear. The situation is made trickier by the fact that the mixtures to be purified contain several species, leading to competition with each other and with the target compound. These aspects when combined together make it impossible to predict the peak shape and position. This explains the importance of developing methodologies based on theoretical models to predict the chromatographic behavior of the target compound under different experimental conditions. These approaches involve the measurement of adsorption isotherms and can help assess the feasibility of the purification process by providing information (*e.g.* maximum loading, affinity for the stationary phase) useful for optimizing large-scale purification. For instance, the retention and isotherm parameters of a compound are strictly affected by the mobile phase composition. The results of investigations of thermodynamic equilibria affecting the separation under batch conditions can then be exploited during process development to design efficient continuous separations. The pharmaceutical industry would most probably benefit from this kind of strategy, which allows the time and the amount of sample necessary for process development to be reduced.

As an example, the behaviors of nociceptin/orphanin FQ and octreotide were modeled under linear and non-linear conditions in the RP mode. Their adsorption isotherms were measured through the so-called inverse method, a powerful numerical approach requiring only a very small amount of analyte, at different isocratic mobile phase compositions. A generalized Langmuir isotherm model including the dependence on the mobile phase composition was then developed to predict the peak shape and retention even under gradient elution.[69,70]

Sporadically, MCSGP operations have also been supported by theoretical modeling to simulate the chromatographic behavior of the system and to predict the process outcome. For instance, Aumann *et al.* reported research used a theoretical approach to simulate the process, by applying a lumped kinetic model and a competitive bi-Langmuir isotherm, to understand the effects of different experimental conditions (feed flow rate, switching times, *etc.*) on the outcome of the separation.[52] For a particularly challenging polypeptide, the same group used process modeling formerly developed to calculate the MCSGP solvent consumption, which was 8 L g^{-1} and corresponded to a 10-fold decrease compared with the corresponding case of a batch process with ideal manual recycling.[71]

15.4 Conclusions and Perspectives

Since the manufacture of ONs and peptides is expected to expand quickly to a large scale, the need to implement existing processes is becoming urgent. Because of the large volumes of solvent waste generated, the bottleneck for the green transition of API production is mainly the downstream processing stage, where the target API contained in the crude mixture is purified from its related impurities. Regulatory authorities have strengthened their requirements consistently by lowering the level of impurities allowed, hence very efficient methods for their removal are required.

Multicolumn continuous chromatographic approaches are becoming increasingly appealing in the field of high-value biopharmaceuticals, especially ONs, proteins, and peptides, since they allow the trade-off typical of batch methods to be alleviated, with economic and environmental repercussions also. This consideration is supported by some recent applications of MCSGP to the purification of ONs and peptides, which demonstrate the boost in the performance achieved with continuous multicolumn chromatography. Especially the productivity and solvent consumption are improved, without a decrease in purity, with a special impact on the footprint of the whole purification process. The extent of the economic gains is, however, product and process specific since it depends on the complexity of the mixture and on the recycling and collection windows chosen. A central thread common to all these applications is, however, that the automation makes the operations more stable and robust, with a consequent reduction in the QC burden, as demanded by Industry 4.0 principles. The design and operation of MCSGP is challenging and requires expert operators. However, Müller-Späth and Bavand conducted an evaluation study which indicated that MCSGP could lead to *ca.* 40% cost savings in downstream processing, compared with batch chromatography scenarios (for a peptide production plant producing around 10 kg of product annually), hence investing in the implementation of continuous chromatographic purification systems can result in significant savings and a reduced payback period (estimated to be around 6–18 months).[72] Recently, one of the leading companies in the field of peptides and ONs, Bachem, entered the MCSGP business on a commercial scale, to move towards the sustainable large-scale production of valuable therapeutics, demonstrating its effectiveness even at a larger scale.[73]

Another strategy to decrease the environmental impact of the purification process is to replace traditional organic solvents with greener alternatives, with lower toxicity and possibly good chromatographic properties. When possible, water-based techniques should be chosen, to avoid organic solvents completely. When this is not possible, ACN, a solvent causing risks to human and environmental health, but which is nevertheless the most often used organic modifier in RP-LC, should be replaced with alcohols or other more eco-friendly mixtures. It has recently been found that a green solvent, namely dimethyl carbonate (DMC), could represent a good alternative for peptide purification owing to its elution strength, which is about three times higher than that of ACN. However, its low miscibility with water limits its applicability, and solutions to overcome this drawback are being sought.

Despite the proven benefits of these innovations, the regulatory authorities are very cautious concerning the approval of new purification methods and make it difficult for these technologies to become established at a fast rate. Therefore, there is still a long way to go before their full integration in pharmaceutical manufacturing procedures. However, even though the path to fully green processes for biomolecules is still long and rough, the attention devoted to sustainability is encouraging and proves that it has become a driving force in all sectors of innovation.

Acknowledgements

The authors would like to thank the National Recovery and Resilience Plan (NRRP), Mission 04 Component 2 Investment 1.5 – NextGenerationEU, Call for tender no. 3277 dated 30/12/2021; Award Number: 0001052 dated 23/06/2022, and the Italian University and Scientific Research Ministry (grant PRIN P2022PTYWP), title: “Design of high-pRofit fostEring bioActive coMpounds through integral valorization of seaWEEDs infesting the MEditerranean sea (DreamWEEDme)”. Chiara De Luca benefits from funding from FSE REACT-EU, within the program PON “Research and Innovation” 2014–2020 (PON R&I), Action IV.6 “Contratti di ricerca su tematiche Green”.

References

1. M. C. Bryan, P. J. Dunn, D. Entwistle, F. Gallou, S. G. Koenig, J. D. Hayler, M. R. Hickey, S. Hughes, M. E. Kopach, G. Moine, P. Richardson, F. Roschangar, A. Steven and F. J. Weiberth, *Green Chem.*, 2018, **20**, 5082.
2. World Health Organization, Biotherapeutic products, https://www.who.int/teams/health-product-policy-and-standards/standards-and-specifications/biotherapeutic-products.
3. L. Wang, N. Wang, W. Zhang, X. Cheng, Z. Yan, G. Shao, X. Wang, R. Wang and C. Fu, *Signal Transduction Targeted Ther.*, 2022, **7**, 48.
4. S. Thakur, A. Sinhari, P. Jain and H. R. Jadhav, *Front. Pharmacol*, 2022, **13**, 1006304.
5. L. Moumné, A. C. Marie and N. Crouvezier, *Pharmaceutics*, 2022, **14**, 260.
6. J. S. Heitmann, T. Bilich, C. Tandler, A. Nelde, Y. Maringer, M. Marconato, J. Reusch, S. Jäger, M. Denk, M. Richter, L. Anton, L. M. Weber, M. Roerden, J. Bauer, J. Rieth, M. Wacker, S. Höber, A. Peter, C. Meisner, I. Fischer, M. W. Löffler, J. Karbach, E. Jäger, R. Klein, H.-G. Rammensee, H. Salih and J. S. Walz, *Nature*, 2022, **601**, 617.
7. K. Nepali, R. Sharma, S. Sharma, A. Thakur and J.-P. Liou, *J. Biomed. Sci.*, 2022, **29**, 65.
8. P. V. Bagwe, P. V. Bagwe, S. S. Ponugoti and S. V. Joshi, *Int. J. Pept. Res. Ther.*, 2022, **28**, 94.
9. G. Lievore, R. Weldon, M. Catani, A. Cavazzini and T. Müller-Späth, *J. Chromatogr. B: Anal. Technol. Biomed. Life Sci.*, 2022, **1209**, 123439.
10. Precedence Research, *Peptide Therapeutics Market – Global Industry Analysis, Size, Share, Growth, Trends, Regional Outlook, and Forecast 2023–2032*, 2023.
11. A. Isidro-Llobet, M. N. Kenworthy, S. Mukherjee, M. E. Kopach, K. Wegner, F. Gallou, A. G. Smith and F. Roschangar, *J. Org. Chem.*, 2019, **84**, 4615.
12. A. Henninot, J. C. Collins and J. M. Nuss, *J. Med. Chem.*, 2018, **61**, 1382–1414.
13. J. L. Lau and M. K. Dunn, *Bioorg. Med. Chem.*, 2018, **26**, 2700.
14. E. Heinzle, A. P. Biwer and C. L. Cooney, *Development of Sustainable Bioprocesses: Modeling and Assessment*, John Wiley & Sons, Ltd, 2007, p. 1.
15. M. Conda-Sheridan and M. Krishnaiah, Protecting groups in peptide synthesis, in *Methods in Molecular Biology*, ed. W. Hussein, M. Skwarczynski, and I. Toth, Humana, 2020, vol. 2103, p. 111.
16. L. Ponchon, G. Beauvais, S. Nonin-Lecomte and F. Dardel, *Nat. Protoc.*, 2009, **4**, 947.
17. L. Ponchon and F. Dardel, *Nat. Methods*, 2007, **4**, 571.
18. M. Catani, C. De Luca, J. Medeiros Garcia Alcântara, N. Manfredini, D. Perrone, E. Marchesi, R. Weldon, T. Müller-Späth, A. Cavazzini, M. Morbidelli and M. Sponchioni, *Biotechnol. J.*, 2020, **15**, e1900226.
19. A. Andrus and W. Bloch, *HPLC of Oligonucleotides and Polynucleotides. HPLC of Macromolecules*, Oxford Academic, Oxford, 1998, p. 141.

20. G. Lievore, D. Bozza, M. Catani, A. Cavazzini, T. Chenet, L. Pasti, L. Ferrazzano, W. Cabri, M. Macis, A. Ricci, C. De Luca and S. Felletti, *Separations*, 2022, **9**, 125.
21. C. De Luca, S. Felletti, G. Lievore, T. Chenet, M. Morbidelli, M. Sponchioni, A. Cavazzini and M. Catani, *Trends Anal. Chem.*, 2020, **132**, 116051.
22. C. De Luca, G. Lievore, D. Bozza, A. Buratti, A. Cavazzini, A. Ricci, M. Macis, W. Cabri, S. Felletti and M. Catani, *Molecules*, 2021, **26**, 4688.
23. D. Bozza, C. De Luca, S. Felletti, M. Spedicato, F. Presini, P. P. Giovannini, M. Carraro, M. Macis, A. Cavazzini, M. Catani, A. Ricci and W. Cabri, *J. Chromatogr. A*, 2024, **1713**, 464530.
24. G. Guiochon, *J. Chromatogr. A*, 2002, **965**, 129.
25. L. Ferrazzano, M. Catani, A. Cavazzini, G. Martelli, D. Corbisiero, P. Cantelmi, T. Fantoni, A. Mattellone, C. De Luca, S. Felletti, W. Cabri and A. Tolomelli, *Green Chem.*, 2022, **24**, 975.
26. J. Becker, C. Manske and S. Randl, *Curr. Opin. Green Sustainable Chem.*, 2022, **1**, 100562.
27. Q. Zhang, H. Lv, L. Wang, M. Chen, F. Li, C. Liang, Y. Yu, F. Jiang, A. Lu and G. Zhang, *Int. J. Mol. Sci.*, 2016, **17**, 2134.
28. T. Stahl, D. Mattern and H. Brunn, *Environ. Sci. Eur.*, 2011, **23**, 38.
29. C. T. Mant, Y. Chen, Z. Yan, T. V. Popa, J. M. Kovacs, J. B. Mills, B. P. Tripet and R. S. Hodges, HPLC analysis and purification of peptides, in *Peptide Characterization and Application Protocols. Methods in Molecular Biology*, ed. G. B. Fields, 2007, vol. 386, p. 3.
30. T. Müller-Späth, L. Aumann, G. Ströhlein, H. Kornmann, P. Valax, L. Delegrange, E. Charbaut, G. Baer, A. Lamproye, M. Jöhnck, M. Schulte and M. Morbidelli, *Biotechnol. Bioeng.*, 2010, **107**, 974.
31. D. Pfister, L. Nicoud and M. Morbidelli, *Continuous Biopharmaceutical Processes: Chromatography, Bioconjugation, and Protein Stability*, Cambridge University Press, 2018.
32. T. P. D. Santos, R. P. Fernandes, R. P. P. L. Ribeiro, C. Peixoto and J. P. B. Mota, *Digital Chem. Eng.*, 2023, **6**, 100081.
33. R. Weldon, J. Lill, M. Olbrich, P. Schmidt and T. Müller-Späth, *J. Chromatogr. A*, 2021, **1663**, 462734.
34. C. De Luca, S. Felletti, G. Lievore, A. Buratti, S. Vogg, M. Morbidelli, A. Cavazzini, M. Catani, M. Macis, A. Ricci and W. Cabri, *J. Chromatogr. A*, 2020, **1625**, 461304.
35. C. De Luca, S. Felletti, D. Bozza, G. Lievore, M. Morbidelli, M. Sponchioni, A. Cavazzini, M. Catani, W. Cabri, M. Macis and A. Ricci, *Ind. Eng. Chem. Res.*, 2021, **60**, 6826.
36. M. Catani and C. De Luca, *Adv. Chem. Eng.*, 2022, **59**, 69.
37. L. Aumann and M. Morbidelli, *Biotechnol. Bioeng.*, 2007, **98**, 1043.
38. A. Tarafder, G. Ströhlein, L. Aumann and M. Morbidelli, *J. Chromatogr. A*, 2008, **1183**, 87.
39. L. Aumann, M. Morbidelli, B. Schenkel and G. Stroehlein, *BioPharm Int.*, 2009, **22**, 46.

40. T. Müller-Späth, G. Ströhlein, O. Lyngberg and D. Maclean, *Chem. Today*, 2013, **31**, 56.
41. N. Ulmer, D. Ristanovic and M. Morbidelli, *Biotechnol. J.*, 2019, **14**, 1800677.
42. F. Steinebach, N. Ulmer, L. Decker, L. Aumann and M. Morbidelli, *J. Chromatogr. A*, 2017, **1492**, 19.
43. S. Vogg, N. Ulmer, J. Souquet, H. Broly and M. Morbidelli, *Biotechnol. J.*, 2019, **14**, 1800732.
44. F. Steinebach, N. Ulmer, M. Wolf, L. Decker, V. Schneider, R. Wälchli, D. Karst, J. Souquet and M. Morbidelli, *Biotechnol. Prog.*, 2017, **33**, 1303.
45. S.-Y. Jing, C. Shi, D. Gao, H.-B. Wang, S.-J. Yao and D.-Q. Lin, *J. Chromatogr. A*, 2023, **1707**, 464292.
46. T. K. Kim, C. Botti, J. Angelo, X. Xu, S. Ghose, Z. J. Li, M. Morbidelli and M. Sponchioni, *Ind. Eng. Chem. Res.*, 2021, **60**, 10764.
47. G. Subramanian, *Biopharmaceutical Production Technology*, Wiley-VCH, 2012.
48. G. Subramanian, in *Continuous processing in pharmaceutical manufacturing*, ed. G. Subramanian, 2005.
49. Purification of PEGlyated proteins with Contichrom®. Application note. www.chromacon.ch
50. ChromaCon – A YMC Company, Highly pure Cannabidiol (CBD) by twin-column chromatography (MCSGP), 2019, https://www.ymcamerica.com/resource/highly-pure-cannabidiol-cbd-by-twin-column-chromatography-mcsgp/
51. L. Aumann and M. Morbidelli, *Biotechnol. Bioeng.*, 2008, **99**, 728.
52. L. Aumann, G. Stroehlein and M. Morbidelli, *Biotechnol. Bioeng.*, 2007, **98**, 1029.
53. ChromaCon – A YMC Company, Purification of a Therapeutic Oligonucleotide Using Twin-Column Chromatography (MCSGP), 2019, https://www.ymcamerica.com/resource/purification-of-a-therapeutic-oligonucleotide-using-twin-column-chromatography-mcsgp/
54. L. Cseri, M. Razali, P. Pogany and G. Szekely, Organic Solvents in Sustainable Synthesis and Engineering, in *Green Chemistry: An Inclusive Approach*, Elsevier, 2018, p. 513.
55. D. Prat, A. Wells, J. Hayler, H. Sneddon, C. R. McElroy, S. Abou-Shehada and P. J. Dunn, *Green Chem.*, 2015, **18**, 288.
56. S. Felletti, M. Spedicato, D. Bozza, C. De Luca, F. Presini, P. P. Giovannini, M. Carraro, M. Macis, A. Cavazzini, M. Catani, A. Ricci and W. Cabri, *J. Chromatogr. A*, 2023, **1712**, 464477.
57. J. A. Kraut and M. E. Mullins, *N. Engl. J. Med.*, 2018, **378**, 270.
58. B. Vanholme, T. Desmet, F. Ronsse, K. Rabaey, F. Van Breusegem, M. De Mey, W. Soetaert and W. Boerjan, *Front. Plant Sci.*, 2013, **4**, 47731.
59. M. Kinoshita, Y. Okamoto and F. Hirata, *J. Am. Chem. Soc.*, 2000, **122**, 2773.
60. M. N. Moore, J. C. Arthur, K. Vansooy and A. N. Scozzari, Processes of purifying oligonucleotides, *WO pat.*, WO2002100873A1, 2001.

61. B. I. Andrews, F. D. Antia, S. B. Brueggemeier, L. J. Diorazio, S. G. Koenig, M. E. Kopach, H. Lee, M. Olbrich and A. L. Watson, *J. Org. Chem.*, 2021, **86**, 49.
62. E. Denton, J. R. Bickler and G. Carlson, Methanol as an Alternative Mobile Phase Solvent for Reversed-Phase Peptide Purification, in *256th ACS National Meeting*, 2018, P187.
63. B. Schäffner, F. Schäffner, S. P. Verevkin and A. Börner, *Chem. Rev.*, 2010, **110**, 4554.
64. B. Lajin and W. Goessler, *J. Anal. At. Spectrom.*, 2021, **36**, 1272.
65. P. D. Boes, S. R. Elleman and N. D. Danielson, *Separations*, 2023, **10**, 70.
66. K. Zhang and X. Liu, *J. Pharm. Biomed. Anal.*, 2016, **128**, 73.
67. F. Gritti and G. Guiochon, *J. Chromatogr. A*, 2014, **1374**, 112.
68. R. Khalaf, N. Forrer, G. Buffolino, A. Butté and M. Morbidelli, *J. Chromatogr. A*, 2015, **1407**, 169.
69. N. Marchetti, F. Dondi, A. Felinger, R. Guerrini, S. Salvadori and A. Cavazzini, *J. Chromatogr. A*, 2005, **1079**, 162.
70. C. De Luca, S. Felletti, M. Macis, W. Cabri, G. Lievore, T. Chenet, L. Pasti, M. Morbidelli, A. Cavazzini, M. Catani and A. Ricci, *J. Chromatogr. A*, 2020, **1616**, 460789.
71. M. Morbidelli, T. Müller-Späth, A. Tarafder, G. Stroehlein and L. Aumann, *BioPharm Int.*, 2007, **22**, 42.
72. T. Müller-Späth and M. Bavand, *Pharm. Eng.*, 2019, https://ispe.org/pharmaceutical-engineering/march-april-2019/purification-synthetic-peptides
73. Bachem multi-column purification overcomes barrier to large scale 'tide' production, https://www.pharmaceutical-networking.com/bachem-multi-column-purification-overcomes-barrier-to-large-scale-tide-production/
74. L. Aumann and M. Morbidelli, *EP Pat.*, 1716900A1, 2006.
75. C. G. Machado, M. P. Winroth and E. H. D. Ribeiro da Silva, Sustainable manufacturing in Industry 4.0: an emerging research agenda, *Int. J. Prod. Res.*, 2019, **58**(5), 1462–1484.

Subject Index